Medicinal Spice and Condiment Crops

Spices are used in small quantities to add flavour, aroma, colour, and texture to food preparation. However, spices and condiments have also been recognized as having medicinal value. They have been used traditionally as curative and preventive agents, with the bioactive compounds obtained from these plants being used in herbal drug formulations in modern times. *Medicinal Spice and Condiment Crops* provides a comprehensive overview of the medicinal properties of spices and condiments, with a focus on the various phytochemicals derived from them.

Key Features:

- Examines spices and condiments with respect to their biology, diversity, distribution, and pharmacological values.
- Explores the scientific validation of the therapeutic potentials of spice and condiment plant species.
- Discusses the bioactive agents involved in alleviating symptoms of cardiovascular disease, cancer, diabetes, ulcers, skin disorders, and neurodegenerative diseases.
- Explains specific medicinal features of spice- and condiment-producing plant species, inspiring research, discovery, and safe herbal drug formulation.

A volume in the *Exploring Medicinal Plants* series, this book is useful for plant scientists, economic botanists, those who work with herbal and complementary medicine, as well as people working in the pharmaceutical industry.

Exploring Medicinal Plants

Series Editor: Azamal Husen, *Wolaita Sodo University, Ethiopia*

Medicinal plants render a rich source of bioactive compounds used in drug formulation and development; they play a key role in traditional or indigenous health systems. As the demand for herbal medicines increases worldwide, supply is declining as most of the harvest is derived from naturally growing vegetation. Considering global interests and covering several important aspects associated with medicinal plants, the Exploring Medicinal Plants series comprises volumes valuable to academia, practitioners, and researchers interested in medicinal plants. Topics provide information on a range of subjects including diversity, conservation, propagation, cultivation, physiology, molecular biology, growth response under extreme environment, handling, storage, bioactive compounds, secondary metabolites, extraction, therapeutics, mode of action, and healthcare practices.

Led by Azamal Husen, PhD, this series is directed to a broad range of researchers and professionals consisting of topical books exploring information related to medicinal plants. It includes edited volumes, references, and textbooks available for individual print and electronic purchases.

Sustainable Uses of Medicinal Plants
Edited by Learnmore Kambizi and Callistus Bvenura

Medicinal Plant Responses to Stressful Conditions
Edited by Arafat Abdel Hamed Abdel Latef

Aromatic and Medicinal Plants of Drylands and Deserts
Ecology, Ethnobiology and Potential Uses

Edited by David Ramiro Aguillón Gutiérrez, Cristian Torres León, and Jorge Alejandro Aguirre Joya

Secondary Metabolites from Medicinal Plants
Nanoparticles Synthesis and their Applications

Edited by Rakesh Kumar Bachheti and Archana Bachheti

Aquatic Medicinal Plants
Edited by Archana Bachheti, Rakesh Kumar Bachheti, and Azamal Husen

Antidiabetic Medicinal Plants and Herbal Treatments
Edited by Azamal Husen

Ethnobotany and Ethnopharmacology of Medicinal and Aromatic Plants
Steps Towards Drugs Discovery

Edited by Mohd Adnan, Mitesh Patel and Mejdi Snoussi

Wild Mushrooms and Health
Diversity, Phytochemistry, Medicinal Benefits, and Cultivation

Edited by Kamal Ch. Semwal, Steve L. Stephenson, and Azamal Husen

Medicinal Roots and Tubers for Pharmaceutical and Commercial Applications
Edited by Rakesh Kumar Bachheti and Archana Bachheti

Pharmacological Aspects of Essential Oils
Current and Future Trends

Edited by Sunita Singh, Pankaj Kumar Chaurasia and Shashi Lata Bharati

Promising Antiviral Herbal and Medicinal Plants
Edited by Nadeem Akhtar, Azamal Husen, Vagish Dwibedi, and Santosh Kumar Rath

Antimalarial Medicinal Plants
Edited by Azamal Husen

Medicinal Spice and Condiment Crops
Edited by Azamal Husen

For more information about this series, please visit: www.routledge.com/Exploring-Medicinal-Plants/book-series/CRCEMP

Medicinal Spice and Condiment Crops

Edited by
Azamal Husen

CRC Press
Taylor & Francis Group
Boca Raton London New York

CRC Press is an imprint of the
Taylor & Francis Group, an **informa** business

First edition published 2024
by CRC Press
2385 NW Executive Center Drive, Suite 320, Boca Raton FL 33431

and by CRC Press
4 Park Square, Milton Park, Abingdon, Oxon, OX14 4RN

CRC Press is an imprint of Taylor & Francis Group, LLC

Library of Congress Cataloging-in-Publication Data

Names: Husen, Azamal, editor.
Title: Medicinal spice and condiment crops / edited by Azamal Husen
Description: First edition | Boca Raton, FL : CRC Press, 2024 | Series: Exploring medicinal plants ; 24 | Includes bibliographical references and index. | Summary: "Spices are used in small quantities to add flavor, aroma, color, and texture to food preparation. However, spices and condiments have also been recognized as having medicinal value. They have been used traditionally as curative and preventive agents, and today the bioactive compounds obtained from these plants are used in herbal drug formulations. Medicinal Spice and Condiment Crops provides a comprehensive overview on the medicinal properties of spices and condiments and various phytochemicals derived from them"-- Provided by publisher.
Identifiers: LCCN 2023046185 (print) | LCCN 2023046186 (ebook) | ISBN 9781032469478 (hardback) | ISBN 9781032480299 (paperback) | ISBN 9781003387046 (ebook)
Subjects: LCSH: Spices--Therapeutic use. | Spice plants--Therapeutic use. | Herbs--Therapeutic use. | Medicinal plants. | Materia medica, Vegetable. | Pharmacognosy.
Classification: LCC RM666.S64 M43 2024 (print) | LCC RM666.S64 (ebook)
LC record available at https://lccn.loc.gov/2023046185
LC ebook record available at https://lccn.loc.gov/2023046186

ISBN: 9781032469478 (hbk)
ISBN: 9781032480299 (pbk)
ISBN: 9781003387046 (ebk)

DOI: 10.1201/9781003387046

Typeset in Times
by Deanta Global Publishing Services, Chennai, India

Dedication

Love you so much, Ammi, for your spicy, delicious food, and for your overall care

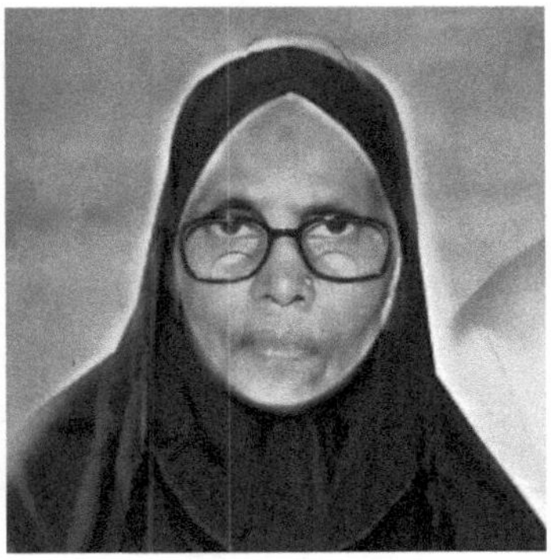

Contents

Preface

Healing through spices and condiments has a long history. Spices are parts of aromatic plants, such as fruit, leaves, seeds, roots, bark, buds, flowers, etc. Spices are used in small quantities to add flavour, aroma, colour, and texture to certain food preparations. However, condiments are used to improve the flavour and quality of food. They are also obtained from aromatic plants, and quite often added to food preparations in higher concentrations than spices. Spices and condiments have no adverse effect on the human body when used in appropriate amounts. Spices and condiments have also been recognized as having some medicinal value since ancient times. They were used traditionally as curative and preventive agents. For instance, ginger has been used for thousands of years in Asian cultures to fight infections. Having become more popular in the Western diet, ginger is often taken in the form of tea for its antibacterial effect. Turmeric powder is used for its anti-inflammatory, antioxidant, and antiviral properties. In different polydrug combinations, turmeric has shown significant antiviral properties and is useful in fighting colds and flu. The dried thread-like parts of the flower (stigmas and styles) obtained from saffron plants are used as a food colouring and flavouring material and a medicine. The dried seeds and oil obtained from fennel are also used as medicine. Dill leaves and seeds are used as a flavouring agent in food and as a fragrance in soaps and cosmetics. They are full of flavonoids, which help reduce the risk of heart disease and stroke. From European folk medicine to Indian and Chinese systems of medicine, onion has a long history of being used for relieving coughs and fighting infections. Garlic is able to kill bacteria and parasites due to the presence of sulphur compounds and bioflavonoids. Similarly, some hot peppers, black pepper seeds, cinnamon barks, clove buds, nutmeg seeds, star anise fruits, black cumin seeds, coriander leaves and seeds, cumin seeds, and mustard seeds, among others, possess several medicinal and pharmacological properties.

The bioactive compounds obtained from spices and condiments are used in various herbal drug formulations for the treatment of cardiovascular disease, cancer, diabetes, ulcers, skin disorders, and neurodegenerative diseases. They are also used as an antimicrobial, anti-inflammatory, spasmolytic, antioxidant, hypoglycaemic, antihepatotoxic, antiallergic, antimigraine, anti-osteoporotic, oestrogenic, androgenic, immunostimulant, antilithic, and anti-insomniac agents. This has also resulted in an increase in the use of spices and condiments as key cooking ingredients as well as a wide variety of herbal medicines and natural dietary supplements. Thus, the claims made in various traditional medical systems must be validated by documenting their therapeutic potential during the course of the development of new drugs. Moreover, the existing challenge lies in defining their benefits, doses, and methods of application, based on scientific grounds.

Taken together, the book in hand is an attempt to address the knowledge and practice gap surrounding the therapeutic qualities of medicinal spice and condiment plant species from around the world. Current trends, technologies, progress reports, processes, and services important to and necessary for the efficient production, use, and understanding of medicinal qualities of spice and condiment plants are highlighted. It is also vital to know the existing sustainable techniques for the postharvest and conservation practices of major spices and condiments to increase the productivity and quality, which will lead to enhancing the profitability for growers. This book provides valuable information to scientists, researchers, and students, working especially on medicinal plants, or in the areas of ethnomedicine, natural products, economic botany, plant biochemistry, biotechnology, pharmacognosy, industrial chemistry, and other allied subjects.

I am grateful to all contributors for readily accepting my invitation, sharing their knowledge in specialized areas of research, and readily adopting suggestions for improving the shape of their contributions. With great pleasure, I also extend my sincere thanks to Randy Brehm, Tom Connelly, and all associates at Taylor & Francis Group, LLC/ CRC Press for their active cooperation. Finally, my special thanks go to Shagufta, Zaara, Mehwish, and Huzaifa for providing their time and overall support to put everything together.

Azamal Husen

About the Editor

Dr Azamal Husen is a Professor at Sankalchand Patel University, Visnagar, India; and Adjunct Professor at Graphic Era (Deemed to be University), Dehradun, Uttarakhand, India. He is also working as a Visiting Professor at University Putra Malaysia, Selangor, Malaysia; and acting as a Foreign Delegate at Wolaita Sodo University, Wolaita, Ethiopia. Previously, he served as Professor and Head of the Department of Biology, University of Gondar, Ethiopia. He also worked as a Visiting Faculty of the Forest Research Institute and the Doon College of Agriculture and Forest at Dehra Dun, India. His research and teaching experience of 25 years encompasses biogenic nanomaterial fabrication and applications; plant responses to nanomaterials; plant adaptation to harsh environments at the physiological, biochemical, and molecular levels; herbal medicine; and clonal propagation for the improvement of tree species.

Dr Husen has conducted several research projects sponsored by various funding agencies, including the World Bank (FREEP), the National Agricultural Technology Project (NATP), the Indian Council of Agriculture Research (ICAR), the Indian Council of Forest Research Education (ICFRE), and the Japan Bank for International Cooperation (JBIC). He received four fellowships from India and a recognition award from the University of Gondar, Ethiopia, for excellent teaching, research, and community service. Dr Husen has been on the editorial board and the panel of reviewers of several reputed journals published by Elsevier, Frontiers Media, Taylor & Francis, Springer Nature, RSC, Oxford University Press, Sciendo, The Royal Society, CSIRO, PLoS, MDPI, John Wiley & Sons, and UPM Journals. He is on the advisory board of Cambridge Scholars Publishing, UK. He is a Fellow of the Plantae group of the American Society of Plant Biologists, and a member of the International Society of Root Research, Asian Council of Science Editors, and INPST. To his credit are over 250 publications, and he is Editor-in-Chief of the *American Journal of Plant Physiology.*

Dr Husen is also working as Series Editor of *Exploring Medicinal Plants* (Taylor & Francis, USA); *Plant Biology, Sustainability, and Climate Change* (Elsevier, USA); and *Smart Nanomaterials Technology* (Springer Nature, Singapore). He achieved the distinguished honour of being recognized as one of the "World's Top 2% Scientists" for 2022, and again for 2023 by Stanford University, USA. This recognition has also been prominently featured in the Elsevier Data Repository.

Contributors

Altero Aguzzi
CREA-Research Centre for Food and Nutrition
Rome, Italy

Mohammad Habiib Abasss Alfaham
Department of Horticultural Science and Agronomy
Science and Research Branch
Islamic Azad University
Tehran, Iran

Ahmad Ali
Department of Life Sciences
University of Mumbai,
Mumbai, India

Hina Alim
Department of Life Sciences
University of Mumbai
Mumbai, India

N. V. Anil Kumar
Department of Chemistry
Manipal Institute of Technology
Manipal Academy of Higher Education
Manipal, India

B. Anusha
Department of Chemistry
Manipal Institute of Technology
Manipal Academy of Higher Education
Manipal, India

Anoja Priyadarshani Attanayake
Department of Biochemistry
Faculty of Medicine
University of Ruhuna
Galle, Sri Lanka

Abdelaali Balahbib
Laboratory of Human Pathologies Biology
Department of Biology
Faculty of Sciences
Mohammed V University in Rabat
Rabat, Morocco

Prairna Balyan
Department of Life Sciences
University of Mumbai
Mumbai, India

Subhrajit Banerjee
Department of Microbiology
St. Xavier's College
Kolkata, West Bengal, India
and
Department of Physiology
Surendranath College
Kolkata, West Bengal, India

Ouafae Benkhnigue
Department of Botany and Plant Ecology
Scientific Institute
University Mohammed V in Rabat
Rabat, Morocco

Chakali Bhargavi
Indira Gandhi Agricultural University, Raipur
Chhattisgarh, India

Mustansir Bhori
School of Biotechnology and Bioinformatics
D Y Patil Deemed to be University
Belapur, Maharashtra, India

Hayette Bouabida
Water and Environment Laboratory
Echahid Cheikh Larbi Tebessi University
Tebessa, Algeria

Abdelhakim Bouyahya
Laboratory of Human Pathologies Biology
Department of Biology
Faculty of Sciences
Mohammed V University in Rabat
Rabat, Morocco

Kolagani Chandramohan
Forest Survey of India
Central Zone
Seminary Hills
Nagpur, Maharashtra, India

Noureddine Chaachouay
Agri-food and Health Research Laboratory, ESEFB
Hassan First University of Settat
Settat, Morocco

Shruti Chattaraj
Department of Botany
Ananda Mohan College
Kolkata, West Bengal, India

Kritika Chauhan
Department of Life Sciences
Graphic Era Deemed to be University
Dehradun, Uttarakhand, India

Anuj Choudhary
Department of Biology and Environmental Sciences
CSKHPKV, Palampur
Himachal Pradesh, India

Siriporn Chuchawankul
Immunomodulation of Natural Products Research Unit
Chulalongkorn University
Bangkok, Thailand

Nadia Dakka
Laboratory of Human Pathologies Biology
Department of Biology
Faculty of Sciences
Mohammed V University in Rabat
Rabat, Morocco

Tuyelee Das
Department of Life Sciences
Presidency University
Kolkata, West Bengal, India

Kruti Dave
Department of Life Science
Ramnarain Ruia Autonomous College
Mumbai, India

Nayomi Deshani De Silva
Department of Biochemistry
Faculty of Medicine
University of Ruhuna
Galle, Sri Lanka

Manik Devgan
Department of Plant Breeding and Genetics
Punjab Agricultural University
Punjab, India

Djemaa Dris
Water and Environment Laboratory
Echahid Cheikh Larbi Tebessi University
Tebessa, Algeria

Alessandra Durazzo
CREA-Research Centre for Food and Nutrition
Rome, Italy

Paolo Gabrielli
CREA-Research Centre for Food and Nutrition
Rome, Italy

Indu Anna George
Department of Life Sciences
University of Mumbai
Maharashtra, India

Sarra Grabsi
Laboratory of Bioactive Molecules and Applications
Echahid Cheikh Larbi Tebessi University
Tebessa, Algeria

Leila Hakimi
Department of Agriculture
Saveh Branch
Islamic Azad University
Saveh, Iran

Azamal Husen
Department of Biotechnology
Smt. S.S. Patel Nootan Science and Commerce College
Sankalchand Patel University
Visnagar, Gujarat, India
and
Department of Biotechnology
Graphic Era (Deemed to be University), Dehradun
Uttarakhand, India
and
Wolaita Sodo University
Wolaita, Ethiopia

Iram Iqbal
Department of Pharmacology
Faculty of Pharmacy
Bahauddin Zakariya University
Multan, Pakistan

Pooja Iyer
School of Biotechnology and Bioinformatics
D Y Patil Deemed to be University
Belapur, Maharashtra, India

Namdev Jagtap
School of Biotechnology and Bioinformatics
D Y Patil Deemed to be University
Belapur, Maharashtra, India

Aayushi Kadam
Department of Food and Human Nutritional Sciences
University of Manitoba
Winnipeg, Manitoba, Canada;
Anatek Services PVT LTD
Sai Chamber,
Mumbai, Maharashtra, India

Kawtar El Kadri
Laboratory of Human Pathologies Biology
Department of Biology
Faculty of Sciences
Mohammed V University in Rabat
Rabat, Morocco

Divya Kapoor
Department of Microbiology
CCS Haryana Agricultural University
Haryana, India

Arshdeep Kaur
Department of Plant Breeding and Genetics
Punjab Agricultural University
Punjab, India

Amit Kumar Kesharwani
Plant Bacteriology Laboratory
Division of Plant Pathology
ICAR-Indian Agricultural Research Institute
New Delhi, India

Salman Khan
Forest Survey of India
Central Zone
Seminary Hills
Nagpur, Maharashtra, India

M. Lakshmikanthan
Department of Chemistry
Manipal Institute of Technology
Manipal Academy of Higher Education
Manipal, India

Muhammad Farhaj Latif
Department of Pharmacology
Faculty of Pharmacy,
Bahauddin Zakariya University
Multan, Pakistan

Atul Loyal
Department of Genetics & Plant Breeding
CCS Haryana Agricultural University
Haryana, India

Massimo Lucarini
CREA-Research Centre for Food and Nutrition
Rome, Italy

Sahil Mehta
Department of Botany
Hansraj College
University of Delhi
New Delhi, India

Shuvadip Mondal
Department of Botany
Fakir Chand College
West Bengal, India

Kuljira Mongkolpobsin
Immunomodulation of Natural Products Research Unit
Chulalongkorn University
Bangkok, Thailand

Mohammad S. Mubarak
Department of Chemistry
The University of Jordan
Amman, Jordan

Sudeshna Nandi
Department of Botany
Vedanta College
Kolkata, West Bengal, India

Samapika Nandy
Department of Botany
Vedanta College
Kolkata, West Bengal, India

Arghya Naskar
Molecular and Applied Mycology and Plant Pathology Laboratory
Department of Botany
University of Calcutta
Kolkata, West Bengal, India

Naser Nasrollahzadeh-Asl
Department of Horticultural Science and Agronomy
Science and Research Branch
Islamic Azad University
Tehran, Iran

Rabia Naz
Department of Pharmacology
Faculty of Pharmacy
Bahauddin Zakariya University
Multan, Pakistan

Sunita Nilkhet
Program in Clinical Biochemistry and Molecular Medicine
Department of Clinical Chemistry
Chulalongkorn University
Bangkok, Thailand

Vedanshi Pal
Department of Life Sciences
Graphic Era Deemed to be University
Dehradun, Uttarakhand, India

Manu Pant
Department of Life Sciences
Graphic Era Deemed to be University
Dehradun, Uttarakhand, India

Swaroopa Amit Patil
Inveniolife Technology PVT LTD
Grow More Tower
Navi Mumbai, Maharashtra, India

Niccolò Pilla
CREA-Research Centre for Food and Nutrition
Rome, Italy

Keddagoda Gamage Piyumi Wasana
Department of Biochemistry
Faculty of Medicine
University of Ruhuna
Galle, Sri Lanka

Manjula Rai
Department of Botany
St. Joseph's College, Darjeeling
West Bengal, India

Shakeelur Rahman
Prakriti Bachao Foundation, Bariatu
Ranchi, Jharkhand, India

Usha Rana
Department of Biology and Environmental Sciences
CSKHPKV, Palampur
Himachal Pradesh, India

Rajeev Rathor
Department of Botany
Hansraj College
University of Delhi
New Delhi, India

Dudipala Ravivarma
Forest College and Research Institute
Mulugu, Siddipet, Telangana, India

Rupa Sanyal
Department of Botany
Bhairab Ganguly College
University of Calcutta
Kolkata, West Bengal, India

Kasturi Sarkar
Department of Microbiology
St. Xavier's College
Kolkata, West Bengal, India

Fatima Saqib
Department of Pharmacology
Faculty of Pharmacy
Bahauddin Zakariya University
Multan, Pakistan

Surjit Sen
Department of Botany
Fakir Chand College
West Bengal, India

Mansi Shah
Centre for Excellence in Marine Studies
University of Mumbai
Maharashtra, India

Rajasekharan Sharika
Immunomodulation of Natural Products Research Unit
Chulalongkorn University
Bangkok, Thailand

Radhika Sharma
Department of Soil Science
Punjab Agricultural University
Punjab, India

Mayur Mukut Murlidhar Sharma
Department of Agriculture & Industry
Kangwon National University
Chuncheon, Republic of Korea

Pankaj Sharma
Department of Microbiology
CCS Haryana Agricultural University
Haryana, India

Nitinkumar S. Shetty
Department of Chemistry
Manipal Institute of Technology
Manipal Academy of Higher Education
Manipal, India

Parames C. Sil
Department of Molecular Medicine
Bose Institute
Kolkata, West Bengal, India

Arundhati Singh
School of Science
The University of Western Australia
Perth, Australia

Tejinder Singh
Department of Agronomy
Punjab Agricultural University
Punjab, India

Tewin Tencomnao
Department of Clinical Chemistry
Faculty of Allied Health Sciences
Chulalongkorn University
Bangkok, Thailand

Kanchanlata Tungare
School of Biotechnology and Bioinformatics
D Y Patil Deemed to be University
Belapur, Maharashtra, India

Sekar Vijayakumar
Marine College
Shandong University
Weihai, PR China

Muqeet Wahid
Department of Pharmacology
Faculty of Pharmacy
Bahauddin Zakariya University
Multan, Pakistan

Lahcen Zidane
Plant, Animal Productions and Agro-industry Laboratory
Department of Biology
Faculty of Sciences
Ibn Tofail University
Kenitra, Morocco

1 Diversity and Geographic Distribution of Spice and Condiment Plant Species

Shakeelur Rahman and Azamal Husen

1.1 INTRODUCTION

The word spice derives from the Latin term *spec*, which means type of appearance. Condiments are generally a combination of different herbs and spices mixed together, prepared in a liquid form and added to food as a sauce. Small quantities of condiments are also used as preservatives and in fermentation (Dubey, 2017). One hundred and nine plant species are used as spices, as documented by the International Standard Organization and Indian Institute of Spices Research, of which more than 75 are found in India and which have been very popular all around the world since ancient times (Mathew, 2013). India is historically known as the 'the home of spices' as the country produces huge quantities of spices and condiments. Spices are cultivated and exported by Indian traders in large amounts all around the world (Singh et al., 2010). The different parts of spice plants, such as roots, rhizomes, stems, barks, leaves, flowers, buds, fruits, seeds, kernels, resins, and gums, are used to provide food taste, aroma, and flavouring, in confectionery, baking, and the processing and preserving of food (Rathore & Shekhawat, 2008). Traditionally, the spices, along with different herbs, have been used in the Indian subcontinent as medicines, in cooking, rituals, magic, and totems, as India is rich in ethnic and cultural diversity (Babu et al., 2013). The different species of spices traded worldwide are mostly produced under tropical climatic conditions. The diverse agroclimatic regions across the Indian subcontinent offer tropical, subtropical, and temperate conditions, suitable for growing almost any kind of plant species. Today, the uses of spices have increased in beverages, cooking, confectionery, liquors and cordials, medicines, perfumery, and cosmetics (Skaria et al., 2005). The different phytoconstituents and volatile essential oils of spices are used in pharmaceutical industries as a result of their antimicrobial, antioxidant, digestive stimulant, tonic, carminative, stomachic, diuretic, and antispasmodic activities, as well as their use in herbal teas and preservatives. Essential oils contain a large number of chemical compounds, and this complex mixture of compounds gives the oils their characteristic fragrances and flavours (Parthasarathy & Kandiannan, 2007; Sugasini et al., 2018).

1.2 DIVERSITY AND GEOGRAPHIC DISTRIBUTION

The different species of spices are grown in various agro-climatic conditions around the world. The production of spices and herbs has been traditionally concentrated in countries such as India, Israel, and Egypt for processing and exporting the commodities. Some of the major spice-producing countries are Bulgaria, Germany, Hungary, China, Malaysia, and Korea. However, India is the largest producer of spices as, out of 109 documented spices, about 75 are produced in that country. India is a suitable country for the cultivation of many spice plants because of its diverse geographical conditions, different soil types, range of soil nutrient and moisture contents, soil temperature, rainfall, and climates (Kodali & Muraleedhar, 2015). The species cultivated in India are grouped into tree,

DOI: 10.1201/9781003387046-1

seed, and herbal spices. The majority of these spices can be grown singly as monocultures or in combination with other crops. It is said that each state of India grows at least 10 different spices, most of them being used for domestic purposes and trade. Kerala is the capital of spice production, and a large number of spices are grown in the state. Kerala is the top producer of pepper (94.1% of the production in India), followed by Karnataka (Dubey, 2017). More than 20 seed spices are grown in India, including coriander, cumin, dill, fenugreek, and fennel seeds. More than 80% of the seed spices grown in India are cultivated in Rajasthan and Gujarat (Joshi & Singh, 2015).

It has been recorded that India produces about 7 million tonnes (Mt) of spices on 3.52 Mha of land. The spices cultivated over the largest area of India are cumin, coriander, fenugreek, and garlic (Meena et al., 2018). True, green, or small cardamom (*Eletteria cardamomum*) is one of the most valuable spices, which is mostly grown in India in the hilly areas of the Western Ghats (Ashok et al., 2020). The climatic conditions of Karnataka, Kerala, and Tamil Nadu are suitable for the cultivation of small cardamom and black pepper (Vijayan et al., 2018). False cardamom, large cardamom, and black cardamom from the related genus *Amomum* are native to Nepal, Sikkim, Bengal, and Southeast Asian countries. African cardamom (*Aframomum danielli*), is native to Southeast Africa, especially Tanzania, Cameroon, Madagascar, and Guinea. Small cardamom is extensively cultivated in Nepal and Sikkim, to a limited extent with large cardamom (*Amomum subulatum*). Cardamom is also known as the 'queen of spices' for its pleasant aroma and taste and is the third most expensive spice after saffron and vanilla (Nair, 2011). Odisha and Tamil Nadu are two states in India renowned for the production of huge amounts of turmeric (Dhanalakshmi et al., 2018). Production of chilli is very high in some Indian states, namely Andhra Pradesh and Karnataka, because of the favorable climatic conditions. Important spices of India and the world are summarized here, with scientific names, common names, family names, and useful parts in Table 1.1. The plants producing a number of commonly used spices and condiments are illustrated in Figure 1.1, and will subsequently be described in different chapters of this book.

TABLE 1.1
Common and Important Spices

Family	Botanical Name	Part Used	Common Name
Acoraceae	*Acorus calamus* L.	Rhizome	Sweet flag, myrtle flag, calamus
Amaryllidaceae	*Allium ascalonicum* L.	Bulb	Shallot
Amaryllidaceae	*Allium cepa* L.	Bulb	Onion
Amaryllidaceae	*Allium cepa* var. *aggregatum* G. Don	Bulb	Potato onion
Amaryllidaceae	*Allium fistulosum* L.	Bulb, leaves	Japanese bunching onion, stony
Amaryllidaceae	*Allium porrum* L.	Bulb, leaves	Leek, winter leek
Amaryllidaceae	*Allium sativum* L.	Bulb	Garlic
Amaryllidaceae	*Allium tuberosum* Rottler.	Bulb	Indian leek
Anacardiaceae	*Buchanania lanzan* Spreng.	Seed	Cuddapah almond
Anacardiaceae	*Mangifera indica* L.	Immature fruit	Mango
Anacardiaceae	*Schinus ole* L.	Fruit	Californian pepper tree
Anacardiaceae	*Schinus terebinthifolia* Raddi	Fruit	Brazilian pepper tree
Annonaceae	*Xylopia aethiopica* (Dunal) A. Rich	Fruit	Negro pepper
Apiaceae	*Anethum graveolens* L.	Leaves	Dill
Apiaceae	*Anethum sowa* Kurz.	Fruit	Indian dill
Apiaceae	*Angelica archangelica* L.	Fruit, petiole,	Garden angelica

(*Continued*)

TABLE 1.1 (CONTINUED)
Common and Important Spices

Family	Botanical Name	Part Used	Common Name
Apiaceae	*Anthriscus cerasifolium* (L.) Hoffm.	Leaf	Chervil
Apiaceae	*Apium graveolens* L.	Seeds, root, leaves	Celery
Apiaceae	*Bunium persicum* (Boiss.) B. Fedtsch	Fruit, bulb	Black caraway
Apiaceae	*Carum bulbocastanum* L.	Fruit, bulb	Black caraway
Apiaceae	*Coriandrum sativum* L.	Leaves, fruits	Coriander
Apiaceae	*Carum copticum* L.	Fruits	Ajowan caraway, blond caraway
Apiaceae	*Cuminum cyminum* L.	Fruits	Cumin
Apiaceae	*Eryngium foetidum* L.	Fruits	Elephant pepper
Apiaceae	*Ferula asa foetida* L.	Rhizome	Asafoetida
Apiaceae	*Foeniculum vulgare* Miller.	Seeds	Fennel
Apiaceae	*Levisticum officinale* Koch.	Leaves	Lovage
Apiaceae	*Pimpinella anisum* L.	Fruits	Aniseed
Asteraceae	*Artemisia dracunculus* L.	Leaves	Tarragon
Brassicaceae	*Armoracia rusticana* Gaertn.	Root	Horseradish
Brassicaceae	*Brassica juncea* (L.) Czernj. et Cosson.	Seeds	Mustard
Brassicaceae	*Brassica nigra* (L.) Koch.	Seeds	Black mustard
Brassicaceae	*Brassica rapa* (L.) Hanelt.	Seed	Field mustard
Brassicaceae	*Sinapis alba* L.	Seeds	White mustard
Caesalpiniaceae	*Tamarindus indica* L.	Fruit	Tamarind
Lamiaceae	*Mentha spicata* L.	Leaves	Spearmint, garden mint
Lamiaceae	*Ocimum bascilicum* L.	Leaves	Sweet basil
Lamiaceae	*Ocimum sanctum* L.	Leaves	Holy basil
Lamiaceae	*Origanum hortensis* L.	Leafy shoots	Marjoram
Lamiaceae	*Origanum vulgare* L.	Leaves	Oregano
Lamiaceae	*Petroselinum crispum* (Mill.) Fuss.	Leaves	Parsley
Lamiaceae	*Rosmarinus officinalis* L.	Leaves	Rosemary
Lamiaceae	*Salvia officinalis* L.	Leaves	Garden sage, Dalmatian sage
Lamiaceae	*Satureja hortensis* L.	Leaves	Summer savory
Lamiaceae	*Thymus vulgaris* L.	Leaves	Thyme
Lauraceae	*Cinnamomum burmanii*	Bark	Indonesian Cassia
Lauraceae	*Cinnamomum cassia* L.	Bark	Cinnamon cassia
Lauraceae	*Cinnamomum loureiroi* Meissn.	Bark	Vietnamese cassia bark
Lauraceae	*Cinnamomum tamala* Nees.	Leaves	Indian cassia bark
Lauraceae	*Cinnamomum verum* Blume.	Leaves	Sri Lankan cinnamon, common
Lauraceae	*Laurus nobilis* L.	Leaves	Bay leaf, true laurel
Lauraceae	*Phoebe cooperiana* U.N Kanjilal ex A. Das	Leaves	Sisir
Malvaceae	*Bombax ceiba* L.	Fruit	Maratimokka, cotton tree
Poaceae	*Cymbopogon citratus* (DC.) Stapf.	Leaves	Lemon grass, West Indian lemon
Poaceae	*Cymbopogon nardus* L.	Leaves	Sri Lankan citronella
Punicaceae	*Punica granatum* L.	Seeds	Pomegranate

(Continued)

TABLE 1.1 (CONTINUED)
Common and Important Spices

Family	Botanical Name	Part Used	Common Name
Ranunculaceae	*Nigella damascena* L.	Seeds	Damas black cumin
Ranunculaceae	*Nigella sativa* L.	Seeds	Black cumin
Rutaceae	*Murraya koenigii* (L.) Spreng.	Leaves	Curry leaf
Rutaceae	*Zanthoxylum armatum* DC.	Fruit	Mountain prickle pepper
Rutaceae	*Zanthoxylum bungee* Planch.	Fruit	Chinese prickly ash pepper
Rutaceae	*Zanthoxylum piperitum* (L.) DC	Fruit	Japanese pepper
Rutaceae	*Zanthoxylum oxyphyllum* Edgew.	Fruit	Wild thorn pepper
Rutaceae	*Zanthoxylum rhetsa* DC.	Fruit	Small thorn pepper
Solanaceae	*Capsicum annuum* L.	Fruit	Wild chilli, capsicum
Solanaceae	*Capsicum frutescens* L.	Fruit	Bird eye chilli
Zingiberaceae	*Kaempferia galanga* L.	Rhizome	Galangal
Zingiberaceae	*Aframomum angustifolium* (Sonn.) K. Schum.	Fruit, seeds	Madagascar cardamom
Zingiberaceae	*Aframomum hanburyi* Schum.	Fruit, seeds	Cameroon cardamom
Zingiberaceae	*Aframomum corrorim* Engl.	Fruit, Seeds	Korarima cardamom

1.3 MEDICINAL PROPERTIES

Most of the spices have medicinal properties which have been used in the treatment of different ailments. More or less all the spices and condiments are aromatic in nature, involving their respective volatile phytochemicals, usually essential oils consisting of terpenoids and oleoresins. The essential oils and oleoresins impart the specific aroma of the spice to the food, act as antioxidants, are active against microbes, preserve food from chemical deterioration, and increase the shelf-life of the food. The spices function as natural preservatives. The quality of the spice is determined by the composition of the respective phytochemicals present in them (Peter, 1998). A number of spices have been identified as having antimicrobial properties. Flavonoids and essential oils present in spices and condiments act against a range of human illnesses, including digestive problems, cardiovascular disease, nervous diseases, and cancer (Johnson, 2000). Spices possess various kinds of phytochemicals which are effective against a large number of diseases. Some examples of spices which contain valuable phytochemicals are black pepper seeds, containing piperine, guineensine, piperamide, and piperolein-B, and hemp seeds, containing cannabinol, tetrahydrocannabinol, cannabidiol, tetrahydrocannabivarin, cannabivarin, and cannabichromene, whereas curry leaves contain α-amorphene, γ-eudesmol, β-pinene, α-terpinolene, limonene, (Z)-β-ocimene, eucalyptol, terpinyl acetate, and other terpenes and sesquiterpenes (Roy, 2019). Similarly, gingerol in ginger, myristin in nutmeg, eugenol in allspice, cinnamon, cassia, and clove, vanillin in vanilla, and curcumin in turmeric are important pharmaceutical compounds with valuable medicinal properties (Nirmal et al., 2000). Capsaicin in chilli is an effective counterirritant used in both pharmaceuticals and cosmetics. Fenugreek, onion, and garlic help to lower cholesterol levels (Kumaravel & Alagusundaram, 2014). Fenugreek seeds are a rich source of soluble fibre which help to reduce blood glucose and lipids and can be used as a food adjuvant in the management of diabetes. Similarly, garlic, onions, and ginger have been found to adjust the process of anti-carcinogenesis favourably (Krishnaswamy, 2008). Traditionally, turmeric is used for wound healing, rheumatic disorders, gastrointestinal symptoms, deworming, rhinitis, and as a cosmetic. Scientific studies have explored its anti-inflammatory, cholekinetic, and antioxidant potentials, with it having beneficial effects on pre-carcinogenic, anti-inflammatory, and anti-atherosclerotic activities. Turmeric and curcumin increase levels

FIGURE 1.1 Illustrations of some common and important plants producing spices and condiments.

of detoxifying enzymes, prevent DNA damage, improve DNA repair, decrease mutation rates and tumour formation, and exhibit antioxidative potential (Manigandan & Saranraj, 2018).

1.4 TRADITIONAL USES

One of the ancient scriptures, Rig Veda, is a good source of information on spices, listing more than one thousand medicinally valuable plants, including spices and condiments (Patil & Dhale, 2013). The traditional use of turmeric is mentioned in the history of the Indus Valley civilization (Velayudhan et al., 2012). Ethnic communities in different parts of India and other countries have considerable knowledge about the formulations and uses of a large number of spice plant species, which have been documented by researchers. Tribal people of Andhra Pradesh, Arunachal Pradesh, Assam, Maharashtra, Jharkhand, Chhatisgarh, and other parts of India have deep knowledge on the use of some native plants as spices and condiments, such as dried fruits of *Tacca leontopetaloides* and tubers of *Decalepis hamiltonii*, *Garcinia pedunculata*, *Hibiscus sabdariffa*, *Houttuynia cordata*, *Neocinnamomum caudatum*, and *Peperomia pellucida* (Reddy, 2007; Bharali et al., 2017). Some of the tribes of Maharashtra have discovered a new spice species, *Guizotia abyssinica*, which is used as a food and flavouring agent (Tayade & Patil, 2010). Some of the aromatic plant species are used as condiments by mixing with pulses and vegetables; these include *Murraya koenigii*, *Artemisia scoparia*, and *Carum carvi* (Bhatia et al., 2018). The Tangkhul tribe of Manipur are heavily dependent on forest-based fuel, food, and medicines. The community uses about 30 plant species as spices. Some of the new species reported by the tribal people are *Allium hookeri*, *Allium tuberosum*, *Costus speciosus*, *Curcuma cassia*, *Elsholtzia blanda*, *Elsholtzia communis*, *Hedychium coronarium*, *Hedychium marginatum*, *Persicaria posumbu*, *Zingiber cassumunar*, and *Zingiber zerumbet* (Salam & Jamir, 2016). The tribal communities of Sikkim use around 14 plant species as spices and preservatives, as well as for medicinal purposes such as antimicrobial and antioxidant activities (Gudade et al., 2015). The people of Kerala and Tamil Nadu have endorsed 11 species as ISO (Salma & Ramakrishanan, 2017). The people of the two states use cardamom tea to obtain relief from anxiety and weakness (Ashok et al., 2020a). The tribal people of the Western Ghats use a large number of spices and condiments as food and medicines, including *Costus speciosus*, *Garcinia gummi-gutta*, *Curcuma neilgherrensis*, *Syzygium palghatense*, *Curculigo orchioides*, and *Cleome monophylla* (Yesodharan & Sujana, 2007). The traditional knowledge system of medicinal plants has gradually been recognized by modern biological science in the development of drugs against various diseases (Sharma & Mujumdar, 2003).

1.5 CONCLUSIONS AND FUTURE REMARKS

Spices are among the important parts of human culture and civilization. Spices and condiments are good natural sources of colours, flavours, aromas, and preservatives of food and beverages. Spices are derived from plant parts, such as bark, buds, flowers, fruits, leaves, rhizomes, roots, seeds, stigmas, styles, or the entire plant tops. Spices are often dried and used in a processed but complete state. The extracts of spices are also used in the form of essential oils and oleoresins. The spices, condiments, and their extracts have wide application in the food and pharmaceutical industries. The most important spices, traditionally traded throughout the world, are products of tropical environments. The major exceptions to this group are the capsicums (chilli peppers, paprika) and coriander, which are grown over a much wider range of tropical and non-tropical environments. In terms of world trade value, the most important spice crops from the tropical regions are pepper, capsicums, nutmeg/mace, cardamom, allspice/pimento, vanilla, cloves, ginger, cinnamon, cassia, and turmeric. Coriander, cumin, mustard, and sesame seeds and the herbs sage, oregano, thyme, bay, and mint are the most important spice crops from non-tropical environments. There are more than 50 spices of global economic and culinary importance. Some other spices, which are used by tribal communities in traditional cooking, have yet to reach any significant level of trade. The introduction of spices to nontraditional areas must be carried out in order to conserve the germplasm. To develop climate-dependent, high-yielding spice varieties, new improved spice crops are to be developed through modern research tools. The introduction of suitable modern hybrids, taking into account the market demand, with a greater emphasis on exploitation of exotic wild species in breeding programmes, needs to be considered. To increase the productivity of spices and condiments, the integration of advanced farming practices with novel technologies and ecofriendly traditional production technologies among the farming community is the current aim.

REFERENCES

Ashok, K., Murugan, M., Dhanya, M.K. & Warkentin, T.D. 2020. Botany, traditional uses, phytochemistry and biological activities of cardamom [*Elettaria cardamomum* (L.) Maton] – A critical review. *Journal of Ethnopharmacology* 246: 112244.

Ashok, K., Pandian, A. & Murugan, M. 2020a.Profiling bioactive flavonoids and carotenoids in select south Indian spices and nuts. *Natural Product Research* 34(9): 1306–1310.

Babu, N., Srivastava, S.K. & Agarwal, S. 2013.Traditional storage practices of spices and condiments in Odisha. *Indian Journal of Traditional Knowledge* 12: 518–523.

Bharali, P., Sharma, C.L., Singh, B. & Sharma, M. 2017. Ethnobotanical studies of spice and condiment plants used by some communities of Assam. *International Journal of Advanced Science and Research* 3(1): 01–11.

Bhatia, H., Sharma, Y.P., Manhas, R.K. & Kumar, K. 2018. Traditionally used wild edible plants of district Udhampur, J&K, India. *Journal of Ethnobiology and Ethnomedicine* 14(1): 73.

Dhanalakshmi, K., Chitra, K. & Menimekalai, R. 2018. Production and economics of turmeric cultivation. *International Journal of Current Microbiology and Applied Sciences* 7(11): 3496–3502.

Dubey, S. 2017. Indian spices and their medicinal value. *Indian Journal of Pharmaceutical Education and Research* 51(3s2): 5330–5332.

Gudade, B.A., Babu, S. & Deka, T.N. 2015. Spices biodiversity and their ethnomedicinal uses by tribal community of Sikkim. *India Vegetos* 28: 141–145.

Johnson, I.T. 2000. Anti-tumour properties. In: *Functional Foods: Concept to Product* (eds. G.R. Gibson & C.M. Williams), 141–166. Cambridge: Woodhead Publishing Ltd.

Joshi, D. & Singh, H.P. 2015. An empirical analysis of growth and instability in major spices in India. *International Journal of Agricultural Science* 7: 440–444.

Kodali, R.K. & Muraleedhar, A. 2015. WSN in spice cultivation. *International Conference on Green Computing and Internet of Things (ICGCIoT)*, 8–10 October, 1173–1177.

Krishnaswamy, K. 2008. Traditional Indian spices and their health significance. *Asia Pacific Journal of Clinical Nutrition* 17: 265–268.

Kumaravel, S. & Alagusundaram, K. 2014. Antimicrobial activity and phytochemical analysis of selected Indian spices. *Journal of Pure and Applied Microbiology* 8: 4131–4136.

Manigandan, M. & Saranraj, P. 2018. Antimicrobial activity of secondary metabolites in an Indian spices – A review. *Life Science Arch* 4(1): 1248–1259.

Mathew, J. 2013. Indigenous aromatic and spice plants described in Van Rheed's Hortus Indici Malabarici. *Indian Journal of Applied Research* 3(11): 30–33.

Meena, M.D., Lal, G., Meena, S.S. & Meena, N.K. 2018. Production and export performances of major seed spices in India during pre and post-WTO period. *International Journal of Seed Spices* 8: 21–30.

Nair, K.P.P. 2011. The agronomy and economy of cardamom (*Elettaria cardamomum* M.): The "Queen of Spices". In: *Agronomy and Economy of Black Pepper and Cardamom* (ed. K.P.P. Nair), 109–366. London: Elsevier

Nirmal, B.K., Ravindran, P.N. & Peter, K.V. 2000. Biotechnology of spices. In: *Biotechnology in Horticulture and Plantation Crops* (eds. K.L. Chadha, P.N. Ravindran & L. Sahijram). New Delhi: Malhotra Publishing House.

Parthasarathy, V.A. & Kandiannan, K. 2007. *Horticulture: Vegetable Science: Spices and Condiments*. Calicut: Indian Institute of Spice Research (ICAR).

Patil, D.A. & Dhale, D.A. 2013. *Spices and Condiments Origin, History and Applications*. New Delhi: Astral International Pvt. Ltd.

Peter, K.V. 1998. Spices research. *Indian Journal of Agricultural Science* 68: 527–532.

Rathore, M.S. & Shekhawat, N.S. 2008. Incredible spices of India: from traditions to cuisine. *American-Eurasian Journal of Botany* 1(3): 85–89.

Reddy, K.N., Pattanaik, C., Reddy, C.S. & Raju, V.S. 2007. Traditional knowledge on wild food plants in Andhra Pradesh. *Indian Journal of Traditional Knowledge* 6: 223–229.

Roy, L. 2019. Phytochemical profile of medicinal Indian spices (Indian medicinal spices). *International Journal of Pharmacy and Biological Science* 10: 105–111.

Salam, S. & Jamir, N.S. 2016. Common spices plant used as medicine by the Tangkhul Tribe of Ukhrul District, Manipur. *International Journal of Science and Research* 6: 22–25.

Salma, S. & Ramakrishanan, L. 2017. Study on most frequently consumed spices in households and the consumer's knowledge and preference to organic spices in Coimbatore district, Tamil Nadu, India. *International Journal of Herbal Medicine* 5: 49–52.

Sharma, P. & Mujumdar, A.M. 2003. Traditional knowledge on plants from Toranmal plateau of Maharashtra. *Indian Journal of Traditional Knowledge* 2: 292–296.

Singh, A.K., Verma, N., Tyagi, V. & Dimree, S. 2010. Indian needs of crop genetic resources setting priorities. *Progressive Agriculture* 10: 1–16.

Skaria, B.P., Joy, P.P., Mathew, G. & Mathew, S. 2005. Spices of Kerala in Taraditional medicine. *National Seminar on Recent Advances in Medicinal and Aromatic Plants*, Meerut (15–17 February 2005).

Sugasini, D., Yalagala, C.R.P. & Kavitha, B. 2018. Indian culinary ethnic spices uses in foods and palate of paradise. *Acta Scientific Nutritional Health* 2: 22–28.

Tayade, S.K. & Patil, D.A. 2010. Ethnomedicinal applications of spices and condiments in Nandurbar district (Maharashtra). *Journal of Ecobiotechnology* 2: 8–10.

Velayudhan, K.C., Dikshit, N. & Nizar, M.A. 2012. Ethnobotany of turmeric (*Curcuma longa* L.). *Indian Journal of Traditional Knowledge* 11: 607–614.

Vijayan, A.K., Pradip, K. & Remashree, A.P. 2018. Small cardamom production technology and future prospects. *International Journal of Agricultural Science* 16: 6943–6948.

Yesodharan, K. & Sujana, K.A. 2007. Wild edible plants traditionally used by the tribes in the Parambikulam Wildlife Sanctuary, Kerala, India. *Natural Product Radiance* 6: 74–80.

2 Biology and Genetic Improvement of Spice and Condiment Plant Species

Shruti Chattaraj, Sudeshna Nandi, and Samapika Nandy

2.1 INTRODUCTION

Spices and condiments are natural plant products, fresh or dried, and have been important parts of cooking since time immemorial. They are traditionally added to food to enhance flavour, taste, and colour, as well as imparting aroma, for the purpose of seasoning and preservation (Asowata-Ayodele et al., 2016; European Spice Association, 2020). The UN Food and Agriculture Organization defines spices and condiments as vegetable products added to food for flavouring, seasoning, and aroma (Food and Agriculture Organization, 2011). They are also used as ingredients in beverages like soups, pharmaceuticals, cosmetics, liquors, and perfumery products. In addition, some of the spices and condiments show important activities like antibacterial, antibiotic, antioxidative, and fungi-static properties (Wendimu et al., 2021). The antioxidative properties of spices enhance the shelf life of food by preserving it from oxidative deterioration. Some of the spices have medicinal value as well. Since ancient times, different roles and applications of spices have been described. Spices are mainly represented by different plant parts such as rhizomes, bark, bulbs, seeds, flowers, fruits, roots, or leaves (Pruthi, 1992).

From prehistoric times, India has been known as the 'Land of Spices' as, here, more than 60 spices are grown out of 109 spice plants reported by the International Organization for Standardization. It is believed that many important spices like black pepper, cardamom, etc. (Ravindran, 2000) originated in the Western Ghats of India, and these Indian spices are used in over 130 countries for the seasoning and flavouring of foods. The crucial spices exported from India to different countries include 10% of cumin and turmeric, 9.5% of coriander, more than 5% of fenugreek and peppers, and 19% of other spices and condiments (Dubey, 2017).

Nowadays with the huge increase in population growth, an intense pressure is developing to enhance the production and yield of the quality spices and condiments to meet the social demands, health requirements, environmental stress, and ecological considerations. The crop improvement approaches try to fulfil these increasing demands by enhancing the yield, quantity, and quality of the crops. But in spice and condiments, the major problem is the lack of disease-resistant varieties, the absence of high-yielding varieties along with the presence of little genetic variability in spice plants. Breeding methods have shown some good results in overcoming these constraints but these methods come with some limitations and drawbacks. The new biotechnological tools came out with some promising results to solve these problems by the development of new breeding lines, transgenic, and novel varieties of spices by recombinant DNA technologies.

DOI: 10.1201/9781003387046-2

2.2 BIOLOGY OF SOME MEDICINALLY IMPORTANT SPICES AND CONDIMENTS

Since ancient times, spices and condiments have played vital roles in the lifestyle of people as important flavouring or seasoning ingredients or as preservatives or colouring agents, or as medicines for the treatment of ailments (Pandey et al., 2016). The molecular basis of the medicinal and therapeutic properties of common kitchen spices are attributed to some active phytochemicals present in the spices and condiments. Some of the commonly used spices with medicinal and therapeutic properties are black pepper, clove, coriander, cinnamon, star anise, nutmeg, cumin, mustard, ginger, etc.

2.2.1 Black Pepper

Black pepper (*Piper nigrum* L.) is an annual vine in the family Piperaceae (Hill, 2004). It is commonly known as the 'King of Spices', and originated in the tropical evergreen forests of the Western Ghats of India. The coastal climate and hilly regions with soils rich in humus are ideal for its cultivation. The plant is a perennial woody evergreen climber, which grows to a height of 10 m or more. The dimorphic branches, i.e., the orthotropic climbing vegetative and the plageotropic fruiting branches, arise from the nodes of the stem. Each node also bears short adventitious roots which help the plant to cling to their supports. The simple, broadly lanceolate leaves are arranged alternately with 2- to 5-cm-long petioles. The inflorescence is a catkin-like spike emerging on the plagiotropic branches opposite upper leaves. About 50 to 120 white to pale yellow unisexual or hermaphrodite flowers develop on the inflorescence with two stamens on either side of the ovary. The stigma bears 3–5 fleshy branches, with a very short style. The bisexual flowers are protogynous. The fruit is a sessile drupe or berry, globose, with a thin, soft pericarp surrounding the seed.

2.2.1.1 Medicinal Properties

Black pepper has good antioxidant properties, relieves coughs and colds, is helpful in treating skin problems, helps in controlling type 2 diabetes, has anti-cancer properties, increases the absorption of essential nutrients, and stimulates digestion.

2.2.2 Cloves

It is one of the most valuable spices of the Myrtaceae family. *Syzygium aromaticum* L. (syn. *Eugenia caryophyllata*) is a tropical evergreen tree with small reddish brown flower buds (Duke, 2002). Basically the dried aromatic flower buds of the clove tree are used as the spice. The word 'clove' is derived from word '*cloy*', a French word meaning 'nail' because of the resemblance of the flower buds of clove with the shape of small irregular nails. Cloves are indigenous to the Indonesia's Moluku Island. It grows well in deep loam, black, and laterite soils rich in humus and in tropical and humid climates.

The clove tree is tropical and evergreen, growing to a height of 8–15 m. The tree trunk is roughly cylindrical and forked with two or three main trunks, with a bushy and dense appearance in the terminal region. The leaves are broad, lanceolate, simple, exstipulate and aromatic due to the presence of oil glands on the lower surface. The younger leaves appear bright pink but, with maturity, become dark green with a glossy upper surface. The inflorescence is a cyme with terminal branching, having 3 to 20 hermaphrodite florets. The flower contains a fleshy hypanthium surrounded by the calyx and four whitish dome-shaped petals above the calyx. Numerous stamens with pale yellow anthers, containing a small pale brown connective oil gland, are found along with a very stout, pale green style dotted with oil glands. The ovary is inferior, two-celled, and multi-ovate, embedded on the hypanthium. The fruit is a fleshy, single-seeded drupe in which the calyx is reduced and forms triangular projections. The dry fruit is appreciated as a spice for its essential oil content. The volatile oil from cloves is rich in eugenol and caryophyllene.

2.2.2.1 Medicinal Properties

The clove shows antioxidative properties, antiseptic, anaesthetic, antibacterial, anti-inflammatory, and anti-cancer activities. The essential oil is used in the treatment of coughs, bronchitis, gastrohelcosis, and cephalalgia, fever, headache, asthma, depression, and indigestion. The clove oil is also used for the treatment of toothache and for dental care.

2.2.3 Cinnamon

The word 'cinnamon' is derived from the Greek *'kinnám ō mon'*, which itself is a Phoenician word. Botanically, cinnamon is *Cinnamomum zeylanicum* (L.) Farw. of the Lauraceae family. Its binomial is derived from Ceylon, the former name of Sri Lanka, as it is mainly present in Sri Lanka, India, Bangladesh, and Nepal (Ballabh et al., 2008). In India, Assam, the Western Ghats of Kerala, and Tamil Nadu are the major producers. Cinnamon, commonly called Dalchini or Dorchini, is one of the oldest known spices in India. The dried bark, mainly as a small piece or powder, is used as spice or condiment due to its strong aroma and sweet taste (Kumar, 2010).

Cinnamon comes from an evergreen, bushy tree of about 6–15 m height. It grows in tropical regions, on sandy loam soils rich in humus. The leaves are ovate or elliptic with a 1- to 2-cm-long stiff petiole. The bases of the leaf are strongly three veined and younger leaves are reddish whereas the matured ones become dark green. The greenish to yellow flowers are present in terminal and axillary panicles at the end of the twigs. The flowers are very small 3 mm in diameter, with small ovate hairy bracts. The stamens are arranged in three whorls with four-celled anthers and hairy filaments; the base of the stamens contains glands. The fruits are dark purple in colour and are one-seeded fleshy berries, 1.5–2 cm long with an enlarged calyx. The tree is valued for its leaves and bark, which have a strong aroma. The seeds of cinnamon contain 3% fixed oil, used in candle making. The leaf oil is used for production of perfumes, toothpastes, hair oil, and soaps. Cinnamic acid from cinnamon is used in flavouring and pharmaceuticals.

2.2.3.1 Medicinal Properties

Cinnamon acts as an antimicrobial, antidiabetic, and antioxidant agent.In addition, immuno-modulatory and wound healing properties have also been reported. It acts against some neurological disorders, like Alzheimer's and Parkinson's diseases.

2.2.4 Ginger

Ginger (*Zingiber officinale* Roscoe) of the Zingiberaceae family is the oldest known spice, valued for its rhizome in both the fresh and dry forms. Ginger originated in Southeast Asia and later spread in the sub-tropical and tropical regions of both hemispheres. Found mainly in humid, partly shaded habitats, it grows on heavy laterite loams to clayey loam soils. Ginger is an herbaceous, monocotyledonous, perennial plant with a well-developed, branched rhizome developed horizontally near the soil surface and an annual pseudostem, formed from leaf sheaths wrapped around one another, which is erect up to about one metre in height. The leaves are linear, lanceolate, and sub-sessile with a thin lamina, rounded base, and a slender tip. The thick rhizome has a brown thin outer skin and a pale-yellow inner material, with many internodes having thin scales. The inflorescence develops from the rhizome and is a cylindrical spike with a cone-like appearance and appressed ovate or elliptical bracts. The flower develops from the axil of bracts. The calyx is tubular and three-toothed, while the corolla has three yellow lobes. Three stamens are opposite the flower labellum or lip, with short, broad filaments and an cream-coloured anther at the tip of each. With two slender free styloids, the trilocular, inferior ovary has multiple ovules per locule.

2.2.4.1 Medicinal Properties

Gingerol in ginger is the main bioactive compound, having antioxidant, anti-inflammatory, anti-nausea and, anti-carcinogenic properties, and is used in the treatment of cardiovascular diseases.

2.2.5 Turmeric

Turmeric (*Curcuma longa* L.) of the ginger family (Zingiberaceae) originated in Southeast Asia. The word 'turmeric' derives from '*terra merita*', meaning 'meritorious earth' in Latin, because of the colour of ground turmeric. The dried rhizome of turmeric is used as spice or flavourant, the yellow-coloured pigment curcumin having a high medicinal value. It is an herbaceous, erect, perennial plant growing annually to reach about 50–100 cm in height. It is cultivated largely in India, and other tropical countries like Sri lanka, China, Thailand, etc. The underground rhizome is brownish and scaly, bearing branches in two rows to form a dense clump. The rhizome appears bright orange in the inside and has a distinctive smell and taste. The leafy erect shoot arising from the rhizome as a pseudostem formed from the leaf sheaths. The leaves have a thin petiole with a broad lanceolate leaf lamina. The flowers develop at the axils of pale green or purple-tinted bracts. The calyx is short and the yellow-orange corolla is tubular at the base and half-cup-shaped at the upper region. It bears two staminodes, each with a short filament, and cylindrical stylodes about 4 mm long. The inferior, trilocular ovary bears a slender style.

2.2.5.1 Medicinal Properties

The medicinal values of turmeric have been known since ancient times and it is used as a stomachic, blood purifier, and antiseptic, and to treat asthma, sore throat, skin disorders, etc. It is used for the preparation of ointments and medicinal oils.

2.2.6 Nutmeg

Nutmeg (*Myristica fragrans* Houtt.) belongs to the Myristicaceae family. Nutmeg is an evergreen tree about 10–20 m tall, with spreading branches and shiny, oval to oblong leaves. The word nutmeg was derived from Latin '*nux muscatus*' which means 'musky nut'. It originated in Indonesia's Banda Islands. The cyme inflorescence bears several small, pale-yellow flowers which are bell-shaped. The tree is valued for its seed (nutmeg) and seed covering (mace), both of which are aromatic and used as spices. The fleshy fruit contains one glossy brown seed with a hard shell or aril.

2.2.6.1 Medicinal Properties

The nutmeg is valued for its antioxidant, anti-diabetic, anti bacterial properties. The volatile oils from nutmeg, like elemicin, safrole, and myristicin, are used for treating joint and muscle pain as they have anti-inflammatory properties. Nutmeg shows carminative and calming effects and is used for treating depression.

2.3 GENETIC IMPROVEMENT OF SPICES AND CONDIMENTS

Spice crop improvement since ancient times has mainly depended on different breeding methods like mass selection, pedigree methods, backcross methods, recurrent selection, and single seed descent methods. These methods aimed to produce high-yielding varieties with broad adaptation and enhanced quality, along with pest and disease resistance. Over the past fifty years, conventional breeding methods have significantly contributed to spice crop improvement but in the recent past, with the population explosion, increasing demand, increasing biotic and abiotic stresses, and decreased area of agricultural land, the output of these methods are becoming limited. The barriers to breeding methods are due to limited gene pool and species barrier, resulting in gene transfer being restricted to a limited range of related plant species. In addition, the conventional methods are time consuming and cumbersome. In the recent past with the advances in biotechnological tools, some promising possibilities have emerged to overcome the above constraints in a shorter period of time. The biotechnological methods, like micropropagation, somaclonal variation, protoplast technology, *in-vitro* cryopreservation, synseed and micro-rhizome technology, molecular markers, and marker-assisted selection, are gaining in importance and showing some positive results in spice crop improvement. Recently, a revolution in the field of plant science research has occurred with the development of new 'omics' technologies. The technologies includes genomics, proteomics, and metabolomics, which enable researchers to identify and study the gene responsible for enhanced productivity and improved quality of the crop. In addition, these technologies help to monitor the components affecting the yield and other properties of crops, and furnish data which could be utilized for the development of improved conventional and genetically modified crops. Here, we discuss some of the strategies used in the development of genetically improved varieties of important spice and condiment crops.

2.3.1 Biotechnological Approaches

2.3.1.1 Micropropagation and Plant Regeneration

Micropropagation is one of the best tissue culture technologies which helps achieve rapid and mass-scale multiplication of some of the important spice crops, which are difficult to multiply by conventional propagation methods (Nirmal Babu & Minoo, 2003). In addition, micropropagation helps to developing disease-free plantlets (Shivakumar, 2014). It is the most efficient and successful method of the tissue culture technologies in terms of commercial application. Micropropagation has been reported in many spice and condiment crops, such as black pepper, cardamom, ginger, turmeric, curcuma, vanilla, capsicum, etc. Furthermore, for genetic modification and improvement of any spice crop species, an efficient protocol for rapid plant propagation is very important.

In black pepper, the successful micropropagation and hardening-off from different explants like shoot, leaf has been reported (Nirmal Babu et al., 2005a, 2012b). Nazeem et al. (2004) reported a viable protocol for *in-vitro* micropropagation of black pepper for commercial large-scale production. The resulting micropropagated and tissue-cultured plants showed superiority when compared with conventionally propagated ones in terms of plant growth in the field, early flower production, plant height, number of leaves, root system size, etc. (Nirmal Babu et al., 2011). In addition, protocols for the rapid multiplication of medicinally valuable species of *Piper* (*Piper longum, Piper betle,* and *Piper chaba*) and endangered *Piper* species (*Piper colubrium* and *Piper barberi*) were reported by Nirmal Babu et al. (2012b). Somatic embryogenesis in black pepper was also reported by different group of researchers from various explants like zygotic embryos (Joseph et al., 1996; Yamuna, 2007), mature leaf tissues (Nirmal Babu et al., 2005a), germinating seed tissue, and from maternal tissues, aas well as by cyclic somatic embryogenesis was reported by Nair and Gupta (2003, 2006), which described plant regeneration at high frequencies. For production of *Piper* varieties with *Phytophthora* foot rot resistance, *in-vitro* screening for resistance to crude culture filtrate and *Phytophthora capsici* metabolite was performed by Shylala et al. (1996). These screening methods helped to identify somaclones with tolerance to *Phytophthora* foot rot.

Rapid *in-vitro* clonal propagation methods for several other spices and condiments like cardamom, ginger, turmeric, different species of *Curcuma*, vanilla, clove, and cinnamon have been standardized and reported by different research groups using various tissues as explants. Some micropropagated spices and the type of explants used are listed in Table 2.1

TABLE 2.1
Some Micropropagated Spices and Condiments with the Type of Explants Used

Spice or Condiment	Explants Used/ Somatic Embryogenesis	Achievements and Salient Features of Micropropagant/Somaclones	Key References
Black pepper and other related species	Shoot tip and leaf. Shoot tip and nodal tissue. Zygotic embryos and mature leaf tissue.	Callus culture derived from shoot tip and leaf explants successfully regenerated into plants. Tissue-cultured plants showed superiority in yield and other characters compared to conventionally developed plants.	Nazeem et al., 2004; Nirmal Babu et al., 2012b
	Cyclic somatic embryogenesis from maternal tissue (integuments).	Standardised protocols for other medicinally important species of *Piper* were also reported.	Sarasan et. al., 1993; Nirmal Babu et al., 2012b
		Direct shoot regeneration in SH and MH media.	Maju & Sonia, 2012
		Somatic emryogenesis resulting in automated micropropagation.	Nirmal Babu et al., 2005a; Yamuna, 2007
		Resulted somatic embryos with enormous potential for automated micropropagation.	Nair & Gupta, 2003, 2006
		The micropropagation protocols helped in induction and identification of *Phytophthora* foot rot tolerant somaclones.	Shylala et al., 1996
Cardamom	Seedling and anther. Somatic embryogenesis.	Plantlets successfully regenerated from callus of seedlings and anthers.	Rao et al., 1982; Nirmal Babu, 1997; Nirmal Babu et al., 2011b
	Rhizome, vegetative bud-derived callus.	Effective protocol for somatic embryogenesis and plant regeneration	Manohari et al., 2008
		Somaclones developed with Katte disease-resistant genotype.	Nirmal babu et al., 1997, Peter et al., 2001
Ginger	Vegetative bud.	Clonal multiplication	Nadgauda et al., 1980;
	Leaf-derived callus, vegetative bud, ovary, anthers, and immature anthers.	Tissue culture helped in development of pathogen-free plant material for superior and high-yielding varieties.	Nirmal Babu et al., 1998; Sharma & Singh, 1997; Rout et al., 2001
		Plantlets regenerated successfully. Somaclones produced. Some somaclones showed promising high yield with rhizome rot resistance.	Nirmal Babu et al., 1992a, 2005a,b; Lincy et al., 2009; Samsudeen etal., 2000; Sumathi, 2007
		Tetraploid ginger induced by in-vitro colchicines treatment and tetraploid somaclone developed with extra bold rhizome.	Adaniya & Shirai, 2001; Smith et al., 2004; Nirmal Babu et al., 2005b; Wang et al., 2010
Turmeric	Callus culture, leaf callus	Rapid and successful micropropagation protocol used for development of disease-free planting material.	Panda et al., 2007; Ghosh et al., 2013
		Organ formation and plantlet regeneration achieved along with the isolation of variants from tissue culture-derived plantlets with high curcumin content.	Nadgauda et al., 1982; Sunitibala et al., 2001; Salvi et al., 2000
		Turmeric clones with root rot resistance were isolated by *in vitro* selection.	Gayatri et al., 2005

(*Continued*)

TABLE 2.1 (CONTINUED)
Some Micropropagated Spices and Condiments with the Type of Explants Used

Spice or Condiment	Explants Used/ Somatic Embryogenesis	Achievements and Salient Features of Micropropagant/Somaclones	Key References
Vanilla	Apical meristem. Callus derived from shoot and seed	Standardized protocol for large-scale micropropagation and multiplication of vanilla for development of genetically stable and disease-free plants.	Minoo, 2002; Minoo et al., 2006; Minoo and Nirmal Babu, 2009
		Plants regenerated successfully.	Minoo, 2002
		Interspecific hybrids between *V. planifolia* and *V. aphylla* were developed by embryo rescue method.	
Cinnamon	Seedling and mature tree explant	Successful micropropagation reported.	Mini et al., 1997; Nirmal Babu et al., 1997
Camphor	Shoot tip, nodal segments, and cotyledonary node	Multiple shoots induction on Woody Plant medium and MS medium.	Huang et al., 1998; Azad et al., 2005
		Cyclic somatic embryogenesis and successful plant regeneration.	Shi et al., 2009, 2010
Nutmeg	Explants from field-grown trees	*In vitro* shoot initiation.	Mallika et al., 1997
	Zygotic embryo	Successful callus formation and plant regeneration was also achieved.	Iyer et al., 2000
		Somatic embryogenesis successfully induced from zygotic embryo.	Nirmal Babu et al., 2016
Clove	Seedling explants	Successful micropropagation , root formation induced in MS medium with IBA or activated charcoal.	Mathew & Hariharan, 1990; Suparman & Blake, 1990
Curry leaf	Nodal cutting, leaves, and immature seeds	Shoot proliferation was reported.	Nirmal Babu et al., 2000; Mathew & Prasad, 2007; Rani et al., 2012
	Intact seedling	Induced direct shoot proliferation with high frequency.	Bhuyan et al., 1997
		Direct and indirect somatic embryogenesis resulted in efficient plant regeneration.	Paul et al., 2011

2.3.1.2 Molecular Characterization

In the recent past, with the development and advancement in the tools and technique of molecular biology, molecular characterization of the crops and spices became much more convenient. Genetic characterization became possible and straightforward, using various molecular markers, along with identification, genetic fingerprinting, cloning of useful genes, and marker-assisted selection of superior genotypes. In different spice and condiment crops, various molecular markers were identified and used for identification and characterization of important traits, for the preparation of genetic maps, to study genetic variability, and for identification of disease-resistant lines, etc. Thomas et al. (2006) used RAPD (Randomly Amplified Polymorphic DNA) and ISSR (Inter-Simple Sequence Repeat) markers for molecular profiling of cardamoms exported from Guatemala, India, and Sri Lanka. RAPD profiling of ginger was reported by various groups (Nirmal Babu et al., 2005a, b; Parthasarathy et al., 2011). AFLP (Amplified Fragment Length Polymorphism) and ISSR marker-based genetic diversity analysis was performed in ginger (Wahyuni et al., 2003; Kizhakkayil & Sasikumar, 2010; Sajeev et al., 2011), whereas AFLP markers were used for characterization of genotypes of some selected cardamoms in Columbia (Tamayo et al., 2007). Nirmal

Babu et al. (2011a) reported the application of various molecular markers like RAPD, ISSR, and RFLP (Restriction Fragment length Polymorphism), as well as PCR (Polymerase Chain Reaction) for studying the interrelationship among and geographical origins of 11 species of cardamom and 96 germplasm accessions from India. For differentiating *Z. officinale* from other *Zingiber* species, SCAR (Sequence Characterized Amplified Region) markers were developed (Chavan et al., 2008). Genetic diversity of *C. longa* was analysed by different research groups using various markers, such as RAPD, ISSR (Jan et al., 2011; Singh et al., 2012), and PCR-based markers (Das et al., 2011; Nirmal Babu et al., 2011d). A characterization and genetic variability study of vanilla hybrids developed between *Vanilla planifolia* and *Vanilla aphylla* was carried out using RAPD and ALFP markers (Minoo et al., 2006), while genetic diversity and interrelationships among different *V. planifolia* clones and related species was carried out using RAPD data (Minoo et al., 2010).

Chloroplast DNA sequences of cinnamon species (*Cinnamomum cassia*, *C. zeylanicum*, *Cinnamomum burmannii* and *Cinnamomum sieboldii*) were analyzed and compared to achieve genetic distinction among these species (Kojoma et al., 2002). For correct identification of the four species, variation in nucleotide sequence at the *trnL-trnF* intergenic spacer (IGS) and *trnL* intron regions was used, along with the SSCP (Single-Strand Conformation Polymorphism) analysis of the PCR products of above nucleotide sequence regions. The use of RAPD and ISSR markers for diversity analysis among different tree spices like cinnamon (Joy & Maridass, 2008), nutmeg (Sheeja et al., 2013), mangosteen (Sulassih & Santosa, 2013), curry leaf (Verma & Rana, 2013), tamarind (Algabal et al., 2011), and allspice (Wadt et al., 2004) was reported by different workers. The phylogenetic relationship study, using sequence data of ITS (Internal Transcribed Spacer) region of several wild *Garcinia* species and *Garcinia mangostana* revealed nucleotide additivity at several sites of the ITS region (Yapwattanaphun et al., 2004). Among different seed and herbal spices, molecular characterization and genetic diversity analyses were reported in different species like cumin, fenugreek, saffron, thyme, garlic, dill, coriander, fennel, etc. using SSR (Simple Sequence Repeats), RAPD, AFLP, and ISSR. Some species- or variety-specific markers like AFLP, RAPD, and SCAR markers were developed for fenugreek (Vinay Kumar et al., 2012), fennel (Agbiotech, 2009), and menthe (Shasany et al., 2005). Microsatellite loci of some of the spices were developed for molecular characterization. Nemati et al. (2012) reported conservation and population studies of saffron (*Crocus sativus*) by developing 12 polymorphic microsatellite loci from a genomic library. Microsatellite markers were developed and characterized from black pepper for studying the genetic diversity of 40 popular genotypes of black pepper from India and 20 varieties from a Brazilian collection (Joy et al., 2007, 2011). The sequence of the rDNA–ITS region was used for phylogenetic analysis of genus Piper (Jaramillo & Manos, 2001). Different molecular markers, such as RAPD, ISSR, or SRAP (Sequence-Related Amplified Polymorphism) were used for genetic diversity and interrelationship analysis of *Piper* species (Sen et al., 2010; Jiang & Liu, 2011; Nirmal Babu et al., 2012b). For the identification of *Phytophthora*-resistant pepper lines, SCAR markers were developed by Sheji et al. (2006). Nirmal Babu et al. (2011a) developed a mapping population for genetic map preparation of *Piper nigrum*, along with genetic fidelity studies of micropropagated black and long pepper plants, using DNA markers.

2.3.1.3 Protoplast Culture

Protoplast culture, followed by somatic hybridization, is one of the best methods for combining the genomes of two sexually incompatible but related species. The plant cell devoid of cell wall is called a 'protoplast'. Thus, genetic improvement of some useful spice and condiment crops became easier by the development of protocols for the isolation and transformation of protoplasts containing genes from disease-resistant and high-yielding species. Recently, various reports were published which describe reliable procedures for protoplast isolation, culture, and somatic hybridization for different spice and condiment crops.

Shaji et al. (1998) reported protoplast isolation from the leaf tissue of black pepper and its successful culture and development up to the microcallus stage. Geetha et al. (2000) isolated cardamom, turmeric, and ginger protoplasts from leaf and cell suspensions, and cultured and developed them up to microcallus stage. In ginger, PEG- (polyethylene glycol-) mediated somatic hybridization of isolated protoplasts from different species was carried out and hybrid identification by RAPD was achieved. The diploid nature of regenerated progenies was analysed by flow cytometry (Guan et al., 2010). In vanilla, both protoplast isolation and PEG-mediated somatic hybridization between *V. planifolia* and *Vanilla andamanica* was reported by Minoo et al. (2008a). Protoplast fusion in *Vanilla* species was performed with the aim of transferring the disease tolerance and natural seed set characters of *V. andamanica* to *V. planifolia.* Among the seed spices, successful protoplast isolation was reported in fenugreek from mesophyll tissue, following which culture and regeneration of shoots was also achieved (Shekhawat & Galston, 1983).

2.3.1.4 Genetic Transformation

Genetic transformation of plants became possible with the development of recombinant DNA technology, along with advances in the tissue culture regeneration technique. Development of transgenic plant became possible, as genetic transformation technique allows transfer of a specific gene from one organism to another, unrelated plant which was never possible with the conventional breeding methods. Now, recombinant DNA technology helps in the introduction of genes with desirable traits like disease resistance, high yield, abiotic stress tolerance, herbicide resistance, etc., without hampering and interrupting the normal functioning of the plant. In plants, both direct and indirect transformation methods are used. For genetic transformation of plants, different methods, like the *Agrobacterium* Ti plasmid vector, microinjection, chemical fusogen-induced protoplast fusion, biolistic bombardment, or electroporation of protoplasts, are used. The different genetic transformation experiments carried out in some of the spice and condiment crops are summarized in Table 2.2

2.3.1.5 Omics, Cloning and Isolation of Candidate Genes

Spice and condiment crops are sources of many important and useful genes, exhibiting various antimicrobial and pharmaceutical qualities. With advances in cloning procedures and the development of various cloning vectors, the isolation, cloning, and characterization of these useful genes became easier. Several disease resistance genes have been isolated from cDNA libraries, using different probes, and cloned. Genes with desirable traits are identified and isolated for application in the crop improvement of spices. Many efforts has been made for the isolation and molecular cloning of genes responsible for improved response to biotic and abiotic stresses in spices such as black pepper, ginger, celery, hot pepper, etc. *Cucumber mosaic virus* (CMV) coat protein gene isolation, sequencing and transfer into black pepper were reported by Bhat et al. (2005) to develop virus-resistant transgenic lines. The suppression subtractive hybridization (SSH) approach was used to identify *Piper colubrinum* resistance genes in fungal pathogens showing differential expression in the presence of salicylic acid (Dicto & Manjusha, 2005). Later, a similar SSH technique was used by Alex et al. (2008) to develop a leaf-specific subtractive cDNA library for tissue-specific gene identification. Two defence-related proteins, namely methyl-glutaryl-CoA reductase and β-1,3-glucanase from *P. colubrinum and P. nigrum* were cloned from cDNA fragments (Girija et al., 2005a, b). The antifungal PR-5 gene homologue, osmotin, was characterized and cloned from a cDNA subtractive library from *P. colubrinum* with the goal that an osmotin gene cloned from a disease-resistant accession could be helpful in the development of more resistant varieties of black pepper by molecular breeding (Mani & Manjula, 2011). Root–knot nematodes, especially *Meloidogyne incognita*, cause huge damage to pepper (*Capsicum annuum* L.) crops. The *Me3* gene of pepper was found to be a useful candidate gene to provide resistance against root–knot nematodes. Yang et al. (2023) reported map-based cloning and functional analysis of *Me3* gene resistance against *M. incognita* and studied its transgenic expression in tobacco and *Arabidopsis*. From *C. longa*, a novel cDNA fragment encoding a phytocystatin gene was cloned and characterized successfully. Phytocystatin is a protease inhibitor, important in plant defence against pathogens and tolerance of plant stress. Phytostatin is gaining importance as a potential candidate in crop protection engineering and synthesis of pharmaceuticals

TABLE 2.2
Genetic Transformation in Some Major Spice and Condiment Crops

Spice or Condiment	Objective of Genetic Improvement	Genetic Transformation Method	Remarks	Key References
Black pepper (*Piper nigrum*)	Genetic diversity conservation; *Phytophthora* foot rot disease-resistant plant development	*Agrobacterium*-mediated gene transfer.	Cyclic secondary somatic embryos developed by Nair and Gupta by micropropagation were transformed with *Agrobacterium* with GUS reporter gene. Transformants were hardened off in the greenhouse. Trangene presence in the transformed plantlets was confirmed by PCR, dot blot and Southern blot hybridization	Nair & Gupta, 2006; Jiby & Bhat, 2011
		Agrobacterium tumefaciens strain EHA 105- mediated transformation.	*Agrobacterium* carrying pCAMBIA 1301 vector with HPT and GUS markers under CaMV 35S promoter was inserted into infecting seedling-derived explants of Panniyur 1 variety of black pepper. Transgenetic nature was analysed by PCR and GUS histochemical assay. Transgenic explants were regenerated into shoots.	Maju & Sonia, 2012
		Agrobacterium-mediated transformation using osmotin gene construct in pGV2260.	Reported successful osmotin gene (a pathogenesis-related protein responsible for *Phytophthora* resistance) transfer to the leaf tissue of black pepper. The resulting transgenic regenerated plants showed decrease spread of foot rot disease caused by *Phytophthora capsici*.	Nirmal Babu et al., 2005, 2013
		In planta transformation	*Piper colubrinum* shows resistance to foot rot-causing pathogen *Phytophthora capsici*. So, the total exogenous DNA of *P. colubrium* was transferred into black pepper variety Panniyur-2 through pollen tube pathway by *in planta* transformation method. The resulting seeds were screened and germinated under both *in vitro* and *ex vitro* conditions, and 39 % of the transformants showed disease tolerance.	Asha & Rajendran, 2010; Krishnan et al., 2015
		A. tumefaciens-mediated transformation.	Virus-resistant transgenic black pepper was developed by transformation with a binary vector pBI121, containing sequences from *Cucumber mosaic virus* (CMV) and *Piper yellow mottle virus* (PYMoV). *A. tumefaciens*-mediated transformation of the binary vector in the embryogenic mass of black pepper varieties resulted in transgenics showing CMV resistance and transgenics with PYMoV sequence which exhibited symptom remission.	Bhat et al., 2014

(*Continued*)

TABLE 2.2 (CONTINUED)
Genetic Transformation in Some Major Spice and Condiment Crops

Spice or Condiment	Objective of Genetic Improvement	Genetic Transformation Method	Remarks	Key References
Cardamom (*Elettaria cardamomum*)	For production of virus free planting material	Biolistic bombardment	A preliminary transformation study in cardamom embryogenic callus using biolistic process and plasmid vector pAHC 25 and promoter Ubi-1 (maize ubiquitin) with GUS reporter gene. Bombarded callus showed transient GUS gene expression.	Nirmal Babu, 1998; Nirmal Babu et al., 2013
			Cloning of coat protein gene of *Cardamom mosaic virus* (CdMV) under ubiquitin promoter in pAHC 17 vector for development of CdMV resistant transgenic cardamom.	Backiyarani et al., 2007
			In vitro studies reported development of transgenic cardamom expressing aprotinin (protease inhibitor) for dealing with *Conogethes punctiferalis* (cardamom shoot and capsule borer).	Josephrajkumar et al., 2006
Ginger (*Zingiber officinale*)	Development of rhizome rot (*Pythium aphanidermatum*) and bacterial wilt (*Ralstonia solanacearum*) disease-resistant plant material and varieties; to develop seed set to induce more genetic variability.	Biolistic bombardment. *Agrobacterium tumefaciens* mediated transformation.	Reported bombardment of embryonic callus with plasmid vector pAHC 25 with promoter Ubi-1 (maize ubiquitin), GUS reporter gene, and BAR (phosphinothricin-acetyl transferase) as a selectable marker. The resulting transformed tissue showed transient expression of GUS.	Nirmal Babu, 1997; Nirmal Babu et al., 1998
			Successful transformation of ginger buds using *Agrobacterium tumefaciens* mediated transformation.	Suma et al., 2008
Turmeric (*Curcuma longa*)	Development of high-yielding curcumin varieties and improvement of physiological traits	Particle bombardment. *Agrobacterium*-mediated transformation.	Shoot-derived callus cultures were transformed with plasmid pAHC25 carrying BAR and GUS genes controlled by the maize ubiquitin promoter. The shoots regenerated from transformants were multiplied and stably developed into plantlets, which showed gluphosinate resistance.	Shirgurkar et al., 2006
			Turmeric inflorescence-derived calli were transformed with plasmid pBISN1 through *Agrobacterium* strain EHA105. The stable expression of the transgene was confirmed by GUS assay, PCR and Southern blot analysis.	He & Gang, 2014

(Continued)

TABLE 2.2 (CONTINUED)
Genetic Transformation in Some Major Spice and Condiment Crops

Spice or Condiment	Objective of Genetic Improvement	Genetic Transformation Method	Remarks	Key References
Vanilla (*Vanilla planifolia*)	Improved disease-free and genetically stable varieties with high yields of vanillin	*Agrobacterium tumefaciens*-mediated transformation	*Vanilla planifolia* genetic transformation using thin shoot tip sections and protocorm-like bodies derived from shoot tips with vector carring nptII and GUS genes controlled by CaMV 35S promoter.	Malabadi & Nataraj, 2007; Ratheesh & Bhat, 2011
Cumin (*Cuminum cyminum*)	Improved abiotic stress response like salinity-tolerant variety development and introduction of geneticvariability.	Particle bombardment.	Using precultured embryo of cumin, stable genetic transformation and plant regeneration was reported.	Singh et al., 2010;
		Agrobacterium -mediated transformation	Using embryo explants of cumin, a transformation experiment was performed with a binary construct and the resulting transformation at the hardening stage showed 1.5% transformation efficiency.	Pandey et al., 2013
Coriander (*Coriandrum sativum*)	Enhance the yield and stability, high essential oil content and stress tolerance.	*Agrobacterium tumefaciens*-mediated transformation	The mutated ethylene receptor ERS1 from *Arabidopsis thaliana* was transformed into coriander to study its role in tissue senescence in heterologous plants. The transgenic coriander was developed by co-cultivating *A. tumefaciens* containing binary vector pCGN1547 carrying the mutated ERS1 gene with hypocotyl segments of coriander. The resulting transgenic coriander showed delayed leaf and flower senescence with a transformation efficiency of 6.6%.	Wang & Kumar, 2004
Fenugreek (*Trigonella foenum-graecum*)	Improved seed production and disease-resistant varieties.	*A. rhizogenes*-induced hairy root transformation	In fenugreek, the hairy root induced by *A. rhizogenes* is ideal for secondary metabolite (diosgenin) production. To investigate the role of hairy root induction, the leaf and stem explants of fenugreek ecotypes (Karaj and Bushehr) were infected with K599 strain of *A. rhizogenes* carrying green fluorescent protein gene (GFP). The stem and leaf explants showed transformation frequency of 81.3% and 71.88%, respectively.	Shahabzadeh et al., 2013
		Agrobacterium tumefaciens.	Trigonellin secondary metabolite production by hairy root cultures of fenugreek was reported in Zanjan and Borazjan, two Iranian masses.	Raheleh et al., 2011
			In vitro crown gall induction in root, hypocotyls, and cotyledon explants of fenugreek was performed using the A281 strain of *A. tumefaciens.*	Khawar et al., 2004

(*Continued*)

TABLE 2.2 (CONTINUED)
Genetic Transformation in Some Major Spice and Condiment Crops

Spice or Condiment	Objective of Genetic Improvement	Genetic Transformation Method	Remarks	Key References
Garlic (*Allium sativum*)	For generating genetic variability and resistance to biotic and tolerance to abiotic stress.	Microprojectile bombardment. Particle bombardment. *A. tumefaciens*-mediated transformation.	A construct carrying *uidA* reporting gene and selectable marker gene *bar* under CaMV35S and Ubi promoters, respectively, was bombarded into leaf, basal plate discs, and embryonic callus of garlic.	Ferrer et al., 2000
			Reported embryogenic callus bombardment with plamid pWRG1515 carrying genes *hpt* and *gusA* resulted in stable transgenics. *GUS* histochemical assay and Southern blot hybridization analysis confirmed the transgenic nature of regenerated plants.	Robledo-Paz et al., 2004
			Chlorsulfuron herbicide-resistant garlic was developed by bombardment of garlic clove apical meristem-derived callus with ALS (acetolactate synthase) gene from chlorsulfuron-resistant *Arabidopsis* mutant. The bombardment was done with tungsten particles coated with recombinant DNA containing CaMV 35S promoter, *GUS* reporter gene, selectable marker hygromycin phosphotransferase (*hpt*).	Park et al., 2002
			Insect (beet armyworm) -resistant garlic was developed by *Agrobacterium*-mediated transformation. The resistant genes *cry1Ca* and H04 from *Bacillus thuringiensis* were transferred along with *gusA* reporter gene and *hpt* selectable marker gene in the garlic cultivar 'Printanor'. The transgenic pants with *cry1Ca* showed complete resistance to beet armyworm.	Zheng et al., 2004
			Efficient protocol for transformation and regeneration was reported in embryo and leaf explants.	Eady et al., 2005; Kenel et al., 2010

(Continued)

TABLE 2.2 (CONTINUED)
Genetic Transformation in Some Major Spice and Condiment Crops

Spice or Condiment	Objective of Genetic Improvement	Genetic Transformation Method	Remarks	Key References
Chilli (*Capsicum annuum*)	Biotic stress resistance and abiotic stress tolerance, especially resistance against viral, bacterial, and fungal diseases.	*Agrobacterium*-mediated genetic transformation. *Agrobacterium* and the particle gun.	Various protocols reported transformation and regeneration of transgenic chilli using different explants such as hypocotyls, cotyledons, leaves, etc.	Liu et al., 1990; Manoharan et al., 1998; Kim et al., 2002; Li et al., 2003; Sobhakumari & Lalithakumari, 2005
		Agrobacterium-mediated transformation *A. rhizogenes*. *Agrobacterium* mediated	Hypocotyle explants from sweet red pepper 'Florinis' and pepper hybrids PO1 and C were transformed with plasmid carrying CaMV 35S promoter controlling *GUS* reporter and *NPTII* selectable marker gene by both particle gun and *A. tumefaciens*-mediated method. More transgenic plants were reported in the case of the particle gun method and they were successfully established in soil.	Nianiou et al., 2002
		transformation.	A callus-induced transformation system was reported using green florescent protein (GFP) as marker for monitoring the early stages of trangene callus development.	Jung et al., 2011
			Hypocotyle explants were transformed with parsley's phenylalanine ammonia-lyase (*PAL*) gene. A13 strain of *A. rhizogenes* harbouring pBI121 recombinant binary vector was used and the transgenic chilli was selected in kanamycin- and cefataxim-e supplemented medium. *PAL*-containg hairy roots of transgenic chilli expressed PAL activity and altered morphology.	Yamakawa et al., 1998
			Virus-resistant transgenic chilli was developed by successful *Agrobacterium*-mediated transformation of *Cucumber mosaic virus* (CMV) and *Tobacco mosaic virus* (TMV) coat protein (*CP*) genes in hypocotyl explants of chili pepper cultivar 8212. The transgenic plants depicted very mild disease severity and showed delayed symptom development.	Cai et al., 2003

(Continued)

TABLE 2.2 (CONTINUED)
Genetic Transformation in Some Major Spice and Condiment Crops

Spice or Condiment	Objective of Genetic Improvement	Genetic Transformation Method	Remarks	Key References
			Transgenic pepper resistant to a new CMV pathotype was developed by successfully transferring coat protein gene from the CMV pathotype.	Lee et al., 2009
			Transformation of chilli plant with *Cre* and *Barnase* genes resulted in altered fertility of transgenic plants. Expression of these genes occurred in specific tissue of the male organ at specific developmental stage of transgene resulted in different fertility level i.e. complete fertility or sterility or partial sterility. The genes were controlled by the Cre/lox system.	Juan-Xu et al., 2009
			Regulation of structural gene biosynthesis in chilli was investigated using gene silencing system induced by *Tobacco rattle virus* (TRV).	Aguilar-Barragán and Ochoa-Alejo, 2014
			Baculovirus chitinase gene transfer in chilli was reported. The resulting transgenic plants depicted *in vitro* inhibition of fungal pathogens (*Colletotrichum capsici* and *Alternaria alternata*) causing anthracnose disease	Mythili et al., 2015
Celery (*Apium graveolens*)	Development of disease, pest, and herbicide resistance.	*A. tumefaciens*-mediated transformation.	Petiole explants of celery cultivar PI257228 were transformed for development of kanamycin-resistant plants.	Catlin et al., 1988
		A. tumefaciens-mediated transformation.	*A. tumefaciens*-mediated transformation of leaf and seedling explants from two cultivars of celery, XP85 and XP166, was performed with *bar* as selectable marker gene. Transgene resulted in gluphosinate herbicide-resistant transgenic celery plants.	Loskutov et al., 2008
		A. tumefaciens-mediated transformation.	Rapid and efficient protocol for *A. tumefaciens*-mediated transformation of leaf, cotyledon, and hypocotyl explants was reported with transformation frequency of 5%, 18% and 16%, respectively. Histochemical GUS assay confirmed the stable inheritance of transgenes.	Song et al., 2007

(*Continued*)

TABLE 2.2 (CONTINUED)
Genetic Transformation in Some Major Spice and Condiment Crops

Spice or Condiment	Objective of Genetic Improvement	Genetic Transformation Method	Remarks	Key References
Onion (*Allium cepa*)	Improvement of the yield and nutritive value, and introduction of useful traits.	*Agrobacterium*-mediated genetic transformation.	Transformation protocol for callus transformation was reported. The resulting transgenic plants were successfully transferred to greenhouse and showed normal ploidy and morphology.	Zheng et al., 2001
		Agrobacterium and biolistic	A new selection system was reported. The system used included *manA* gene from *Escherichia coli* which encodes phosphomannose isomerase (*pmi*). The transgenic plant harbouring *manA* genes detoxify mannose-6-phosphate by changing it into fructose-6-phosphate.	Aswath et al., 2006
		Agrobacterium-mediated transformation.	*Agrobacterium*-mediated transformation of epidermal explants of onion resulted in transient expression of GUS reporter gene in transformed cells.	Cheng et al., 2009
		Particle bombardment.	Onion epidermal layers were bombarded with a construct carrying *GUS* reporter gene under CaMV 35S promoter control on plasmid pWRG2426. Transient GUS expression was observed within 24 hours of bombardment.	Sandhu & Gosal, 2009
		Agrobacterium-mediated transformation	Different factors influencing the *Agrobacterium*-mediated transformation and *in vitro* regeneration of onion was studied.	Mythili et al., 2012
		In planta transformation by agroinfiltration.	A highly efficient *in planta* transient transformation in living normal epidermal cells of onion was reported with a transformation efficiency of 43.87%. The binary vector pLPGM202 contained modified GFP gene and kanamycin (Kar) resistance gene.	Xu et al., 2014
		Agrobacterium-mediated transformation.		
		Agrobacterium-mediated transformation.		
		Agrobacterium-mediated genetic transformation.	Callus derived from meristem plate and twin-scale leaves was transformed using binary vector pCAMBIA 1301 with *GUS* marker gene along with various concentrations of acetosyringone. Maximum callus induction was reported in MS medium supplemented with B5 vitamins At 150 μM concentration of acetosyringone, maximum GUS gene expression was observed.	Malla et al., 2015

(*Continued*)

TABLE 2.2 (CONTINUED)
Genetic Transformation in Some Major Spice and Condiment Crops

Spice or Condiment	Objective of Genetic Improvement	Genetic Transformation Method	Remarks	Key References
			Glyphosate and phosphinothricin herbicide-tolerant transgenic onion was produced with transformation frequency of 0.9 percent. The construct used for transformation carried a CaMV 35S promoter and GFP gene from pBin m GFP ER.	Eady et al., 2003a, 2003b
			Beet armyworm (*Spodoptera exigua*) -resistant shallots were developed by transformation of *Bacillus thuringiensis*-derived *cry1Ca* or a *H04* hybrid gene in Tropix and Kuning cultivars of shallot. The hybrid gene was controlled by chrysanthemum Rubisco SSU promoter and CaMV 35S promoter controlling HPT. The transgenic plants developed showed high expression of the transferred genes and also armyworm resistance.	Zheng et al., 2005

(Chan et al., 2014). Molecular cloning of the flavone synthase I gene was reported from total RNA and cDNA libraries of cumin by Yousefi et al. (2019). As flavones are useful plant metabolites with high pharmaceutical value, having a cloned flavone synthase gene could help in crop genetic improvement. Rostiana et al. (2022) reported the molecular analysis and cloning of a bacterial wilt resistance gene from ginger. Improved response to bacterial wilt is a bottleneck in ginger cultivation, as it can damage more than 70% of the crop.

Nilam et al. (2020) reported the transcriptome sequencing of coriander from seeds, leaves, and flowers for the identification of genes and SSR marker development, which is useful for breeding and genetic diversity studies in coriander. Choudhary et al. (2019) studied the transcriptome profile of coriander to determine the genes associated with the resistance to stem gall development in coriander. Analysis of the fenugreek transcriptome was reported by Ciura et al. (2018). They used the RNA-sequencing technique to analyse defence-related changes in the transcriptome, i.e., the production of steroidal sapogenins and other secondary metabolites following the application of cholesterol, methyl jasmonate, and squalene. These results will help in understanding the biosynthethic pathway of steroidal saponins and diosgenin. Transcriptome analysis in black pepper helped our understanding of piperine biosynthesis and genetic changes and evolution associated with this pathway. The data obtained from this analysis will help in biological and agronomic research in black pepper (Hu et al., 2019).

2.4 CONCLUSION AND FUTURE PERSPECTIVES

Spices and condiments are produced from very important plant crops which play a pivotal role in the life of mankind as seasoning and flavouring agents in foods, in cosmetic and perfumes, and, most importantly, in the pharmaceutical industry due to their medicinal properties. Now with the increase in demand for production of crops of high-quality spices and condiments, with the ever-growing population and natural resource exploitation, there is a huge demand to enhance the quality and production as quickly as possible. But overcoming these problems with conventional breeding seems impossible as combining or inducing genetic variation and producing improved varieties of crops is very difficult and time consuming. In the recent past, considerable progress has been made in the biotechnological and genomics field for the development of tools for characterization, to introduce genetic variation for crop improvement, enhance potential yields, and increase disease resistance and enhance crop health. Biotechnological approaches like tissue culture, micropropagation, molecular characterization by developing various molecular markers, cloning and isolation of candidate genes, etc., has shown great potential for rapid and efficient genetic improvement of the spice and condiment crops. To date, different research groups have reported the isolation and cloning of genes responsible for disease resistance, for improved biotic and abiotic stress response, and other agronomically important traits in major spice and condiment crops. Many rapid and efficient transformation strategies are available in various spice crops. Information on large-scale production of secondary metabolites by hairy root cultures is emerging as a promising approach. All the information and the data obtained from the application of biotechnological strategies to the analysis of spice and condiment crop will be of enormous value in the genetic improvement of spices and condiments for increasing the quality and quantity of the crops.

LIST OF ABBREVIATIONS

AFLP: Amplified Fragment Length Polymorphism
ISSR: Inter-Simple Sequence Repeat
ITS: Internal Transcribed Spacer
PCR: Polymerase Chain Reaction
PEG: polyethylene glycol
RAPD: Random Amplified Polymorphic DNA

RFLP: Restriction Fragment Length Polymorphism
SCAR: Sequence Characterized Amplified Region
SRAP: Sequence-Related Amplified Polymorphism
SSCP: Single-strand Conformation Polymorphism
SSR: Simple Sequence Repeats
SSH: Suppression Subtractive Hybridization

REFERENCES

Adaniya, S. & Shirai, D. 2001. *In vitro* induction of tetraploid ginger (*Zingiber officinale* Roscoe) and its pollen fertility and germinability. *Scientia Horticlturae,* 88:277–287.

Agbiotech. 2009. WWW.agbiotech.net/ Accessed on 12 September 2009.

Aguilar-Barragán, A. & Ochoa-Alejo, N. 2014. Virus-induced silencing of MYB and WD40 transcription factor genes affects the accumulation of anthocyanins in chilli pepper fruit. *Biologia Plantarum,* 58(3):567–574.

Alex, S. M., Dicto, J., Purushothama, M. G. & Manjula, S. 2008. Differential expression of metallothionein type-2 homologues in leaves and roots of Black pepper (*Piper nigrum* L). *Genetics and Molecular Biology,* 31(2):551–554.

Algabal, A. Q. A. Y., Papanna, N. & Simon, L. 2011. Amplified fragment length polymorphism marker-based genetic diversity in tamarind (*Tamarindus indica*). *International Journal of Fruit Science,* 11(1):1–16.

Asha, S. & Rajendran, P. C. 2010. Putative transgenic plants through *in planta* transformation against *Phytophthora* foot rot in black pepper (*Piper nigrum* L.). *Asian Journal of Biological Sciences,* 4:135–141.

Asowata-Ayodele, A. M., Afolayan, A. J. & Otunola, G. A. 2016. Ethnobotanical survey of culinary herbs and spices used in the traditional medicinal system of Nkonkobe Municipality, Eastern Cape, South Africa. *South African Journal of Botany,* 104:69–75.

Aswath, C. R., Mo, S. Y., Kim, D. H. & Park, S. W. 2006. *Agrobacterium* and biolistic transformation of onion using non-antibiotic selection marker phosphomannose isomerase. *Plant Cell Reports,* 25(2):92–99.

Azad, M. A. K., Yokota, S., Ishiguri, F., Yahara, S. & Yoshizawa, N. 2005. Large-scale clonal propagation of *Cinnamomum camphora* (L.) Nees and Eberm. *Bulletin of the Utsunomiya University Forests,* 41:101–109.

Backiyarani, S., Manohari, C., Jebasingh, T., Jacob, T. & Usha, R. 2007. Cloning of coat protein gene of Kursupara isolate of *Cardamom mosaic virus* for developing transgenic virus-resistant cardamom. In: *Recent trends in horticultural biotechnology,* Vol. II. ICAR national symposium on biotechnological interventions for improvement of horticultural crops: issues and strategies, Kerala, India, 10–12 January, 2005 (pp. 693–698). New India Publishing Agency.

Ballabh, B., Chaurasia, O. P., Ahmed, Z. & Singh, S. B. 2008. Traditional medicinal plants of cold desert Ladakh used against kidney and urinary disorders. *Journal of Ethnopharmacology,* 118(2):331–339.

Bhat, A. I., Haresh, P. S. & Madhubala, R. 2005. Sequencing of coat protein gene of an isolate of *Cucumber mosaic virus* infecting black pepper in India. *Jurnal of Plant Biochemistry and Biotechnololy,* 14(1):37–40.

Bhat, A. I., Revathy, K. A., Sasi, S. & Jiby, M. V. 2014. Production and evaluation of transgenic black pepper expressing *Cucumber mosaic virus* and *Piper yellow mottle virus* sequences. Paper presented at the VIROCON 2014. Tamil Nadu Agricultural University, Coimbatore.

Bhuyan, A. K., Pattnaik, S. & Chand, P. K. 1997. Micropropagation of curry leaf tree (*Murraya koenigii* (L.) Spreng.) by axillary proliferation using intact seedlings. *Plant Cell Reports,* 16(11):779–782.

Cai, W. Q., Fang, R. X., Shang, H. S., Wang, X., Zhang, F. L., Li, Y. R., Zhang, J. C., Cheng, X. Y., Wang, G. L. & Mang, K. Q. 2003. Development of CMV- and TMV-resistant transgenic chili pepper: Eld performance and biosafety assessment. *Molecular Breeding,* 11(1):25–35.

Catlin, D., Ochoa, O., McCormick, S. A. & Quiros, C. F. 1988 Celery transformation by *Agrobacterium tumefaciens*: Cytological and genetic analysis of transgenic plants. *Plant Cell Reports,* 7(2):100–103.

Chan, S. N., Abu Bakar, N., Mahmood, M., Ho, C.-L. & Shaharuddin, N. A. 2014. Molecular cloning and characterization of novel phytocystatin gene from turmeric, *Curcuma longa. BioMed Research International,* 2014:Article ID 973790, 9 pages.

Chavan, P., Warude, D., Joshi, K. & Patwardhan, B. 2008. Development of SCAR (sequence-characterized amplifi ed region) markers as a complementary tool for identification of ginger (*Zingiber officinale* Roscoe) from crude drugs and multicomponent formulations. *Biotechnology and Applied Biochemistry,* 50(1):61–69.

Cheng, Y. Q., Yang, J., Xu, F. P., An, L. J., Liu, J. F. & Chen, Z. W. 2009. Transient expression of minimum linear gene cassettes in onion epidermal cells *via* direct transformation. *Applied Biochemistry and Biotechnology*, 159(3):739–749.

Choudhary, S., Naika, M. B., Sharma, R., Meena, R. D., Singh, R. & Lal, G. 2019. Transcriptome profiling of coriander: A dual purpose crop unravels stem gall resistance genes. *Journal of Genetics*, 98(1):1–11.

Ciura, J., Szeliga, M., Grzesik, M. & Tyrka, M. 2018. Changes in fenugreek transcriptome induced by methyl jasmonate and steroid precursors revealed by RNA-Seq. *Genomics*, 110:267–276.

Das, A., Kesari, V., Satyanarayana, V. M., Parida, A. & Rangan, L. 2011. Genetic relationship of *Curcuma* species from Northeast India using PCR-based markers. *Molecular Biotechnolology*, 49(1):65–76.

Dicto, J. & Manjusha, S. 2005. Identification of elicitor induced PR5 gene homologue in *Piper colubrinum* link by suppression subtractive hybridization. *Current Science*, 88(4):25.

Dubey, S. 2017. Indian spices and their medicinal value. *Indian Journal of Pharmaceutical Education and Research*, 51(3s2):s330–s332.

Duke, J. 2002. *CRC Handbook of Medicinal Spices*. CRC Press, Boca Raton, FL, 360p.

Eady, C., Davis, S., Catanach, A., Kenel, F. & Hunger, S. 2005. *Agrobacterium tumefaciens*-mediated transformation of leek (*Allium porrum*) and garlic (*Allium sativum*). *Plant Cell Reports*, 24(4):209–215.

Eady, C., Davis, S., Farrant, F., Reader, J. & Kenel, F. 2003a. *Agrobacterium tumefaciens*-mediated transformation and regeneration of herbicide resistant onion (*Allium cepa*) plants. *Annals of Applied Biology*, 142(2):213–217.

Eady, C. C., Reader, J., Davis, S. & Dale, T. 2003b. Inheritance and expression of introduced DNA in transgenic onion plants (*Allium cepa*). *Annals of Applied Biology*, 142(2):219–224.

ESA (European Spice Association) 2020. *List of culinary herbs and spices adopted by 1*. European Spice Association (ESA) TC, p. 2. Accessed on 30 August 2020 from https://www.esa-spi ces.org/index-esa .html/publications-esa.

FAO (Food and Agriculture Organization) 2011. *Spices and Herbs for Home and Market, FAO Diversification Booklet No. 20th*. Rome. Accessed on 25 August 2020 from http://www.fao. org/3/i2476e/i2476e00.pdf.

Gayatri, M. C., Roopadarshini, V., Kavyashree, R. & Kumar, C. S. 2005. Encapsulation and regeneration of aseptic shoot buds of turmeric (*Curcuma longa* L.). *Plant Cell Biotechnology and Molecular Biology*, 6(3/4):89–94.

Geetha, S. P., Nirmal Babu, K., Rema, J., Ravindran, P. N. & Peter, K. V. 2000. Isolation of protoplasts from cardamom (*Elettaria cardamomum* Maton.) and ginger (*Zingiber offi cinale* Rosc.). *Journal of Spices and Aromatic Crops*, 9(1):23–30.

Ghosh, A., Chatterjee, P. & Ghosh, P. 2013. A protocol for rapid propagation of genetically true to type Indian turmeric (*Curcuma longa* L.) through *in vitro* culture technique. *Advances in Applied Science Research*, 4(3):39–45.

Girija, D., Beena, P. S. & Nazeem, P. A. 2005a. Molecular cloning of a cDNA fragment encoding the defense relatednprotein β-1, 3-glucanase in black pepper (*Piper nigrum* L.). *Proceedings of the Kerala Science Congress*. KFRI, Peechi, pp. 81–82.

Girija, D., Beena, P. S., Nazeem, P. A. & Puroshothama, M. G. 2005b. Molecular cloning of cDNA fragment encoding hydroxy methyl glutaryl CoA reductase in *Piper colubrinum. Proceedings of the National Symposium on Biotechnological Interventions for Improvement of Horticultural Crops: Issues and Strategies*. Kerala Agricultural University, Thrissur, pp. 303–306.

Guan, Q., Guo, Y., Wei, Y., Meng, F. & Zhang, Z. 2010. Regeneration of somatic hybrids of ginger via chemical protoplast fusion. *Plant Cell, Tissue and Organ Culture*, 102(3):279–284.

He, R. & Gang, D. R. 2014. Somatic embryogenesis and *Agrobacterium*-mediated transformation of turmeric (*Curcuma longa*). *Plant Cell, Tissue and Organ Culture*, 116(3):333–342.

Hill, T. 2004. *The Contemporary Encyclopedia of Herbs & Spices*. John Wiley & Sons Inc., New York, 432p.

Huang, L. C., Huang, B. L. & Murashige, T. 1998. A micropropagation protocol for *Cinnamomum camphora*. *In Vitro Cellular and Developmental Biology – Plant*, 34(2):141–146.

Iyer, R. I., Jayaraman, G., Gopinath, P. M. & Sita, G. L. 2000. Direct somatic embryogenesis in zygotic embryos of nutmeg (*Myristica fragrans* Houtt.). *Tropical Agriculture*, 77(2):98–105.

Jan, H. U., Rabbani, M. A. & Shinwari, Z. K. 2011. Assessment of genetic diversity of indigenous turmeric (*Curcuma longa* L.) germplasm from Pakistan using RAPD markers. *Journal of Medicinal Plant Research*, 5(5):823–830.

Jaramillo, M. A. & Manos, P. S. 2001. Phylogeny and patterns of floral diversity in the genus *Piper* (Piperaceae). *American Journal of Botany*, 88(4):706–716.

Jiang, Y. & Liu, J. P. 2011. Evaluation of genetic diversity in *Piper* spp using RAPD and SRAP markers. *Genetics and Molecular Research*, 10(4):2934–2943.

Jiby, M. V. & Bhat, A. I. 2011. An efficient Agrobacterium-mediated transformation protocol for black pepper (*Piper nigrum* L.) using embryogenic mass as explants. *Journal of Crop Science and Biotechnology*, 14(4):2, 47–254.

Joseph, B., Joseph, D. & Philip, V. J. 1996. Plant regeneration from somatic embryos in black pepper. *Plant Cell, Tissue and Organ Culture*, 47(1):87–90.

Josephrajkumar, A., Chakrabarty, R. & Thomas, G. 2006. Midgut proteases of the cardamom shoot and capsule borer *Conogethes punctiferalis* (Lepidoptera: Pyralidae) and their interaction with aprotinin. *Bulletin of Entomological Research*, 96(1):91–98.

Joy, N., Abraham, Z. & Soniya, E. V. 2007. A preliminary assessment of genetic relationships among agronomically important cultivars of black pepper. *BMC Genetics*, 8(1):42.

Joy, N., Prasanth, V. P. & Soniya, E. V. 2011. Microsatellite based analysis of genetic diversity of popular black pepper genotypes in South India. *Genetica*, 139(8):1033–1043.

Joy, P. & Maridass, M. 2008. Inter species relationship of *Cinnamomum* species using RAPD marker analysis. *Ethnobotanical Leaflets*, 12:476–480.

Juan-Xu, L., Yi-Xun, Y., Jian-Jun, L., Guo-Ju, C., Bi-Hao, C. & China, P. R. 2009. Study on *Agrobacterium*-mediated transformation of pepper with *barnase* and *Cre* genes. *Agricultural Sciences in China*, 8(8):947–955.

Jung, M., Shin, S. H., Park, J. M., Lee, S. N., Lee, M. Y., Ryu, K. H., Paek, K. Y. & Harn, C. H. 2011. Detection of transgene in early developmental stage by GFP monitoring enhances the efciency of genetic transformation of pepper. *Plant Biotechnology Reporter*, 5:157–167.

Kenel, F., Eady, C. & Brinch, S. 2010. Efficient *Agrobacterium tumefaciens*-mediated transformation and regeneration of garlic (*Allium sativum*) immature leaf tissue. *Plant Cell Reports*, 29(3):223–230.

Khawar, K. M., Onarici, G., Cocu, S., Erisen, S., Sancak, S. C. & Ozcan, S. 2004. *In vitro* crown galls inducted by *Agrobacterium tumefaciens* strain A281 (p TiBo542) in *Trigonella foenum-graecum*. *Biologia Plantarum*, 48(3):441–444.

Kim, J. Y., Jung, M., Kim, H. S., Lee, Y. H., Choi, S. H., Lim, Y. P., Min, B. W., Yang, S. G. & Harn, C. H. 2002. A new selection system for pepper regeneration by mannose. *Journal of Plant Biotechnology*, 4:129–134.

Kizhakkayil, J. & Sasikumar, B. 2010. Genetic diversity analysis of ginger (*Zingiber officinale* Rosc.) germplasm based on RAPD and ISSR markers. *Scientia Horticulturae*, 125(1):73–76.

Kojoma, M., Kurihara, K., Yamada, K., Sekita, S., Satake, M. & Iida, O. 2002. Genetic identifi cation of cinnamon (*Cinnamomum* spp.) based on the trnL-trnF chloroplast DNA. *Planta Medica*, 68(1):94–96.

Krishnan, A., Mahadevan, C., Mani, T. & Sakuntala, M. 2015. Virus-induced gene silencing (VIGS) for elucidation of pathogen defense role of serine/threonine protein kinase in the non-model plant *Piper colubrinum* Link. *Plant Cell, Tissue and Organ Culture*, 122(2):269–283.

Kumar, K., Abdulkhadar, Rangaswamy, P. & Irrwlappom. 2010. *Introduction to Spices, Plantation Crops, Medicinal and Aromatic Plants*. Oxford and IBH Publishing Co, New Delhi.

Lee, Y. H., Jung, M., Shin, S. H., Lee, J. H., Choi, S. H., Her, N. H., Lee, J. H. & Harn, C. H. 2009. Transgenic peppers that are highly tolerant to a new CMV pathotype. *Plant Cell Reports*, 28(2):223–232.

Li, D., Zhao, K., Xie, B., Zhang, B. & Luo, K. 2003. Establishment of highly efficient transformation system for pepper (*Capsicum annuum* L.). *Plant Cell Reports*, 21(8):785–785.

Lincy, A. K., Remashree, A. B. & Sasikumar, B. 2009. Indirect and direct somatic embryogenesis from aerial stem explants of ginger (*Zingiber officinale* Rosc.). *Acta Botanica Croatica*, 68(1):93–103.

Hu, L., Xu, Z., Wang, M., Fan, R., Yuan, D., Wu, B., Wu, H., Qin, X., Yan, L., Tan, L. and Sim, S., Li, W. Saski, C. A., Daniell, H., Wendel, J. F., Lindsey, K., Zhang, X., Hao, C. & Jin, S. 2019. The chromosome-scale reference genome of black pepper provides insight into piperine biosynthesis. *Nature Communications*, 10:4702.

Liu, W., Parrot, W. A., Hilderbrand, D. P., Collins, G. B. & Williams, E. G. 1990. *Agrobacterium* induced gall formation in bell pepper (*Capsicum annuum* L.) and formation of shoot like structures expressing introduced genes. *Plant Cell Reports*, 9(7):360–364.

Loskutov, A. V., Song, G. Q. & Sink, K. C. 2008. Evaluation of methods for celery (*Apium graveolens* L.) transformation using *Agrobacterium tumefaciens* and the *bar* gene as a selectable marker. *In Vitro Cell Development Biology-Plant*, 44(4):239–245.

Maju, T. T. & Soniya, E. V. 2012. *In vitro* regeneration system for multiplication and transformation in *Piper nigrum* L. *International Journal of Medicinal and Aromatic Plants*, 2(1):178–184.

Malabadi, R. B. & Nataraja, K. 2007. Genetic transformation of *Vanilla planifolia* by *Agrobacterium tumefaciens* using shoot tip sections. *Research Journal of Botany*, 2(2):86–94.

Malla, A., Srinivasan, B., Shanmugaraj, B. M. & Ramalingam, S. 2015. Micropropagation and DNA delivery studies in onion cultivars of Bellary, CO3. *Journal of Crop Science and Biotechnology*, 18(1):37–43.

Mallika, V. K., Rekha, K., Marymol, M., Manjula, M. & Vikraman Nair, R. 1997. *In vitro* shoot initiation from explants of fi eld grown trees of nutmeg (*Myristica fragrans* Houtt.). In: Edison, S., Ramana, K.V., Sasikumar, B., Nirmal Babu, K. & Santhosh, J.E. (eds) *Biotechnology of Spices, Medicinal and Aromatic Crops*. Indian Society for Spices, Calicut, pp. 29–34.

Mani, T. & Manjula, S. 2011. Optimization of Agrobacterium-mediated transient gene expression and endogenous gene silencing in *Piper colubrinum* Link. by vacuum infi ltration. *Plant Cell, Tissue and Organ Culture*, 105(1):113–119.

Manoharan, M., Vidya, C. S. S. & Sita, G. L. 1998. *Agrobacterium*-mediated genetic transformation in hot chilli (*Capsicum annuum* L. var. Pusa jwala). *Plant Science*, 131(1):77–83.

Manohari, C., Backiyarani, S., Jebasingh, T., Somanath, A. & Usha, R. 2008. Efficient plant regeneration in small cardamom (*Elettaria cardamomum* Maton.) through somatic embryogenesis. *Indian Journal of Biotechnology*, 7:407–409.

Mathew, D. & Prasad, M. C. 2007. Multiple shoot and plant regeneration from immature leaflets of *in vitro* origin in curry leaf (*Murraya koenigii* Spreng). *Indian Journal of Plant Physiology*, 12(1):18–22.

Mathew, M. K. & Hariharan, M. 1990. *In vitro* multiple shoot formation in *Syzygium aromaticum*. *Annals of Botany*, 65(3):277–279.

Mini, P. M., John, C. Z., Samsudeen, K., Rema, J., Nirmal Babu, K. & Ravindran, P. N. 1997. Micropropagation of *Cinnamomum verum* (Bercht and Presl.). In: Edison, S., Ramana, K. V., Sasikumar, B., Nirmal Babu, K. & Santhosh, J. E. (eds) *Biotechnology of Spices, Medicinal and Aromatic Crops*. Indian Society for Spices, Calicut, pp. 35–38.

Minoo, D. & Nirmal Babu, K. 2009. Micropropagation and *in vitro* conservation of vanilla (*Vanilla planifolia* Andrews). In: Jain, S. M. & Saxena, P. K. (eds) *Springer protocols, Methods in Molecular Biology 547, Protocols for In Vitro Cultures and Secondary Metabolite analysis of Aromatic and Medicinal Plants*. The Humana Press/Springer, New York, pp. 129–138.

Minoo, D. 2002. *Seedling and Somaclonal Variation and Their Characterization in Vanilla*. PhD thesis. Calicut University, Kerala.

Minoo, D., Geetha, S. P., Nirmal Babu, K. & Peter, K. V. 2008a. Isolation and fusion of protoplasts in Vanilla species. *Current Science*, 94(1):115–120.

Minoo, D., Nirmal Babu, K. & Grisoni, M. 2010. Biotechnological applications. In: Eric, O. & Michel, G. (eds) *Vanilla*. CRC Press, Boca Raton, pp. 51–73.

Minoo, D., Nirmal Babu, K., Ravindran, P. N. & Peter, K. V. 2006. Interspecific hybridization in vanilla and molecular characterization of hybrids and selfed progenies using RAPD and AFLP markers. *Scientia Horticulturae*, 108(4):414–422.

Mythili, J. B., Rajeev, P. R. & Chethana, B. S. 2012. Factors influencing *in vitro* regeneration and *Agrobacterium*-mediated transformation of onion 'Arka Niketan'. *Acta Horticulturae*, 929:417–422.

Mythili, J. B., Rashmi, H. J., Suneetha, C., Saiprasad, G. V. S., Rajeev, P. R., Naveena, C., Anand, L., Ganeshan, G. & Mahamood, Riaz. 2015. Transgenic chili possessing *Baculovirus* chitinase gene exhibits *in vitro* fungal inhibition. *Journal of Crop Improvement*, 29(2):159–187.

Nadgauda, R. S., Khuspe, S. S. & Mascarenhas, A. F. 1982. Isolation of high-curcumin varieties of turmeric from tissue culture. In: Iyer, R. D. (ed). *Proceedings of the V Annual Symposium on Plantation Crops*. CPCRI, Kasargod, pp. 143–144.

Nadgauda, R. S., Kulkarni, D. B., Mascarenhas, A. F. & Jaganathan, V. 1980. Development of plantlets from tissue cultures of ginger. *Proceedings of the Annual Symposium on Plantation Crops*. CRCRI, Kasargod, pp. 143–147.

Nair, R. R. & Dutta Gupta, S. 2003. Somatic embryogenesis in black pepper (*Piper nigrum* L.): 1. Direct somatic embryogenesis from the tissues of germinating seeds and ontogeny of somatic embryos. *The Journal of Horticultural Science and Biotechnology*, 78:416–421.

Nair, R. R. & Dutta Gupta, S. 2006. High-frequency plant regeneration through cyclic secondary somatic embryogenesis in black pepper (*Piper nigrum* L.). *Plant Cell Reports*, 24(12):699–707.

Nazeem, P. A., Augustin, M., Rathy, K., Sreekumar, P. K., Rekha, C. R., Shaju, K. V., Peter, K. V., Girija, D. & Kesavachandran, R. 2004. A viable protocol for large scale *in vitro* multiplication of black pepper (*Piper nigrum* L.). *J Plant Crops*, 32:163–168.

Nemati, Z., Zeinalabedini, M., Mardi, M., Pirseyediand, S. M., Marashi, S. H. & Nekoui, S. M. K. 2012. Isolation and characterization of a first set of polymorphic microsatellite markers in saffron, *Crocus sativus* (Iridaceae). *American Journal of Botany*, 99(9):340–343.

Nianiou, M., Zambounis, K. A. & Tsaftaris, A. 2002. Development of pepper transgenic plants via *Agrobacterium* and biolistic transformation. *Acta Horticulturae*, 579:83–87.

Nilam, J. T., Rasmieh, H., Feba, J., Nimita, G. U., Abhijeeta, K. N., Rukam, S. T. & Balaji, A. G. 2020. Transcriptome landscaping for gene mining and SSR marker development in coriander (*Coriandrum sativum* L.). *Genomics*, 112(2):1545–1553.

Nirmal Babu, K. & Minoo, D. 2003. Commercial micropropagation of spices. In: Chandra, R. & Misra, M. (eds) *Micropropagation of Horticultural Crops*. International Book Distributing Company, Lucknow, p. 345.

Nirmal Babu, K., Anu, A., Remasree, A. B. & Praveen, K. 2000. Micropropagation of curry leaf tree *Murraya koenigii* (L.) *Spreng. Plant Cell, Tissue and Organ Culture*, 61(3):199–203.

Nirmal Babu, K., Asha, S., Saji, K. V. & Parthasarathy, V. A. 2011a. Black pepper. In: Singh, H. P., Parthasarathy, V. A., & Nirmal Babu, K. (eds) *Advances in Horticulture Biotechnology*, vol 3, Molecular Markers and Marker-Assisted Selection – Fruit Crops, Plantation Crops and Spices. Westville Publishing House, New Delhi, pp. 247–260.

Nirmal Babu, K., George, J. K., Anandaraj, M., Venugopal, M. N., Nair, R. R. et al. 2005a. Improvement of selected spices through biotechnology tools – Black pepper, Cardamom, Ginger, Vanilla. Final report, DBT, Government of India, p. 111.

Nirmal Babu, K., George, J. K., Bhat, A. I., Prasath, D. & Parthasarathy, V. A. 2011. Tropical spices. In: Singh, H. P., Parthasarathy, V. A. & Nirmal Babu, K. (eds) *Advances in Horticulture Biotechnology*, vol 5, Gene Cloning and Transgenics. Westville Publishing House, New Delhi, pp. 529–542.

Nirmal Babu, K., Minoo, D., Geetha, S. P., Samsudeen, K., Rema, J., Ravindran, P. N. & Peter, K. V. 1998. Plant biotechnology – Its role in improvement of spices. *The Indian Journal of Agricultural Science*, 68(8):533–547.

Nirmal Babu, K., Minoo, D., Geetha, S. P., Samsudeen, K., Rema, J., Ravindran, P. N. & Peter, K. V. 1998. Plant Biotechnology- Its role in improvement of Spices. *Indian Journal of Agricultural Sciences*, 68(8):533–547.

Nirmal Babu, K., Minoo, D., Rahul, P. R., Anupama, K., Peter, K. V. & Sarma, Y. 2016. *Biotechnological Approaches in Improvement of Spices: A Review*. DOI: 10.1007/978-81-322-2283-5_25.

Nirmal Babu, K., Nair, R. R., Saji, K. V. & Parthasarathy, V. A. 2012b. Biotechnology. In: Singh, H. P., Parthasarathy, V. A., Srinivasan, V. & Saji K. V. (eds) *Piperaceae*. Westville Publishing House, New Delhi, pp. 57–81.

Nirmal Babu, K., Ravindran, P. N. & Peter, K. V. (eds) 1997. *Protocols for Micropropagation of Spices and Aromatic Crops*. Indian Institute of Spices Research, Calicut, p. 35.

Nirmal Babu, K., Samsudeen, K. & Ratnambal, M. J. 1992a. *In vitro* plant regeneration from leaf-derived callus in ginger, *Zingiber officinale* Rosc. *Plant Cell, Tissue and Organ Culture*, 29(2):71–74.

Nirmal Babu, K., Samsudeen, K. & Ravindran, P. N. 1996. Biotechnological approaches for crop improvement in ginger, *Zingiber officinale* Rosc. In: Ravishanker G. A. & Venkataraman, L. V. (eds) *Recent Advances in Biotechnological Applications on Plant Tissue and Cell Culture*. Oxford IBH Publishing Co., New Delhi, pp. 321–332.

Nirmal Babu, K., Samsudeen, K., Minoo, D., Geetha, S. P. & Ravindran, P. N. 2005b. Tissue culture and biotechnology of ginger. In: Ravindran, P. N. & Nirmal Babu, K. (eds) *Ginger – The Genus Zingiber*. CRC Press, Boca Raton, pp. 181–210.

Nirmal Babu, K., Suraby, E. J., Cissin, J., Minoo, D., Pradeepkumar, T., Parthasarathy, V. A. & Peter, K. V. 2013. Status of transgenics in Indian spices. *Journal of Tropical Agriculture*, 51:1–14.

Nirmal Babu, K., Usha Rani, T. R. & Parthasarathy, V. A. 2011b. Cardamom. In: Singh, H. P., Parthasarathy, V. A., Nirmal K (eds) *Advances in Horticulture Biotechnology*, vol 3, Molecular Markers and Marker-Assisted Selection – Fruit Crops, Plantation Crops and Spices. Westville Publishing House, New Delhi, pp. 261–268.

Nirmal Babu, K. 1997. *In vitro studies in ginger (Zingiber officinale* Rosc.*)*. Ph.D thesis. University of Calicut, Calicut, Kerala, India.

Rostiana, O., Syahid, S. F., Arlianti, T. & Supriadi 2022. Molecular analysis of gene from ginger and *Zingiber zerumbet* encoding resistance to bacterial wilt. *IOP Conference Series: Earth and Environmental Science*, 974(1):01205.

Panda, M. K., Mohanty, S., Subudhi, E., Acharya, L. & Nayak, S. 2007. Assessment of genetic stability of micropropagated plants of *Curcuma longa* L. by cytophotometry and RAPD analysis. *Internationl Journal of Integrative Biology*, 1(3):189–195.

Pandey, R., Tiwari, R. K. & Shukla, S. S. 2016. Omics: A newer technique in herbal drug standardization and quantification. *Journal of Young Pharmacista*, 8(2):76–81.

Pandey, S., Mishra, A., Patel, M. K. & Jha, B. 2013. An efficient method for *Agrobacterium*-mediated genetic transformation and plant regeneration in cumin (*Cuminum cyminum* L.). *Applied Biochemistry and Biotechnology*, 71(1):1–9.

Park, M. Y., Yi, N. R., Lee, H. Y., Kim, S. T., Kim, M., Park, J. H., Kim, J. K., Lee, J. S., Cheong, J. J. & Choi, Y. D. 2002. Generation of chlorsulfuron-resistant transgenic garlic plants (*Allium sativum* L.) by particle bombardment. *Molecular Breeding*, 9(3):171–181.

Parthasarathy, V. A. & Nirmal Babu, K. 2011. Ginger. In: Singh HP, Parthasarathy VA, Nirmal Babu K (eds) *Advances in Horticulture Biotechnology*, vol 3, Molecular Markers and Marker Assisted Selection – Fruit Crops, Plantation Crops and Spices. Westville Publishing House, New Delhi, pp. 279–290.

Paul, S., Dam, A., Bhattacharyya, A. & Bandyopadhyay, T. K. 2011. An efficient regeneration system via direct and indirect somatic embryogenesis for the medicinal tree *Murraya koenigii*. *Plant Cell, Tissue and Organ Culture*, 105(2):271–283.

Peter, K. V., Ravindran, P. N., Nirmal Babu, K., Venugopal, M. N., Geetha, S. P. & Benny, D. 2001. *Production of Somaclones and Somatic Hybrids of Cardamom (Elettaria Cardamomum* Maton*) for High Yield and Resistance to Diseases*. ICAR Project report. Indian Institute of Spices Research, Calicut, p. 120.

Pruthi, J. S. 1992. *Spices and Condiments*. National Book Trust, New Delhi.

Raheleh, A., Hasanloo, T. & Khosroshahli, M. 2011. Evaluation of trigonelline production in *Trigonella foenum-graecum* hairy root cultures of two Iranian masses. *Plant Omics Journal*, 4(7):408–412.

Rani, U., Sharma, M. M., Ismail, N. & Batra, A. 2012. *In vitro* plant regeneration from immature seeds of *Murraya koenigii* L. *Indian Journal of Biotechnology*, 11(1):108–110.

Rao, S. N. K., Narayanaswamy, S., Chacko, E. K. & Doraiswamy, R. 1982. Regeneration of plantlets from callus of *Elettaria cardamomum* Maton. *Proceedings of the Indian Academy of Science (Plant Sci)*, 91(1):37–41.

Ravindran, P. N. 2000. *Black Pepper: Piper nigrum, Series: Medicinal and Aromatic Plants- Industrial Profiles*. Harwood Academic Publishers, Amsterdam, Netherlands.

Retheesh, S. T. & Bhat, A. I. 2011. Genetic transformation and regeneration of transgenic plants from protocorm-like bodies of vanilla (*Vanilla planifolia* Andrews) using *Agrobacterium tumefaciens*. *Journal of Plant Biochemistry and Biotechnology*, 20(2):262–269.

Robledo-Paz, A., Cabrera-Ponu, J. L., Villalobos-Aranbula, V. M., Herrer-Estrella, L. & Jofre-Garfias, A. E. 2004. Genetic transformation of garlic (*Allium sativum* L) by particle bombardment. *Horticulture Science*, 39(6):1208–1211.

Sajeev, S., Roy, A. R., Langrai, B., Pattanayak, A. & Deka, B. C. 2011. Genetic diversity analysis in the traditional and improved ginger (*Zingiber officinale* Rosc.) clones cultivated in North-East India. *Scientia Horticulturae*, 128(3):182–188.

Salvi, N. D., George, L. & Eapen, S. 2000. Direct regeneration of shoots from immature inflorescence cultures of turmeric. *Plant Cell Tissue and Organe Culture*, 62(3):235–238.

Samsudeen, K., Nirmal Babu, K., Minoo, D. & Ravindran, P. N. 2000. Plant regeneration from anther-derived callus cultures of ginger (*Zingiber officinale* Rosc.). *Journal of Horticultural Science and Biotechnology*, 75(4):447–450.

Sandhu, J. S. & Gosal, S. S. 2009. A simple and rapid method for optimizing particle bombardment parameters through transient expression of gus reporter gene in onion epidermal layer cells. *Crop Improvement*, 36:6–8.

Sen, S., Skaria, R. & Muneer, P. A. 2010. Genetic diversity analysis in *Piper* species (Piperaceae) using RAPD markers. *Molecular Biotechnology*, 46(1):72–79.

Shahabzadeh, Z., Heidari, B. & Hafez, R. F. 2013. Induction of transgenic hairy roots in *Trigonella foenum-graceum* co-cultivated with *Agrobacterium rhizogenes* harboring a GFP Gene. *Journal of Crop Science and Biotechnolgy*, 16(4):263–268.

Shaji, P., Anandaraj, M. & Sharma, Y. R. 1998. Comparative study of protoplast isolation and development in *Piper nigrum* (black pepper) and *P. colubrinum*. In: Mathew NM, Jacob CK (eds) *Developments in Plantation Crops Research*. Allied Publishers, New Delhi, pp. 51–53.

Sharma, T. R. & Singh, B. M. 1997. High frequency *in vitro* multiplication of disease-free *Zingiber officinale* Rosc. *Plant Cell Reports*, 17(1):68–73.

Shasany, A. K., Darokar, M. P., Dhawan, S., Gupta, A. K. et al. 2005. Use of RAPD and AFLP markers to identify inter- and intraspecific hybrids of mentha. *Oxford Journal of Heredity*, 96(5):542–549.

Sheeja, T. E., Sabeesh, C., Shabna, O. V., Shalini, R. S. & Krishnamoorthy, B. 2013. Genetic diversity analysis of *Myristica* and related genera using RAPD and ISSR markers. *Journal of Spices and Aromatic Crops*, 22(1):38–46.

Sheji, C., Smitha, K. S., George, R. S., Bhat, I. & Anandaraj, M. 2006. Development of SCAR marker for locating *Phytophthora* resistance in black pepper (*Piper nigrum* L). Southern Zone Meeting. Indian Phytopathological Society, Kasargod.

Shekhawat, N. S. & Galston, A. W. 1983. Mesophyll protoplasts of fenugreek (*Trigonella foenum-graecum*): Isolation, culture and shoot regeneration. *Plant Cell Reports*, 2(3):119–121.

Shi, X., Dai, X., Liu, G. & Bao, M. 2009. Enhancement of somatic embryogenesis in camphor tree (*Cinnamomum camphora* L.): Osmotic stress and other factors affecting somatic embryo formation on hormone-free medium. *Trees*, 23(5):1033–1042.

Shi, X., Dai, X., Liu, G., Zhang, J., Ning, G. & Bao, M. 2010. Cyclic secondary somatic embryogenesis and effi cient plant regeneration in camphor tree (*Cinnamomum camphora* L.). *In Vitro Cellular and Developmental Biology – Plant*, 46(2):117–125.

Shirgurkar, M. V., Naik, V. B., Arnold, S. V., Nadgauda, R. S. & Clapham, D. 2006. An efficient protocol for genetic transformation and shoot regeneration of turmeric (*Curcuma longa* L.) via particle bombardment. *Plant Cell Reports*, 25(2):112–116.

Shivakumar, N. & Agrawal, P. 2014. Callus induction and regeneration from adventitious buds of *Zingiber officinale. Asian Journal of Microbiology, Biotechnology and Environmental Sciences*, 16(4):881–885.

Shylala, M. R., Nair, S. G., Nazism, P. A., Millikan, V. K. & Mathew, M. K. 1996. *In vitro* screening of black pepper for tolerance to *Phytophthora capsici. J Plant Crops*, 24:171–178.

Singh, N., Mishra, A., Joshi, M. & Jha, B. 2010. Microprojectile bombardment mediated genetic transformation of embryo axes and plant regeneration in cumin (*Cuminum cyminum* L.). *Plant Cell, Tissue and Organ Culture*, 103(1):1–6.

Singh, S., Panda, M. K. & Nayak, S. 2012. Evaluation of genetic diversity in turmeric (*Curcuma longa* L) using RAPD and ISSR markers. *Industrial Crops and Products*, 37(1):284–291.

Smith, M. K., Hamill, S. D., Gogel, B. J. & Severn-Ellis, A. A. 2004. Ginger (*Zingiber officinale*) autotetraploid with improved processing quality produced by an *in vitro* colchicines treatment. *Australian Journal of Experimental Agriculture*, 44(10):1065–1072.

Sobhakumari, V. P. & Lalithakumari, D. 2005. High-frequency shoot regeneration and *Agrobacterium*-mediated DNA transfer in red chilli (*Capsicum annuum* L.). *Plant Cell Biotechnology and Molecular Biology*, 6:9–16.

Song, G. Q., Loskutov, A. V. & Sink, K. C. 2007. Highly efficient *Agrobacterium tumefaciens*-mediated transformation of celery (*Apium graveolens* L.) through somatic embryogenesis. *Plant Cell, Tissue and Organ Culture*, 88(2):193–200.

Sulassih, S. & Santosa, E. 2013. Phylogenetic analysis of mangosteen (*Garcinia mangostana* L.) and its relatives based on morphological and inter simple sequence repeat (issr) markers. *SABRAO Journal of Breeding and Genetics*, 45(3):478–490.

Suma, B., Keshavachandran, R. & Nybe, E. V. 2008. *Agrobacterium tumefaciens* mediated transformation and regeneration of ginger (*Zingiber officinale* Rosc.). *Journal of Tropical Agriculture*, 46:38–44.

Sumathi, V. 2007. *Studies on Somaclonal Variation in Zingiberaceous Crops.* PhD thesis. University of Calicut, Kerala.

Suparman, U. & Blake, J. 1990. Studies on tissue culture of clove tree plant. *Indonesian Journal of Crop Science*, 5(2):67–75.

Tamayo, A. C. 2007. Characterizacion de genotipos seleccionados de cardamomom (*Elettaria cardamomum* L. Maton) (Zingiberaceae) in Colombia. *Tierra Tropical*, 3(2):233–241.

Thomas, E., Kizhakkayil, J., Zachariah, T. J., Syamkumar, S. & Sasikumar, B. 2006. Comparative quality characterization and molecular profiling of Indian, Sri Lankan and Guatemalan cardamoms. *Journal of Food, Agriculture and Environment*, 4(2):129–133.

Verma, S. & Rana, T. S. 2013. Genetic relationships among wild and cultivated accessions of curry leaf plant (*Murraya koenigii* (L.) Spreng.), as revealed by DNA fingerprinting methods. *Molecular Biotechnology*, 53(2):139–149.

Wadt, L. H. D. O., Ehringhaus, C. & Kageyama, P. Y. 2004. Genetic diversity of " Pimenta Longa" genotypes (Piper spp., Piperaceae) of the Embrapa Acre germplasm collection. *Genetics and Molecular Biology*, 27(1):74–82.

Wahyuni, S., Xu, D. H., Bermawie, N., Tsunematsu, H. & Ban, T. 2003. Genetic relationships among ginger accessions based on AFLP marker. *J Bioteknologi Pertanian*, 8:60–68.

Wang, Y. & Kumar, P. P. 2004. Heterologous expression of Arabidopsis ERS1 causes delayed senescence in coriander. *Plant Cell Reports*, 22(9):678–683.

Wang, Z. M., Niu, Y., Song, M. & Tang, Q. L. 2010. Tetraploid of *Zingiber officinale* Roscoe. *in vitro* inducement and its morphology analysis. *China Vegetables*, 4:13.

Wendimu, A., Tekalign, W. & Asfaw, B. 2021. A survey of traditional medicinal plants used to treat common human and livestock ailments from Diguna Fango district, Wolaita, southern Ethiopia. *Nordic Journal of Botany*, 39(5):1–20.

Xu, K., Huang, X., Wu, M., Wang, Y., Chang, Y., Liu, K., et al. 2014. A rapid, highly efficient and economical method of Agrobacterium-mediated *in planta* transient transformation in living onion epidermis. *PLOS ONE*, 9(1):e83556.

Yamakawa, T., Sekiguchi, S., Kodama, T., Smith, S. M. & Yeoman, M. M. 1998. Transformation of chilli pepper (*Capsicum frutescens*) with a phenylalanine ammonia-lyase gene. *Plant Biotechnology*, 15(4):189–193.

Yamuna, G. 2007. *Studies on Cryopreservation of Spice Genetic Resources*. PhD thesis. Calicut University, Kerala.

Yang, L., Hongyi,C., Jian, L., Yuhong, Y., Yan, L., Bingyan, X., Jianlong, Z. & Zhenchuan, M. 2023. Molecular cloning and functional analysis of the pepper resistance gene Me3 to root-knot nematode. *Horticultural Plant Journal*, 9(1):133e144.

Yousefi, K., Razmju, F. & Osmani, A. A. 2019. Molecular cloning of flavone synthase I gene from cumin (*Cuminum cyminum*). *International Journal of Science and Research*, 8(12):1924–1929.

Zheng, S. J., Henken, B., Ahn, Y. K., Krens, F. A. & Kik, C. 2004. The development of a reproducible *Agrobacterium tumefaciens* transformation system for garlic (*Allium sativum* L.) and the production of transgenic garlic resistant to beet armyworm (*Spodoptera exigua* Hübner). *Molecular Breeding*, 14(3):293–307.

Zheng, S. J., Khrustaleva, L., Henken, B., Soari, E., Jacobsen, E., Kik, C. & Krens, F. A. 2001. *Agrobacterium tumefaciens*-mediated transformation of *Allium cepa* L.: The production of transgenic onions and shallots. *Molecular Breeding*, 7(2):101–115.

3 Sustainable Techniques for the Harvest and Conservation of Spice and Condiment Plant Species

Sudeshna Nandi, Shruti Chattaraj, and Samapika Nandy

3.1 INTRODUCTION

Biodiversity furnishes goods and services that comfort human life (Bini et al., 2005), yet biodiversity is decreasing at an unparalleled rate, which has injurious effects on human health. The quest to continue the potential benefits of biodiversity has led to the advancement of multilateral environmental agreements (Stanley et al., 2012). In recent times, recognition has been devoted to the sustainable use of natural resources and the conservation of biological diversity (Niesenbaum, 2019). By far, non-timber forest products are among the most studied groups and rural people depend on them for their livelihood and trade (Heubach et al., 2011). The non-timber forest products are vital in meeting medicinal needs, providing sources of income and ensuring food security (Meinhold & Darr, 2019; Husen et al., 2021). Forest products include different taxonomic and functional groups, including spices and condiments. Spices and condiments are natural plant mixtures used for complementing foods either in whole or ground forms. Spices are basically fragrant, pungent, and aromatic dried parts of the plants, such as bark, roots, leaves, stems, flowers, bud, fruit, seeds, etc., usually utilized in minute quantities, whereas condiments are the herbs mostly added in higher quantities as compared with spices (D'Souza et al., 2017). Since antiquity, spices and condiments have been a vital part of human history, plant products, and human nutrition, and have been considered to be essential to the culinary arts to elevate the taste and savour of cooked foods (Balkrishna et al., 2023). Along with elevating taste or flavouring, spices and condiments are widely used as colouring agents, preservatives, and as essential oils in many pharmaceutical and chemical industries. Some spices and condiments were traditionally used as curative and preventive agents, boosting the immune system if consumed appropriately (D'Souza et al., 2017; Husen, 2021). In addition, they also played a notable role in the economy of exporting and importing countries of the world, as well in the development of most cultures around the globe (Thomas & Sanil, 2019; Husen, 2023).

Reports have stated that there are some particular phytochemical compounds which are responsible for the colouring, aromatic, and pungent qualities of spices and condiments. These phytochemical compounds have medicinal activities including antimicrobial, antimutagenic, antioxidant, anticarcinogenic, anti-inflammatory, anti-arthritic, hypoglycaemic, and antidiabetic properties, which are beneficial in treating several diseases (Dubey, 2017). The antimicrobial and antifungal activity of the natural constituents of spices even help in inhibiting both food spoilage microorganisms (food preservation) as well as human pathogens (food safety) (Gottardi et al., 2016). Some widely accepted spices, which are renowned for their taste in every cookhouse of the world and for healing and medication, are as follows: asfoetida, carom seeds, curry leaves, fenugreek, bay leaf,

DOI: 10.1201/9781003387046-3

capsicum, cardamom, garlic, mint, clove, coriander, cinnamon, black paper, cumin, nutmeg, fennel, nigella, ginger, mustard, etc. (Muggeridge, 2001). The incorporation of spices in meals has several advantageous effects like preventing colds and influenza, stimulating the secretion of saliva, promoting digestion, and even suppressing nausea and vomiting (Sultana et al., 2010). The International Organization for Standardization (ISO) has listed 109 spices in the globe, among which 63 spices are grown in India (Shukla & Yadav, 2018). Since various wild spices are under threat of extinction, action must be taken to domesticate or cultivate these crops, rather than collect from the wild, for their sustainable conservation and also to meet demand from all around the globe. The present review aims to provide an exhaustive summary of most of the relevant and sustainable techniques to harvest and conserve the spices and condiments from plant species.

3.2 HISTORY OF SPICES AND CONDIMENTS

Records for spices and condiments date back to ancient and medieval times, when spices and condiments were the most prized elements of trade (Purseglove et al., 1969). The first century AD, in Rome, represented the first expansion of the use of spices and condiments in food (Rosengarten, 1969). The utilization of spices expanded from the Middle East to the eastern Mediterranean and Europe, as well as in Asian countries (Purseglove et al., 1981). Similarly, condiments were also popular in ancient China, India, Italy, and Greece, where the condiments were used to make food more appetizing (Farrell, 1990). Some of the widely used spices and condiments used to flavour food or beverages include turmeric, cloves, cardamom, black pepper, coriander, cinnamon, fenugreek, mustard, ginger, onion, and garlic, possessing a broad spectrum of biological functions, whereas the production of tree spices like cinnamon, clove, nutmeg, and allspice appeared to be very restricted (Sachan et al., 2018). In recent times, the consumption and export of spices and condiments are increasing every day all around the world, pointing to the fact that there is a sizeable opportunity for increasing the growing area and improving the harvesting and storage processes for these spices and condiments to increase the quantity and quality of the produce.

3.3 HARVESTING AND POSTHARVEST HANDLING TECHNIQUES OF PLANT-BASED SPICES AND CONDIMENTS

Depending on the requirements of the end users of the product, the harvesting stage is controlled; if the produce, like ginger, is for extraction, harvesting is recommended 7–8 months after planting, whereas, for home use, harvesting at 8–10 months is suggested (Girma et al., 2008). During harvesting, cautionary steps should be taken to make sure that no contaminating matter gets blended with the harvested plant materials. The harvested materials need to be carried promptly to a drying shed if harvesting occurs under moist conditions. All major equipment, including harvesters, cutting devices, and other equipment, should be properly maintained, cleaned, and calibrated to lessen contamination from soil and other materials (Douglas et al., 2005). According to Hiranya (2013), a lack of understanding of varieties, irrigation, washing, and storage were the major factors undermining the quality of spices in the export chain.

3.3.1 Pre-cooling and Trimming

To remove the field heat from spices like ginger and to better withstand transportation, pre-cooling is carried out. Trimming involves the removal of pest-infested parts, undesirable stems, diseased roots and other parts with a sharp knife to prevent decay during storage and supply (Sharma, 2009). According to the findings of Yiljep et al. (2005) and Belay et al. (2012), the damaged, injured, and cut ends, if not removed carefully, might serve as initiation points for the growth of moulds and rots, as well as source of inoculums for the spread of pathogenic fungi and bacteria. Likewise, the quality and shelf life of spices like ginger might be negatively affected due to improper storage in a room without any cooling provisions, resulting in thereby decay and shrinkage of ginger rhizomes, loss of quantity, and infection by diseases (Casimero et al., 2021). Cold storage might overcome the problem and maintain the quality.

3.3.2 Washing and Cleaning

In order to get highest market price for high-quality product, washing is one of the vital postharvest handling practices (Belay et al., 2012). Fresh rhizomes of ginger and turmeric are washed to remove dirt and soil, and pressure washing is approved as it is more effective and tends to lower the microbial loads, although washing can also be achieved by soaking and spraying equipment. After harvesting, cleaning is also necessary to minimize losses from the growth of moulds, decay, and microbial contamination. During postharvest handling operations, hot air dryers or mechanical washers can help to reduce contamination from dust (Abate, 2013). If rhizomes were not washed properly, the physical appearance deteriorates, affecting the quality of the produce during transportation, storage, and packaging, thereby increasing the postharvest losses during this period. A lack of appreciation of the washing practices and their significance for export markets are among the prime reasons for the economic losses suffered by spice traders (Hiranya, 2013).

3.3.3 Curing, Sorting, and Grading

Curing is the next operation in the processing of spices, which helps to heal the damaged, scratched, and wounded tissues of the produce by keeping them under shade at a suitable temperature for 3–4 days to achieve the dry product. High temperature and relative humidity favour periderm formation. Curing aims to increase the produce shelf life and control the attack by and spread of diseases (Acharya, 2012). Through the process of curing, the vitality of fresh rhizomes, like ginger and turmeric, was maintained in such a way that it can slow down the postharvest decaying, starch was gelatinized, they were handled to preclude raw odour, lowering the drying time and disseminating the pigment consistently so that it provided a more uniformly coloured produce. After the extraction of rhizomes from the ground, curing must be finished in about ten days to achieve the firmest, most usable product. Long-duration types of spices have a medium curing percentage (21–24%), medium-duration types achieve the lowest curing percentage (14–20%), and short-duration types have the highest curing percentage (24–26%) (Hordofa et al., 2020). A variety of spices, including saffron, cinnamon, cassia, and vanilla beans, are cured by various specialized techniques in order to generate the characteristic flavour or aroma for which they are valued (Momin & Jamir, 2021). Grading is performed based on the size, colour, and freshness, while sorting is executed to remove the decayed, diseased, and damaged spices (Hordofa et al., 2020).

3.3.4 Threshing

Threshing is the procedure of the separation of fruits or seeds from unwanted flower stems or plant stalks. Separation is done by the use of winnowing, even manually, with hand-utilized sieves and screens, or by mechanical shakers and sorters. Seed spices are threshed by beating with wooden and bamboo sticks (Momin & Jamir, 2021) or by the use of engine-driven threshing machines.

3.3.5 Packaging

Other chief postharvest practices include adequate packaging to diminish postharvest losses and to maintain the quality of the spices. Packaging is considered to be an essential procedure through which the intrinsic feature of the spices is conserved and damage caused by microbial infection or insect infestation is prevented (Gautam and Acharya., 2009; Zakir et al., 2018).

3.3.6 Storage

High-quality storage is a crucial step in postharvest handling operations with cool stores and warehousing facilities linked to postharvest crop management. Under adverse environmental conditions, spices rapidly deteriorate and hence should be conserved in well-kept storage facilities. The storehouses need to be properly ventilated with devices to check humidity and temperature in order to eliminate mould and insect attacks. The rooms or storage places need fumigation before storage. Utilizing appropriate postharvesting techniques and storage procedures, spices like ginger, turmeric, pepper, clove, nutmeg, and chillies, can be conveniently stored for several months (Girma et al., 2016).

There are a few spices available, like black cumin, which need intensive labour during collection and harvesting as the capsules (fruits) at maturity tend to shatter. The mature fruits do not need much attention as they are self-preserving.

3.4 CONSERVATION OF SPICES AND CONDIMENTS

The most important key to food security and sustainable agricultural development is the use of plant genetic resources. The plant genetic resources are mainly constituted of populations of cultivars. Maintenance of plant genetic resources is one of the major and most important techniques for conservation of the particular species. Spices are propagated naturally by the rhizomes or bulbs, depending on the spice. Farmers propagate the spices by stem cuttings. Conservation of germplasm can thus be achieved by using most suitable or diverse explants (Chaudhury et al., 2013). The traditional ecologists prefer *in-situ* conservation over *ex-situ* conservation as some of the useful traits are the result of specific environmental factors, but this approach is not applicable in case of critically endangered or threatened species which demands different conservation strategies. In that case, conservation of only a small portion of the germplasm is enough to maintain the existence of the species and prevent the extinction (Nirmal Babu et al., 2012). Spices are the most important plants and represent a precious jewel in the chest of Asia, with India being one of the countries with most spice plant species. India has given priority to the conservation and characterization of different spices through different scientific research programmes.

3.4.1 Importance of Germplasm Conservation

Seed storage is one of the oldest and sometimes most effective methods of conservation. But this process has its own limitations. Crops with recalcitrant seed or where the crop needs to be conserved urgently because of serious threats to its existence cannot be conserved by only seed storage. This must be augmented by *in-vitro* conservation (Nirmal Babu et al., 2012). Seeds have a high probability of being affected by a pathogen or pest. If this happened, then the entire conservation programme will be undermined; in *in-vitro* conservation under sterile conditions, this possibility can be minimized. Hence, *in-vitro* conservation and cryo-preservation are important supplementary strategies.

Most of the spices are propagated by vegetative methods and they are conserved in field gene banks. Conservation of this type of crop in a field gene bank increases the risk of loss of accessions by bacteria, viruses, or other pathogens, or natural disasters such as storms, and hence the conserved collection may be damaged or even destroyed (Nirmal Babu et al., 2012). Conservation by means of *in vitro* or cryopreservation is much safer in all aspects although it is more expensive than conservation at the field level.

3.4.2 Methods

Several methods have been practised for *in vitro* conservation. The most important method of conservation of plants with recalcitrant seeds is plant tissue culture. This technique is considered as the base line technique of the cryopreservation and *in vitro* conservation. The plant regenerated through these techniques can also be used for different studies, such as experiments to examine physiological and biochemical processes, cell and callus lines developed for *in vitro* synthesis of valuable secondary metabolites and other important compounds (Nirmal Babu et al., 1999; Peter et al., 2002; Minoo, 2002, Nirmal Babu et al., 2005).

Generally, callus culture is much easier than cryopreservation. The callus is a heterogeneous mass of cells, with a high volume and a slow growth rate. These features constrain the cryopreservation of plant tissue cultures. Nag and Street (1975) first successfully cryopreserved the plant cell. This method of cryopreservation is not popular in the case of spices but it has slowly taken hold. Umetsu et al. (1995) conserved fennel embryonic cell lines at 4°C.

Long-term conservation of plant germplasm is the primary topic of concern nowadays, and cryopreservation is the only method to solve this problem. Generally, liquid nitrogen is used for this technique. In liquid nitrogen, all the cellular and physiological processes become too slow to have any deteriorative effect (Kartha, 1985). Over 100 different spices have been conserved using this technique. Cellular division stops when plant germplasm is stored in liquid nitrogen (–196°C). All the metabolic and physiological processes stop at this ultralow temperature, which lowers the probability of genetic instability or loss of the accession due to contamination or human error during the subculturing processes. Generally, the vegetatively propagated plants or elite genotype plants are conserved better by this technique (Withers, 1985; Kartha, 1985). Care must be taken during this technique, as tissue damage may occur at this very low temperature.

3.5 CONCLUSIONS AND RECOMMENDATIONS

Agriculture remains the pillar of economy for many countries, with traditional procedures that have been passed down through generations. Spices and condiments, being cash crops, have a massive potential for the economies of developing countries by elevating the purchasing power of the smallholder farmers, creating and extending employment convenience and dispersal of income, and thereby aiding in lowering poverty and establishing food security. Spices and condiments have become a natural and necessary component of people's daily nutrition; beyond their role in conferringexerting flavour to food, they also affect many advantageous physiological effects, including pharmaceutical ones. The normal intake of spices and condiments has proved to be safe, as well as leading to several beneficial effects. They bear multiple health beneficial properties and there is also a probability of synergy among them in their mechanism of action. Thereby, a diet rich in spices is compatible with a healthy life.

The worldwide demand for processed foods has also become a fundamental part of the modern way of human life which has geared up the research into spices and condiments. For spices and condiments harvested from wild populations, with restricted abundance and slow growth rates, destructive harvesting of wild-grown plants usually results in exhaustion of resource and extinction of species. Therefore, utilization of improved sustainable harvesting practices and conservation of spices and condiments should be considered, and must be formulated to boost their productivity and survival. Root and whole-plant harvesting is considered to be more destructive than collecting flowers, buds, or leaves. Postharvest handling practices are other major concerns for spice growers as most of the freshly harvested spice crops are highly perishable, with high moisture content and highly susceptible to microbial contamination. Hence, refined contemporary processing methods, relevant postharvest handling strategies, and suitable conservation methods are required to extend the mean shelf life and quality of spices and condiments to meet international standards.

REFERENCES

Abate, S. 2013. Protection of ginger. In: Fikre, T. & Kifle, A. (eds.) *Ginger (Zingiber officinale* Roscoe) *Production, Postharvest Handling, Processing and Marketing: A Com-prehensive Extension Package Manual*. Hawassa, Ethiopia, 139.

Acharya, B. 2012. *Sustainable Ginger Production, Protective Storage and Dried Ginger Processing Technology. Protective Storage of Ginger.* Nepal Agricultural Research Council, National Ginger Research Program.

Balkrishna, A., Joshi, B., Srivastava, A. & Bisht, S. 2023. *An Insight on Nutraceutical Influences of Spice and Condiments: Serve as a Flavor and Immunity Enhancer from Ages. Herbs and Spices - New Advances.* https://doi.org/10.5772/intechopen.108749.

Belay, B., Kebede, A., Tameru, A. & Fasil, A. 2012. Isolation, characterization and identification of post-havest spoilage fungi of ginger (*Zingiber officinale*) at HadaroTunto and Boloso-Bombae, Southern Ethiopia. *International Journal of Life Sceinces* 1(2): 19–27.

Bini, L., Alexandre, J., Diniz-Filho, F., Carvalho, P., Pinto, M. P. & Rangel, T. F. L. 2005. Lomborg and the litany of biodiversity crisis: What the peer-reviewed literature says. *Conservation Biology* 19(4): 1301–1305.

Casimero, N. G., Quevedo, M. A. & Binas, E. E. 2021. Postharvest quality and shelf life of ginger (*Zingiber officinale*) as affected by different storage methods. *International Journal for Innovative Engineering and Management Research* 10(4): 467–479.

Rekha, C. & Malik, S. K. 2013. Conservation of spices and tree borne oil seed crops. In: Normah, M., Chin, H. & Reed, B. (eds.) *Conservation of Tropical Plant Species.* Springer, New York.

Douglas, M., Heyes, J. & Small, F. 2005. *Herbs, Spices and Essential Oils: Postharvest Operations in Developing Countries.* NZ Institute for Crop and Food Research Ltd New Zealand, FAO.

D'Souza, S. P., Chavannavar, S. V., Kanchanashri, B. & Niveditha, S. B. 2017. Pharmaceutical perspectives of spices and condiments as alternative antimicrobial remedy. *Journal of Evidence-Based Complementary and Alternative Medicine* 22(4): 1002–1010. https://doi.org/10.1177/2156587217703214.

Dubey, S. 2017. Indian spices and their medicinal value. *Indian Journal of Pharmaceutical Education and Research* 51(3): 330–331.

Endrias, G. & Asfaw, K. 2011. Production, processing and marketing of ginger in Southern Ethiopia. *Journal of Horticulture and Forestry* 3(7): 207–213.

Farrell, K. T. 1990. *Spices, Condiments and Seasonings* (2nd ed.). Aspen Publishers, MA. ISBN 9780834213371.

Gautam, K. G. J. & Acharya., B. 2009. *Ginger and Turmeric Production, Improved Technology.* Nepal Agricultural Research Council, Ginger Research Program.

Girma, H., Digafe, T., Edossa, E., Belay, Y. & Weyessa, G. 2008. *Spices Research Achievements EIAR.*

Girma, H., Haimanot, M. & Habtewold, K. 2016. *In Search of Spices and Herbs in Ethiopia. Zinjibel in Ethiopia.* Tepi National Spices Research Center.

Gottardi, D., Bukvicki, D., Prasad, S. & Tyagi, A. K. 2016. Beneficial effects of spices in food preservation and safety. *Frontiers in Microbiology* 7: 1394. https://doi.org/10.3389/fmicb.2016.01394.

Heubach, K., Wiitg, R., Nuppenau, E. & Hahn, K. 2011. The economic importance of nontimber forest products (NTFPs) to livelihood maintenance of rural west African communities: A case study from northern Benin. *Ecological Economics* 70(11): 1991–2001.

Hiranya, B. S. 2013. *An Assessment of Pre & Post-harvest Factors Affecting Quality of Ginger in the Export Chain: A Case Study of Salyan District and Nepalgunj City.* Mid-Western Development Region. Larenstein University of Applied Sciences, Larenstein, Neatherlands.

Hordofa, T. S., Tolossa, T. T. & Yildiz, F. 2020. Cultivation and postharvest handling practices affecting yield and quality of major spices crops in Ethiopia: A review. *Cogent Food and Agriculture* 6(1): 1788896.

Husen, A., Bachheti, R. K. & Bachheti, A. 2021. *Non-Timber Forest Products (Food, Healthcare and Industrial Applications).* Springer Nature Switzerland AG, Gewerbestrasse 11, 6330. https://doi.org/10.1007/978-3-030-73077-2.

Husen, A. 2021. *Traditional Herbal Therapy for the Human Immune System.* CRC Press, Boca Raton, FL. https://doi.org/10.1201/9781003137955.

Husen, A. 2023. *Antidiabetic Medicinal Plants and Herbal Treatments.* CRC Press, Boca Raton, FL. https://doi.org/10.1201/b23347.

Kartha, K. K. 1985. *Cryopreservation of Plant Cells and Organs.* Kartha, K. K. (ed.). CRC Press, Boca Raton, FL, 243–267.

Meinhold, K. & Darr, D. 2019. The processing of non-timber forest products through small and medium enterprises-A review of enabling and constraining factors. *Forests* 10(11): 1026. https://doi.org/10.3390/f10111026.

Minoo, D. 2002. *Seedling and Somaclonal Variation and Their Characterization in Vanilla.* Ph.D. Thesis. Calicut University, Kerala, India.

Divakaran, M., Nirmal Babu, K. & Peter, K. V. 2011. Cryopreservation of pollen and inter specific hybridization in important orchid species *Vanilla planifolia* and V. *aphylla.* National Consultation for Production and Utilisation of Orchids 19–21 February.

Momin, M. C. & Jamir, A. R. 2021. Harvesting, Handling, and processing of Spices. In: Surajit Mitra (ed.) *Postharvest Management of Horticultural Crops.* JAYA Publishing House, New Delhi, 458–484.

Muggeridge, M., Lion, Foods, Clay, M. & European Spices Association 2001. Quality specifications for herbs and spices. In: Peter, K. V. *Handbook of Herbs and Spices* (3rd ed.). Woodhead Publishing Limited, Cambridge, 1–311.

Nag, K. K. & Street, H. E. 1975. Freeze preservation of cultured plant cells: I. the pretreatment phase. *Physiologia Plantarum* 34(3): 254–260.

Niesenbaum, R. A. 2019. The integration of conservation, biodiversity, and sustainability. *Sustainability* 11(17): 4676.

Nirmal Babu, K., Geetha, S. P., Minoo, D., Ravindran, P. N. & Peter, K. V. 1999. In vitro conservation of germplasm. In: Ghosh, S. P. (ed.) *Biotechnology and its Application in Horticulture* (Ed.) Narosa Publishing House, New Delhi, 106–129.

Nirmal Babu, K., Minoo, D., Geetha, S. P., Ravindran, P. N. & Peter, K. V. 2005. Advances in biotechnology of spices and herbs. *Ind J Bot Res* 1(2): 155–214.

Nirmal Babu, K., Yamuna, G., Praveen, K., Minoo, D., Ravindran, P. N. & Peter, K. V. 2012. Cryopreservation of spices genetic resources. In: Katkov, Igor (ed.) *Current Frontiers in Cryobiology*, 2012 March 9. San Diego, California, 457–484.

Patel, A., Tiwari, S., Pandey, N., Gupta, D. & Prasad, S. M. 2020. Role of spices beyond a flavouring agent: The antioxidant and medicinal properties. In: Mishra, N. (ed.) *Ethnopharmacological Investigation of Indian Spices.* IGI Global, Hershey, PA, 5–35.

Peter, K. V., Ravindran, P. N., Nirmal Babu, K., Sasikumar, B., Minoo, D., Geetha, S. P. & Rajalakshmi, K. 2002. *Establishing In Vitro Conservatory of Spices Germplasm.* ICAR Project report. Indian Institute of Spices Research, Calicut, Kerala, 131.

Purseglove, J. W., Brown, E. G. & Rosengarten, F. J. 1969. *The Book of Spices* (2nd ed.). Rosengarten, F. (ed.). CRC press. Washington DC, PA, 489.

Purseglove, J. W., Brown, E. G., Green, C. L. & Robbins, S. R. J. 1981. Spices Vol. 2 (pp. 447–813). Longman Group Ltd.

Sachan, A. K., Kumar, S., Kumari, K. & Singh, D. 2018. Medicinal uses of spices used in our traditional culture: Worldwide. *Journal of Medicinal Plants Studies* 6: 116–122.

Sharma, B. P. 2009. *Diseases and Ginger and Their Management: Post-harvest Consideration of Ginger to Control Diseases.* Nepal Agricultural Research Council, Ginger Research Program, Lalitpur, Nepal.

Shukla, A. & Yadav, N. 2018. Role of Indian spices in Indian history. *International Journal of Management Research and Reviews* 8(11): 1–6.

Stanley, D., Voeks, R. & Short, L. 2012. Is non-timber forest product harvest sustainable in the less developed world? A systematic review of the recent economic and ecological literature. *Ethnobiology and Conservation* 1: 1–39.

Sultana, S., Ripa, F. A. & Hamid, K. 2010. Comparative antioxidant activity study of some commonly used spices in Bangladesh. *Pakistan Journal of Biological Sciences* 13(7): 340–343. https://doi.org/10.3923/pjbs.2010.340.343.

Rosengarten Jr, F. 1969. The book of spices. The Book of Spices.

Thomas, L. & Sanil, P. C. 2019. Competitiveness in spice export trade from India: A review. *Journal of Spices & Aromatic Crops* 28: 1–19. https://doi.org/10.25081/josac.2019.v28.i1.5738.

Umetsu, H., Wake, H., Saitoh, M., Yamaguchi, H. & Shimomura, K. 1995. Characteristics of cold preserved embryogenic suspension cells in fennel *Foeniculum vulgare* Miller. *Journal of Plant Physiology* 146(3): 337–342.

Withers, L. A. 1985. Cryopreservation of cultured cells and meristems. In: Vasil, I. K. (ed.) *Cell Culture and Somatic Cell Genetics of Plants, Cell Growth, Nutrition, Cyto-Differentiation and Cryopreservation.* Academic Press, Orlando, FL, 253–316.

Yiljep, G. A., Fumen, A. & Ajisegiri, E. A. 2005. The effects of peeling, splitting and drying on ginger quality and oil/oleoresin content. *Agricultural Engineering International: The CIGR E Journal* VII: 1–8.

Zakir, M., Hailemichael, G., Seyoum, M. &Addisu, M. 2018. Status of ginger (*Zingiber officinale*) research and production challenges and future prospects in Ethiopia: A review. *Academia Journal of Agricultural Research* 6(5): 276–290.

4 Tissue Culture of Medicinal Spice and Condiment Crops for Mass Propagation

Arundhati Singh, Vedanshi Pal, Kritika Chauhan, Manu Pant, and Azamal Husen

4.1 INTRODUCTION

Spices and condiments have been an integral part of local food since prehistoric times and are added to enhance the flavour and taste of the food. Condiments are generally added in greater quantity than spices. In addition to flavour, spices and condiments are also valued for their medicinal value. Consumption of these medicinally active plants has been known to prevent age-related and metabolic disorders (Tapsell et al. 2006; Kaefer et al. 2008). The therapeutic properties of spices and condiments have attracted researchers to identify their secondary metabolites and evaluate their medicinal properties. Methods have been standardized for extracting the phytochemicals from different species (Wilson 1995). These extracts are further used in deciphering these species' medicinal properties and as a sensory character of the food (Martinez et al. 2015). Over the years, spices have been increasingly used in medicinal preparations, food preservatives, cosmetic and skin-care products, and even in pesticide formulations (Adeeyo et al. 2021). A large number of plants fall under the category of medicinal spices and condiments. Some of them include plant species from the genera *Curcuma*, *Zingiber*, *Allium*, *Piper*, *Cinnamomum*, *Syzygium*, *Coriandrum*, *Trigonella*, etc. These plants have been shown to have a wide range of medicinal properties, like antifungal, antimicrobial, antibacterial, antiviral, anticancer, antidiabetic, antioxidant and anti-inflammatory activities (Husen 2021, 2022a, 2022b; 2023; Rahman & Husen 2021, 2022, 2023a, 2023b). These properties are generally attributed to some phytochemical constituent(s) of the plant species (Figure 4.1)

Despite the known benefits of medicinal spices and condiments, little attention has been paid to their conservation and propagation methods. Extensive usage of these plants in traditional food and medicines, associated with their increased utilization in herbal preparations, has aggravated their non-scientific collection from natural stands in the wild. The decline in suitable growth habitats, lack of planned plantations, and inadequate scientific propagation methods has furthered the problem of the unavailability of superior-quality raw material in sufficient quantities. This scenario calls for concerted efforts towards large scale propagation of medicinal spices and condiments. Plant tissue culture techniques have been recognized as a potential method for the mass production of a wide variety of plants through organogenic different pathways. These methods have also been tried for *in vitro* production of medicinal spices. The tissue culture-generated plants offer the benefits of mass production in limited time and space, seasonal independence, and the generation of disease-free, superior-quality plants. The present chapter highlights some of the commonly used medicinally important spices and condiments, their medicinal values, the bioactive compounds conferring these therapeutic properties to these plants, and the tissue culture strategies for large-scale production of these plant species.

DOI: 10.1201/9781003387046-4

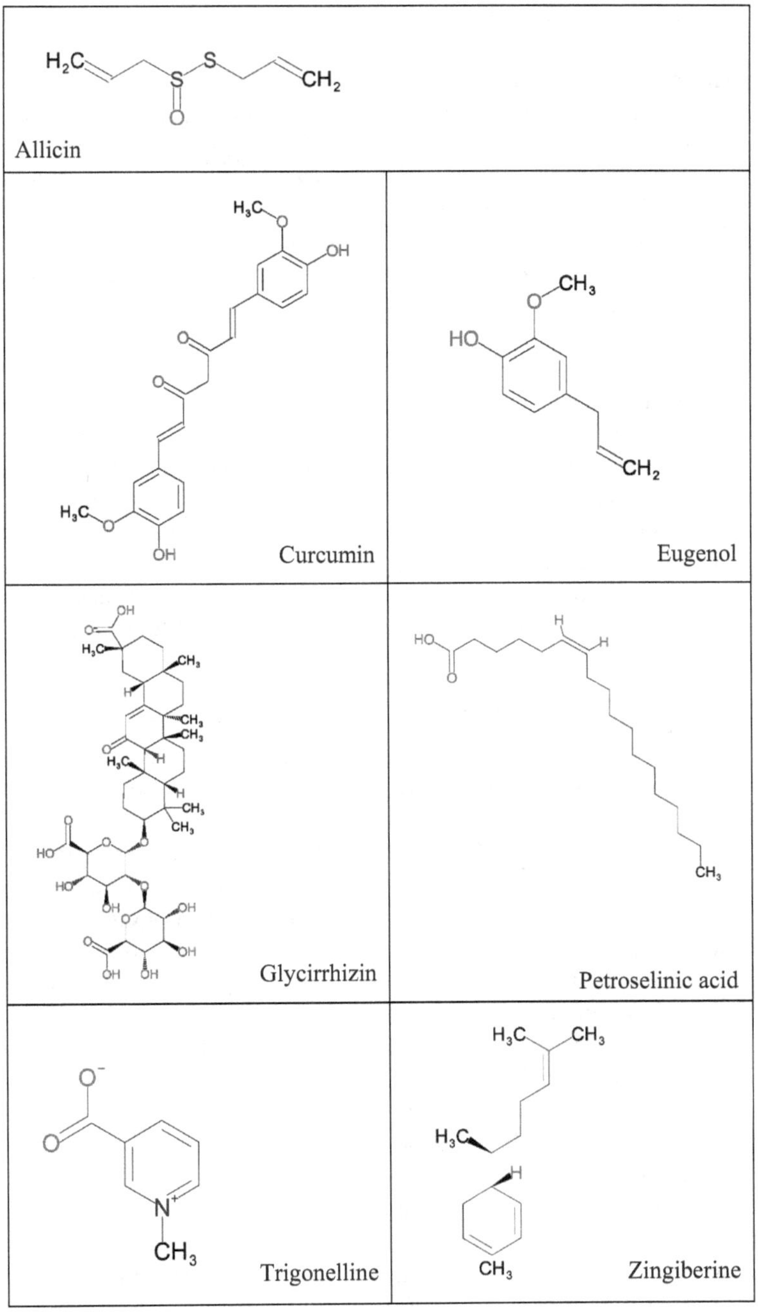

FIGURE 4.1 Structures of some bioactive ingredients of medicinal spices and condiments.

4.1.1 *Allium Sativum* (Garlic)

Garlic is a modified stem, more commonly known as a bulb. It belongs to the family Amaryllidaceae and is cultivated worldwide. Garlic bulb has been used in cooking since ancient times and is well established to have health benefits (Rivlin 2001; Cao et al. 2014). It has been shown to possess anti-allergic activities by inhibiting antibody-mediated histamine production, further leading to the modulation of airway allergic responses (Zare et al. 2009). Garlic is known to have anticancer activity. It activates metabolizing enzymes, further inhibiting the accumulation of reactive oxygen species, thereby preventing tumour formation and DNA damage (Zhang et al. 2020). Garlic inhibits NF-κB signalling, bringing it to its anti-inflammatory and anti-arthritic activities (Ban et al. 2009). It also performs immunomodulatory effects by modulating cytokine production (Arreola et al. 2015). Garlic oils have antifungal activities that work efficiently against *Candida albicans* and *Penicillium funiculosum* (Li et al. 2016). The crude or fresh extracts have been reported to extend to antibacterial properties. Consumption of garlic is known to prevent food poisoning, by positively influencing gut health (Rimoldi et al. 2020).

Tissue culture studies on garlic have been performed using different shoot and root explants. Bhojwani et al. (1980) reported *in vitro* proliferation from shoot buds on B5 medium containing 0.5 mg/l 2-IP and 0.1 mg/l NAA. Shoot multiplication from shoot tip cultures of garlic was reported by Nagakubo et al. (1993) on LS medium supplemented with equimolar concentrations of IAA and IBA. A study by Lee et al. (2009) confirmed the superiority of MS medium for somatic embryo induction in garlic when fortified with 2,4-D and BAP. Shoot tips are preferable explants for shoot proliferation in garlic. With this explant, MS medium containing cytokinin (BAP) and auxin (IAA) has been reported to give optimal results (Bekheet 2004; Ayed et al. 2018). In addition to the shoot segments, garlic's basal part and root apices have also been known to respond to *in vitro* treatments. Callus induction from root apices on B5 medium with 2,4-D, NAA, and IBA (Metwally et al. 2014) and basal segments of bulbs on MS medium containing kinetin and 2,4-D (Bekheet 2004) have been reported. The efficiency of the basal part of garlic for shoot organogenesis has also been studied when inoculated on MS medium supplemented with cytokinin (Mubarrat 2016)

4.1.2 *Cinnamomum Verum* (Cinnamon)

Cinnamon belongs to the family Lauraceae and is an evergreen tree native to Sri Lanka and southern India. The plant bears green-coloured flowers and single-seeded berries as fruits (Jakhetia et al. 2010). *C. verum* is known for its essential oils and active ingredients like cinnamic acid, cinnamaldehyde, etc. (Tung et al. 2008). Cinnamaldehyde accounts for the characteristic taste and aroma (Singh et al. 2007). Cinnamon is widely valued as a natural antioxidant that accounts for its therapeutic properties (Shahidi et al. 1992). The antimicrobial effect of cinnamon on *Helicobacter pylori* has been documented (Tabak et al. 1999). It has been reported to have anticancer, antidiabetic, anti-inflammatory, and antimicrobial properties (Matan et al. 2006; Kim et al. 2006; Subash et al. 1292007; Jayaprakasha et al. 2007).

Tissue culture studies on cinnamon have been performed using shoots, nodes, and leaves as explants. *In-vitro* proliferation on MS medium using shoot tips was studied by Soulange et al. (2007); the response observed was the breaking of bud with the supplementation of MS medium with 1mg/l BAP and 2mg/l TDZ, wherein shoot terminals regenerated roots, which was achieved on coir dust medium containing IBA and IAA (2000 ppm) (Rema et al. 1993). Muralidharan et al. (2001) showed the superiority of kinetin on axillary bud induction by supplementing the MS medium with 0.1mg/l MS medium, and the explants used were young twigs, bearing nodes. MS medium can be used with a combination of auxin (2,4-D) with cytokinin (BAP or Kn) to induce callus from leaf tissues (Muralidharan et al. 2001). The *in-vitro* seedling can be induced by using an embryonic axis with a half-cotyledon portion as an explant while the half-strength MS medium was supplemented with 1.5 mg/l BAP + 0.2 mg/l IAA + 1g/l activated charcoal as an additive (Subasinghe et al. 2016).

4.1.3 *Curcuma Longa* (Turmeric)

C. longa (family Zingiberaceae) is a native of tropical South Asia. The medicinal and therapeutic properties have been utilized for ages in ancient medical systems like Ayurveda, Unani, and Siddha. Chemical constituents, like turmerones, zingiberene, β-sesquiphellandrene, terpinolene, and a number of sesquiterpenoids are found in the rhizome, leaves, and flower oil (Chane-Ming et al. 2002).

Turmeric leaf oil has been reported to be a potent natural antioxidant (Priya et al. 2012). The main active compound is curcumin, which has been reported to alter several physiological processes and has properties that include anti-inflammatory, antioxidant, antibacterial, antidepressant, antinociceptive, and pain-relieving activites (Banafshe et al. 2014; Gopinath et al. 2018;Lopresti et al. 2021). Anticancer activity of curcumin has also been established. Curcumin has been demonstrated to enhance the effects of chemotherapy and radiotherapy to check the spread of cancer, protect healthy tissues from oxidative stress, and control several pathways leading to the death of glioblastoma cells (Walker et al. 2021). Research reports indicate that gargling with turmeric while receiving radiation therapy for head and neck cancer can significantly reduce the intensity and severity of mucositis (Rao et al. 2014).

The turmeric extract known as turmeric oleoresin has hypoglycaemic properties and is frequently used as a flavouring and colouring agent. Consumption of turmeric oleoresin and essential oil has been reported to check the accumulation of belly fat and elevated blood sugar levels (Honda et al. 2006).

Considering the benefits of turmeric, *C. longa* has been widely propagated through tissue culture techniques (Figure 4.2). Most of the studies have been performed using rhizomes and shoots as explants. MS medium is most optimal for the micropropagation of *C. longa*. In a study by Naz et al. (2009), MS medium containing 3 mg/l BAP resulted in shoot induction from rhizome segments when cultured *in vitro*. Young vegetative buds were also used as explants, resulting in shoot organogenesis when medium was supplemented with the cytokinin BA and the auxin NAA (Salvi et al. 2002). In *vitro* shoot proliferation on MS medium containing 6-benzyl adenine (BA) (2.0 mg/l)

FIGURE 4.2 *In vitro* culture of turmeric.

has been reported (Rahman et al. 2004; Anik et al. 2018). Shoot organogenesis can also be observed by using a shoot tip as an explant and supplementing the MS medium with NAA (1 mg/l) and BA (2,3 mg/l) (Jala et al. 2012). Another study reported callus induction from vegetative buds on LSBM (Linsmaier and Skoog's medium) supplemented with 2,4-D (Roopadarshini et al. 2010).

4.1.4 *Coriandrum Sativum* (Coriander)

Coriander is an annual medicinal herb belonging to the Apiaceae family. It possesses bioactive components, marking its importance in nutrition and cooking. The essential oil extracted from the seeds contains linalool, a terpene alcohol, in high concentrations, whereas lipids isolated from aerial parts of the plant contain high levels of petroselinic acid (Sahib et al. 2013). In addition, several other phytochemicals are known to be present in coriander (Table 4.1)

TABLE 4.1
Phytoconstituents of Some Important Medicinal Herbs and Spices

Phytoconstituents	Key References
Coriandrum sativum	
Phytosterols, polyphenols, gallic acid, bornyl acetate, thymol, apigenin, rutin, luteolin, pinocembrin, pseudobaptigenin, quercetin, kaempferol dimethyl ether, galangin-5-methyl ether, baicalein trimethyl ether, pinobanksin-3-*O*-pentenoate, pinobanksin-3-*O*-phenylpropionate, apigenin-7-*O*-glucuronoide, pinobanksin-5-methylether-3-*O*-acetate, pectolinarigenin, apigenin-C-glucoside, kaempferol-3-7-dimethyl ether-3-*O*-glucoside, and pinobanksin-3-*O*-pentanoate	Matsubara et al. 2011; Reilla et al. 2012; Sun et al. 2016; Hussain et al. 2018
Cuminum cyminum	
Bornyl acetate, α-terpinene, and γ-terpinene, *p*-mentha-1,4-dien-7-al, cumin aldehyde, β-pinene, safranal, and *p*-cymene were the main components of the oils; iron, manganese, potassium, zinc, vital amino acids, proteins, and other unsaturated fatty acids	Li et al. 2004; Iacobellis et al. 2005; Bettaieb et al. 2010; Sahana et al. 2011
Foeniculum vulgare	
α-Thujone, (Z)-anethole and (E)-anethole fenchone and 50 to 60% anethole, anisic acid, fenchone, methyl chavicol, anethole, phellandrene, anisic acid, camphene, limonene, and flavonoids	Badgujar et al. 2014; Ghiasvand et al. 2021
Glycyrrhiza glabra	
Ocimum sanctum	
Flavonoids, alkaloids, terpenoids, tannins, steroids, saponins, and glycosides	Varsha et al. 2013
flavonoids, volatile oils, including eugenol and β-caryophyllene, and a variety of additional substances found in a fixed oil.	Das et al. 2006; Harsha et al. 2020
Syzygium aromaticum	
3-Allyl-6-methoxyphenyl, also known as *m*-eugenol (69.44%), eugenol acetate (10.79%), 4-hydroxy-4-methylpentan-2-one, also known as tyranton (7.78%), caryophyllene (6.80%), 1,4,7-cycloundecatriene, 1,5,9,9-tetramethyl-, Z, Z,Z and small amounts (<1%) of other components .	Uddin et al. 2017
Trigonella foenum-graecum	
Linoleic acid (54.13%), palmitic acid (16.21%), pinene (4.56%), 4-pentyl-1-(4-propylcyclohexyl)-1-cyclohexene (3.87%), and linoleic acid methyl ester (3.19%).	Akbari et al. 2019
Zingiber officinale	
Zingeberene, curcumene, and farnesene. The aromatic compounds include gingerols (in fresh ginger), shagaols (in dried ginger), paradols, and zingerone.	Shukla & Singh 2007; Zick et al. 2008

Owing to its biochemical composition, coriander is known to have hypolipidaemic and hyperglycaemic effects and serves as an antimicrobial, anti-inflammatory, anticonvulsant, anticancer, analgesic, antihypertensive, and antioxidant agent. (Larbi et al. 2015). The seeds are extensively used to treat digestive problems such as diarrhoea, bloating, nausea, and indigestion by inducing the liver to secrete more bile juice and activating digestive enzymes (Leung et al. 1996; Platel et al. 2004). The seeds are also used to treat rheumatoid arthritis, relieving pain and inflammation. Methanolic extracts of the coriander fruit are rich in antioxidant properties, favouring its use as a food additive (Msaada et al. 2017).

Tissue culture studies on coriander have been done using vegetative parts, fruits, or seeds as explants. SH medium (Schenk and Hilderbrandt medium) containing NAA and gibberellic acid was reported to be optimal for shoot induction from coriander seeds (Stephen et al. 1998). In contrast, Dias et al. (2011) showed that a half-strength MS medium supplemented with BAP and IBA was most suitable for seed germination *in vitro.* The suitability of half-strength MS medium was further supported by Ali et al. (2017), who obtained the best results for seed-based shoot proliferation on the medium containing BAP and NAA. Other combinations shown to trigger shoot induction from seedling explants are modified MS medium with kinetin or kinetin plus IAA (Kataeva et al. 1993), and MS medium with BAP and IBA (Dias et al. 2010). Vegetative parts and fruits as explants were reported to induce shoot organogenesis on MS medium supplemented with BAP (Barros et al. 2012). Hassan (2019) showed that callus could be induced from seeds inoculated onto an MS medium containing BAP and the auxin 2,4-D.

4.1.5 *Cuminum Cyminum* (Cumin)

Cumin is a small slender annual plant with greyish-brown-coloured seeds, which are frequently used as a spice in food preparation. The seeds have also been valued for their medicinal benefits. They have been largely used in Ayurvedic medicine to treat dyspepsia, diarrhoea, and jaundice (Dhandapani et al. 2002). Conventional healers have also explored their role as a stimulant, carminative, and astringent (Thippeswamy et al. 2005). There is a long history of the usage of cumin in the treatment and control of hypertension, indigestion, sleep disturbance, neurological disorders, gastrointestinal issues, and dermal problems (Sahana et al. 2011; Kalaivani et al. 2013). The evidence also demonstrates cumin's value as an antioxidant, which may neutralise hydroxyl, peroxyl, and DPPH free radicals and prevent them from mediating lipid peroxidation (Thippeswamy et al. 2005). *C. cyminum* oil has been shown to have strong antibacterial and antifungal activities (Hajlaoui et al. 2010). Cumin has been demonstrated to be effective in treating diabetes, affecting insulin metabolism, and helping overweight persons to lose weight (Taghizadeh et al. 2015; Keihan et al. 2016). According to a recent study, cumin essential oil nanoemulsion has been proposed to be an effective plant-based therapy for cancer treatment and to overcome antimicrobial resistance (Nirmala et al. 2020).

Tissue culture studies on cumin have been mainly focused on seed-based explants. Cotyledonary explants have been shown to give rise to embryonic callus when inoculated onto MS medium supplemented with IAA and kinetin (Shukla et al. 1997). Hypocotyl explants have been used for shoot induction through callogenesis or somatic embryogenesis. Studies have shown that, whereas MS medium containing thidiazuron (TDZ) could induce somatic embryogenesis, medium supplemented with BAP was beneficial for adventitious shoot formation, while RS (Rimler and Shotts) medium containing 2,4-D, NAA, BAP, and kinetin was beneficial for the development of flowering shoots (Dave et al. 1995; Tawfik et al. 2001; Khavari et al. 2001). For callus induction from hypocotyl segments, MS medium containing BAP has been shown to give optimal results (Jha et al. 1982). In contrast, Gamborg's medium (B5 medium) has been demonstrated to be superior to other nutrient media for embryo culture of *C. cyminum.* B5 medium supplemented with different concentrations of cytokinins (BAP/kinetin), and auxins (NAA/IAA) have been reported to induce shoots or callus from embryo explants (Ebrahimie et al. 2003). B5 medium has also been shown to induce organogenesis in cultured nodal explants when supplemented with BAP and NAA (Ebrahimie et al. 2007). The effectiveness of TDZ in B5 medium in inducing indirect shoot organogenesis and that of BAP and NAA in inducing indirect organogenesis from seedlings of *C. cyminum* under *in vitro* conditions has been shown (Kahrizi et al. 2013).

4.1.6 *Foeniculum Vulgare* (Fennel)

Fennel is a medicinal spice (family Apiaceae) valued for its nutritional, medicinal, and aroma qualities. The essential oils isolated from the plant contain several phytoactive ingredients (Table 4.1), which have excellent lipid peroxidation inhibitory and radical scavenging activities. Fennel seed powder has been used as a feed component to induce rapid growth in chickens, owing to its essential oils (Badgujar et al. 2014; Ghiasvand et al. 2021). Fennel extract is used as an oestrogenic agent to cure liver toxicity caused by carbon tetrachloride (Özbek et al. 2004), to treat idiopathic hirsutism in women, and to relieve pain and inflammation (Choi et al. 2004). The vitamin C content of the plant is known to impart anti-obesity value to fennel (Bhadra 2021). A recent study (Suleiman et al. 2022) has confirmed the anticancer potential of fennel extract *in vitro* against MCF-7 (breast tumour cell line) and HEPG-2 (hepatocellular carcinoma cell line).

In vitro propagation studies on fennel have confirmed the superiority of MS medium for plantlet regeneration from explants. MS medium supplemented with 2,4-D, TDZ and BAP in amounts $\leq$ 1.0 mg/l are optimal in inducing callus from seed explants (Abd-Allah et al. 2015). Hypocotyls have been the most utilized starting material for tissue culture-based propagation of fennel. Varying combinations of plant growth regulators like 1 mg/l kinetin + 0.5 mg/l 2,4- D (Saxena et al. 2012), 0.1 mg/l BAP + 0.1 mg/l NAA + 0.5 mg/l IAA (Tabibazar et al. 2020), 1mg/l 2,4-D + 0.5mg/l BAP or 1 mg/l 2,4-D + 0.5 mg/l BAP have been used for callus induction in different studies (Saremirad et al. 2021). In one study (Dwivedi et al. 2020), cotyledonary nodes of fennel have been shown to proliferate when inoculated on MS medium supplemented with 1mg/l BAP and 29.91 μM ZnSO4, the latter working as an additive. Leaf explants can also be used for indirect organogenic differentiation into shoots when cultured on half-strength MS medium containing 2,4-D and BAP (Torabi et al. 2014) or NAA and BAP (Asree et al. 2020).

4.1.7 *Glycyrrhiza Glabra* (Licorice)

Glycyrrhiza glabra is a perennial herb of the family Fabaceae that is valued as a medicinal herb (Sharma et al. 2017). It has been an essential ingredient in traditional medicines to treat various disorders, from the common cold and coughs to hepatitis to more severe conditions like cancer and SARS (Dhingra et al. 2004; Saxena et al. 2005). Plant roots are known to be the source of a wide array of phytochemicals (Table 4.1) with glycyrrhizin as the main active ingredient. Glycyrrhizin is 50 times sweeter than refined sugar and is widely used in foods, tea, tonic, extract, tincture, or decoctions in both conventional and alternative medicine (Sharma et al. 2017). Licorice extract has also been shown to have a memory-boosting effect, and to improve renal function in mice (Fukai et al. 2003; Dhingra et al. 2004). The aqueous root extract has been shown to have vital anti-angiogenic and anticancer activities (Sheela et al. 2006). The roots are a possible source of antioxidants and urease inhibitors, which may be a novel lead in developing new drugs to treat ulcers (Lateef et al. 2012). The hydroxyl radical scavenging and antibacterial properties of the root extract have been demonstrated, making it a promising medication for preventing bacterial infections and scavenging hydroxyl radicals produced during the cancer-forming process (Varsha et al. 2013). Licorice root can treat bacterial infections and remove hydroxyl radicals.

Several studies have been conducted for the *in vitro* propagation of *G. glabra* through different explants. Nodal explants have been utilized for axillary bud induction and subsequent plantlet development. MS medium supplemented with cytokinins like BAP, kinetin (Thengane et al. 1998; Arya et al. 2009) or cytokinin (BAP), and auxin (NAA) (Yadav et al. 2012) has been shown to induce shoot organogenesis from cultured nodal explants. Somatic embryogenesis in *G. glabra* has also been reported from cotyledonary, stem, and leaf explants. Wawrosch et al. (2009) showed that embryos could be induced from cotyledons under *in vitro* conditions when inoculated on MS medium supplemented with TDZ. Stems as the explant have been utilized to induce somatic embryos on an MS medium containing 2,4-D and BAP (Parsaeimehr et al. 2009). Leaf segments were shown to give rise to somatic embryos on MS medium containing either BAP or NAA in concentrations of less than 1 mg/l (Kakutani et al. 1999). Caulogenic differentiation in *G. glabra* has also been reported from stem explants (MS medium + 2 mg/l BAP + 0.5 mg/l 2,4-D + 50 mg/l ascorbic acid (Jaiswal et al. 2021) and leaf explants on MS + 0.5 mg/l 2,4-D) (Neelima et al. 2017).

4.1.8 *Ocimum Sanctum* (Holy Basil)

Holy basil, a member of the family Lamiaceae, , is popular in the ancient systems of medicine to treat coughs, colds, stomach discomfort, skin illnesses, arthritis, severe eye ailments, measles, and diarrhoea (Parasuraman et al. 2015; Mondal et al. 2011). In the Indian subcontinent, *O. sanctum* is known as 'tulsi' and is honoured as the mother medicine in Indian traditional history. Numerous pharmacological effects of its extract include hypoglycaemia, immunomodulation, anticoagulant, antistress, analgesia, antipyrexia, anti-ulcerogenic, antihypertensive, anti-inflammatory, chemo-protective, CNS depressive, radioprotective, antitumour, and antibacterial activities, owing to its phytochemical ingredients (Table 4.1). The biocidal properties of plant extracts have been used in preparing products like mouthwash (Jain et al. 2022). An alcoholic extract of holy basil has been demonstrated to affect immunity in experimental trials, confirming its analgesic and antistress potential (Khanna et al. 2003; Ravindran et al. 2005; Mondal et al. 2011). Leaf extracts of the plant have been shown to have hepatoprotective properties (Chattopadhyay et al. 1992). In male and female rats, the leaf extract of *O. sanctum* has been shown to have an infertility effect by impacting on the sperm's ability to mature physiologically (Srinivasulu et al. 2017; Poli et al. 2019). The findings can be used for effective drug formulations.

For mass propagation of *O. sanctum* through tissue culture technology, different parts of the plant have been utilised as the starting material (Figure 4.3). Young inflorescences were shown to develop callus when cultured on MS medium containing either 2,4-D or TDZ (Singh et al. 1999). Nodal explants have been extensively used to achieve shoot organogenesis. MS medium containing BAP and NAA (Jamal et al. 2016), BAP and GA_3 (Pattnaik et al. 1996; Shukla et al. 2021), or BAP alone (Deepa 2019) are some of the medium combinations that have been reported to induce organogenic differentiation. Leaf segments have been used for callus-mediated plantlet regeneration. B5 medium + zeatin + 2,4-D (Manjudevi et al. 2022), MS medium + NAA + kinetin (Shilpa et al. 2010) or MS medium + picloram + BAP + IAA (Mishra 2015) are effective in inducing callus formation from leaf segments. In a recent study, Kishan et al. (2022) showed the effectiveness of 2,4-D and BAP in MS medium to induce callus from nodal segments of *O. sanctum.*

4.1.9 *Piper Nigrum* (Black Pepper)

Black pepper belongs to the family Piperaceae, is a native of Southeast Asia and is one of the most widely used spices (Govindarajan 1977). It is often referred to as the king of spices, and is beneficial in the human diet and used as a medicine, preservative, perfume, and insecticide (Shrinivasan 2007). *P. nigrum* is well known for its potent antioxidant properties that have long been tested in mice and guinea pigs (Atal et al. 1985; Singh et al. 1986). The main phytochemical, piperine, is responsible for inhibiting and quenching free radicals and reactive oxygen species and inhibiting the peroxidation of lipids (Mittal et al. 2000). In traditional systems and home therapies, black pepper is extensively used to treat coughs, colds, fever, and digestion-related disorders.

Many studies have been done on the mass propagation of *P. nigrum.* Verma et al. (2019) excised shoot apical meristems from black pepper for axillary shoot proliferation on MS medium

FIGURE 4.3 Nodal explant culture of holy basil (*Ocimum sanctum*).

supplemented with 5 μm 6-BAP. Shoot tips have been used to induce shoot proliferation on MS medium containing 1.5 mg/l BAP (Philip et al. 1992). Nisar et al. (2011) used petiole as an explant to induce callogenesis on MS medium with 0.5 mg/l BA. Micropylar tissue from germinating seeds was cultured on plant growth regulator-free SH medium without light. This gave rise to primary somatic embryos from which cyclic secondary somatic embryogenesis occurred (Nair et al. 2006). Rahman et al. (2007) showed that callus induction could be observed from immature zygotic embryos of black pepper on MS medium containing 2 mg/l IAA and 0.5 mg/l kinetin. Nodal and internodal rings have been reported to give rise to adventitious shoot bud differentiation on B5 medium supplemented with 10 μm 6- benzyladenine Shoot tip apices were used to initiate shoot cultures on MS medium where 4.5 mg/l BAP, 5.0 mg/l BAP, or 4 mg/l BAP were the best cytokinin in combination with 1 mg/l IAA (Bhat et al. 1995).

Leaves of black pepper could be observed to induce callus culture on MS medium containing 0.5 mg/l BA, 3 mg/l 2,4-D, and 1.5 mg/l IBA. Callogenesis could also occur from the stem or the shoot tip using MS medium supplemented with 1.5 mg/l BA. Shoot regeneration from callus was reported on MS medium with 0.5 mg/l BA. Root regeneration could also be observed from callus on a nutrient medium containing indole-3-butyric acid at a 1.5 mg/l concentration (Hussain et al. 2011).

4.1.10 *Syzygium Aromaticum* (Clove)

S. aromaticum is a member of the Myrtaceae family and is a well-known spice. The unopened flower buds found in clusters are collected and dried to form complete clove buds (Nurdjannah et al. 2012). Clove's potent ability to prevent several degenerative diseases and its strong antioxidant activity is linked to its chemical components (Astuti et al. 2019) (Table 4.1). Clove-derived sesquiterpenes are purported to have anticarcinogenic properties (Miyazawa et al. 2001); The primary volatile component of clove essential oil, eugenol, is produced by hydrodistilling primarily the buds and leaves of the plant (Kasmatou et al. 2012). The essential oil has been used to treat wounds and burns, controlling toothaches, and relieving pain during dental procedures since ancient times (Sarrami et al. 2002). It has also been shown to have strong antifungal activity (Ahmad et al. 2005) and the potential to control biofilm development and microbial growth in food and food products (Hu et al. 2018)

Cloves have been propagated through tissue culture on a variety of nutrient media. Mathew et al. (1990) could obtain shoot induction from shoot tip, nodal, and cotyledonary segments on MS and B5 medium containing BAP, NAA, and coconut milk. Suparman et al. (1990) showed that Woody Plant Medium (WPM) could be effectively used for shoot proliferation from nodal segments of the clove plant, when supplemented with BAP and NAA. The superiority of modified MS medium containing NAA and kinetin in inducing shoot development from different explants was confirmed by Hongsaranon et al. (1992). In more recent studies, callus induction from leaves and hypocotyl segments of cloves was reported on MS medium containing 2,4-D , or 2,4-D + BAP + NAA (Rasud et al. 2020).

4.1.11 *Trigonella Foenum-Graecum* (Fenugreek)

Fenugreek (from the family Fabaceae) is one of the oldest and most promising therapeutic herbs. It is known to have a wide range of medicinal benefits, like diabetes control by reducing high blood sugar in the urine, renal hypertrophy, polyuria, and low glomerular filtration rate (Shetty et al. 2009), reducing high cholesterol, and treating wounds, inflammation, and digestive problems (Khoja et al. 2011), and having analgesic and anti-inflammatory effects (Younesy et al. 2014). The antioxidant, antimycotic, and antibacterial properties of fenugreek seeds are well reported (Kaviarasan et al. 2004; Bhatia et al. 2006; Varadarajan et al. 2015, Kaviarasan et al. 2007). The aqueous extract of fenugreek has been reported to have immunomodulatory properties (Bin-Hafeez et al. 2003). The Indian system of food and medicine is known to extensively use plant seeds and leaves (Akbari et al. 2019; Patel et al. 2014). The seed oil is rich in chemicals (Table 4.1) that confer the odour, bitter flavour, and medicinal properties to the plant seeds (Sulieman et al. 2008; Akbari et al. 2019). The anti-hyperglycaemic chemical GII isolated from fenugreek seed aqueous extract is reported to have high antidiabetic activity (Moorthy et al. 2010). The polysaccharide galactomannan found in fenugreek gum is used in food industry to check surface activity (Brummer et al. 2003; Youssef et al. 2009). More recently, the effectiveness of fenugreek extracts against the fungi *Candida albicans* and *Aspergillus niger* has been documented (Kulkarni et al. 2020).

Considering fenugreek's wide-ranging medicinal applications, it has been vastly studied for mass propagation using tissue culture technology, mainly through the use of seed and stem explants. Oncina et al. (2002) used leaf, stem, and root segments of fenugreek for callus initiation on MS medium containing NAA and supplemented with coconut water. Cotyledonary nodes have been reported to give rise to shoots on MS medium with TDZ (Aasim et al. 2010). Afshari et al. (2011) used stem segments to generate callus on a medium containing 2,4-D and kinetin. Hypocotyl embryos were used to develop somatic embryos on a medium containing BAP and NAA. Hypocotyl and cotyledons of fenugreek have also been reported to initiate callus on a medium containing kinetin with NAA or 2,4-D (Elnour et al. 2015; Tavangar et al. 2021). In another study, the effectiveness of NAA alone in inducing callus from hypocotyl segments of fenugreek was reported (Valizadeh et al.,2018). Ahmed et al. (2019) showed that direct shoot regeneration could be achieved from cultured stem segments of fenugreek on an MS medium containing BAP. In a recent study, fenugreek seeds were shown to respond differently under different conditions of plant growth regulators (Agha et al. 2022). The study reported that, although MS medium containing BAP (2.0 mg/l) and 2,4-D (0.5 mg/l) was optimal for germination, medium supplemented with BAP (2.0 mg/l), 2,4-D (2.0 mg/l) and NAA (0.5 mg/l) could induce callus from inoculated seeds.

4.1.12 *Zingiber Officinale* (Ginger)

Ginger (family Zingiberaceae) is a perennial plant used as medicines a flavouring food agent, and a spice (Sharma 2017). The aromatic compounds of the plants are responsible for the hot and spicy flavour when consumed (Ali et al. 2008). The components of the volatile oil that confer medicinal value on ginger are listed in Table 4.1.

Throughout the centuries, ginger has been known to carry strong medicinal properties for treating morning sickness in expectant mothers, migraine, arthritis, ulcer, constipation, diabetes, and stroke (Ali et al. 2008; Sharma 2017). Ginger has been reported to be an effective anticancer medicine. It has been shown that ginger helps inhibit various pathways that lead to cancer, like initiation, promotion, and progression of different stages (Shukla & Singh 2007; Kundu et al. 2009). Other studies show that ginger is effective against skin, breast, and colon cancers (Ryan & Morrow 2010).

Several researchers have propagated ginger by *in vitro* techniques. In the majority of the studies, MS medium has been shown to be the best for micropropagation of ginger. Dogra et al. (1994) reported that rhizome buds could be used for development of rooted shoots on MS medium supplemented with cytokinin (BAP) and auxin (NAA). Rout et al. (2001) studied the effect of growth regulators on plantlet regeneration *via* meristem tip culture. They confirmed the positive effect of adenine sulphate when used in conjunction with cytokinin (BAP) and auxin (IAA). LSBM medium containing high concentrations of BAP was reported to be favourable for multiple shoot development from vegetative buds of ginger (Kavyashree 2009). Lincy and Sasikumar (2010) showed that half-strength MS medium containing TDZ and IBA could be used for adventitious shoot and root regeneration from stem explants of ginger.

4.2 CONCLUSION

Plant tissue culture technology has been used for large-scale propagation of several medicinal plants, including those which are commonly known as spices and condiments. The medical benefits of these plants have become better understood thanks to advances in research, which has resulted in their increased use in the pharmaceutical and food industries. Elevated demands have led to increased pressure on the crops in the field while also creating a gap between demand and supply of the plant raw material. Therefore, the *in-vitro* propagation approach can be used to meet this expanding demand. From a small explant, tissue culture enables the production of superior-grade plant material. For *in-vitro* cultivation of medicinal spices, a variety of explants, including nodes, leaves, seeds, shoot-tip meristems, etc., have been used successfully. More focused attention is required for large-scale production of such important medicinal spices and condiments and their effective translation to the field. The *in-vitro* protocols can be further used for enhancing secondary metabolite production by these species and their subsequent exploitation.

REFERENCES

Aasim, M., Hussain, N., Umer, E. M., Zubair, M., Hussain, S. B., Saeed, S. H., Rafique, T. S. & Sancak, C. 2010. In-vitro shoot regeneration of fenugreek (Trigonellafoenum-graecum L.) using different cytokinins. *African Journal of Biotechnology* 42: 7174–7179.

Abd-Allah, R. H., Zeid, E. M. A., Zakaria, M. M. & Eldahmy, S. I. 2015. Micropropagation of wild fennel (Foeniculum vulgare var. vulgare) via organogenesis and somatic embryogenesis. *Journal of Developmental Biology and Tissue Engineering* 7: 1–10.

Adeeyo, A. O., Edokpayi, J. N., Alabi, M. A., Msagati, T. A. & Odiyo, J. O. 2021. Plant active products and emerging interventions in water potabilisation: Disinfection and multi-drug resistant pathogen treatment. *Clinical Phytoscience* 7(1): 1–16.

Afshari, E., Ranjbar, G. A., Kazemitabar, S. K., Riasat, M. & Poshtmasari, H. K. 2011. Callus induction, somatic embryogenesis and plant regeneration in fenugreek (*Trigonella foenum-Graecum* L.). *Iranian Journal of Medicinal and Aromatic Plants* 27(1): 147–160.

Agha, H. M., Radzun, K. A., Sidik, N. J. & Jawad, A. H. 2022. Callus induction of fenugreek *trigonella foenum-graecum* via auxin combined with cytokinins hormones, and assessment of toxicity via brine shrimp assay. *Journal of Asian Scientific Research* 12(1): 12–27.

Ahmad, N., Alam, M. K., Shehbaz, A., Khan, A., Mannan, A., Hakim, S. R., Bisht, D. & Owais, M. 2005. Antimicrobial activity of clove oil and its potential in the treatment of vaginal candidiasis. *Journal of Drug Targeting* 13(10): 555–561.

Ahmed, A. O. M. 2019. *A Studies on Propagation of Fenugreek (Trigonella foenum-graecum L.) Using Tissue Culture* (Doctoral dissertation, Sudan University of Science and Technology).

Akbari, S., Abdurahman, N. H., Yunus, R. M., Alara, O. R. & Abayomi, O. O. 2019. Extraction, characterization and antioxidant activity of fenugreek (*Trigonella -foenum- graecum*) seed oil. *Materials Science for Energy Technologies* 2(2): 349–355.

Ali, B. H., Blunden, G., Tanira, M. O. & Nemmar, A. 2008. Some phytochemical, pharmacological and toxicological properties of ginger (*Zingiber officinale* Roscoe): A review of recent research. *Food and Chemical Toxicology* 46(2): 409–420.

Ali, M., Mujib, A., Tonk, D. & Zafar, N. 2017. Plant regeneration through somatic embryogenesis and genome size analysis of *Coriandrum sativum* L. *Protoplasma* 254(1): 343–352.

Anik, S. H. 2018. *In vitro Propagation of Turmeric (Curcuma longa L.)* (Doctoral dissertation, Department of Biotechnology, Sher-e-Bangla Agricultural University, Dhaka-1207).

Arreola, R., Quintero-Fabián, S., López-Roa, R. I., Flores-Gutiérrez, E. O., Reyes-Grajeda, J. P., Carrera-Quintanar, L. & Ortuño-Sahagún, D. 2015. Immunomodulation and anti-inflammatory effects of garlic compounds. *Journal of Immunology Research* 2015: 401630.

Arya, S., Rathi, N. & Arya, I. D. 2009. Micropropagation protocol for Glycyrrhiza glabra L. *Phytomorphology* 59(1): 71–76.

Asree, H. J. & Khirallah, A. A. 2020. Induced callus of fennel plant (Foeniculum vulgare mill.) in *Invitro. Plant Archives* 20(1): 2949–2952.

Astuti, R. I., Listyowati, S. & Wahyuni, W. T. 2019. Life span extension of model yeast Saccharomyces cerevisiae upon ethanol derived-clover bud extract treatment. In: *IOP Conference Series: Earth and Environmental Science* (Vol. 299, No. 1, p. 012059). IOP Publishing.

Atal, C. K., Dubey, R. K. & Singh, J. 1985. Biochemical basis of enhanced drug bioavailability by piperine: Evidence that piperine is a potent inhibitor of drug metabolism. *Journal of Pharmacology and Experimental Therapeutics* 232(1): 258–262.

Ayed, C., Bayoudh, C., Rhimi, A., Mezghani, N., Haouala, F. & AL MohandesDridi, B. 2018. *In vitro* propagation of Tunisian local garlic (*Allium sativum* L.) from shoot tip culture. *Journal of Horticulture and Postharvest Research* 1(2): 75–86.

Badgujar, S. B., Patel, V. V. & Bandivdekar, A. H. 2014. *Foeniculum vulgare* mill, a review of its botany, phytochemistry, pharmacology, contemporary application and toxicology. *BioMed Research International* 2014: 842674.

Ban, J. O., Oh, J. H., Kim, T. M., Kim, D. J., Jeong, H. S., Han, S. B. & Hong, J. T. 2009. Anti-inflammatory and arthritic effects of thiacremonone, a novel sulfurcompound isolated from garlic via inhibition of NF-κB. *Arthritis Research & Therapy* 11(5): 1–13.

Banafshe, H. R., Hamidi, G. A., Noureddini, M., Mirhashemi, S. M., Mokhtari, R. & Shoferpour, M. 2014. Effect of curcumin on diabetic peripheral neuropathic pain: Possible involvement of opioid system. *European Journal of Pharmacology* 723: 202–206.

Barros, L., Duenas, M., Dias, M. I., Sousa, M. J., Santos-Buelga, C. & Ferreira, I. C. 2012. Phenolic profiles of *in vivo* and *in vitro*-grown grown *Coriandrum sativum* L. *Food Chemistry* 132(2): 841–848.

Bekheet, S. A. 2004. Micropropagation of garlic (*Allium sativum* L.). *Annals of Agricultural Scieces* 49: 223–232.

Bettaieb, I., Bourgou, S., Wannes, W. A., Hamrouni, I., Limam, F. & Marzouk, B. 2010. Essential oils, phenolics, and antioxidant activities of different parts of cumin (*Cuminum cyminum L.*). *Journal of Agricultural and Food Chemistry* 58(19): 10410–10418.

Bhadra, P., Banerjee, P. & Deb, A. 2021. *Fennel: Functional Food for Obesity*. New Delhi Publishers, 82–97. https://doi.org/10.30954/ anthnutraceuticals.882.

Bhat, S. R., Chandel, K. P. S. & Malik, S. K. 1995. Plant regeneration from various expiants of cultivated *Piper* species. *Plant Cell Reports* 14(6): 398–402.

Bhatia, K., Kaur, M., Atif, F., Ali, M., Rehman, H., Rahman, S. & Raisuddin, S. 2006. Aqueous extract of *Trigonella foenum-graecum* L. ameliorates additive urotoxicity of buthionine sulfoximine and cyclophosphamide in mice. *Food and Chemical Toxicology* 44(10): 1744–1750.

Bhojwani, S. S. 1980. *In vitro* propagation of garlic by shoot proliferation. *Scientia Horticulturae* 13(1): 47–52.

Bin-Hafeez, B., Haque, R., Parvez, S., Pandey, S., Sayeed, I. & Raisuddin, S. 2003. Immunomodulatory effects of fenugreek (*Trigonella foenum graecum* L.) extract in mice. *International Immunopharmacology* 3(2): 257–265.

Brummer, Y., Cui, W. & Wang, Q. 2003. Extraction, purification and physicochemical characterization of fenugreek gum. *Food Hydrocolloids* 17(3): 229–236.

Cao, H. X., Zhu, K. X., Fan, J. G. & Qiao, L. 2014. Garlic-derived allyl sulfides in cancer therapy. *Anti-Cancer Agents in Medicinal Chemistry (Formerly Current Medicinal Chemistry-Anti-Cancer Agents)* 14(6): 793–799.

Chane-Ming, J., Vera, R., Chalchat, J. C. & Cabassu, P. 2002. Chemical composition of essential oils from rhizomes, leaves and flowers of *Curcuma longa* L. from Reunion Island. *Journal of Essential Oil Research* 14(4): 249–251.

Chattopadhyay, R. R., Sarkar, S. K., Ganguly, S., Medda, C. & Basu, T. K. 1992. Hepatoprotective activity of *Ocimum sanctum* leaf extract against paracetamol included hepatic damage in rats. *Indian Journal of Pharmacology* 24: 163–165.

Choi, E. M. & Hwang, J. K. 2004. Antiinflammatory, analgesic and antioxidant activities of the fruit of Foeniculum vulgare. *Fitoterapia* 75(6): 557–565.

Das, S. K. & Vasudevan, D. M. 2006. Tulsi: The Indian holy power plant. *Natural Product Radiance* 5(4): 279–283.

Dave, A. N. J. U. & Batra, A. M. L. A. 1995. Role of protein metabolism constituents in somatic embryo formation in cumin. *Indian Journal of Plant Physiology* 38: 25–27.

de Oliveira Santos, M. V., Nascimento, L. E., Praxedes, É. A., Borges, A. A., Silva, A. R., Bertini, L. M. & Pereira, A. F. 2019. Syzygium aromaticum essential oil supplementation during *in vitro* bovine oocyte maturation improves parthenogenetic embryonic development. *Theriogenology* 128: 74–80.

Deepa, M. K. 2019. *Invitro* studies of ocimum sanctum using various plant growth regulators. *ZENITH International Journal of Multidisciplinary Research* 9(1): 252–257.

Dhandapani, S., Subramanian, V. R., Rajagopal, S. & Namasivayam, N. 2002. Hypolipidemic effect of *Cuminum cyminum* L. on alloxan-induced diabetic rats. *Pharmacological Research* 46(3): 251–255.

Dhingra, D., Parle, M. & Kulkarni, S. K. 2004. Memory enhancing activity of *Glycyrrhiza glabra* in mice. *Journal of Ethnopharmacology* 91(2–3): 361–365.

Dias, M. I., Barros, L., Sousa, M. J. & Ferreira, I. C. 2011. Comparative study of lipophilic and hydrophilic antioxidants from *in vivo* and *in vitro* grown *Coriandrum sativum. Plant Foods for Human Nutrition* 66(2): 181–186.

Dias, M. I., Cardoso, S., Martins, A. & Sousa, M. J. 2010. *In vitro* culture of *Coriandrum sativum* L. In: *XXVIII International Horticultural Congress on Science and Horticulture for People (IHC2010): A New Look at Medicinal and 925*, 271–274.

Dogra, S. P., Koria, B., Nad, N. & Sharma, P. P. 1994. *In vitro* clonal propagation of ginger (*Zingiber officinale* Rosc.). *The Horticulture Journal* 7: 45–50.

Dwivedi, P., Amin, D. & Sharma, A. 2020. Effect of differential concentration of micronutrient copper and zinc on *in vitro* morphogenesis of Foeniculum vulgare Mill. *Plant Physiology Reports* 25: 178–184.

Ebrahimie, E., Habashi, A. A., Ghareyazie, B., Ghannadha, M. & Mohammadie, M. 2003. A rapid and efficient method for regeneration of plantlets from embryo explants of cumin (*Cuminum cyminum*). *Plant Cell, Tissue and Organ Culture* 75: 19–25.

Ebrahimie, E., Naghavi, M. R., Hosseinzadeh, A., Behamta, M. R., Mohammadi-Dehcheshmeh, M., Sarrafi, A. & Spangenberg, G. 2007. Induction and comparison of different *in vitro* morphogenesis pathways using embryo of cumin (Cuminum cyminum L.) as a model material. *Plant Cell, Tissue and Organ Culture* 90: 293–311.

ElNour, M. E., Ali, A. M. & Saeed, B. E. A. E. 2015. Antimicrobial activities and phytochemical screening of callus and seeds extracts of fenugreek (*Trigonella foenum-graecum*). *International Journal of Current Microbiology and Applied Sciences* 4(2): 147–157.

Fukai, T., Satoh, K., Nomura, T. & Sakagami, H. 2003. Preliminary evaluation of antinephritis and radical scavenging activities of glabridin from *Glycyrrhiza glabra*. *Fitoterapia* 74(7–8): 624–629.

Ghiasvand, A. R., Khatibjoo, A., Mohammadi, Y., Gharaei, M. A. & Shirzadi, H. 2021. Effect of fennel essential oil on performance, serum biochemistry, immunity, ileum morphology and microbial population, and meat quality of broiler chickens fed corn or wheat-based diet. *British Poultry Science* 2021: 1–11.

Gopinath, H. & Karthikeyan, K. 2018. Turmeric: A condiment, cosmetic and cure. *Indian Journal of Dermatology, Venereology and Leprology* 84(1): 16–21.

Govindarajan, V. S. & Stahl, W. H. 1977. Pepper—Chemistry, technology, and quality evaluation. *CRC Critical Reviews in Food Science and Nutrition* 9(2): 115–225.

Hajlaoui, H., Mighri, H., Noumi, E., Snoussi, M., Trabelsi, N., Ksouri, R. & Bakhrouf, A. 2010. Chemical composition and biological activities of Tunisian *Cuminum cyminum* L. essential oil: A high effectiveness against Vibrio spp. strains. *Food and Chemical Toxicology* 48(8–9): 2186–2192.

Harsha, M., Kumar, K. M., Kagathur, S. & Amberkar, V. S. 2020. Effect of *Ocimum sanctum* extract on leukemic cell lines: A preliminary *in-vitro* study. *Journal of Oral and Maxillofacial Pathology* 24(1): 93–98.

Hassan, S. F. 2019. Shoots regeneration in callus tissue of coriander plant (*Coriandrum sativum* L.) *in vitro*. *Plant Archives* 19: 1203–1205.

Honda, S., Aoki, F., Tanaka, H., Kishida, H., Nishiyama, T., Okada, S., Matsumoto, I., Abe, K. & Mae, T. 2006. Effects of ingested turmeric oleoresin on glucose and lipid metabolisms in obese diabetic mice: A DNA microarray study. *Journal of Agricultural and Food Chemistry* 54(24): 9055–9062.

Hongsaranon, Y. & Piriyaprin, S. 1992. In-vitro propagation of Syzygium aromaticum (L.) Merr. and Perry. In *Proceeding of Seminar on Medicinal Plants and Herbs Group*. Department of Agriculture, Bangkok, 156–166. http://biozoojournals.ro/bihbiol/cont/v7n1/bb.131107.Kahrizi.pdf.

Hu, Q., Zhou, M. & Wei, S. 2018. Progress on the antimicrobial activity research of clove oil and eugenol in the food antisepsis field. *Journal of Food Science* 83(6): 1476–1483.

Husen, A. 2021. *Traditional Herbal Therapy for the Human Immune System*. CRC Press, Boca Raton, FL. https://doi.org/10.1201/9781003137955.

Husen, A. 2022a. *Herbs, Shrubs and Trees of Potential Medicinal Benefits*. CRC Press, Boca Raton, FL. https://doi.org/10.1201/9781003205067.

Husen, A. 2022b. *Exploring Poisonous Plants: Medicinal Values, Toxicity Response and Other Potential Uses*. CRC Press, Boca Raton, FL. https://doi.org/10.1201/b23017.

Husen, A. 2023. *Antidiabetic Medicinal Plants and Herbal Treatments*. CRC Press, Boca Raton, FL. https://doi.org/10.1201/b23347.

Hussain, A., Naz, S., Nazir, H. & Shinwari, Z. K. 2011. Tissue culture of black pepper (*Piper nigrum* L.) in Pakistan. *Pakistan Journal of Botany* 43(2): 1069–1078.

Hussain, F., Jahan, N., Rahman, K. U., Sultana, B. & Jamil, S. 2018. Identification of hypotensive biofunctional compounds of *Coriandrum sativum* and evaluation of their angiotensin-converting enzyme (ACE) inhibition potential. *Oxidative Medicine and Cellular Longevity* 2018. Article ID 4643736 | https://doi.org/10.1155/2018/4643736.

Iacobellis, N. S., Lo Cantore, P., Capasso, F. & Senatore, F. 2005. Antibacterial activity of *Cuminum cyminum L.* and *Carum carvi L.* essential oils. *Journal of Agricultural and Food Chemistry* 53(1): 57–61.

Jain, A., Singh, V., Lukram, A., Chatterjee, S., Khan, A. M. & Dawar, G. 2022. Antibacterial efficacy of manuka honey, *Oocimum sanctum, Ccurcuma longa* and 0.2% chlorhexidine mouthwash on the level of streptococcus mutans and lactobacillus acidophilus-A randomized controlled trial. *Indian Journal of Dental Research* 33(2): 169.

Jaiswal, N., Verma, Y. & Misra, P. 2021. High frequency *in vitro* callogenesis and plant regeneration of Glycyrrhiza glabra L. *Vegetos* 34(3): 495–504.

Jakhetia, V., Patel, R., Khatri, P., Pahuja, N., Garg, S., Pandey, A. & Sharma, S. 2010. Cinnamon: A pharmacological review. *Journal of Advanced Scientific Research* 1(02): 19–23.

Jala, A. 2012. Effects of NAA BA and Sucrose on Shoot Induction and Rapid Micropropagation by trimming shoot of *Curcuma longa* L. *Science & Technology Asia*: 54–60.

Jamal, M. A. H. M., Sharif, I. H., Shakil, M. M., Rahman, A. R. B., Banu, N. A., Islam, M. R. & Nazmuzzaman, M. 2016. *In vitro* regeneration of a common medicinal plant, *Ocimum sanctum* L. for mass propagation. *African Journal of Biotechnology* 15(24): 1269–1275.

Jayaprakasha, G. K., Negi, P. S., Jena, B. S. & Rao, L. J. M. 2007. Antioxidant and antimutagenic activities of *Cinnamomum zeylanicum* fruit extracts. *Journal of Food Composition and Analysis* 20(3–4): 330–336.

Jha, T. B., Roy, S. C. & Mitra, G. C. 1982. *In vitro* culture of (*Cuminum cyminum* L.): Regeneration of flowering shoots from calli of hypocotyls and leaf explants. *Plant Cell, Tissue and Organ Culture* 2: 11–14.

Kaefer, C. M. & Milner, J. A. 2008. The role of herbs and spices in cancer prevention. *The Journal of Nutritional Biochemistry* 19(6): 347–361.

Kahrizi, D. & Soorni, J. 2013. Study on shoot regeneration and somatic embryogenesis in cumin (*Cuminum cyminum* L.) landraces. *Biharean Biologist* 7(1): 37–41.

Kakutani, K., Ozaki, K., Watanabe, H. & Tomoda, K. 1999. Somatic embryogenesis and plantlet regeneration on several species of licorice. *Plant Biotechnology* 16(3): 239–241.

Kalaivani, P., Saranya, R. B., Ramakrishnan, G., Ranju, V., Sathiya, S., Gayathri, V., … Thanikachalam, S. 2013. *Cuminum cyminum*, a dietary spice, attenuates hypertension via endothelial nitric oxide synthase and NO pathway in renovascular hypertensive rats. *Clinical and Experimental Hypertension* 35(7): 534–542.

Kamatou, G. P., Vermaak, I. & Viljoen, A. M. 2012. Eugenol—from the remote Maluku Islands to the international market place: A review of a remarkable and versatile molecule. *Molecules* 17(6): 6953–6981.

Kataeva, N. V. & Popowich, E. A. 1993. Maturation and rejuvenation of *Coriandrum sativum* shoot clones during micropropagation. *Plant Cell, Tissue and Organ Culture* 34(2): 141–148.

Kaviarasan, S., Naik, G. H., Gangabhagirathi, R., Anuradha, C. V. & Priyadarsini, K. I. 2007. *In vitro* studies on antiradical and antioxidant activities of fenugreek (*Trigonella foenum- graecum*) seeds. *Food Chemistry* 103(1): 31–37.

Kaviarasan, S., Vijayalakshmi, K. & Anuradha, C. V. 2004. Polyphenol-rich extract of fenugreek seeds protect erythrocytes from oxidative damage. *Plant Foods for Human Nutrition* 59: 143–147.

Kavyashree, R. 2009. An efficient *in vitro* protocol for clonal multiplication of ginger- var. Varada. *Indian Journal of Biotechnology* 8: 328–331.

Keihan, G. S., Gharib, M. H., Momeni, A., Hemati, Z. & Sedighin, R. 2016. A comparison between the effect of *cuminum cyminum* and vitamin E on the level of leptin, paraoxonase 1, Hbalc and oxidized LDL in diabetic patients. *International Journal of Molecular and Cellular Medicine* 5(4): 229–235.

Khanna, N. & Bhatia, J. 2003. Antinociceptive action of *Ocimum sanctum* (Tulsi) in mice: Possible mechanisms involved. *Journal of Ethnopharmacology* 88(2–3): 293–296.

Khavari, K. S. & Bagheri, A. 2001. Evaluation of Iranian cumin landraces and their purification to obtain high yielding cultivars. Research Report, Ferdowsi University of Mashhad and Research Organization of Ministry of Jehad Keshavarzi, Iran, 40

Khoja, K. K., Shaf, G., Hasan, T. N., Syed, N. A., Al-Khalifa, A. S., Al-Assaf, A. H. & Alshatwi, A. A. 2011. Fenugreek, a naturally occurring edible spice, kills MCF-7 human breast cancer cells via an apoptotic pathway. *Asian Pacific Journal of Cancer Prevention* 12(12): 3299–3304.

Kim, S. H., Hyun, S. H. & Choung, S. Y. 2006. Anti-diabetic effect of cinnamon extract on blood glucose in db/db mice. *Journal of Ethnopharmacology* 104(1–2): 119–123.

Kishan, A. S., Paul, M. & Reddy, J. 2022. *In vitro* micropropagation of *Ocimum sanctum* L.*International Journal of Botany* 7(1): 225–229.

Kulkarni, M., Hastak, V., Jadhav, V. & Date, A. A. 2020. Fenugreek leaf extract and its gel formulation show activity against Malassezia furfur. *Assay and Drug Development Technologies* 18(1): 45–55.

Kundu, J. K., Na, H. K. & Surh, Y. J. 2009. Ginger-derived phenolic substances with cancer preventive and therapeutic potential. *Forum of Nutrition* 61: 182–192.

Laribi, B., Kouki, K., M'Hamdi, M. & Bettaieb, T. 2015. Coriander (*Coriandrum sativum* L.) and its bioactive constituents. *Fitoterapia* 103: 9–26.

Lateef, M., Iqbal, L., Fatima, N., Siddiqui, K., Afza, N., Zia-ul-Haq, M. & Ahmad, M. 2012. Evaluation of antioxidant and urease inhibition activities of roots of Glycyrrhiza glabra. *Pak J Pharm Sci* 25(1): 99–102.

Lee, S. Y., Kim, H. H., Kim, Y. K., Park, N. I. & Park, S. U. 2009. Plant regeneration of garlic (*Allium sativum* L.) via somatic embryogenesis. *Scientific Research and Essay* 4(13): 1569–1574.

Leung, A. Y. & Foster, S. 1996 *Encyclopedia of Common Natural Ingredients Used in Food, Drugs, and Cosmetics* (2nd ed). John Wiley & Sons, Inc., Michigan.

Li, R. & Jiang, Z. T. 2004. Chemical composition of the essential oil of *Cuminum cyminum L.* from China. *Flavour and Fragrance Journal* 19(4): 311–313.

Li, W. R., Shi, Q. S., Dai, H. Q., Liang, Q., Xie, X. B., Huang, X. M., Zhao, G. Z. & Zhang, L. X. 2016. Antifungal activity, kinetics and molecular mechanism of action of garlic oil against *Candida albicans. Scientific Reports* 6: 22805.

Lincy, A. & Sasikumar, B. 2010. Enhanced adventitious shoot regeneration from aerial stem explants of ginger using TDZ and its histological studies. *Turkish Journal of Botany* 34(1): 21–29.

Lopresti, A. L., Smith, S. J., Jackson-Michel, S. & Fairchild, T. 2021. An investigation into the effects of a curcumin extract (Curcugen®) on osteoarthritis pain of the knee: A randomised, double-blind, placebo-controlled study. *Nutrients* 14(1): 41.

Manjudevi, M., Thirugnanasampandan, R. & Vishnupriya, B. 2022. *In vitro* propagation of Ocimum sanctum L., Ocimum canum Sims., and Ocimum tenuiflorum L., and evaluation of antioxidant, MMP-9 down regulation of eugenol and camphor. *South African Journal of Botany* 151: 208–217.

Martínez-Graciá, C., González-Bermúdez, C. A., Cabellero-Valcárcel, A. M., Santaella-Pascual, M. & Frontela-Saseta, C. 2015. Use of herbs and spices for food preservation: Advantages and limitations. *Current Opinion in Food Science* 6: 38–43.

Matan, N., Rimkeeree, H., Mawson, A. J., Chompreeda, P., Haruthaithanasan, V. & Parker, M. 2006. Antimicrobial activity of cinnamon and clove oils under modified atmosphere conditions. *International Journal of Food Microbiology* 107(2): 180–185.

Mathew, M. K. & Hariharan, M. 1990. *In vitro* multiple shoot regeneration in Syzygium aromaticum. *Annals of Botany* 65(3): 277–279.

Matsubara, E., Fukagawa, M., Okamoto, T., Ohnuki, K., Shimizu, K. & Kondo, R. 2011. (-)-Bornyl acetate induces autonomic relaxation and reduces arousal level after visual display terminal work without any influences of task performance in low-dose condition. *Biomedical Research* 32(2): 151–157.

Metwally, E. I., El-Denary, M. E., Dewir, Y. H. & Naidoo, Y. 2014. *In vitro* propagation of garlic (*Allium sativum* L.) through adventitious shoot organogenesis. *African Journal of Biotechnology* 13(38): 3892–3900.

Mishra, T. 2015. Protocol establishment for multiplication and regeneration of 'Holy Basil'(Ocimum sanctum Linn). An important medicinal plant with high religious value in India. *Journal of Medicinal Plants Studies* 3(4): 16–19.

Mittal, R. & Gupta, R. L. 2000. *In vitro* antioxidant activity of piperine. *Methods and Findings in Experimental and Clinical Pharmacology* 22(5): 271–274.

Miyazawa, M. & Hisama, M. 2001. Suppression of chemical mutagen-induced SOS response by alkylphenols from clove (Syzygium aromaticum) in the Salmonella typhimurium TA1535/pSK1002 umu test. *Journal of Agricultural and Food Chemistry* 49(8): 4019–4025.

Mondal, S., Varma, S., Bamola, V. D., Naik, S. N., Mirdha, B. R., Padhi, M. M., Mehta, N. & Mahapatra, S. C. 2011. Double-blinded randomized controlled trial for immunomodulatory effects of Tulsi (*Ocimum sanctum* Linn.) leaf extract on healthy volunteers. *Journal of Ethnopharmacology* 136(3): 452–456.

Moorthy, R., Prabhu, K. M. & Murthy, P. S. 2010. Anti-hyperglycemic compound (GII) from fenugreek (*Trigonella foenum-graecum* Linn.) seeds, its purification and effect in diabetes mellitus. *Indian Journal of Experimental Biology* 48(11): 1111–1118.

Msaada, K., Ben Jemia, M., Salem, N., Bachrouch, O., Sriti, J., Tammar, S., Bettaieb, I. et al. 2017 Antioxidant activity of methanolic extracts from three coriander (*Coriandrum sativum* L.) fruit varieties. *Arabian Journal of Chemistry* 10: S3176–S3183.

Mubarrat, F. 2016. *In vitro Propagation of Garlic (Allium sativum L.)* (Doctoral dissertation, Department of Biotechnology, Sher-e-Bangla Agricultural University).

Muralidharan, E. M. 2001. KFRI research report no. 211 micropropagation of important rare and endangered tree species of Western ghats [Final report of project no. KFRI 253/96]. Genetics Division Kerala Forest Research Institute Peechi, Kerala.

Nagakubo, T., Nagasawa, A. & Ohkawa, H. 1993. Micropropagation of garlic through *in vitro* bulblet formation. *Plant Cell, Tissue and Organ Culture* 32(2): 175–183.

Nair, R. R. & Dutta Gupta, S. 2006. High-frequency plant regeneration through cyclic secondary somatic embryogenesis in black pepper (*Piper nigrum* L.). *Plant Cell Reports* 24(12): 699–707.

Naz, S. H. A. G. U. F. T. A., Ilyas, S., Javad, S. U. M. E. R. A. & Ali, A. 2009. *In vitro* clonal multiplication and acclimatization of different varieties of turmeric (*Curcuma longa* L.). *Pakistan Journal of Botany* 41(6): 2807–2816.

Neelima, R., Rao, N. N., Seema, D., Sarita, A. & Arya, I. D. 2017. Optimization of media constituents for regeneration via callus cultures of *Glycyrrhiza glabra* L. an endangered plant. *Biochemical and Cellular Archives* 17(1): 389–398.

Nirmala, M. J., Durai, L., Rao, K. A. & Nagarajan, R. 2020. Ultrasonic nanoemulsification of *Cuminum cyminum* essential oil and its applications in medicine. *International Journal of Nanomedicine* 15: 795.

Nisar, A., Bin, G., Hina, F., Bilal, H. A., Chun-Zhao, L., Tariq, M. & Zabta, K. S. 2011. Feasible plant regeneration in black pepper from petiole explants. *Journal of Medicinal Plants Research* 5(18): 4590–4595.

Nurdjannah, N. & Bermawie, N. 2012. Cloves. In: *Handbook of Herbs and Spices.* Woodhead Publishing, 197–215.

Oncina, R., Del Río, J. A., Gomez, P. & Ortuno, A. 2002. Effect of ethylene on diosgenin accumulation in callus cultures of *Trigonella foenum-graecum* L. *Food Chemistry* 76(4): 475–479.

Özbek, H., Ugras, S., Bayram, I., Uygan, I., Erdogan, E., Öztürk, A. & Huyut, Z. 2004. Hepatoprotective effect of Foeniculum vulgare essential oil: A carbon-tetrachloride induced liver fibrosis model in rats. *Scandinavian Journal of Laboratory Animal Science* 31(1): 9–17.

Parasuraman, S., Balamurugan, S., Christapher, P. V., Petchi, R. R., Yeng, W. Y., Sujithra, J. & Vijaya, C. 2015. Evaluation of antidiabetic and antihyperlipidemic effects of hydroalcoholic extract of leaves of *Ocimum tenuiflorum* (Lamiaceae) and prediction of biological activity of its phytoconstituents. *Pharmacognosy Research* 7(2): 156.

Parsaeimehr, A. 2009. Producing friable callus for suspension culture in Glycyrrhiza glabra. *Advances in Environmental Biology*: 125–129.

Patel, S., Shah, D. B. & Panchal, H. J. 2014. De novo RNA seq assembly and annotation of *Trigonella foenum-gGraecum L* (SRR066197). *Legume Genomics and Genetics* 5(7): 1–7

Pattnaik, S. & Chand, P. K. 1996. *In vitro* propagation of the medicinal herbs *Ocimum americanum* L. syn. O. canum Sims.(hoary basil) and Ocimum sanctum L.(holy basil). *Plant Cell Reports* 15: 846–850.

Philip, V. J., Joseph, D., Triggs, G. S. & Dickinson, N. M. 1992. Micropropagation of black pepper (Piper nigrum Linn.) through shoot tip cultures. *Plant Cell Reports* 12(1): 41–44.

Platel, K. & Srinivasan, K. 2004. Digestive stimulant action of spices: A myth or reality? *Indian Journal of Medical Research* 119(5): 167.

Poli, V. & Challa, C. 2019. A comparative study of eugenol and *Ocimum sanctum* Linn. leaf extract on the antifertility effect in female albino rats. *Journal of the Chinese Medical Association* 82(3): 231–234.

Priya, R., Prathapan, A., Raghu, K. G. & Menon, A. N. 2012. Chemical composition and *in vitro* antioxidative potential of essential oil isolated from *Curcuma longa L.* leaves. *Asian Pacific Journal of Tropical Biomedicine* 2(2): S695–S699.

Rahman, M. M., Amin, M. N., Jahan, H. S. & Ahmed, R. 2004. *In vitro* regeneration of plantlets of *Curcuma longa* Linn. A valuable spice plant in Bangladesh. *Asian Journal of Plant Sciences* 3(3): 306–309.

Rahman, M. A., Khondaker, I. A. & Joarder, O. I. 2007. *In vitro* callus induction and plant regeneration from immature zygotic embryos of black pepper (*Piper nigrum* L.). *Journal of the Bangladesh Agricultural University* 5(452-2018-3945): 87–90.

Rahman, S. & Husen, A. 2021. Potential role of medicinal plants in the cure of liver and kidney diseases. In: *Non-Timber Forest Products* (Eds. Husen, A., Bachheti, R. K. & Bachheti, A.) Springer International Publishing AG, Gewerbestrasse, Cham, 229–254. https://doi.org/10.1007/978-3-030-73077-2_10.

Rahman, S. & Husen, A. 2022. *Asparagus racemosus* (Shatavari) and *Dioscorea villosa* (Wild) yams). In: *Herbs, Shrubs, and Trees of Potential Medicinal Benefits* (Ed. Husen, A.). CRC Press, Boca Raton, FL, 95–112.

Rahman, S. & Husen, A. 2023a. *Trigonella foenum-gGraecum* L.: A useful medicinal plant for diabetes and other diseases. In: *Antidiabetic Medicinal Plants and Herbal Treatments* (Ed. Husen, A.). CRC Press, Boca Raton, FL, 145–153.

Rahman, S. & Husen, A. 2023b. Role of green chiretta [*Andaphis paniculata* (Burm. F.)] Nees in diabetes and other ailments. In: *Antidiabetic Medicinal Plants and Herbal Treatments* (Ed. Husen, A.). CRC Press, Boca Raton, FL, 169–176

Rao, S., Dinkar, C., Vaishnav, L. K., Rao, P., Rai, M. P., Fayad, R. & Baliga, M. S. 2014. The Indian spice turmeric delays and mitigates radiation-induced oral mucositis in patients undergoing treatment for head and neck cancer: An investigational study. *Integrative Cancer Therapies* 13(3): 201–210.

Rasud, Y. & Bustamana, B. 2020. *In vitro* callus induction from clove (Syzigium aromaticum L.) leaves on medium containing various auxin concentrations. *Jurnal Ilmu Pertanian Indonesia* 25(1): 67–72.

Ravindran, R., Devi, R. S., Samson, J. & Senthilvelan, M. 2005. Noise-stress-induced brain neurotransmitter changes and the effect of *Ocimum sanctum* (Linn.) treatment in albino rats. *Journal of Pharmacological Sciences* 98(4): 354–360.

Rema, J. & Krishnamoorthy, B. 1993. Rooting response of elite cinnamon (*Cinnamomum verum* Bercht & Presl.) lines. *Journal of Spices and Aromatic Crops* 2: 21–25.

Riella, K. R., Marinho, R. R., Santos, J. S., Pereira-Filho, R. N., Cardoso, J. C., Albuquerque-Junior, R. L. C. & Thomazzi, S. M. 2012. Anti-inflammatory and cicatrizing activities of thymol, a monoterpene of the essential oil from *Lippia gracilis*, in rodents. *Journal of Ethnopharmacology* 143(2): 656–663.

Rimoldi, S., Torrecillas, S., Montero, D., Gini, E., Makol, A., Valdenegro, V., Izquierdo, M. & Terova, G. 2020. Assessment of dietary supplementation with galactomannan oligosaccharides and phytogenics on gut microbiota of European sea bass (*Dicentrarchus labrax*) fed low fishmeal and fish oil- based diet. *PLOS One* 15(4): e0231494.

Rivlin, R. S. 2001. Historical perspective on the use of garlic. *The Journal of Nutrition* 131(3): 951S–954S.

Roopadarshini, V. 2010. High frequency shoot multiplication and callus regeneration of turmeric. *International Journal of Biotechnology and Biochemistry* 6(5): 723–734.

Rout, G. R., Palai, S. K., Samantaray, S. & Das, P. 2001. Effect of growth regulator and culture conditions on shoot multiplication and rhizome formation in ginger (*Zingiber officinale* Rosc.) *in vitro. In vitro Cellular & Developmental Biology-Plant* 37: 814–819.

Ryan, J. L. & Morrow, G. R. 2010. Ginger. *Oncology Nurse Edition* 24(2): 46–49.

Sahana, K., Nagarajan, S. & Rao, L. J. M. 2011. Cumin (*Cuminum cyminum L.*) seed volatile oil: Chemistry and role in health and disease prevention. In: *Nuts and Seeds in Health and Disease Prevention.* Academic Press, 417–427.

Sahib, N. G., Anwar, F., Gilani, A. H., Hamid, A. A., Saari, N. & Alkharfy, K. M. 2013. Coriander (*Coriandrum sativum* L.): A potential source of high-value components for functional foods and nutraceuticals- a review. *Phytotherapy Research* 27(10): 1439–1456.

Salvi, N. D., George, L. & Eapen, S. 2002. Micropropagation and field evaluation of micropropagated plants of turmeric. *Plant Cell, Tissue and Organ Culture* 68(2): 143–151.

Saremirad, A., Azarkish, R. & Abbasi, S. 2021. Optimization callus induction in fennel (Foeniculum vulgare M.). *Journal of Medicinal Plants Biotechnology* 6(2): 1–18.

Sarrami, N., Pemberton, M. N., Thornhill, M. H. & Theaker, E. D. 2002. Adverse reactions associated with the use of eugenol in dentistry. *British Dental Journal* 193(5): 257–259.

Saxena, S. 2005. Glycyrrhiza glabra: Medicine over the millennium. *Natural Product Radiance* 4(5): 358–367.

Saxena, R., Kothari, P., Rathore, S. & Ishan-Ulla-Khan, R. 2012. Organogenesis in fennel (Foeniculum vulgare Mill.). *International Journal of Seed Spices* 2(2): 1–4.

Shahidi, F., Janitha, P. K. & Wanasundara, P. D. 1992. Phenolic antioxidants. *Critical Reviews in Food Science and Nutrition* 32(1): 67–103.

Sharma, Y. 2017. Ginger (*Zingiber officinale*) - An elixir of life a review. *The Pharmaceutical Innovation* 6(11, Part A): 22.

Sheela, M. L., Ramakrishna, M. K. & Salimath, B. P. 2006. Angiogenic and proliferative effects of the cytokine VEGF in Ehrlich ascites tumor cells is inhibited by *Glycyrrhiza glabra. International Immunopharmacology* 6(3): 494–498.

Shetty, A. K. & Salimath, P. V. 2009. Reno-protective effects of fenugreek (*Trigonella foenum- greacum*) during experimental diabetes. *e-SPEN, the European e-Journal of Clinical Nutrition and Metabolism* 4(3): e137–e142.

Shilpa, K., Selvakkumar, C., Senthil, A. K. & Lakshmi, B. S. 2010. *In vitro* root culture of Ocimum sanctum L. and evaluation of its free radical scavenging activity. *Plant Cell, Tissue and Organ Culture (PCTOC)* 101: 105–109.

Shukla, M. R., Kibler, A., Turi, C. E., Erland, L. A., Sullivan, J. A., Murch, S. J. & Saxena, P. K. 2021. Selection and micropropagation of an elite melatonin rich Tulsi (Ocimum sanctum L.) germplasm line. *Agronomy* 11(2): 207.

Shukla, M. R., Subhash, N., Patel, D. R. & Patel, S. A. 1997. *In vitro* studies in cumin (*Cuminum cyminum* L.). In: *Proceedings of the National Seminar on Biotechnology of Spices and Aromatic Plants* (Eds. Edison, S., Ramana, K. V., Sasikumar, B., Babu, K. N. & Eapen, S. J.), 45–48.

Shukla, Y. & Singh, M. 2007. Cancer preventive properties of ginger: A brief review. *Food and Chemical Toxicology* 45(5): 683–690.

Singh, G., Maurya, S., DeLampasona, M. P. & Catalan, C. A. 2007. A comparison of chemical, antioxidant and antimicrobial studies of cinnamon leaf and bark volatile oils, oleoresins and their constituents. *Food and Chemical Toxicology* 45(9): 1650–1661.

Singh, J., Dubey, R. K. & Atal, C. K. 1986. Piperine-mediated inhibition of glucuronidation activity in isolated epithelial cells of the guinea-pig small intestine: Evidence that piperine lowers the endogeneous UDP-glucuronic acid content. *Journal of Pharmacology and Experimental Therapeutics* 236(2): 488–493.

Singh, N. K. & Sehgal, C. B. 1999. Micropropagation of 'Holy Basil'(Ocimum sanctum Linn.) from young inflorescences of mature plants. *Plant Growth Regulation* 29: 161–166.

Singh, S., Taneja, M. & Majumdar, D. K. 2007. Biological activities of *Ocimum sanctum* L. fixed oil—An overview. *Indian Journal of Experimental Biology* 45(3): 403–412.

Soulange, J. G., Ranghoo-Sanmukhiya, V. M. & Seeburrun, S. D. 2007. Tissue culture and RAPD analysis of *Cinnamonum camphora* and *Cinnamomum verum*. *Biotechnology* 6(2): 239–244.

Srinivasan, K. 2007. Black pepper and its pungent principle-piperine: A review of diverse physiological effects. *Critical Reviews in food Science and Nutrition* 47(8): 735–748.

Srinivasulu, K. & Changamma, C. 2017. A study on the effect of *Ocimum sanctum* (Linn.) Leaf extract and ursolic acid on spermatogenesis in male rats. *Indian Journal of Pharmaceutical Sciences* 79(1): 158–163.

Stephen, R. & Jayabalan, N. 1998. *In vitro* flowering and seed setting formation of coriander (*Coriandrum sativum* L.). *Current Science* 74(3): 195–197.

Subash Babu, P., Prabuseenivasan, S. & Ignacimuthu, S. 2007. Cinnamaldehyde—A potential antidiabetic agent. *Phytomedicine* 14(1): 15–22.

Subasinghe, S., Hettiarachchi, C. S. & Iddagoda, N. 2016. *In-vitro* propagation of cinnamon (*Cinnamomum verum* Presl.) using embryos and *in vitro* axillary buds. *Journal of Advanced Agricultural Technologies* 3(3): 164–169.

Sulieman, A. M. E., Ali, A. O. & Hemavathy, J. 2008. Lipid content and fatty acid composition of fenugreek (*Trigonella foenum-graecum* L.) seeds grown in Sudan. *International Journal of Food Science and Technology* 43(2): 380–382.

Suleiman, W. B. & Helal, E. E. H. 2022. Chemical constituents and potential pleiotropic activities of Foeniculum vulgare (Fennel) ethanolic extract; *in vitro* approach. *Egyptian Journal of Chemistry* 65(7): 617–626.

Sun, G., Zhang, S., Xie, Y., Zhang, Z. & Zhao, W. 2016. Gallic acid as a selective anticancer agent that induces apoptosis in SMMC-7721 human hepatocellular carcinoma cells. *Oncology Letters* 11(1): 150–158.

Suparman, U. & Blake, J. 1990. Studies on tissue culture of clove tree plant. *Indonesian Journal of Crop Science* 5(2): 67–75.

Tabak, M., Armon, R. & Neeman, I. 1999. Cinnamon extracts' inhibitory effect on Helicobacter pylori. *Journal of Ethnopharmacology* 67(3): 269–277.

Tabibazar, S., Aharizad, S., Uliaie, E. D., Nojadeh, M. S. & Kosari-Nasab, M. 2020. Effect of plant regulations on callus essential Oil content of fennel (Foeniculum Vulgare Miller) populations. *Journal of Biochemical Technology* (2): 141–145.

Taghizadeh, M., Memarzadeh, M. R., Asemi, Z. & Esmaillzadeh, A. 2015. Effect of the *Cumin cyminum* L. intake on weight loss, metabolic profiles and biomarkers of oxidative stress in overweight subjects: A randomized double-blind placebo-controlled clinical trial. *Annals of Nutrition and Metabolism* 66(2–3): 117–124.

Tapsell, L. C., Hemphill, I., Cobiac, L., Patch, C. S., Sullivan, D. R., Fenech, M., Roodenrys, S., Keogh, J. B., Clifton, P. M., Williams, P. G., Fazio, V. A. & Inge, K. E. 2006. Health benefits of herbs and spices: The past, the present, the future. *The Medical Journal of Australia* 185(S4): S1–S24.

Tavangar, A., Karami, L., Hedayat, M. & Gholamreza, A. B. D. I. 2021. Effect of salinity and drought stress on morphological and biochemical properties of two Iranian fenugreek (*Trigonella foenum-graecum*) populations. *Notulae Botanicae Horti Agrobotanici Cluj-Napoca* 49(2): 12038–12038.

Tawfik, A. A. & Noga, G. 2001. Adventitious shoot proliferation from hypocotyl and internodal stem explants of cumin. *Plant Cell, Tissue and Organ Culture* 66(2): 141–147.

Thengane, S. R., Kulkarni, D. K. & Krishnamurthy, K. V. 1998. Micropropagation of licorice (Glycyrrhiza glabra L.) through shoot tip and nodal culture. *In Vitro Cellular & Developmental Biology - Plant* 34: 331–334.

Thippeswamy, N. B. & Naidu, K. A. 2005. Antioxidant potency of cumin varieties—Cumin, black cumin and bitter cumin—On antioxidant systems. *European Food Research and Technology* 220(5): 472–476.

Torabi, S., Far, M. S., Omidi, M., Gharakhanlou, H. & Saffar, Z. N. 2014. A survey of the best growing conditions of *in vitro* culture in fennel different genotypes for mass production. *International Journal of Biosciences* 5(4): 7–17.

Tung, Y. T., Chua, M. T., Wang, S. Y. & Chang, S. T. 2008. Anti-inflammation activities of essential oil and its constituents from indigenous cinnamon (*Cinnamomum osmophloeum*) twigs. *Bioresource Technology* 99(9): 3908–3913.

Uddin, M. A., Shahinuzzaman, M., Rana, M. S. & Yaakob, Z. 2017. Study of chemical composition and medicinal properties of volatile oil from clove buds (Eugenia caryophyllus). *International Journal of Pharmaceutical Sciences and Research* 8(2): 895.

Valizadeh, M. 2018. *In vitro* regeneration in medicinal plant fenugreek (*Trigonella foenum-graecum* L.). *Journal of Plant Physiology and Breeding* 8(2): 43–51.

Varadarajan, S., Narasimhan, M., Malaisamy, M. & Duraipandian, C. 2015. *In vitro* anti-mycotic activity of hydro alcoholic extracts of some Indian medicinal plants against fluconazole resistant Candida albicans. *Journal of Clinical and Diagnostic Research: JCDR* 9(8): ZC07.

Varsha, S., Agrawal, R. C. & Sonam, P. 2013. Phytochemical screening and determination of anti-bacterial and anti-oxidant potential of *Glycyrrhiza glabra* root extracts. *Journal of Environmental Research and Development* 7(4A): 1552.

Verma, V. M. 2019. Black pepper: Health benefits, *in vitro* multiplication, and commercial cultivation. In: *Medicinal Plants, From Farm to Pharmacy* (Eds. Joshee, N., Dhekney, S. & Parajuli, P.). Springer, Cham, r111–127.

Walker, B. C., Adhikari, S. & Mittal, S. 2021. Therapeutic potential of curcumin for the treatment of malignant gliomas. In: *Gliomas* (Eds. Debinski, W.). Exon Publications, Brisbane, April 30. Chapter 8 Available from: https://www.ncbi.nlm.nih.gov/books/NBK570704/ https://doi.org/10.36255/exonpublications.gliomas.2021.chapter8.

Wawrosch, C., Kazianka, C., Küchler, V., Lämmermayer, K., Winter, M. & Kopp, B. 2009. *In vitro* multiplication of Glycyrrhiza glabra L. through somatic embryogenesis. *Planta Medica* 75(09): PJ28.

Wilson, R. 1995. Aromatherapy for vibrant health & beauty. In: *A Guide to Understanding and Using Aromatherapy for Vibrant Health and Beauty*. Avery Publishing, Garden City Park.

Yadav, K. & Singh, N. 2012. Factor influencing *in vitro* plant regeneration of licorice (Glycyrrhiza glabra L.). *Iranian Journal of Biotechnology* 10(2): 161–167

Younesy, S., Amiraliakbari, S., Esmaeili, S., Alavimajd, H. & Nouraei, S. 2014. Effects of fenugreek seed on the severity and systemic symptoms of dysmenorrhea. *Journal of Reproduction and Infertility* 15(1): 41–48.

Youssef, M. K., Wang, Q., Cui, S. W. & Barbut, S. 2009. Purification and partial physicochemical characteristics of protein free fenugreek gums. *Food Hydrocolloids* 23(8): 2049–2053.

Zare, A., Farzaneh, P., Pourpak, Z., Hassan, Z. & Moin, M. 2009. Aged garlic extract modulates allergic airway inflammation in Balb/c mice. In: *Allergy* (Vol. 64). Wiley-Balckwell Publishing, Malden, 311–311.

Zhang, Y., Liu, X., Ruan, J., Zhuang, X., Zhang, X. & Li, Z. 2020. Phytochemicals of garlic: Promising candidates for cancer therapy. *Biomedicine & Pharmacotherapy* 123: 109730.

5 Ethnobotany of Spice and Condiment Plant Species

Kolagani Chandramohan, Salman Khan, Dudipala Ravivarma, Chakali Bhargavi, and Azamal Husen

5.1 INTRODUCTION

Spices and condiments are used for flavouring food and beverages. Since at least 4000 BC, local Chinese and Indian populations have used herbs and spices in traditional medical practices. The use of spices was associated in popular culture with magic, religion, tradition, and cultural ceremonies. Spices are defined as 'plant parts like seed, bark, berries, buds, floral parts, fruits, kernels, leaves, rhizome, latex, etc. that impart strong flavour or aroma to food or drinks', among other definitions (Rathore & Shekhawat, 2008; Ramkumar & Karuppusamy, 2021). The meaning of spice is derived from Latin word '*spec*', meaning 'appearance or kind' (Dubey, 2017).

These aromatic plant products are used whole, broken, or in ground form to season food rather than for their nutrient value (FAO, 2005; Ramkumar & Karuppusamy, 2021). On the other hand, condiments are often a mixture of herbs and spices combined in a liquid form and used to complement dishes and are typically applied in greater quantities than spices (Dubey, 2017; D'Souza et al., 2017; Ramkumar & Karuppusamy, 2021). Unlike spices, they have no toxic effect on the human body (D'Souza et al., 2017). However, according to the International Organization for Standardization, both spices and condiments can be grouped under one roof, termed as 'spices and condiments', regardless of how they are used for flavouring and seasoning, including in the preparation of meat, baking, confections, and other forms of food processing and preservation (Ramkumar & Karuppusamy, 2021).

5.2 HISTORY OF SPICES AND CONDIMENTS

The history of spices started in ancient and medieval times when spices were the most valuable trade items. The use of spices became popular and spread through the Middle East to the eastern Mediterranean and Europe. The trading of spices was carried out from China, Indonesia, and India and the spices were originally transported overland by donkey or camel caravans. Around 2000 BC, the trade in spices spread from East Asia to South Asia and the Middle East, starting with cinnamon and black pepper and ending with herbs and pepper in South Asia. Egyptians employed herbs in the process of mummification, and their need for rare spices and plants fueled global trade (Garland, 1972). For 5000 years, Arab middlemen controlled the trade in spices, until India and Eastern countries were discovered as spice-producing countries. During those days, worldwide exploration in search of spices was undertaken by Columbus, Vasco de Gama, and Sir Francis Drake. Then, London became the world's major spice market as a result of the British East India Company and the British Empire, which dominated for 200 years.

In the Indian region, Spices and herbs such as black pepper, cinnamon, turmeric, and cardamom were used by Indians for thousands of years for both culinary and health purposes. Cardamon and turmeric, which are both indigenous to India, were cultivated as early as the eighth century BC in the gardens of Babylon. In the book 'Sushruta Samhita', Sushruta (around the fourth century BC)

DOI: 10.1201/9781003387046-5

used white mustard and other aromatic plants in bed sheets to ward off malignant spirits and also applied sesame on wounds as an antiseptic (Duke, 2002). Medical writings of Charaka (between 100 BCE and 200 CE) and Sushruta II (6th century BCE) referenced spices and herbs. Sushruta also used spices and herbs, such as cinnamon, cardamom, ginger, turmeric, and pepper, for healing purposes. Spices such as cardamom, ginger, black pepper, cumin, and mustard seed were included in ancient herbal medicines for different types of health benefits. In Ayurvedic medicine, spices such as cloves and cardamom were wrapped in betel nut leaves and chewed after meals to increase the flow of saliva and aid digestion (Rathore & Shekhawat, 2008).

Since ancient times, trading on spices has been a preeminent business. Throughout the world, spices and herbs are frequently used in cuisine, to improve flavour and add new tastes (Grivetti, 2017). The United States Department of Agriculture (USDA) reports revealed that the consumption of spices in the US has increased exponentially over the course of the past half-century, with spices like ginger and chili pepper frequently being part of food to ensure the best taste. The use of spices in global cuisine, as well as in the world economy, has increased faster than ever before (USDA Report, 2021). There is also a renewed interest in the health benefits of spices and herbs. Recent research revealed that 5–10% of adults in the United States were using botanical supplements, such as herbal products and spices, for health benefits (Clarke et al., 2015). Spices and herbs can help flavour foods when concentrations of added sugars, saturated fat, and sodium are decreased, and they also can add to the enjoyment of nutrient-dense foods, dishes, and meals that reflect specific cultures (USDHHS and USDA, 2015).

Since ancient times, people have used condiments to enhance the flavour of the food. Salt and vinegar have been used since ancient times, both as a preservative and to enhance the flavour of food. Vinegar was used as a medicine as well as a food but, throughout the centuries, other sauces and spices became popular as a result of trade empire expansion. The Romans liked condiments and they made many sauces for their own food. The Romans also grew mustard and they introduced it into the parts of Europe they conquered. They also made mint sauce. In the Middle Ages, mustard was a popular condiment in Europe. In the sixteenth, seventeenth, and eighteenth centuries, new condiments, like pesto sauce from Italy, chutney from India, soy sauce from China, and mayonnaise were invented. In response to civilization of people and increasing demand, the production of condiments and their related products increased (Clarke et al., 2015).

5.3 IMPORTANCE OF SPICES AND CONDIMENTS

One of the most promising developments for spices in modern times is that scientific evidence is accumulating that supports the anecdotal health benefits touted by our ancestors. Research shows that culinary spices and herbs may have beneficial effects in areas such as heart health, cognition, and weight management, as well as in improving diet quality by making healthier foods more acceptable to consumers. The body of scientific evidence is ever expanding, supporting the wisdom of our ancestors throughout the ages. Spice sales and marketing generates significant foreign cash thanks to their widespread demand. Only 6–7% of India's total exports are spices, which are sent to 130 different nations. Due to the enormous domestic demand, the remaining products are consumed in the Indian market. Along with other spices, India is currently one of the world's top producers of ginger and turmeric (Ray, 2007; Babu et al., 2013). Typically, spices are utilized directly or indirectly in a variety of sectors, including cosmetics, perfumes, pharmaceuticals, and medicine. Antioxidants, antimicrobials, antibiotics, and preservatives are the key properties of spices. Some of the spices are high in potassium, iron, vitamins, and trace metals (Babu et al., 2013). Spices are also used for colouring the foodstuffs (Parthasarathy & Kandiannan, 2007). Numerous leafy spices are occasionally used to make "herbal" teas that promote general wellness. Spices are well known for their ability to stimulate the digestive system in addition to improving the flavour and taste of meals (Sugasini et al., 2018; Ramkumar & Karuppusamy, 2021).

Many people choose to use spices as medication in preference to conventional medicinal drugs since they are vital remedies. Many spices gained prominence for their therapeutic value due to the development of scientifically validated knowledge on their phytochemistry, the therapeutic effects of their bio-active principles, and their mechanisms of action. Examples of these properties include antioxidant and antiseptic potential, singlet oxygen quenching, carminative action, hypolipidaemic effect, antidiabetic, antilithogenic, and anti-inflammatory properties, antimutagenic effects, and decreased induction and advancement of cancer cells. As varied as the morphology of the plant parts used to make spices is their application (Divakaran et al., 2018; Figure 5.1).

FIGURE 5.1 A. *Brassica juncea*; B. *Capparis spinosa*; C. *Cinnamomum verum*; D. *Punica granatum*; E. Cultivation of *Piper nigrum*; F. *Zingiber montanum*; G. *Piper longum*.

5.4 DISTRIBUTION OF SPICES AND CONDIMENTS

The global spice and condiment market was estimated at USD 36.91 billion in 2021 and it is projected to exceed USD 59.5 billion by 2030. India is one of the top spice-producing countries in the world, accounting for more than one-third of global production, producing more than 75 spices, including cardamom, chilli, coriander, and cumin seeds. Turkey and China are the second largest spice-producing countries in the world. India is known as the 'Home of Spices' and the 'Land of Spices' and produces such a large range of spices due to the diverse agroclimatic regions across the subcontinent which offer tropical, subtropical, and temperate conditions, suitable to grow almost all such species (Singh et al., 2011). India contributes about 20–25% of the world trade in Spices. In India, 52–60 spices are being cultivated (Ramkumar & Karuppusamy, 2021). There is a good export demand for Indian spices because of their high quality with the maximum content of essential oils, oleoresins and active principles. Kerala, Andhra Pradesh, Odisha, Gujarat, Maharashtra, West Bengal, Karnataka, Tamil Nadu, and Madhya Pradesh are the largest states in terms of spice production, while every state and union territory of the nation grows at least a few spice crops (Naik, 2007; Babu et al., 2013). Spices produced from tree-based crops are referred to as "tree spices." Tamil Nadu, Kerala, Andhra Pradesh, Maharashtra, and Karnataka are the key states that cultivate tree spices. There are 17 tree spices that are widely cultivated in India, including allspice, tamarind, garcinia, clove, nutmeg, and cinnamon (Sugasini et al., 2018). Whole spice, ground spice, oil, and oleoresin are used to create the economically significant products. Clove crops are grown around the Western Ghats in Kerala, Karnataka, and the Kanyakumari and Nilgiris regions of Tamil Nadu. In Kerala and a few isolated locations in Tamil Nadu and Karnataka, nutmeg growing is more prevalent. In the gardens of Kerala, Tamil Nadu, Karnataka, West Bengal, and Odisha, all tree spices were grown sporadically (Divakaran et al., 2018; Ramkumar & Karuppusamy, 2021). Kerala in India is ideally suited for the growing of spices. Black pepper, cardamom, ginger, turmeric, clove, garcinia, nutmeg, and mace are important spices that are grown. Pepper, cardamom, ginger, and turmeric make up over 95% of the entire production; although ginger and turmeric predominate on the plains, pepper and cardamom are grown in the high mountains (Sugasini et al., 2018; Ramkumar & Karuppusamy, 2021).

5.5 ETHNOBOTANICAL IMPORTANCE OF SPICES AND CONDIMENTS

Ethnobotany is a multidisciplinary science involving the traditional use of plants by human beings (Husen, 2020, 2023; Rahman & Husen, 2021, 2022, 2023a,2023b). Many Indian spice plants have been investigated for their beneficial use as antioxidants or source of antioxidants, using presently available experimental techniques (Ali et al., 2008). This plant-based traditional knowledge has become a recognized tool in the search for new sources of drugs, pharmaceuticals, and nutraceuticals (Sharma & Mujumdar, 2003). The Rig Veda, one of the ancient Hindu scriptures, is the earliest literary record in India on spices, listing more than one thousand healing plant species, including spices and condiments. In the modern world, spices and condiments provide the traditional value in curing various diseases, improving health awareness and relaxation from stress, in addition to flavouring the food.

To manufacture condiments and other foods unique to tribal people, they employ various unusual plants. These include the dried fruits of *Tacca leontopetaloides* and the tubers of *Decalepis hamiltonii*, which are rare in distribution (Reddy et al., 2007). Fifty-eight plant species of 29 families are being used by people of the Nkonkobe Municipality of Eastern Cape Province, South Africa, as spices (Abiola et al., 2016). Spices are being used in the treatment of various diseases by Maonan people in China (Hong et al., 2015). The significance of spices in cancer prevention has been revealed (Keafer & Milner, 2008). Some of the plant species are known as new source of spice plants such as *Acmella paniculata*, *Dendrocalamus hamiltonii*, *Etlingera linguiformis*, *Gynura procumbens*, *Hydrocotyle sibthorpioides*, *Magnolia oblonga*, *Oenanthe javanica*, *Paederia foetida*, *Perilla frutescens*, *Persicaria nepalensis*, *Phyllostachys bambusoides*, and *Spilanthes acmella*, which are used by tribes such as the Adi, Apatani, Nyishi, Galo, Khampti, Khowa, Mishmi, Idu, Taroan, Momba, Sherdukpen, Singpho, Hrusso, Tagin, and Khamba (Bharali et al., 2017). Some Assamese ethnic tribal groups, including the Ahom, Deori, Mishing, and Sonowal-kachari, use 51 species of spices and condiments, including newly discovered spice species such as *Garcinia pedunculata*, *Hibiscus sabdariffa*, *Houttuynia cordata*, *Neocinnamomum caudatum*, and *Peperomia pellucida* (Bharali et al., 2017). A new spice species, *Guizotia abyssinica*, is also included for use in flavouring food (Tayade & Patil, 2010; Ramkumar & Karuppusamy, 2021). The state of Sikkim, India is inhabited by various ethnic tribes such as the Lepcha, Bhutis, Limboo, and Nepalese, who mostly construct their tiny, remote communities on hills and hilltops. It is crucial to preserve this biological legacy of knowledge and investigate new resources (Rai et al., 1998). *Allium hookeri*, *Allium tuberosum*, *Costus speciosus*, *Curcuma cassia*, *Elsholtzia blanda*, *Elsholtzia communis*, *Hedychium coronarium*, *Hedychium marginatum*, *Persicaria posumbu*, *Zingiber cassumunar*, and *Zingiber zerumbet* are important spices used by Hmar tribes (Salam & Jamir, 2016). Babu et al. (2013) assessed the status of traditional practices of farmers, and women in particular, for storage of spices and condiments especially for ginger (*Zingiber officinale*), turmeric (*Curcuma longa*), chilli (*Capsicum annuum*), onion (*Allium cepa*), garlic (*Allium sativum*), and coriander (*Coriandrum sativum*). Observations revealed that a large number of farmers still practice the traditional storage system. The tea, known as 'Elaichi tea', which has been used to treat fatigue and depression. Beta-carotene is found in high concentrations in cardamom capsules (Ashokkumar et al., 2020). In traditional medicine, daily consumption of cardamom with a tablespoon of honey improves eyesight (Singh & Singh, 1996). However, some people believe that excessive uses of cardamom capsules may cause impotency (Nair, 2006; Ramkumar & Karuppusamy, 2021). Some wild plants, such as *Zanthoxylum armatum*, were recorded as being used as a spice (Singh, 2017). The tribal groups Kadar, Malasars, Mudhuvan, Kurumbans, Kaatunaickens, Thodar, Kanis, and Paliyars, as well as other tribal groups, still adhere to a traditional way of life that closely depends on forest plants for all of their food and livelihood needs. Numerous local species of spices and condiments are noted, such as *Costus speciosus*, *Garcinia gummi-gutta*, *Curcuma neilgherrensis*, *Syzygium palghatense*, *Curculigo orchioides*, and *Cleome paniculata*, which are used to preserve, and add flavour and colour to food products (Yesodharan & Sujana, 2007; Figure 5.2).

Detailed information, such as the name of the species, chemical constituents, and ethnobotanical importance of spices and condiments are provided in Table 5.1

Figure 5.3 shows members of the family Lamiaceae which aremostly used as spices and condiments, followed by members of the Apiaceae and Zingiberaceae.

5.6 CONSERVATION AND SUSTAINABLE UTILIZATION OF WILD RELATIVES OF SPICES AND CONDIMENTS

India has a high level of diversity in spices such as black pepper, cardamom, ginger, turmeric, cinnamon, tamarind, and garcinia. The other important spices include coriander, fennel, fenugreek, cumin, nutmeg, clove, and vanilla. In traditional medicinal systems or folklore systems, in addition to spices and condiments, some of the plants and plant products are used for preservation and flavouring of foods. These spice or condiment plants are also used locally to heal various ailments. India is the land of spices, where several spice genera originated in the diversity hot spots of either the Himalayas or the Western Ghats. The majority of *Zingiber* species originated from the Western Ghats of India, where they have traditional importance in various systems of medicine (Sabu, 2006). Nearly 25% of spice and condiment plants are reported to grow wild and are collected from local vegetation. Several local spices are wild relatives of a known valuable spice, for example, in the genera *Allium*, *Alpinia*, *Cinnamomum*, *Curcuma*, *Hedychium*, *Zanthoxylum*, etc. Wild germplasm of cultivated spice plants and wild relatives of current spice species are very important for the

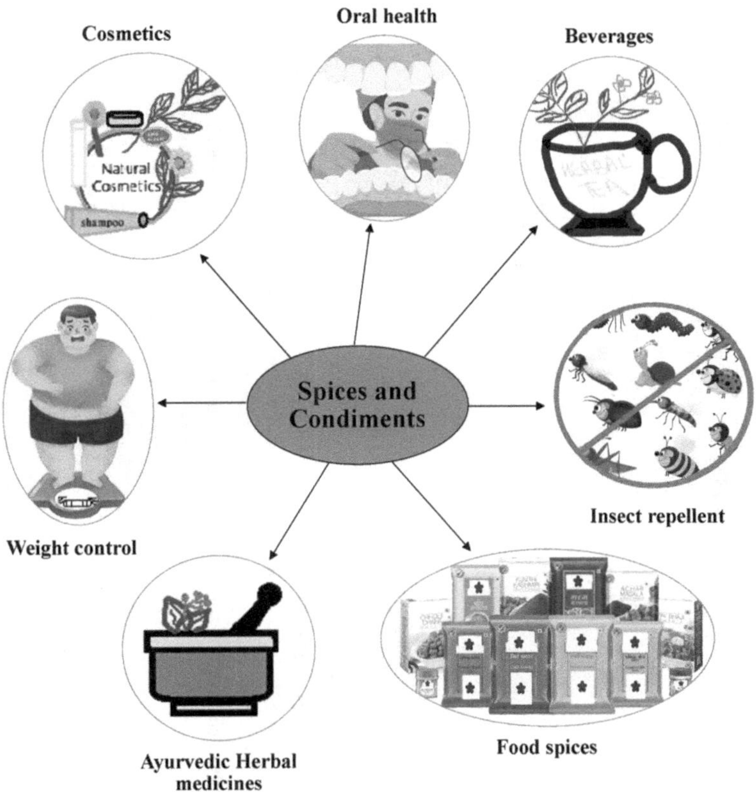

FIGURE 5.2 Importance of spices and condiments in various uses.

improvement and development of high-quality spice varieties (Pedapati et al., 2014). Many research centres are actively engaged in the collection, cultivation, and agronomic development of wild relatives of currently usable spices and condiments. Due to pressure caused by the increasing interest in herbal medicine, collection and conservation of valuable wild spice species in Indian region have become a necessity protect the germplasm of indigenous plant species.

5.7 CONCLUSION AND FUTURE REMARKS

Spices and condiments are sources of phytochemicals with specific health advantages. As food additives, spices are used singly or in combination to provide flavour, colour, and scent. Many spices have been shown to have medicinal benefits for treating wounds, coughs and colds, fever, hyperglycaemia, and hyperlipidaemia, according to traditional knowledge prevalent in nations like India. It can be concluded that species from the Lamiaceae family are mostly used as spices and condiments followed by the families Apiaceae and Zingiberaceae, where mostly fruits are used, followed by leaves. These spices confer a wide range of health advantages, according to evaluations of their nutritional and medicinal uses. Considering the immense health advantages of these spices, additional study is required to confirm their effectiveness in treating a variety of ailments, to discover novel ingredients with health benefits, and further refine existing spices. There is a need to increase the productivity to meet the domestic requirement and for export of valued spices and condiments. It has been observed that technological interventions and possible methods for yield increase and plant protection measures need to improve as a result of focused research. Reintroducing the germplasm from wild areas of the diversity centres of crops such as *Cinnamomum*, nutmeg, clove, allspice, and *Vanilla* will improve our genetic stocks for future utilization. Explorations to locate new spices

TABLE 5.1
List of Major Spices and Condiments Used by People and Their Ethnobotanical Use

Name of Family	Name of Species	Popular Name/ Trade Name	Part Used	Chemical Constituents	Ethnobotanical Uses	Key References
Acoraceae	*Acorus calamus* L.	Sweet flag, myrtie flag, calamus, flag root	Rhizome	Alpha-asarone, β-asarone	Thermogenic, anti-constipation, emmenagogue, intellect promoting, emetic, carminative, stomachic, expectorant, sudoriphic, antipyretic, resusticative, tranquilising, sedative, nervine tonic.	Yadav et al., 2019; Spice Borad of India, 2019
Amaryllidaceae	*Allium ascalonicum* L.	Shallot	Bulb	Dipropyl trisulphide, propylmethyl trisulphide	Boils, dysentery	Bastaki et al., 2021
Amaryllidaceae	*Allium cepa* L.	Onion	Bulb	Quercetin, apigenin, dipyridyl disulphide, rutin, quercetin-4-glucoside	Earache, boils	Alam et al., 2022; Salam & Jamir, 2016
Amaryllidaceae	*Allium cepa* var. *aggregatum*	Potato onion	Bulb	Quercetin, apigenin, dipyridyl disuphfide, rutin, quercetin-4-glucoside	Earache, boils	Alam et al., 2022
Amaryllidaceae	*Allium fistulosum* L.	Japanese bunching onion, stony leek	Bulb, leaves	Quercetin, apigenin, dipyridyl disuphide, rutin, quercetin-4-glucoside	Common cold, influenza, arthritis, headache, abdominal pain, constipation, dysentery	Alam et al., 2022
Amaryllidaceae	*Allium sativum* L.	Garlic	Bulb	Allicin, allyl alcohols	Anticholesterol, antifungal, tonic, rubefacient, stimulant, thermogenic, aphrodisiac, used to treat coughs, asthma, cardiopathy	Alam et al., 2022; Spice Board of India, 2019
Amaryllidaceae	*Allium tuberosum* Rottler	Indian leek	Bulb	Dipropyl trisulphide, Propylmethyl trisulphide	Urinary tract stones	Bastaki et al., 2021
Anacardiaceae	*Schinus molle* L.	Californian pepper tree	Fruit	α-Phellandrene, α- pinene, β-myrcene	Anti-depressant, anti-spasmodic, astringent, purgative, healing, stimulant, tonic and for treating blennorrhagia, bronchitis,	Abrha & Unnithan, 2014

(Continued)

TABLE 5.1 (CONTINUED)
List of Major Spices and Condiments Used by People and Their Ethnobotanical Use

Name of Family	Name of Species	Popular Name/ Trade Name	Part Used	Chemical Constituents	Ethnobotanical Uses	Key References
Apiaceae	*Anethum graveolens* L.	Dill	Leaves	d-Carvone	Carminative, stomachic, antipyretic	Ramkumar & Karuppusamy, 2021; Spice Board of India, 2019
Apiaceae	*Anethum sowa* Kurz.	Indian dill	Fruit	d-Carvone	Carminative, stomachic, antipyretic	Ramkumar & Karuppusamy, 2021; Spice Board of India, 2019
Apiaceae	*Apium graveolens* L.	Celery	Fruit, bulb	phenol, toluene, butylated hydroxyl toluene	Stimulant, tonic, diuretic, carminative, emmenagogue, anti-inflammatory	Adelakun et al., 2021; Spice Board of India, 2019
Apiaceae	*Bunium persicum* (Boiss.) Fedtsch.	Black caraway	Fruit, bulb	d-Carvone, carvone derivatives	Stomachic, carminative, anthelmintic, lactagogue, adjuvant/ corrective for treating nausea	Ramkumar & Karuppusamy, 2021; Spice Board of India, 2019
Apiaceae	*Carum bulbocastanum* L.	Black caraway	Fruit, bulb	d-Carvone, carvone derivatives	Stomachic, carminative, anthelmintic, lactagogue, adjuvant/ corrective for treating nausea	Ramkumar & Karuppusamy, 2021; Spice Board of India, 2019
Apiaceae	*Coriandrum sativum* L.	Coriander	Leaves, fruits	d-Linalool, C10–C14-2-alkenals	Carminative, diuretic, tonic, stimulant, stomachic, refrigerant, aphrodisiac, analgesic, anti-inflammatory	Ramkumar & Karuppusamy, 2021; Spice Board of India, 2019
Apiaceae	*Carum copticum* L.	Ajowan caraway, blond caraway	Fruits	d-Carvone, carvone derivatives	Digestive, antispasmodic, stimulant, carminative, expectorant	Ramkumar & Karuppusamy, 2021; Spice Board of India, 2019

(*Continued*)

TABLE 5.1 (CONTINUED)
List of Major Spices and Condiments Used by People and Their Ethnobotanical Use

Name of Family	Name of Species	Popular Name/ Trade Name	Part Used	Chemical Constituents	Ethnobotanical Uses	Key References
Apiaceae	*Cuminum cyminum* L.	Cumin	Fruits	Cuminaldehyde, *p*-1,3-mentha-dienal	Digestive, carminative, astringent, anti-inflammatory, anticonstipation, diuretic, revulsive, galactogogue	Ramkumar & Karuppusamy, 2021; Spice Board of India, 2019
Apiaceae	*Ferula assa- foetida* L.	Asafoetida	Rhizome exudates	2-butyl-1-propenyl disulphide, (1-methylthiopropyl) 1-propenyl, di-and 2-butyl-3-methylthioallyl disulphide	Asthma, gastrointestinal disorders, intestinal parasites	Iranshahy & Iranshahi, 2011
Apiaceae	*Foeniculum vulgare* Miller.	Fennel	Seeds	(E)-Anethole, fenchone	Stimulant, carminative, stomachic, emmenagogue, refrigerant, cardiac stimulant, anti-emetic, aphrodisiac, anthelmintic	Ramkumar & Karuppusamy, 2021; Spice Board of India, 2019
Apiaceae	*Pimpinella anisum* L.	Aniseed	Fruits	(E)-Anethole, methyl chavicol	Mild expectorant, stimulant, carminative, diuretic, diaphoretic	Ramkumar & Karuppusamy, 2021; Spice Board of India, 2019
Asteraceae	*Artemisia dracunculus* L.	Tarragon	Leaves	Methyl chavicol, anethole	Aperient, stomachic, stimulant, febrifuge	Ramkumar & Karuppusamy, 2021; Spice Board of India, 2019
Brassicaceae	*Armoracia rusticana* Gaertn.	Horse radish	Root	Sinigrin, sinigrin- derived allyl isothiocyanate and diallylsulphide	Thermogenic, appetizing, digestive, stomachic, laxative, anti-inflammatory, anodyne, refreshing, antibacterial	Yasmeen et al., 2020; Spice Board of India, 2019

(*Continued*)

TABLE 5.1 (CONTINUED)
List of Major Spices and Condiments Used by People and Their Ethnobotanical Use

Name of Family	Name of Species	Popular Name/ Trade Name	Part Used	Chemical Constituents	Ethnobotanical Uses	Key References
Brassicaceae	*Brassica juncea* (L.) Czernj. et Cosson.	Mustard	Seeds	Allyl isothiocyanate	Thermogenic, anodyne, anti-inflammatory, carminative, digestive, anthelmintic, sudoriphic, tonic, emetic, used to treat vomiting, abdominal colic, dyspepsia, flatulence, skin diseases	Ramkumar & Karuppusamy, 2021; Spice Board of India, 2019 (Figure 5.1 A)
Brassicaceae	*Brassica nigra* (L.) Koch.	Black mustard	Seeds	Allyl isothiocyanate	Thermogenic, anodyne, anti inflammatory, carminative, digestive, anthelmintic, sudoriphic, tonic, emetic, used to treat vomiting, abdominal colic, dyspepsia, flatulence, skin diseases	Ramkumar & Karuppusamy, 2021; Spice Board of India, 2019
Brassicaceae	*Brassica rapa* (L.) Hanelt.	Field mustard	Seed	Allyl isothiocyanate	Thermogenic, anodyne, anti inflammatory, carminative, digestive, anthelmintic, sudoriphic, tonic, emetic, used to treat vomiting, abdominal colic, dyspepsia, flatulence, skin diseases	Ramkumar & Karuppusamy, 2021; Spice Board of India, 2019
Brassicaceae	*Sinapis alba* L.	White mustard	Seeds	Allyl isothiocyanate	Thermogenic, anodyne, anti-inflammatory, carminative, digestive, anthelmintic, sudoriphic, tonic, emetic, used to treat vomiting, abdominal colic, dyspepsia, flatulence, skin diseases	Ramkumar & Karuppusamy, 2021; Spice Board of India, 2019
Caesalpiniaceae	*Tamarindus indica* L.	Tamarind	Fruit	Phenolic compounds, cardiac, malic acid, uronic acid, tartaric acid,	Refrigerant, digestive, carminative, laxative, antiscerbutic, febrifuge, ophthalmic, useful in gastropathy, to treat datura poisoning, alcoholic intoxication, scabies, constipation.	Pal & Mukherjee, 2020; Spice Board of India, 2019

(Continued)

TABLE 5.1 (CONTINUED)
List of Major Spices and Condiments Used by People and Their Ethnobotanical Use

Name of Family	Name of Species	Popular Name/ Trade Name	Part Used	Chemical Constituents	Ethnobotanical Uses	Key References
Capparidaceae	*Capparis spinosa* L.	Caper bud	Flower bud	Rutin, palmitic acid	Diuretic, aspirant, expectorant, emmenagogue, tonic and in scurvy	Mollica et al., 2019; Spice Board of India, 2019 (Figure 5.1B)
Clusiaceae	*Garcinia gummi-gutta* (L.) Robson.	Malabar tamarind, camboge	Fruit	Garcinol, isogarcinol, xanthones	Astringent, digestive, thermogenic, to treat constipation, haemorrhoids, diarrhoea	Anju & Rameshkumar, 2016; Spice Board of India, 2019
Clusiaceae	*Garcinia indica* (Thouars) Choisy.	Kokam	Fruit	Hydroxy citric acid, garcinol	Cooling, anthelmintic, cardiotonic, astringent, emollient, useful in treating piles, dysentery, heart complaints, bilious affections	Ananthakrishnan & Rameshkumar, 2016; Spice Board of India, 2019
Cupressaceae	*Juniperus communis* L.	Juniper berry (technically, seed cone)	Fruit	α-Pinene, β-pinene, apigenin, sabinene, β-sitosterol, campesterol	Carminative, stimulant, diuretic, to treat dropsy, leucorrhoea, urino-genital disorders.	Bais et al., 2014; Spice Board of India, 2019
Fabaceae	*Trigonella foenum-graecum* L.	Fenugreek	Leaves, seeds	Trigonelline, linoleic, palmitic acids	Carminative, tonic, aphrodisiac, emollient, antibacterial, used to treat vomiting, fever, anorexia, colonitis.	Divakaran et al., 2018; Spice Board of India, 2019
Illiciaceae	*Illicium verum* Hook.f.	Star anise	Fruit	Anethole, α-pinene, phellandrene, *p*-cymene, limonene	Astringent, carminative, deodorant, expectorant, digestive	D'Souza et al., 2017; Spice Board of India, 2019
Iridaceae	*Crocus sativus* L.	Saffron	Stigma, style	Safranal	Stimulant, tonic, stomachic, aphrodisiac, anodyne, anti spasmodic, emmenagogue, diuretic, laxative, used to treat bronchitis, fever, epilepsy, skin diseases, decolouration of skin	Ramkumar & Karuppusamy, 2021; Spice Board of India, 2019
Lamiaceae	*Hyssopus officinalis* L.	Hyssop	Leaves	Pinocamphone, isopinocamphone	Stimulant, carminative, pectoral, used to treat nervous disorders, toothache, pulmonary and uterine troubles.	Baj et al., 2018; Spice Board of India, 2019

(Continued)

TABLE 5.1 (CONTINUED)
List of Major Spices and Condiments Used by People and Their Ethnobotanical Use

Name of Family	Name of Species	Popular Name/ Trade Name	Part Used	Chemical Constituents	Ethnobotanical Uses	Key References
Lamiaceae	*Lippia graveolens* Kunth.	Mexican oregano	Leaves, terminal shoot	Carvacrol, thymol	Stimulant, carminative, stomachic, diuretic, diaphoretic and emmenagogue	Ramkumar & Karuppusamy, 2021; Spice Board of India, 2019
Lamiaceae	*Mentha arvensis* L.	Mint, field mint	Leaves	1-Menthol, menthone,	Stimulant, stomachic, carminative, antiseptic, digestive, antispasmodic, contraceptive, used to treat vomiting, skin diseases, amenorrhoea, dental caries	Ramkumar & Karuppusamy, 2021; Spice Board of India, 2019
Lamiaceae	*Mentha piperita* L.	Pepper mint	Leaves	1-Menthol, menthone, menthofuran	Stimulant, stomachic, carminative, antiseptic, digestive	Ramkumar & Karuppusamy, 2021; Spice Board of India, 2019
Lamiaceae	*Ocimum basilicum* L.	Sweet basil	Leaves	Methyl chavicol, linalool, methyl eugenol	Stomachic, anthelmintic, diaphoretic, expectorant, antipyretic, carminative	Ramkumar & Karuppusamy, 2021; Spice Board of India, 2019
Lamiaceae	*Ocimum sanctum* L.	Holy basil	Leaves	Oleanolic acid, ursolic acid, rosmarinic acid, eugenol, carvacrol, linalool,	Stomachic, anthelmintic, diaphoretic, expectorant, antipyretic, carminative,	Ramkumar & Karuppusamy, 2021; Spice Board of India, 2019
Lamiaceae	*Origanum hortensis* L.	Marjoram	Leafy shoots	E & T sabinene hydrates, 1-terpinen-4-ol	Carminative, expectorant, tonic, astringent	Ramkumar & Karuppusamy, 2021; Spice Board of India, 2019
Lamiaceae	*Origanum vulgare* L.	Oregano	Leaves	Carvacrol, Thymol	Stimulant, carminative, stomachic, diuretic, diaphoretic and emmenagogue	Ramkumar & Karuppusamy, 2021; Spice Board of India, 2019

(Continued)

TABLE 5.1 (CONTINUED)
List of Major Spices and Condiments Used by People and Their Ethnobotanical Use

Name of Family	Name of Species	Popular Name/ Trade Name	Part Used	Chemical Constituents	Ethnobotanical Uses	Key References
Lamiaceae	*Petroselinum crispum* (Miller) Nyman.	Parsley	Leaves	Apiole	Stimulant, diuretic, carminative, emmenagogue, antipyretic, anti- inflammatory, emetic, aphrodisiac, alexipharmic, refrigerant	Ramkumar & Karuppusamy, 2021; Spice Board of India, 2019
Lamiaceae	*Rosmarinus officinalis* L.	Rosemary	Leaves	Verbenone, 1-8-cineole, camphor, linanool	Astringent, nervine tonic, stomachic, antibacterial, protistocidal, rubefacient	Ramkumar & Karuppusamy, 2021; Spice Board of India, 2019
Lamiaceae	*Salvia officinalis* L.	Garden sage, Dalmatian sage	Leaves	Salvial-4 (14)-en-1-one, linalool	Mild tonic, astringent, carminative, deodorant, insecticidal,, antipyretic, used to treat gingivitis, dentifice, as mouthwash, gargles	Ramkumar & Karuppusamy, 2021; Spice Board of India, 2019
Lamiaceae	*Satureja hortensis* L.	Summer savory	Leaves	Carvacrol	Antispasmodic, anticatarrhal, astringent, carminative, laxative, diuretic, stomachic, sudoriphic, vermifuge	Ramkumar & Karuppusamy, 2021; Spice Board of India, 2019
Lamiaceae	*Thymus vulgaris* L.	Thyme	Leaves	Thymol, carvacrol	Antispasmodic, carminative, emmenagogue, anthelmintic, spasmodic, laxative, stomachic, tonic, vermifuge	Ramkumar & Karuppusamy, 2021; Spice Board of India, 2019
Lauraceae	*Cinnamomum burmanii* Blume.	Indonesian cassia	Bark	Cinnamaldehyde, eugenol	Astringent, stimulant, carminative, germicidal and for treating nausea and vomiting	Ramkumar & Karuppusamy, 2021; Spice Board of India, 2019
Lauraceae	*Cinnamomum cassia* L.	Cinnamon cassia	Bark	Cinnamaldehyde, eugenol	Astringent, stimulant, carminative, germicidal and for treating nausea and vomiting	Ramkumar & Karuppusamy, 2021; Spice Board of India, 2019

(*Continued*)

TABLE 5.1 (CONTINUED)
List of Major Spices and Condiments Used by People and Their Ethnobotanical Use

Name of Family	Name of Species	Popular Name/ Trade Name	Part Used	Chemical Constituents	Ethnobotanical Uses	Key References
Lauraceae	*Cinnamomum loureiroi* Nees.	Vietnamese cassia bark	Bark	cinnamate, cinnomyl-acetate	Astringent, stimulant, carminative, germicidal and for treating nausea and vomiting	Ramkumar & Karuppusamy, 2021; Spice Board of India, 2019
Lauraceae	*Cinnamomum tamala* Nees.	Indian cassia bark	Leaves	cinnamaldehyde, cinnamate, cinnomyl-acetate	Astringent, stimulant, carminative, germicidal and for treating nausea and vomiting	Ramkumar & Karuppusamy, 2021; Spice Board of India, 2019
Lauraceae	*Cinnamomum verum* Blume.	Sri Lankan cinnamon, common cinnamon	Leaves	cinnamaldehyde, cinnamate, cinnomyl-acetate, β-caryophyllene	Astringent, diuretic, carminative, aphrodisiac, deodorant, expectorant, febrifuge, stomachic	Ramkumar & Karuppusamy, 2021; Spice Board of India, 2019 (Figure 5.1C)
Lauraceae	*Laurus nobilis* L.	Bay leaf, true laurel	Leaves	1,8-Cineole	Stimulant in sprains, narcotic and in veterinary medicine	Ramkumar & Karuppusamy, 2021; Spice Board of India, 2019
Myristicaceae	*Myristica argentea* Warb.	Papuan nutmeg	Kernel and mace	Sabinene, α-pinene, myristicin	anticonvulsant, antiseptic, astringent, sweat, thermogenic	Ramkumar & Karuppusamy, 2021; Spice Board of India, 2019
Myristicaceae	*Myristica fragrans* Houtt.	Nutmeg	Kernel and mace	Sabinene, α-pinene, myristicin	Astringent, sweat, thermogenic, aphrodisiac, anti- inflammatory, anodyne, deodorant, digestive, expectorant, narcotic, anticonvulsant, antiseptic, constipating	Ramkumar & Karuppusamy, 2021; Spice Board of India, 2019

(*Continued*)

TABLE 5.1 (CONTINUED)
List of Major Spices and Condiments Used by People and Their Ethnobotanical Use

Name of Family	Name of Species	Popular Name/ Trade Name	Part Used	Chemical Constituents	Ethnobotanical Uses	Key References
Myrtaceae	*Syzygium aromaticum* Thunb.	Clove	Flower bud	Eugenol, eugenyl acetate	Refrigerant, ophthalmic, digestive, carminative, stomachic, stimulant, antispasmodic, antibacterial, expectorant, rubefacient, aphrodisiac, appetize enhancer, emollient	Ramkumar & Karuppusamy, 2021; Spice Board of India, 2019
Myrtaceae	*Pimenta dioica* (L.) Merr.	Allspice	Leaves	Eugenol, caryophyllene	Stimulant, digestive, carminative, anodyne against rheumatism and neuralgia	Ramkumar & Karuppusamy, 2021; Spice Board of India, 2019
Myrtaceae	*Pimenta racemosa* (Miller) J. Moore.	West Indian bay	Leaves	Geranial, methylchavicol/ methyleugenol	Carminative, used to treat colic, diarrhoea	Ramkumar & Karuppusamy, 2021; Spice Board of India, 2019
Orchidaceae	*Vanilla planifolia* Andrews.	Vanilla	Fruit	Vanillin, *p*-OH-benzyl-methyl ether	Aphrodisiac	Ramkumar & Karuppusamy, 2021; Spice Board of India, 2019
Orchidaceae	*Vanilla tahitensis* J. Moore.	Tahitian vanilla	Fruit	Vanillin, *p*-OH-benzyl-methyl ether	Aphrodisiac	Ramkumar & Karuppusamy, 2021; Spice Board of India, 2019
Orchidaceae	*Vanilla pompona* Schiede.	Pompona vanilla	Fruit	Vanillin, *p*-OH-benzyl-methyl ether	Aphrodisiac	Ramkumar & Karuppusamy, 2021; Spice Board of India, 2019
Papaveraceae	*Papaver somniferum* L.	Poppy, blue maw	Seeds	Morphine, codeine, thebaine, porphyroxine	Expectorant, sudoriphic, sedative, nervine tonic, constipating, aphrodisiac	Masihuddin et al., 2018; Spice Board of India, 2019

(*Continued*)

TABLE 5.1 (CONTINUED)
List of Major Spices and Condiments Used by People and Their Ethnobotanical Use

Name of Family	Name of Species	Popular Name/ Trade Name	Part Used	Chemical Constituents	Ethnobotanical Uses	Key References
Piperaceae	*Piper guineense* Schum.	West African pepper	Fruits	Piperine	Purgative, stomachic, digestive, emollient	Zaveri et al., 2010; Spice Board of India, 2019
Piperaceae	*Piper longum* L.	Long pepper	Fruits	Piperine	Expectorant, thermogenic, diuretic, tonic, purgative, stomachic, digestive, emollient, antiseptic	Zaveri et al., 2010; Spice Board of India, 2019 (Figure 5.1E)
Piperaceae	*Piper mullesua* Buch.-Ham. ex D.Don.	Wild pepper	Fruits	Piperine	Anthelmintic, carminative, alterant, antiperiodic, diuretic, digestive, emmenagogue, rubefacient, stimulant, stomachic, used to treat fever, asthma, cough, dyspepsia, flatulence, arthritis	Zaveri et al., 2010; Spice Board of India, 2019
Piperaceae	*Piper nigrum* L.	Black pepper	Fruits	Piperine, S-3-carene, 5-caryophyllene	Anthelmintic, carminative, alterant, antiperiodic, diuretic, digestive, emmenagogue, rubefacient, stimulant, stomachic, used to treat fever, asthma, cough, dyspepsia, flatulence, arthritis	Ramkumar & Karuppusamy, 2021; Spice Board of India, 2019 (Figure 5.1G)
Punicaceae	*Punica granatum* L.	Pomegranate	Seeds	Ellagic acid ellagitannins), punicic acid, flavonoids, anthocyanidins, anthocyanins	Astringent, cooling, tonic, aphrodisiac, laxative, diuretic, cardiotonic, used to treat pectoral diseases	Garachh et al., 2012; Spice Board of India, 2019 (Figure 5.1D)
Ranunculaceae	*Nigella damascena* L.	Damas black cumin	Seeds	Pinene, cuminaldehyde, *p*-l,3-mentha-dienal	Liver diseases, the expulsion of gases, diarrhoea and the promotion of menstruation in women	Ramkumar & Karuppusamy, 2021; Spice Board of India, 2019
Ranunculaceae	*Nigella sativa* L.	Black cumin	Seeds	Pinene, cuminaldehyde, *p*-l,3-mentha-dienal	Liver diseases, the expulsion of gases, diarrhoea and the promotion of menstruation in women	Ramkumar & Karuppusamy, 2021; Spice Board of India, 2019

(*Continued*)

TABLE 5.1 (CONTINUED)
List of Major Spices and Condiments Used by People and Their Ethnobotanical Use

Name of Family	Name of Species	Popular Name/ Trade Name	Part Used	Chemical Constituents	Ethnobotanical Uses	Key References
Rutaceae	*Murraya koenigii* (L.) Sprengel.	Curry leaf	Leaves	*cis*-Ocimene, β-caryophyllene	Astringent, anthelmintic, febrifuge, stomachic, appetizing, carminative, constipating, anti inflammatory, antiseptic	Rana et al., 2004; Spice Board of India, 2019
Rutaceae	*Zanthoxylum armatum* DC.	Mountain prickle pepper	Fruit	Linalool, limonene, β-fagarine, magnoflorine, nitidine	stomachic, carminative and anthelmintic	Verma et al., 2021; Okagu et al., 2021
Rutaceae	*Zanthoxylum bungee* Planch.	Chinese prickly ash pepper	Fruit	Nitidine, skimminiadine, hyperoside	Sinusitis	Okagu et al., 2021; Salam & Jamir, 2016
Solanaceae	*Capsicum annuum* L.	Wild chilli, capsicum	Fruit	Capsaicin, dihydrocapsaicin	Digestive, thermogenic, carminative, stimulant, cardiotonic, antipyretic, serdorific, rubefacient and sialagogue.	Ramkumar & Karuppusamy, 2021; Spice Board of India, 2019
Solanaceae	*Capsicum frutescens* L.	Bird eye chilli	Fruit	Capsaicin, dihydrocapsaicin	Digestive, thermogenic, carminative, stimulant, cardiotonic, antipyretic, serdorific, rubefacient, sialagogue.	Ramkumar & Karuppusamy, 2021; Spice Board of India, 2019
Zingiberaceae	*Aframomum angustifolium* Schum.	Madagascar cardamom	Fruit, seeds	α-Terpinyl acetate, 1-8-cineole, linalool	Hypnotic, appetite stimulant, astringent, bowels, tonic to heart, liver	Ramkumar & Karuppusamy, 2021; Spice Board of India, 2019
Zingiberaceae	*Aframomum hanburyi* Schum.	Cameroon cardamom	Fruit, seeds	α-Terpinyl acetate, 1-8-cineole, linalool	Stimulant, tonic, diuretic, carminative, digestive, expectorant, cardiotonic	Ramkumar & Karuppusamy, 2021; Spice Board of India, 2019
Zingiberaceae	*Aframomum corrorim* Engl.	Korarima cardamom	Fruit, Seeds	α-Terpinyl acetate, 1-8-cineole, linalool	Carminative, tonic, headache, stomachache, sore throat	Ramkumar & Karuppusamy, 2021; Spice Board of India, 2019

(Continued)

TABLE 5.1 (CONTINUED)
List of Major Spices and Condiments Used by People and Their Ethnobotanical Use

Name of Family	Name of Species	Popular Name/ Trade Name	Part Used	Chemical Constituents	Ethnobotanical Uses	Key References
Zingiberaceae	*Alpinia calcarata* Roscoe.	Galangal	Rhizome	Hydroquinone, pyranone	Asthma digestive, vulnerary, febrifuge, stimulant, depurative, used to treat skin diseases, rheumatism	Singh et al., 2020; Spice Board of India, 2019
Zingiberaceae	*Alpinia galanga* (L.) Willd.	Greater galangal	Rhizome	Carotol, eucalyptol, 5-hydroxymethylfurfural	Carminative, expectorant, digestive, vulnerary, febrifuge, stimulant, depurative, used to treat skin diseases, rheumatism, asthma, wounds, fever, haemorrhoids.	Singh et al., 2020; Spice Board of India, 2019
Zingiberaceae	*Curcuma amada* Roxb.	Mango ginger	Rhizome	Curcumin, demethoxycurcumin, bisdemethoxycurcumin, ferulic acid, gallic acid	Digestive, carminative, emollient, appetizer, stomachic, rubefacient, anodyne, expectorant, anthelmintic, stimulant	Mahadevi & Kavitha, 2020; Spice Board of India, 2019
Zingiberaceae	*Curcuma longa* L.	Turmeric	Rhizome	Turmerone, zingiberene, 1,8-cineole	Thermogenic, emollient, anodyne, anti- inflammatory, vulnerary, depurative, antiseptic, appetite stimulant, carminative, expectorant, stomachic, anthelmintic, stimulant, ophthalmic, tonic, used to treat skin diseases, dyspepsia, asthma, coughs, bronchitis, inflammations, ulcers, worms, skin discolouration	Ramkumar & Karuppusamy, 2021; Spice Board of India, 2019
Zingiberaceae	*Zingiber montanum* (Koenig.) Link.	Wild ginger, mountain ginger	Rhizome	Gingerol, shogaol, neral, geranial	Digestive, carminative, emollient, appetite stimulant, stomachic, rubefacient, anodyne, expectorant, anthelmintic, stimulant	Ramkumar & Karuppusamy, 2021; Spice Board of India, 2019 (Figure 5.1F)
Zingiberaceae	*Zingiber officinale* Roscoe.	Ginger	Rhizome	Gingerol, shogaol, neral, geranial	Digestive, carminative, emollient, appetizer, stomachic, rubefacient, anodyne, expectorantanthelmintc, stimulant.	Ramkumar & Karuppusamy, 2021; Spice Board of India, 2019

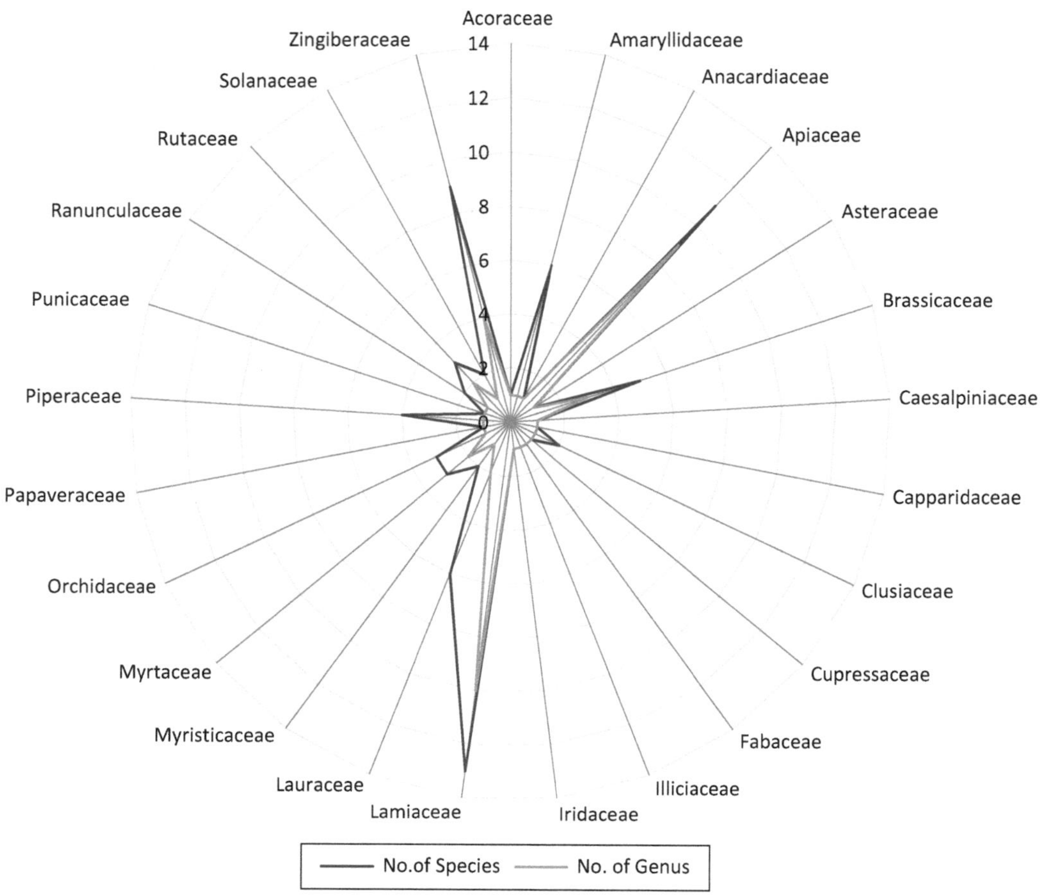

FIGURE 5.3 Family-wise distribution of major spices and condiments.

and condiments from wild localities need to be carried out to help in diversifying Indian spice production. To supplement the climate-dependent high-yielding spice varieties, new improved spice crops suited for Indian climates need to be developed through modern research methods. High-yielding hybrids, sustainable utilization of exotic wild species, and advanced farming technologies are required to achieve these ends.

ACKNOWLEDGEMENTS

The authors are greatly indebted to Shri. Anoop Singh IFS, Director General, Forest Survey of India, Dehradun and Shri. Chaturbhuja Behera IFS, Regional Director, Forest Survey of India Central Zone, Nagpur for their support and the use of facilities.

REFERENCES

Abiola, M., Asowata, A., Anthony, J. A. & Gloria, A. O. 2016. Ethnobotanical survey of culinary herbs and spices used in the traditional medicinal system of Nkonkobe Municipality, Eastern Cape, South Africa. *South African Journal of Botany* 104: 69–75.

Abrha, L. H. & Unnithan, C. R. 2014. Chemical composition and antibacterial activity of essential oil of *Schinus molle. Unique Journal of Pharmaceutical and Biological Sciences* 2(1): 9–12.

Adelakun, E. A., Achuenu, C., Ngurukwem, E. C. & Chuwang, D. J. 2021. Gas chromatography mass spectrometry Analysis of phthalide composition of celery oil from *Apium graveolens* L. cultivated in Nigeria. *Journal of Chemical Society of Nigeria* 46(2): 245–254.
Alam, A., Al Arif Jahan, A., Bari, M. S., Khandokar, L., Mahmud, M. H., Junaid, M. & Haque, M. A. 2022. Allium vegetables: Traditional uses, phyto constituents, and beneficial effects in inflammation and cancer. *Critical Reviews in Food Science and Nutrition*: 63(23): 6580–6614.
Ali, S. S., Kasoju, N., Luthra, A., Singh, A., Sharanabasava, H., Sahu, A. & Bora, U. 2008. Indian medicinal herbs as sources of antioxidants. *Food Research International* 41(1): 1–15.
Ananthakrishnan, R. & Rameshkumar, K. B. 2016. Phytochemicals and bioactivities of *Garcinia indica* (Thouars) Choisy - A review. *Diversity of Garcinia Species in the Western Ghats: Phytochemical Perspective* 142–150.
Anju, V. & Rameshkumar, K. B. 2016. Phytochemicals and bioactivities of *Garcinia gummi-gutta* (L.) N. Robson – A review. *Diversity of Garcinia Species in the Western Ghats: Phytochemical Perspective* 151–161.
Ashokkumar, K., Pandian, A., Murugan, M., Dhanya, M. K., Sathyan, T., Sivakumar, P. & Warkentin, T. D. 2020. Profiling bioactive flavonoids and carotenoids in select south Indian spices and nuts. *Natural Product Research* 34(9): 1306–1310.
Babu, N., Srivastava, S. K. & Agarwal, S. 2013. Traditional storage practices of spices and condiments in Odisha. *Indian Journal of Traditional Knowledge* 12(3): 518–523.
Baj, T., Korona-Głowniak, I., Kowalski, R. & Malm, A. 2018. Chemical composition and microbiological evaluation of essential oil from *Hyssopus officinalis* L. with white and pink flowers. *Open Chemistry* 16(1): 317–323.
Bais, S., Gill, N. S., Rana, N. & Shandil, S. 2014. A phytopharmacological review on a medicinal plant: *Juniperus communis. International Scholarly Research Notices.*
Bastaki, S. M., Ojha, S., Kalasz, H. & Adeghate, E. 2021. Chemical constituents and medicinal properties of *Allium* species. *Molecular and Cellular Biochemistry* 476(12): 4301–4321.
Bharali, P., Sharma, M., Sharma, C. L. & Singh, B. 2017. Ethnobotanical survey of spices and condiments used by some tribes of Arunachal Pradesh. *Journal of Medicinal Plants Studies* 5(1): 101–109.
Clarke, T. C., Black, L. I., Stussman, B. J., Barnes, P. M. & Nahin, R. L. 2015. Trends in the use of complementary health approaches among adults: United States, 2002–2012. *National Health Statistics Reports.* Hyattsville, MD: National Center for Health Statistics.
D'Souza, S. P., Chavannavar, S. V., Kanchanashri, B. & Niveditha, S. B. 2017. Pharmaceutical perspectives of spices and condiments as alternative antimicrobial remedy. *Journal of Evidence-Based Complementary and Alternative Medicine* 22(4): 1002–1010.
Divakaran, M., Jayasree, E., Nirmal Babu, K. & Peter, K. V. 2018. Legacy of Indian spices: Its production and processing. *Indian Spices: The Legacy, Production and Processing of India's Treasured Export* 13–30.
Dubey, S. 2017. Indian spices and their medicinal value. *Indian Journal of Pharmaceutical Education and Research* 51(3): S330–332.
Duke, J. A. 2002. *Handbook of Medicinal Spices.* CRC Press. Boca Raton.
Eyob, S., Appelgren, M., Rohloff, J., Tsegaye, A. & Messele, G. 2008. Traditional medicinal uses and essential oil composition of leaves and rhizomes of korarima (*Aframomum corrorima* (Braun) PCM Jansen) from southern Ethiopia. *South African Journal of Botany* 74(2): 181–185.
FAO. 2005. *Spices and Essential Oils: Post-Harvest Operations in Developing Countries*, by Douglas. J. Heyes and B. Smallfield, Rome.
Garachh, D., Patel, A., Chakraborty, M. & Kamath, J. V. 2012. Phytochemical and pharmacological profile of *Punica granatum*: An overview. *International Research Journal of Pharmacy* 3(2): 65–68.
Garland, S. 1972. *The Herbs and Spices Book.* London: Frances Lincoln Publishers.
Grivetti, L. E. 2017. Herbs, spices, and flavoring agents Part 5: Fusion of herbs and spices in the 19th-century and into the 21st-century North American culture. *Nutrition Today* 52(1): 44–51.
Hong, L., Guo, Z., Huang, K., Wei, S., Liu, B., Meng, S. & Long, S. 2015. Ethnobotanical study on medicinal plants used by Maonan people in China. *Journal of Ethnobiology and Ethnomedicine* 11: 32–65.
Husen, A. 2020. *Exploring Medicinal Plants* (Book Series). Boca Raton, FL: CRC Press.
Husen, A. 2023. *Antidiabetic Medicinal Plants and Herbal Treatments.* Boca Raton, FL: CRC Press. https://doi.org/10.1201/b23347.
Iranshahy, M. & Iranshahi, M. 2011. Traditional uses, phytochemistry and pharmacology of asafoetida (*Ferula assa-foetida* oleo-gum-resin)-A review. *Journal of Ethnopharmacology* 134(1): 1–10.
Kaefer, C. M. & Milner, J. A. 2008. The role of herbs and spices in cancer prevention. *The Journal of Nutritional Biochemistry* 19: 347–361.

Mahadevi, R. & Kavitha, R. 2020. Phytochemical and pharmacological properties of *Curcuma amada*: A review. *International Journal of Research in Pharmaceutical Sciences* 11(3): 3546–3555.

Masihuddin, M., Jafri, M. A., Siddiqui, A. & Chaudhary, S. 2018. Traditional uses, phytochemistry and pharmacological activities of *Papaver somniferum* with special reference of unani medicine an updated review. *Journal of Drug Delivery and Therapeutics* 8(5): 110–114.

Mollica, A., Stefanucci, A., Macedonio, G., Locatelli, M., Luisi, G., Novellino, E. & Zengin, G. 2019. Chemical composition and biological activity of *Capparis spinosa* L. from Lipari Island. *South African Journal of Botany* 120: 135–140.

Naik, D. 2007. Spices need priority for its research and development in Orissa. In: *National Symposium on Spices and Aromatic Crops*, 75–82.

Nair, K. P. 2006. The agronomy and economy of cardamom (*Elettaria cardamomum* Maton): The "queen of spices". *Advances in Agronomy* 91: 179–471.

Nurdjannah, N. & Bermawie, N. 2012. *Handbook of Herbs and Spices* (2nd edition). Indonesia Agency for Agriculture, 197–211.

Okagu, I. U., Ndefo, J. C., Aham, E. C. & Udenigwe, C. C. 2021. Zanthoxylum species: A review of traditional uses, phytochemistry and pharmacology in relation to cancer, infectious diseases and sickle cell anemia. *Frontiers in Pharmacology* 12: 713090.

Pal, D. & Mukherjee, S. 2020. Tamarind (*Tamarindus indica*) seeds in health and nutrition. In: Victor R. Preedy & Ronald Ross Watson, *Nuts and Seeds in Health and Disease Prevention*, pp. 171–182. Academic Press, United Kingdom.

Parthasarathy, V. A. & Kandiannan, K. 2007. *Horticulture: Vegetable Science: Spices and Condiments.* Indian Institute of Spice Research (ICAR), Calicut, India.

Pedapati, A., Tyagi, V., Singh, A. K., Verma, N., Yadav, S. K., Singh, S. P., Singh, S. & Pratibha, B. 2014. Spices and condiments: Status of genetic resources and setting priorities for introduction in India. *Environment & Ecology* 32: 1051–1056.

Phuyal, N., Jha, P. K., Raturi, P. P. & Rajbhandary, S. 2019. *Zanthoxylum armatum* DC: Current knowledge, gaps and opportunities in Nepal. *Journal of Ethnopharmacology* 229: 326–341.

Rahman, S. & Husen, A. 2022. *Asparagus racemosus* (Shatavari) and *Dioscorea villosa* (Wild yams). In: *Herbs, Shrubs, and Trees of Potential Medicinal Benefits* (Ed. Husen, A.). Boca Raton, FL: CRC Press.

Rahman, S. & Husen, A. 2021. Potential role of medicinal plants in the cure of liver and kidney diseases. In: *Non-timber Forest Products* (Ed. Husen, A., Bachheti, R. K., Bachheti, A.). Cham: Springer International Publishing AG, pp. 229–254. https://doi.org/10.1007/978-3-030-73077-210.

Rahman, S. & Husen, A. 2023a. *Trigonella foenum-graecum* L.: A useful medicinal plant for diabetes and other diseases. In: *Antidiabetic Medicinal Plants and Herbal Treatments* (Ed. Husen, A.). Boca Raton, FL: CRC Press.

Rahman, S. & Husen, A. 2023b. Role of green chiretta [*Andrographis paniculata* (Burm. F.) Nees in diabetes and other ailments. In: *Antidiabetic Medicinal Plants and Herbal Treatments* (Ed. Husen, A.). Boca Raton, FL: CRC Press.

Rai, P. C., Sarkar, A., Bhujel, R. B. & Das, A. P. 1998. Ethnomedicinal studies in some fringe areas of Sikkim and Darjeeling Himalaya. *Journal of Hill Research* 11(1): 1–19.

Ramkumar, T. R. & Karuppusamy, S. 2021. Plant diversity and ethnobotanical knowledge of spices and condiments. In: S.K. Upadhyay & S.P. Singh, *Bioprospecting of Plant Biodiversity for Industrial Molecules*, 231–260. John Wiley & Sons Ltd, UK.

Rana, V. S., Juyal, J. P. & Blazquez, M. A. 2004. Chemical constituents of the volatile oil of *Murraya koenigii* leaves. *International Journal of Aromatherapy* 14(1): 23–25.

Rathore, M. S. & Shekhawat, N. S. 2008. Incredible spices of India: from traditions to cuisine. *American-Eurasian Journal of Botany* 1(3): 85–89.

Ray, D. P. 2007. Spices cultivation in eastern and north eastern regions-A perspective. In: *Souvenir & Abstracts of National Symposium on Spices and Aromatic Crops, Bhubaneshwar*, pp. 25–26.

Reddy, K. N., Pattanaik, C., Reddy, C. S. & Raju, V. S. 2007. Traditional knowledge on wild food plants in Andhra Pradesh. *Indian Journal of Traditional Knowledge* 6(1): 223–229.

Sabu, M. 2006. *Zingiberaceae and Costaceae of South India.* Calicut, Kerala: Indian Association for Angiosperm Taxonomy.

Salam, S. & Jamir, N. S. 2016. Common spices plant used as medicine by the Tangkhul tribe of Ukhrul district, Manipur, India. *International Journal of Scientific and Research Publications* 6(7): 22–25.

Sharma, P. P. & Mujumdar, A. M. 2003. Traditional knowledge on plants from Toranmal Plateau of Maharashtra. *Indian Journal of Traditional Knowledge* 2(3): 292–296.

Singh, A. G. 2017. Ethnomedicinally important plants used as spices and condiments in the rupandehi district, West Nepal. *Life Sciences Leaflets* 85: 64–71.

Singh, A. K., Verma, N., Tyagi, V. & Dimree, S. 2011. Indian needs of crop genetic resources setting priority. *Progressive Agriculture* 11(1): 1–16.

Singh, S., Sahoo, B. C., Kar, S. K., Sahoo, A., Nayak, S., Kar, B. & Sahoo, S. 2020. Chemical constituents analysis of *Alpinia galanga* and *Alpinia calcarata*. *Research Journal of Pharmacy and Technology* 13(10): 4735–4739.

Singh, V. B. & Singh, K. 1996. *Spices*. New Delhi. India: New Age International Ltd.

Spice Board of India 2019. Spice Catalogue-https://www.indianspices.com/spices-development/spice-catalogue.html, Accessed on 02.11.2022.

Sugasini, D., Yalagala, P. C., Kavitha, B., Kasthuri, T., Vijayalakshmi, Y., Kumar, P. K. & Kumar, S. 2018. Indian culinary ethnic spices uses in foods are palate of paradise. *Acta Scientific Nutritional Health* 2(8): 22–28.

Tayade, S. K. & Patil, D. A. 2010. Ethnomedicinal applications of spices and condiments in Nandurbar district (Maharashtra). *Journal of Ecobiotechnology* 2(8): 8–10.

USDHHS & USDA. 2015. *U.S. Department of Health and Human Services and U.S. Department of Agriculture. 2015–2020: Dietary Guidelines for Americans* (8th edition).

United States Department of Agriculture. 2021. Food availability (per capita) data system. *Spices: Supply and Use*. https://www.ers.usda.gov/data-products/food-availability-per-capita-data-system/ Accessed January 2021.

Verma, K. K., Kumar, B., Raj, H. & Sharma, A. 2021. A review on chemical constituents, traditional uses, pharmacological studies of *Zanthoxylum armatum* (Rutaceae). *Journal of Drug Delivery and Therapeutics* 11(2): 136–142.

Yadav, D., Srivastava, S. & Tripathi, Y. B. 2019. Acorus calamus: A review. *International Journal of Scientific Research in Biological Sciences* 6(4): 62–67.

Yasmeen, A., Qayyum, F., Mughal, A. A. & Amjad, O. 2020. Phytochemical and biological investigation of *Armoracia rusticana*. *International Journal of Pharmacy & Integrated Health Sciences* 1(1): 58–68.

Yesodharan, K. & Sujana, K. A. 2007. Wild edible plants traditionally used by the tribes in the Parambikulam Wildlife Sanctuary, Kerala, India. *Indian Journal of Natural Products and Resources* 6(1): 74–80.

Zaveri, M., Khandhar, A., Patel, S. & Patel, A. 2010. Chemistry and pharmacology of *Piper longum* L. *International Journal of Pharmaceutical Sciences Review and Research* 5(1): 67–76.

6 Nutritional and Chemical Values of Spices and Condiment Plant Species

Keddagoda Gamage Piyumi Wasana, Nayomi Deshani De Silva, and Anoja Priyadarshani Attanayake

6.1 INTRODUCTION

From prehistoric times, spices and condiments are used to intensify the flavour and taste of food. In the Middle Ages, strange food was described using the word 'species', and, based on this word, the term 'spices' originated (Sachan et al., 2018). In general, spices are aromatic, pungent, or seasoning agents and are used in small quantities compared to condiments (D'Souza et al., 2017). Condiments are agents that are added to complement food during eating or/and cooking (D'Souza et al., 2017). In addition to adding flavour to food, the colour of condiments can easily mask unpleasant sensory characteristics. Condiments are available in the form of a single component or a mixture of ingredients. Depending on the physical nature, plant-based condiments can be divided into three phases; solid (e.g., pepper, mustard), semi-solid (e.g., tomato paste, soybean paste, chutney), and liquid (e.g., honey, chili sauce, chili oil, ketchup) (Chavasit and Photi, 2018). Nevertheless, in consonance with the International Organization for Standardization, there is no definite explanation for spices and condiments and, therefore, these two terms are combined together. In general, spices and condiments are not ranked as food because of their low nutritional value (Boelens, 2017); the same fact is evident in the definition of spices given by the US Food and Drug Administration (US FDA, 2017).

Spices and condiments are derived from several parts of plants, such as leaves, bark, root, fruits, seeds, rhizomes, and buds. Some spices and condiments that are famous for their flavour are curry leaves (*Murraya koenigii* L. Spreng; family: Rutaceae), black paper (*Piper nigrum* L.; family: Piperaceae), garlic (*Allium sativum* L.; family: Amaryllidaceae), turmeric (*Curcuma longa* L.; family: Zingiberaceae), clove (*Syzygium aromaticum* L.; family: Myrtaceae), ginger (*Zingiber officinale* Rosc.; family: Zingiberaceae), carom seeds (*Trachyspermum ammi* L.; family: Apiaceae), capsicum (*Capsicum annuum* L.; family: Solanaceae), coriander (*Coriandrum sativum* L.; family: Apiaceae), cinnamon (*Cinnamomum verum* J.Presl.; family: Lauraceae), cumin (*Cuminum cyminum* L.; family: Apiaceae), dill (*Anethum graveolens* L.; family: Apiaceae), fennel (*Foeniculum vulgare* Mill.; family: Apiaceae), fenugreek (*Trigonella foenum-graecum* L.; family: Fabaceae), mint (*Mentha piperita* L.; family: Lamiaceae), mustard (*Brassica nigra* L.; family: Brassicaceae), etc.

In general, spices and condiments are included in curries, salads, cakes, sausages, pickles, and alcoholic beverages. In modern society, the economic significance of spices and condiments has been raised to a higher level, since they are employed not only as food flavour agents but also because of their nutritional and medicinal value. Spices and condiments play important roles as medicines in the management of asthma, fever, skin diseases, vomiting, diarrhoea, constipation, ulcers, coughs, cancer, cardiovascular disease, etc. (Dubey, 2017). One of the drivers of their use as medicines is that most spices and condiments possess bioactive compounds with declared health benefits (Hussein and El-Anssary, 2018). It remains a fact that the nutrition and chemical value of

DOI: 10.1201/9781003387046-6

spices and condiments is the major determinant of their importance. Therefore, it is worth highlighting the nutritional and chemical value of spice and condiment plant species.

6.2 HISTORICAL PERSPECTIVE AND DISTRIBUTION

The awareness and use of plants or herbs as spices and condiments is not a current aspect or experience, and, its practice is as old as human history. Although to begin with, rare spices were used for culinary purposes as a sign of wealth in Rome, their application was developed to exaggerate the taste for food (Gupta, 2010). Over time, the use of spices spread to the Eastern Mediterranean and Europe through the Middle East. In the origin, spices were transported from India, China, and Indonesia by donkey or camel caravans. Until the discovery of the sea route to India and other spice-producing countries, the spice trade was controlled by Arab middlemen. In reality, Christopher Columbus, as a spice seeker, was able to find the shortest way to India and thereby facilitate the trade in spices from India (García-Casal et al., 2016). Vasco da Gama, who was exhilarated by pepper, also helped to find the path to India (García-Casal et al., 2016). Currently, India is perceived as the most important country for spices; more importantly, about 50 of the the 86 spices produced around the world are grown in India (García-Casal et al., 2016; Padakatti & Meti, 2020). Furthermore, Asia is reputed as the main leader in the production of spices such as cinnamon, pepper, cloves, nutmeg, and ginger (Gottardi et al., 2016). European countries produce bay leaves, celery leaves, chives, coriander, dill tips, thyme, and watercress, while, in the Americas, a wide distribution of ginger, nutmeg, and chili peppers can be seen (Gottardi et al., 2016).

In addition to spices, condiments were used to enhance food items long ago. In particular, Romans liked to use plant-based condiments and they prepared many sauces for their food, among which, mint sauce was the most common. In the Middle Ages, Europe invented mustard as a condiment. Subsequently, during the sixteenth to eighteenth centuries, several condiments came to market. Italy was the mother country of pesto sauce, while the use of condiments such as chutney and soy sauce came from India and China, respectively. With the Industrial Revolution in the ninteenth century, condiments began to be mass-produced in factories, and today these condiments are bestselling all around the world.

6.3 NUTRITIONAL VALUE OF SPICE AND CONDIMENT PLANT SPECIES

On the basis of nutrient content, spice and condiment plant species are not considered to be significant even though they are widely utilized in the world. However, spices and condiments have been observed to be good sources of several nutritional components, particularly in the diets of rural populations (Pradeep et al., 1993). To date, the nutritional value of several spices and condiments has been scientifically assessed.

The most exceptional spice, turmeric rhizome is widely used by Sri Lankans and Indians for their cooking as a food colouring agent. As revealed by the analysis of nutritional composition, 100 g of turmeric showed the presence of total carbohydrate, dietary fiber, protein, and total fat at 69.9, 21, 8, and 10 g, respectively (Balakrishnan, 2007). Minerals, including calcium, phosphorus, sodium, potassium, and iron, were found in turmeric at concentrations of 0.2 g, 0.26 g, 10 mg, 2.5 g, and 47.5 mg, respectively, per 100 g (Balakrishnan, 2007). The high protein content of turmeric is one of the reasons for its great consumption, and Asians have been reported to consume turmeric at the rate of 200 to 1000 mg/day (Polasa et al., 1991) or 160 to 440 g/person/year (Krishnaswamy, 1996). The presence of fibre in turmeric rhizomes is reported to detoxify consumers' digestive tract by eliminating harmful threats and is disclosed to be an absorption agent of extreme cholesterol levels (Zhang et al., 2021). Also, fibre adds bulk to a food portion and thereby it prevents excessive consumption of starchy food, reducing the risk of diseases like diabetes mellitus (Okwu & Josiah, 2006). Due to the significant iron content of turmeric, it is also recommended for iron-deficient anaemia (Mughal, 2019).

Similarly, nutrition and proximate analysis carried out on garlic reported that it could be ranked as a protein- and carbohydrate-rich spice (Yusuf et al., 2018). The use of garlic as a seasoning food agent by Asians, Africans, and Europeans dates back more than 5,000 years. According to Yusuf et al. (2018), garlic contained moisture, crude protein, crude fat, crude fibre, ash, and carbohydrate at 4.55, 15.33, 0.72, 2.10, 4.08, and 73.22 mg/100 g respectively.

In addition, garlic contains significant amount of allicin and capsaicin (Dashak et al., 2001). The crude protein content of garlic, at 15.33 mg/100g, and it is not considered a predominant source of protein. The low moisture content in the dry matter of all spices and condiment plant species, including garlic, also gives rise to extended shelf life with limited deterioration, especially due to microbial contamination. Furthermore, garlic is reported to contain minerals including potassium, calcium, phosphorus, iron, magnesium, zinc, and manganese at 10.19, 26.30, 10.19, 5.29, 0.001, 0.34, and 0.001 mg/100 g, respectively (Yusuf et al., 2018). These minerals are involved in bone and teeth formation and enzymatic activities, particularly those involved in protection against oxidative stress and those involved in protein synthesis (Boulfia et al., 2021). Furthermore, the mineral content of garlic makes it a potential regulator of non-communicable diseases such as heart disease, diabetes mellitus, and hypertension (DiNicolantonio et al., 2018; Hassanin et al., 2021). Importantly, zinc is an established mineral in terms of antioxidant potency, as it is directly involved in superoxide dismutase (Boulfia et al., 2021).

The highly aromatic tropical clove tree is widely grown in Sri Lanka, Indonesia, India, and Brazil. Cloves are often added as a flavouring agent in the making of pickles, meat curries, and syrups. In addition, cloves could be served as a dietary supplement due to their potent nutritional profile, which contribute to the high economic value of clove plant buds. According to Liu and Liu (2021), clove buds had total carbohydrate, crude protein, crude fibre, crude fat, and ash concentrations of 86.75, 39.8, 111.72, 123.58, and 34.17 mg/g respectively. Minerals include 14.08 g/kg potassium, 6.89 g/kg calcium, 4.00 g/kg sodium, 3.41 g/kg magnesium, and 0.54 g/kg iron in cloves with further evidence supporting its antidiabetic, antimicrobial, antiviral, anti-obesity, and neuroprotective properties (Liu & Liu, 2021; Xue et al., 2022). According to their nutritional profiles, cinnamon, coriander, fennel, curry leaves, black pepper, fenugreek, and ginger are excellent carbohydrate sources (Pradeep et al., 1993; Parthasarathy et al., 2008; Dubey, 2017). Compared to these spice and condiment plant species, a proximate and nutritional analysis revealed that cumin has a low nutritional value (Dubey, 2017).

The herbs chives (*Allium schoenoprasum* L.; family: Amaryllidaceae), and fennel (*Foeniculum vulgare* Mill.; family: Apiaceae) are reported to have linoleic and α-linolenic acids in their nutrient composition analysis (Pereira et al., 2015). The interesting fact is that linoleic and α-linolenic acids are the precursors of omega-6 and omega-3 fatty acids which cannot be synthesized in the human body and therefore must be obtained from diet. Overall, plant-based herbs are rich in a wide range of sugars, with the main functions of signalling and being involved with physiological processes including cell adhesion, and defensive actions (Pereira et al., 2015).

From a global public health perspective, food fortification is one of the most cost-effective interventions. Beyond the fortification of staple foods, the fortification of alternative products with potent nutritional profiles, including condiments, has been suggested as an additional strategy by which to improve micronutrient intake. The fortification of condiments, as possible candidate vehicles for micronutrients, is a new strategy with several advantages, including cost-effectiveness, feasibility, sensory acceptability, and especially the targeting of subpopulations (García-Casal et al., 2016). The condiment, tomato ketchup, has been reported to provide several micronutrients, particularly beta-carotene, and vitamins A and C, and therefore tomato ketchup is used as a vehicle for fortification by several civilizations (Chavasit & Photi, 2018). Beta-carotene (pro-vitamin A) in tomato ketchup contributes approximately 4% of vitamin A, according to the recommended dietary intake values in Thailand (Chavasit & Photi, 2018). Consequently, the use of condiments conquers monotony in the diet and improves its quality expressly in terms of micronutrients, with such commonly used products conceivably reaching some people in rural areas where most staple foods, which are produced locally, are unfortified.

Fermentation is one of the oldest preservative methods and is used in the production of plant-based condiments, and this method is accompanied by several benefits in terms of nutritional content. Well-known fermented condiments are iru or locust beans (*Parkia biglobosa* (Jacq.) R.Br. ex G.Don; family: Fabaceae) and ogiri or melon seeds (*Citrullus lanatus* (Thunb.) Matsum. & Nakai; family: Cucurbitaceae). Interestingly, during the fermentation process of iru and ogiri, their nutritional value increased, especially in terms of crude protein content, by 30.6 and 18.4%, respectively (Omafuvbe et al., 2004). In addition, the fermentation of iru and ogiri increased reducing sugar, α-amylase, and free amino acid concentrations while decreasing total sugar content (Omafuvbe et al., 2004). Furthermore, the presence of vitamins A, E, and D of iru and ogiri upon fermentation is an additional advantage over using the seasoning salt or sugar, which are the main agents that tend to replace the local preservative method in fermentation (Omafuvbe et al., 2004). This fact also means that the preparation methods of condiments affect their nutritional composition. Supporting this observation, the fermented soybean condiment enhanced its nutritional profile as evidenced by the significantly higher content of isoflavone (0.096–0.132 % w/w) than that of unfermented soybean (0.068 %) (Iheukwumere & Iheukwumere, 2022). Consequently, the fermentation of plant-based condiments causes a reduction in dry matter parallel to the increase in vitamin, mineral, and protein contents when measured on a dry weight basis (Olasupo & Okorie, 2019). In addition to the improvement in the nutritional profile, an improvement in flavour, texture, antioxidant activity, antimicrobial activity, probiotic function, digestibility, and degradation of undesirable compounds and antinutritional factors are found in the fermented condiments compared with that of the fresh materials (Olasupo & Okorie, 2019). Because of all these findings, several communities in the world use fermented plant species condiments as dietary supplements in addition to their use as flavouring agents (Ogueke et al., 2010).

The vast majority of the studies mentioned that spice and condiment plant species have a strong nutritional value which plays an important role in our daily lifestyle in addition to the delicious preparation of our food.

6.4 CHEMICAL VALUES OF SPICE AND CONDIMENT PLANT SPECIES

The research community has classified organic compounds produced by the plant kingdom, into three encompassing groups: primary metabolites that are directly involved with plant growth, secondary metabolites that intercede interactions between plants and the environment, and hormones that regulate metabolism (Erb & Kliebenstein, 2020). The secondary metabolites play a prominent role in determining the chemical value of plant species, including spices and condiments. Secondary metabolites mainly consist of phenolics, alkaloids, saponins, and terpenes, and basically, these groups are classified according to their basic chemical skeletons (Hussein and El-Anssary, 2018). The main chemical families of secondary metabolites include monoterpenes, monoterpenoids, and phenylpropanoids (El-Zaeddi et al., 2016). Additionally, alkaloids, aldehydes, alcohols, esters, sesquiterpenes, and sesquiterpenoids could be found in small amounts (El-Zaeddi et al., 2016).

Analysis of the composition of spices and condiments, especially in terms of their secondary metabolites, has begun. Secondary metabolites of spice and condiment plant species have been scientifically shown to exert potent biological effects, providing solid research evidence for their applications in traditional healthcare systems for the treatment of a number of diseases from time immemorial. In modern medicine, spices and condiments provide lead compounds for the production of drugs to manage various diseases from migraine to cancer. Importantly, the chemical profiles of essential oils have shown potent effects against the top ten killer diseases, including ischemic heart disease and stroke, diarrhoeal disease, diabetes mellitus, tuberculosis, and lung cancer (Fokou et al., 2020).

To date, several secondary metabolites of spice, condiment and other plant species, which are determinants of chemical value, have been identified, isolated, and scientifically evaluated to assess their efficacy and safety and thus introduce lead molecules for drug development for the treatment of various diseases (Husen, 2020, 2021, 2022a,b; 2023; Rahman & Husen, 2021, 2022, 2023a,b). From this point on, the chemical value of the secondary metabolites of well-known spices and condiments is revealed.

Phenolics, which are universally identified as the "wonder drug lead of life", are the largest group of secondary metabolites of spices and condiment plant species. Phenolics are the key contributing factors to the colour, taste, and flavour of spices and condiments. According to the basic chemical skeleton, phenolic compounds could be classified into different groups, namely simple phenolics, polyphenols, tannins, flavonoids, coumarins, stilbenes, chromones, and xanthones. To date, several phenolic compounds have been isolated from spices and condiments. Curcumin is the main chemical value determinant of turmeric, other than volatile oils with the sesquiterpenes turmerone and zingiberene. Curcumin is a polyphenol that has been used as a food-cum-medicine in Ayurveda and traditional healthcare systems in addition to its practice as a flavouring and food colouring agent (Gera et al., 2017). The chemical structure of curcumin plays a prominent role in its identification and, in fact, curcumin exhibits keto-enol tautomerism. Therefore, curcumin can exist in the keto form in natural and acidic media, whereas the enol form of curcumin, as the predominant form, exists in an alkaline medium. In the 1815s, curcumin was first isolated, and, in later studies, its biological importance in terms of antidiabetic, antioxidant, anti-inflammatory, anticancer, neuroprotective, and hepatoprotective efficacy, together with safety, was proven via pre-clinical and clinical studies (Chainani-Wu, 2003; Sharma et al., 2004; Rahmani et al., 2016; Li et al., 2018; Peng et al., 2018; Salehi et al., 2020; Sharifi-Rad et al., 2020). With these promising results, scientists have developed several formulations of curcumin to manage several ailments (Meena et al., 2017; Salehi et al., 2020). Curcumin exerts several mechanisms of action, including the scavenging of reactive oxygen species and inhibition of pro-inflammatory transcription factors, pro-inflammatory cytokines, and prostaglandin production in activated microglia and astrocytes, etc. (Cianciulli et al., 2016; Yang et al., 2017). Having these benefits, together with the poor bioavailability of curcumin, makes it a drug of great potential, albeit with pros and cons (Anand et al., 2007). After oral administration, the hydrophobic nature of curcumin triggers a poor gastrointestinal tract absorption rate and rapid metabolism (Hewlings & Kalman, 2017; Sharifi-Rad et al., 2020). However, the concern of poor bioavailability can be overcome by adding a bioavailability enhancement agent that creates a curcumin complex. All these prospects of curcumin have been ascribed to its chemical structure (**1**, Figure 6.1) (Wright, 2002; Priyadarsini et al., 2003). Similarly, the chemical value of phenolic compounds such as eugenol (**2**, Figure 6.2), gingerols (**3**, Figure 6.2), kaempferol (**4**, Figure 6.2), 4-hydroxycinnamic acid (**5**, Figure 6.2), and 4-hydroxybenzoic acid (**6**, Figure 6.2) collected from spices and condiments such as cloves, ginger, fenugreek seeds, cinnamon, and black pepper, respectively, have been documented (Yashin et al., 2017; Singh & Yadav, 2022).

Alkaloids are a group of structurally diverse secondary metabolites that comprise at least one nitrogen atom, and therefore alkaloids can act as bases in acid-base reactions. Based on this

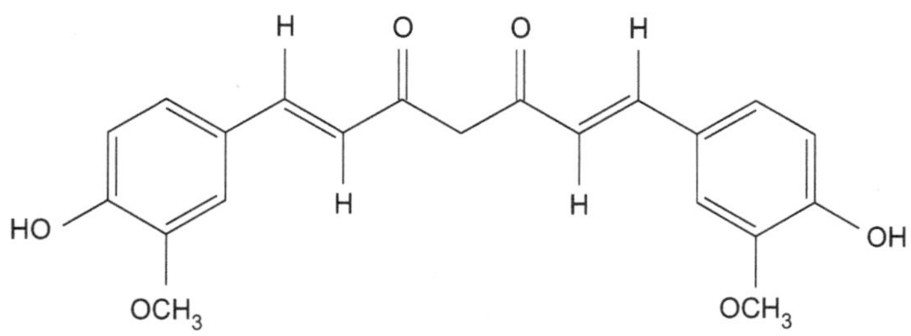

FIGURE 6.1 Chemical structure of curcumin (**1**).

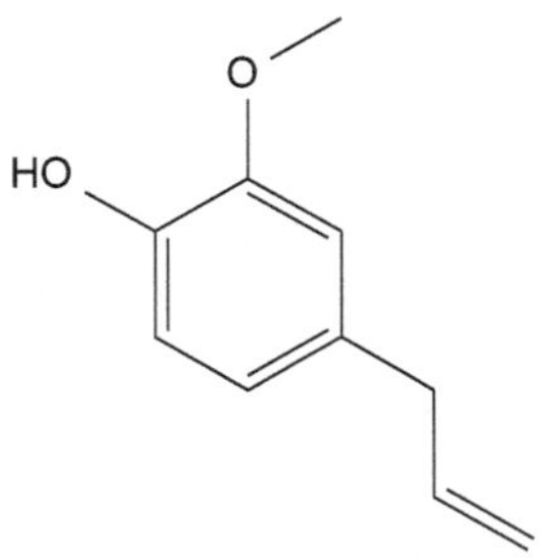

eugenol (**2**)

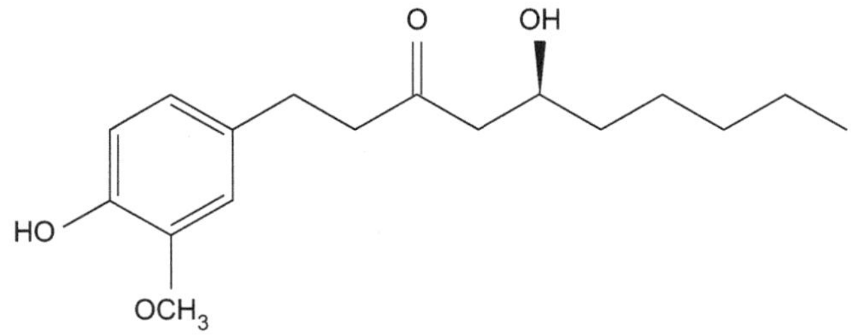

gingerol (**3**)

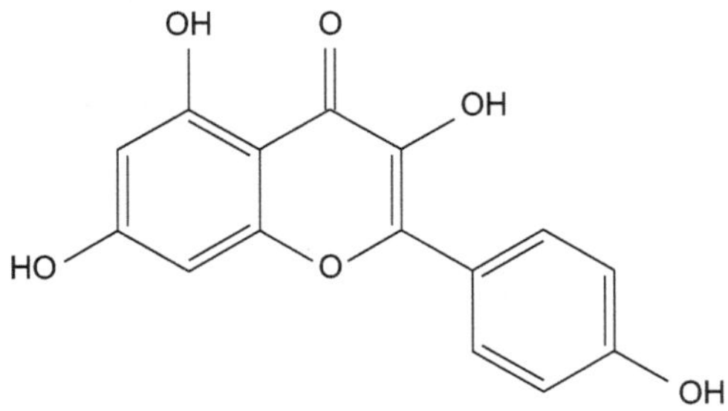

kaempferol (**4**)

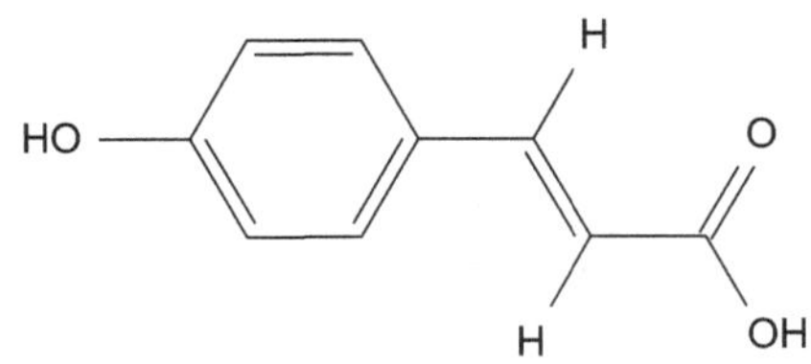

4-hydroxycinnamic acid (**5**)

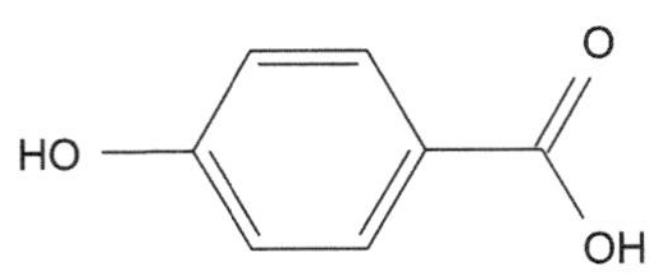

4-hydroxybenzoic acid (**6**)

FIGURE 6.2 Some phenolic compounds from plant-based spices and condiments (**2–6**).

alkalinity, the name alkaloid originated from 'alkali-like' compounds. Alkaloids can be classified into three forms: true alkaloids, psedoalkaloids, and protoalkaloids, based on their molecular precursor (Dey et al., 2020). According to another comprehensive classification, importantly based on the heterocyclic nucleus in structures, alkaloids in spices and condiments can be classed as tropane, pyrrolizidine, piperidine, quinoline, isoquinoline, indole, steroidal, imidazole, purine, or pyrrolidine alkaloids (Dey et al., 2020). It is notable that the alkaloid concentration of some spices and condiments increases just before the formation of the seed and then gradually decreases when the seeds are ripening, indicating the importance of alkaloids in the seeds. Several alkaloids of spices and condiments have recently been identified as pharmacologically active agents, mainly because of their vast chemical structural diversity. In addition to their pharmacological potency, the chemical value of alkaloids makes spices and condiments natural pesticides (Kurek, 2016).

Of all the spices, black pepper is known as the 'king of spices' or 'black gold' since it has a distinctive pungent flavor. The pungent flavor of pepper is mainly due to the main alkaloid, piperine, and multifaceted essential oil content. The content of piperine varies among different species of the family, Piperaceae; however, certain studies reported that the highest content of piperine is up to 9% present in black pepper (Chopra et al., 2016; Tiwari et al., 2020). Piperine is reported to have a multifaceted biological profile, including anti-inflammatory, antioxidant, anti-ulcer, anticancer, antimicrobial, and anti-infective activities (Rauscher et al., 2000; Bang et al., 2009; Stojanović-Radić et al., 2019; Tavares et al., 2011). Parallel to these biological activities, piperine is reputed to have the ability to cause increased bioavailability (Chopra et al., 2016; Gorgani et al., 2017). It is reported that piperine is associated with an increase of 2000% in the bioavailability of curcumin (Shoba et al., 1998). In the piperine molecule (**7**, Figure 6.3), the conjugated aliphatic chain between piperidine and the 5-(3,4-methylenedioxyphenyl) moiety makes it a unique and excellent molecule to exhibit optimal attributes for its therapeutic activities (Tiwari et al., 2020). Naturally, black pepper contains four isomeric forms of piperine, namely piperine, isopiperine, chavicine, and isochavicine (Tiwari et al., 2020). Importantly, piperine is the only isomer that has pungency, compared with its other geometrical isomers (Ravindran, 2000). Although the transformation of chavicine to piperine is possible, it destroys the potent pungency of piperine, further indicating the uniqueness of the chemical arrangement of piperine (Kozukue et al., 2007).

All these attestations highlight the stimulating chemical value of piperine as a prominent alkaloid in spices. Other than piperine, several alkaloids such as piperanine, piperettine, piperylin A, piperolein B, and pipericine have been identified from black pepper (Hirasa & Takemasa, 1998). Analyzing other spices and condiment plant species, such as fenugreek, curry leaves, and garlic, has led to the identification of protoberberine alkaloids such as trigonelline (**8**, Figure 6.4), karapinchamine A (**9**, Figure 6.4), and allicin (**10**, Figure 6.4), respectively (Nakamura et al., 2013; Iqbal, 2019; Tudu et al., 2022).

Saponins are compounds that possess hydrophobic-hydrophilic asymmetry through a polycyclic aglycone moiety with either a steroid or a triterpenoid attached to a carbohydrate unit. This hydrophobic-hydrophilic asymmetry of saponins lowers the surface tension and, more importantly, they can act as natural micellar systems (Peng et al., 2018; Hussein and El-Anssary, 2018). Spices and condiments are rich in saponins, as evidenced by several studies (Elbandy et al., 2009; Ali et al., 2018; Zhang & Kitts, 2021). Structure-based activities of saponins isolated from spices and condiments relate to numerous therapeutic potentials such as anti-inflammatory, antimicrobial, anti-obesity, anti-ageing, etc. (Elbandy et al., 2009; D'Souza et al., 2017; Sharma et al., 2023).

Terpenes are the metabolites responsible for the fragrance, taste, and pigment of spices and condiments (Singh & Sharma, 2015). Terpenes are an arrangement of isoprene units that are naturally occurring, and volatile and, therefore, give off a scent or a taste as an anti-feedant defence mechanism (Singh & Sharma, 2015). The chemical value of terpenes is driven by stereospecific carbocation cyclization, rearrangement, and elimination reactions, and therefore a single terpene could be able to give rise to multiple products with vastly different pharmacological activities (Boncan et al., 2020). So far, a small number of terpenes have been isolated from spices and condiments. Isolated terpenes from spices and condiments exert potent antioxidant, anti-inflammatory, anticancer, digestive, and diuretic activities, have become recent trends in healthy foods, and have opened doors for several medical investigations (Cox-Georgian et al., 2019).

In general, the plants or herbs used as spices and condiments are aromatic and pungent (Achinewu et al., 1995), and therefore, they are prominently contain essential oils (Green et al., 2012). Essential oils, which are volatile secondary metabolite complexes, mainly including terpenoids and phenolics, are produced by all parts of spices and condiment plant species, generally located in temperate to warm countries, either via secretion or excretion during the metabolic process. Due to the volatility of essential oil, they evaporate when they are exposed to the air, facilitating that distillation of essential oil of spices and condiments. Coriander seeds, as a well-known spice, consist of

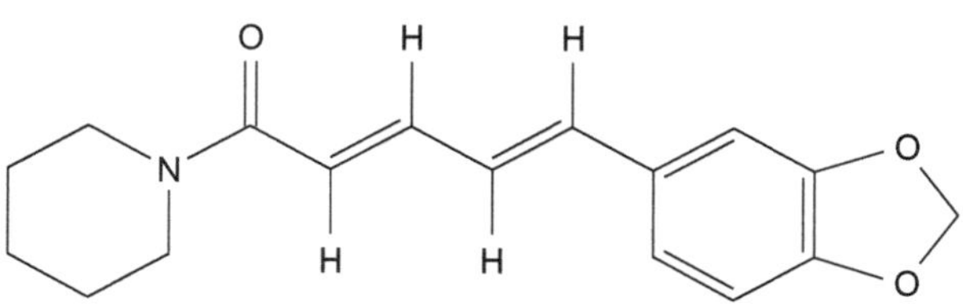

FIGURE 6.3 Chemical structure of piperine (**7**).

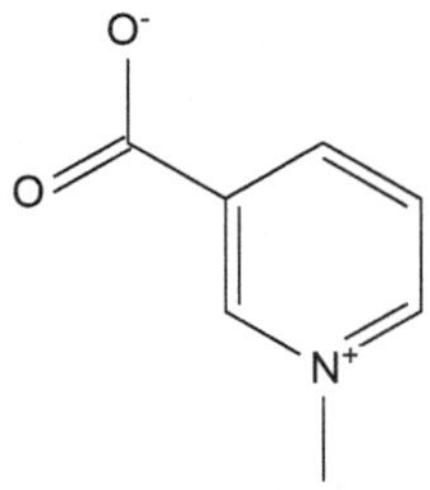

trigonelline (**8**)

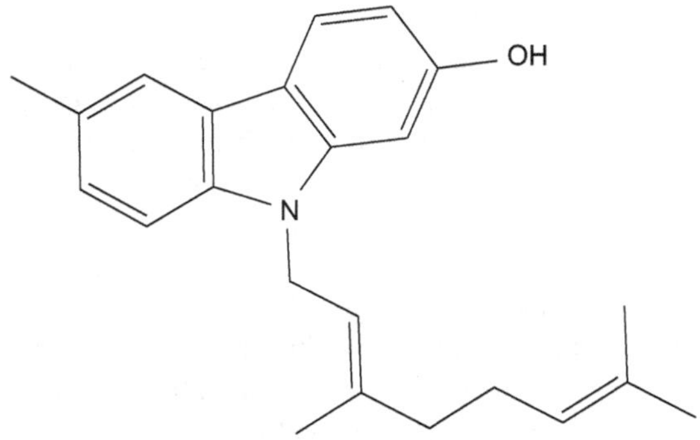

karapinchamine A (**9**)

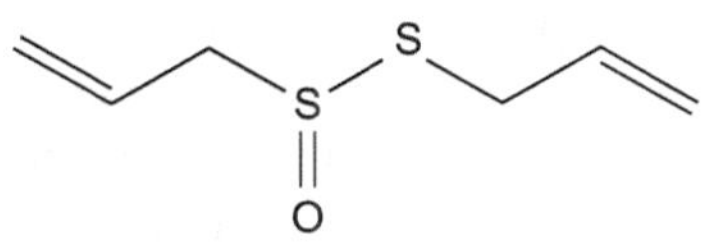

allicin (**10**)

FIGURE 6.4 Some alkaloids from plant-based spices and condiments (**8–10**).

0.2–2.6% essential oil with compounds such as linalool, neryl acetate, γ-terpinene, and α-pinene (Nejad Ebrahimi et al., 2010). Aliphatic aldehydes, having an unpleasant odor, are the main components of volatile oil from the fresh coriander herb (Potter & Fagerson, 1990). According to the origin of coriander, the yield of essential oil and its composition varies (Mandal & Mandal, 2015). This fact could be rationalized, as coriander seeds originating from Tunisia, Canada, Egypt, Turkey, and India consisted of essential oils at 0.37, 0.44, 0.31, 0.43, and 0.82%, respectively (Telci et al., 2006; Romeilah et al., 2010; Sriti et al., 2011).

The Food and Drug Administration (FDA) committee classifies clove essential oil as a safe agent with a potent chemical value that could be used in perfumes, cosmetics, sanitary products, medicines, and foods (Haro-González et al., 2021). As revealed in several studies, clove contains approximately 15 to 20 % wt. of essential oil (Golmakani et al., 2017; Haro-González et al., 2021). To date, scientific investigators have identified at least 30 compounds from clove essential oil and, among these compounds, eugenol is the major compound, accounting for at least 50 % by weight (Golmakani et al., 2017). The spectacular skeletal formula of euginol is the reason for its ultimate biological potential as an antioxidant, with antiviral, and anti-inflammatory activity (Khalil et al., 2017; Banerjee et al., 2020). Similarly, the essential oil content of spices and condiments, including curry leaves, ginger, fennel, and turmeric, with the main compounds of linalool, zingiberene, anethole, and ar-turmerone, respectively (Liju et al., 2011; Rajendran et al., 2014; Abdellaoui et al., 2020; Wang et al., 2020), has been evaluated. These observations have led to the notion that all spices and condiments contain essential oil profiles with diverse chemical values and, therefore, in modern science, extraction of essential oils from spices and condiments to develop value-added products has become an incredibly interesting area. Commonly, essential oil isolated from spices and condiments are used to relieve body pain and slow down the development of ischemic heart disease, stroke, diarrhoeal disease, depressive disorders, and several adverse health conditions that result from oxidative stress and inflammation (El-Zaeddi et al., 2016; Fokou et al., 2020; Orellana-Paucar and Machado-Orellana, 2022).

Importantly, it is notable that there are synergistic effects in spices and condiments, and therefore their combined effects make impacts that are more significant than the sum of their separate impacts. The chemical value of spices and condiments is the main determinant of their synergistic effect. For example, the mixture of cumin, fenugreek, and mustard spices has shown higher antimicrobial activity than the sum of their separate antimicrobial activities (Das et al., 2012). The antimicrobial activity of aqueous extracts of cloves, oregano, and cinnamon was tested for antimicrobial activity separately and as a combination, and, as a result, it was reported that the mixture of the three extracts had the strongest antimicrobial activity due to the synergistic activity of the antimicrobial compounds present in the extracts (Babuskin et al., 2014). Furthermore, the chemical value of spices is reported through the synergistic activity, as the combination of clove and cinnamon aqueous extracts with oxygen-deficient modified atmospheric packaging reduces the growth and development of microbes until the end of the storage period without changing the organoleptic properties of poultry meat (Sivarajan et al., 2017). Similarly, due to the synergistic activity, the condiment of pesto sauce, which is a combination of several plant-based spices such as garlic, olive oil, and pine nuts, may be an excellent pharmacological agent. Compared with spices alone, in general, condiments are a mixture of plant-based species and, therefore, the benefits of the synergistic effect can be readily achieved.

According to all these aforementioned pieces of evidence, it could be implicated that a keen understanding of the chemical value of spices and condiments leads to an improved interest in their possible medicinal value, and therefore, it could be stated that the chemical value of spice and condiment plant species is the mirror image of their medicinal value. Owing to these chemical/medicinal values, spices and condiments could be used as candidates in food supplements and medicines, or even as superfoods (Aprotosoaie et al., 2013; Mandal & Mandal, 2015). However, the chemical profile of secondary metabolites can change from plant to plant even in the same species due to several reasons, such as abiotic and biotic factors, extraction methods, post-harvest treatments, and product storage conditions (Khazaie et al., 2008; Fokou et al., 2020).

6.5 CONCLUSION AND FUTURE REMARKS

Despite the use of spices and condiments as adjuvants to enhance the flavour and taste of food, its nutritional value provides additional value to consumers in terms of macro- and micronutrients. In parallel to that, the chemical value of spices and condiment plant species, in the sense of secondary metabolites, plays a vital role in the control of various diseases. The active secondary metabolites of spices and condiments have been isolated, identified, and subjected to determine their efficacy and safety in the management of disease conditions. Accordingly, the nutritional and chemical values of spices and condiments are known to have several beneficial effects on both health promotion and disease prevention. It is far-reaching to further assess the nutritional and chemical value of spices and condiment plant species to introduce value-added agents/products.

In general, several condiments include high sugar and/or salt, especially for the preservative process. High salt and/or sugar intake is a notable risk factor for non-communicable diseases including diabetes mellitus, cardiovascular disease, and cancer. Therefore, necessary actions should be taken in the future to reduce the sugar and salt content of condiments as much as possible by replacing them with value-added preservatives. Furthermore, complicated technologies and improved logistics are required to maintain acceptable nutritional and chemical values of spices and condiments in marketable value-added products. Moreover, it is important to study the bioavailability of bioactive chemicals in spice and condiment plant species and thereby assess the ultimate availability of their nutritional and chemical value for utilization after digestion, absorption, and circulation around the body.

REFERENCES

Abdellaoui, M., Derouich, M. & El-Rhaffari, L. 2020. Essential oil and chemical composition of wild and cultivated fennel (*Foeniculum vulgare* Mill.): A comparative study. *South African Journal of Botany*, 135: 93–100.

Achinewu, S.C., Aniena, M.I. & Obomanu, F.G. 1995. Studies on spices of food value in the South eastern states of Nigeria 1: Antioxidant properties. *Journal Africain du Cancer Medicinal Plant*, 18: 135–139.

Ali, M., Nas, F.S., Yahaya, A. & Ibrahim, I.S. 2018. Phytochemical screening of some spices used as condiment in Kano, North Western Nigeria. *Open Access Journal of Pharmaceutical Research*, 2: 000150.

Anand, P., Kunnumakkara, A.B., Newman, R.A. & Aggarwal, B.B. 2007. Bioavailability of curcumin: Problems and promises. *Molecular Pharmaceutics*, 4(6): 807–818.

Aprotosoaie, A.C., Spac, A., Cioancă, O., Trifan, A., Miron, A. & Hăncianu, M. 2013. The chemical composition of essential oils isolated from sweet fennel fruits available as herbal tea products. *Revista Medico-Chirurgicala a Societatii de Medici si Naturalisti din Iasi*, 117(3): 819–824.

Babuskin, S., Babu, P.A.S., Sasikala, M., Sabina, K., Archana, G., Sivarajan, M. & Sukumar, M. 2014. Antimicrobial and antioxidant effects of spice extracts on the shelf life extension of raw chicken meat. *International Journal of Food Microbiology*, 171: 32–40.

Balakrishnan, K.V. 2007. Postharvest technology and processing of turmeric. In: Ravindran P.N., Nirmal Babu K., Sivaraman K. (eds.) *Turmeric: The Genus Curcuma*. CRC Press, Boca Raton. http://doi.org/10.1201/9781420006322

Banerjee, K., Madhyastha, H., Sandur, R., Manikandanath, N.T., Thiagarajan, N. & Thiagarajan, P. 2020. Anti-inflammatory and wound healing potential of a clove oil emulsion. *Colloids and Surfaces, B: Biointerfaces*, 193: 111102.

Bang, J.S., Oh, D.H., Choi, H.M., Sur, B.J., Lim, S.J., Kim, J.Y., Yang, H.I., Yoo, M.C., Hahm, D.H. & Kim, K.S. 2009. Anti-inflammatory and antiarthritic effects of piperine in human interleukin 1β-stimulated fibroblast-like synoviocytes and in rat arthritis models. *Arthritis Research and Therapy*, 11(2): 1–9.

Boelens, M.H. 2017. Spices and condiments II. In: *Volatile Compounds in Foods and Beverages*. Routledge, New York. http://doi.org/10.1201/9780203734285

Boncan, D.A.T., Tsang, S.S., Li, C., Lee, I.H., Lam, H.M., Chan, T.F. & Hui, J.H. 2020. Terpenes and terpenoids in plants: Interactions with environment and insects. *International Journal of Molecular Sciences*, 21(19): 7382.

Boulfia, M., Lamchouri, F., Senhaji, S., Lachkar, N., Bouabid, K. & Toufik, H. 2021. Mineral content, chemical analysis, *in vitro* antidiabetic and antioxidant activities, and antibacterial power of aqueous and organic extracts of *Moroccan Leopoldia* comosa (L.) Parl. bulbs. *Evidence-Based Complementary and Alternative Medicine*, 2021: 9932291.

Chainani-Wu, N. 2003. Safety and anti-inflammatory activity of curcumin: A component of tumeric (*Curcuma longa*). *The Journal of Alternative and Complementary Medicine*, 9(1): 161–168.

Chavasit, V. & Photi, J. 2018. Condiments and sauces. In: *Food Fortification in a Globalized World*. Academic Press. http://doi.org/10.1016/B978-0-12-802861-2.00015-8

Chopra, B., Dhingra, A.K., Kapoor, R.P. & Prasad, D.N. 2016. Piperine and its various physicochemical and biological aspects: A review. *Open Chemistry Journal*, 3(1): 75–96.

Cianciulli, A., Calvello, R., Porro, C., Trotta, T., Salvatore, R. & Panaro, M.A. 2016. PI3k/Akt signalling pathway plays a crucial role in the anti-inflammatory effects of curcumin in LPS-activated microglia. *International Immunopharmacology*, 36: 282–290.

Cox-Georgian, D., Ramadoss, N., Dona, C. & Basu, C. 2019. Therapeutic and medicinal uses of terpenes. *Medicinal Plants: From Farm to Pharmacy*: 333–359. http://doi.org/10.1007/978-3-030-31269-5_15

Das, S., Anjeza, C. & Mandal, S. 2012. Synergistic or additive antimicrobial activities of Indian spice and herbal extracts against pathogenic, probiotic and food-spoiler micro-organisms. *International Food Research Journal*, 19: 1185–1191.

D'Souza, S.P., Chavannavar, S.V., Kanchanashri, B. & Niveditha, S.B. 2017. Pharmaceutical perspectives of spices and condiments as alternative antimicrobial remedy. *Journal of Evidence-Based Complementary and Alternative Medicine*, 22(4): 1002–1010.

Dashak, D.A., Dawang, M.L. & Lucas, N.B. 2001. An assessment of the proximate chemical composition of locally produced spices known as dadawa basso and dadawa kalwa from three markets in Plateau State of Nigeria. *Food Chemistry*, 75(2): 231–235.

Dey, P., Kundu, A., Kumar, A., Gupta, M., Lee, B.M., Bhakta, T., Dash, S. & Kim, H.S. 2020. Analysis of alkaloids (indole alkaloids, isoquinoline alkaloids, tropane alkaloids). *Recent Advances in Natural Products Analysis*, 2020: 505–567.

DiNicolantonio, J.J., Liu, J. & O'Keefe, J.H. 2018. Magnesium for the prevention and treatment of cardiovascular disease. *Open Heart*, 5(2): e000775.

Dubey, S. 2017. Indian spices and their medicinal value. *Indian Journal of Pharmaceutical Education and Research*, 51(3s2): 330–331.

Elbandy, M., Kang, O.H., Kwon, D.Y. & Rho, J.R. 2009. Two new anti-inflammatory triterpene saponins from the Egyptian medicinal food black cumin (seeds of *Nigella sativa*). *Bulletin of the Korean Chemical Society*, 30(8): 1811–1816.

El-Zaeddi, H., Martínez-Tomé, J., Calín-Sánchez, Á., Burló, F. & Carbonell-Barrachina, Á.A. 2016. Volatile composition of essential oils from different aromatic herbs grown in Mediterranean regions of Spain. *Foods*, 5(2): 41.

Erb, M. & Kliebenstein, D.J. 2020. Plant secondary metabolites as defenses, regulators, and primary metabolites: The blurred functional trichotomy. *Plant Physiology*, 184(1): 39–52.

Fokou, J.B.H., Dongmo, P.M.J. & Boyom, F.F. 2020. Essential oil's chemical composition and pharmacological properties. In: El-Shemy, H.A. (ed.) *Essential Oils-Oils of Nature*. IntechOpen. http://doi.org/10.5772/intechopen.77673

García-Casal, M.N., Peña-Rosas, J.P. & Malavé, H.G. 2016. Sauces, spices, and condiments: Definitions, potential benefits, consumption patterns, and global markets. *Annals of the New York Academy of Sciences*, 1379(1): 3–16.

Gera, M., Sharma, N., Ghosh, M., Lee, S.J., Min, T., Kwon, T. & Jeong, D.K. 2017. Nanoformulations of curcumin: An emerging paradigm for improved remedial application. *Oncotarget*, 8(39): 66680–66698.

Golmakani, M.T., Zare, M. & Razzaghi, S. 2017. Eugenol enrichment of clove bud essential oil using different microwave-assisted distillation methods. *Food Science and Technology Research*, 23(3): 385–394.

Gorgani, L., Mohammadi, M., Najafpour, G.D. & Nikzad, M. 2017. Piperine—The bioactive compound of black pepper: from isolation to medicinal formulations. *Comprehensive Reviews in Food Science and Food Safety*, 16(1): 124–140.

Gottardi, D., Bukvicki, D., Prasad, S. & Tyagi, A.K. 2016. Beneficial effects of spices in food preservation and safety. *Frontiers in Microbiology*, 7: 1394.

Green, B.O., Nworgu, F.C. & Obazee, M.N. 2012. Spices and food condiments in Niger-Delta region of Nigeria. *African Journal of Biotechnology*, 11: 14468–14473.

Gupta, M. 2010. Pharmacological properties and traditional therapeutic uses of important Indian spices: A review. *International Journal of Food Properties*, 13(5): 1092–1116.

Haro-González, J.N., Castillo-Herrera, G.A., Martínez-Velázquez, M. & Espinosa-Andrews, H. 2021. Clove essential oil (*Syzygium aromaticum* L. Myrtaceae): Extraction, chemical composition, food applications, and essential bioactivity for human health. *Molecules*, 26(21): 6387.

Hassanin, N., Elkishki, M.M. & Fawzy, L.H. 2021. Trace elements and their relation to diabetes mellitus and obesity. *Journal of Recent Advances in Medicine*, 2: 128–132.

Hewlings, S.J. & Kalman, D.S. 2017. Curcumin: A review of its effects on human health. *Foods*, 6(10): 92.

Hirasa, K. & Takemasa, M. 1998. *Spice Science and Technology*. CRC Press, Boca Raton. http://doi.org/10.1201/9780367800451

Hussein, R.A. & El-Anssary, A.A. 2018. Plants secondary metabolites: The key drivers of the pharmacological actions of medicinal plants, herbal medicine. In: Builders P.F. (ed.) *Herbal Medicine*. IntechOpen. http://doi.org/10.5772/intechopen.76139

Husen, A. 2020. *Exploring Medicinal Plants* (Book Series). CRC Press, Boca Raton, FL.

Husen, A. 2021. *Traditional Herbal Therapy for the Human Immune System*. CRC Press, Boca Raton, FL. https://doi.org/10.1201/9781003137955

Husen, A. 2022a. *Herbs, Shrubs and Trees of Potential Medicinal Benefits*. CRC Press, Boca Raton, FL. https://doi.org/10.1201/9781003205067

Husen, A. 2022b. *Exploring Poisonous Plants: Medicinal Values, Toxicity Response and Other Potential Uses*. CRC Press, Boca Raton, FL. https://doi.org/10.1201/b23017

Husen, A. 2023. *Antidiabetic Medicinal Plants and Herbal Treatments*. CRC Press, Boca Raton, FL. https://doi.org/10.1201/b23347

Iheukwumere, C.M. & Iheukwumere, I.H. 2022. Nutritive and antinutrient values of soybean condiments produced from indigenous fermenters. *IPS Applied Journal of Nutrition, Food and Metabolism Science*, 1(1): 1–5.

Iqbal, H.A.K. 2019. A new method for the isolation and purification of trigonelline as hydrochloride from *Trigonella foenum-graecum* L. *Baghdad Science Journal*, 16(3): 550–554.

Khalil, A.A., Rahman, U., Khan, M.R., Sahar, A., Mehmood, T. & Khan, M. 2017. Essential oil eugenol: Sources, extraction techniques and nutraceutical perspectives. *RSC Advances*, 7(52): 32669–32681.

Khazaie, H.R., Nadjafi, F. & Bannayan, M. 2008. Effect of irrigation frequency and planting density on herbage biomass and oil production of thyme (*Thymus vulgaris*) and hyssop (*Hyssopus officinalis*). *Industrial Crops and Products*, 27(3): 315–321.

Kozukue, N., Park, M.S., Choi, S.H., Lee, S.U., Ohnishi-Kameyama, M., Levin, C.E. & Friedman, M. 2007. Kinetics of light-induced cis– trans isomerization of four piperines and their levels in ground black peppers as determined by HPLC and LC/MS. *Journal of Agricultural and Food Chemistry*, 55(17): 7131–7139.

Krishnaswamy, K. 1996. Indian functional foods: Role in prevention of cancer. *Nutrition Reviews*, 54: S127–S131.

Kurek, J. 2016. Alkaloids-their importance in nature and for human life. In: Kurek J. (ed.). http://doi.org/10.5772/intechopen.73336

Li, X., Ma, S., Yang, P., Sun, B., Zhang, Y., Sun, Y., Hao, M., Mou, R. & Jia, Y. 2018. Anticancer effects of curcumin on nude mice bearing lung cancer A549 cell subsets SP and NSP cells. *Oncology Letters*, 16(5): 6756–6762.

Liju, V.B., Jeena, K. & Kuttan, R. 2011. An evaluation of antioxidant, anti-inflammatory, and antinociceptive activities of essential oil from *Curcuma longa*. L. *Indian Journal of Pharmacology*, 43(5): 526–531.

Liu, J.G. & Liu, H.P. 2021. Total nutrient analysis of various parts of *Eugenia caryophyllata*. *Research Biotica*, 43: 357–362.

Mandal, S. & Mandal, M. 2015. Coriander (*Coriandrum sativum* L.) essential oil: Chemistry and biological activity. *Asian Pacific Journal of Tropical Biomedicine*, 5(6): 421–428.

Meena, R., Kumar, S., Gaharwar, U.S. & Rajamani, P. 2017. PLGA-CTAB curcumin nanoparticles: Fabrication, characterization and molecular basis of anticancer activity in triple negative breast cancer cell lines (MDA-MB-231 cells). *Biomedicine and Pharmacotherapy*, 94: 944–954.

Mughal, M.H. 2019. Turmeric polyphenols: A comprehensive review. *Integrative Food, Nutrition and Metabolism*, 6(6). http://doi.org/10.15761/IFNM.1000269

Nakamura, S., Nakashima, S., Oda, Y., Yokota, N., Fujimoto, K., Matsumoto, T., Ohta, T., Ogawa, K., Maeda, S., Nishida, S. & Matsuda, H. 2013. Alkaloids from Sri Lankan curry-leaf (*Murraya koenigii*) display melanogenesis inhibitory activity: Structures of karapinchamines A and B. *Bioorganic and Medicinal Chemistry*, 21(5): 1043–1049.

Nejad Ebrahimi, S., Hadian, J. & Ranjbar, H. 2010. Essential oil compositions of different accessions of *Coriandrum sativum* L. from Iran. *Natural Product Research*, 24(14): 1287–1294.

Ogueke, C.C., Nwosu, J.N., Owuamanam, C.I. & Iwouno, J.N. 2010. Ugba, the fermented African Oilbean Seeds; its production, chemical composition, preservation, safety and health benefits. *Pakistan Journal of Biological Sciences*, 13(10): 489–496.

Okwu, D.E. & Josiah, C. 2006. Evaluation of the chemical composition of two Nigerian medicinal plants. *African Journal of Biotechnology*, 5: 357–361.

Olasupo, N.A. & Okorie, P.C. 2019. African fermented food condiments: Microbiology impacts on their nutritional values. *Frontiers and New Trends in the Science of Fermented Food and Beverages*, 1: 1–20.

Omafuvbe, B.O., Falade, O.S., Osuntogun, B.A. & Adewusi, S.R. 2004. Chemical and biochemical changes in African locust bean (*Parkia biglobosa*) and melon (*Citrullus vulgaris*) seeds during fermentation to condiments. *Pakistan Journal of Nutrition*, 3(3): 140–145.

Orellana-Paucar, A.M. & Machado-Orellana, M.G. 2022. Pharmacological profile, bioactivities, and safety of turmeric oil. *Molecules*, 27(16): 5055.

Padakatti, T. & Meti, R. 2020. Indian spices: Traditional and medicinal use. *International Journal of Home Science*, 6: 42–44.

Parthasarathy, V.A., Chempakam, B. & Zachariah, T.J. (eds.) 2008. *Chemistry of Spices*. CABI, Wallingford, Oxfordshire.

Peng, S., Li, Z., Zou, L., Liu, W., Liu, C. & McClements, D.J. 2018. Improving curcumin solubility and bioavailability by encapsulation in saponin-coated curcumin nanoparticles prepared using a simple pH-driven loading method. *Food and Function*, 9(3): 1829–1839.

Peng, X., Dai, C., Liu, Q., Li, J. & Qiu, J. 2018. Curcumin attenuates on carbon tetrachloride-induced acute liver injury in mice via modulation of the Nrf2/HO-1 and TGF-β1/Smad3 pathway. *Molecules*, 23(1): 215.

Pereira, C., Barros, L. & Ferreira, I.C. 2015. A comparison of the nutritional contribution of thirty-nine aromatic plants used as condiments and/or herbal infusions. *Plant Foods for Human Nutrition*, 70(2): 176–183.

Polasa, K., Sesikaran, B., Krishna, T.P. & Krishnaswamy, K. 1991. Turmeric (*Curcuma longa*)-induced reduction in urinary mutagens. *Food and Chemical Toxicology*, 29(10): 699–706.

Potter, T.L. & Fagerson, I.S. 1990. Composition of coriander leaf volatiles. *Journal of Agricultural and Food Chemistry*, 38(11): 2054–2056.

Pradeep, K.U., Geervani, P. & Eggum, B.O. 1993. Common Indian spices: Nutrient composition, consumption and contribution to dietary value. *Plant Foods for Human Nutrition*, 44(2): 137–148.

Priyadarsini, K.I., Maity, D.K., Naik, G.H., Kumar, M.S., Unnikrishnan, M.K., Satav, J.G. & Mohan, H. 2003. Role of phenolic OH and methylene hydrogen on the free radical reactions and antioxidant activity of curcumin. *Free Radical Biology and Medicine*, 35(5): 475–484.

Rahman, S. & Husen, A. 2022. *Asparagus racemosus* (Shatavari) and *Dioscorea villosa* (Wild yams). In: Husen, A. (ed.) *Herbs, Shrubs, and Trees of Potential Medicinal Benefits*. CRC Press, Boca Raton, FL.

Rahman, S. & Husen, A. 2021. Potential role of medicinal plants in the cure of liver and kidney diseases. In: Husen, A., Bachheti, R.K. & Bachheti, A. (eds.) *Non-timber Forest Products*. Springer International Publishing AG, Cham, pp. 229–25. https://doi.org/10.1007/978-3-030-73077-2_10

Rahman, S. & Husen, A. 2023a. *Trigonella foenum-Graecum* L.: A Useful Medicinal Plant for Diabetes and Other Diseases. In: Husen, A. (ed.) *Antidiabetic Medicinal Plants and Herbal Treatments*. CRC Press, Boca Raton, FL.

Rahman, S. & Husen, A. 2023b. Role of Green chiretta [*Andaphis paniculata* (Burm. F.)] Nees in diabetes and other ailments. In: Husen, A. (ed.) *Antidiabetic Medicinal Plants and Herbal Treatments*. CRC Press, Boca Raton, FL.

Rahmani, S., Asgary, S., Askari, G., Keshvari, M., Hatamipour, M., Feizi, A. & Sahebkar, A. 2016. Treatment of non-alcoholic fatty liver disease with curcumin: A randomized placebo-controlled trial. *Phytotherapy Research*, 30(9): 1540–1548.

Rajendran, M.P., Pallaiyan, B.B. & Selvaraj, N. 2014. Chemical composition, antibacterial and antioxidant profile of essential oil from *Murraya koenigii* (L.) leaves. *Avicenna Journal of Phytomedicine*, 4(3): 200–214.

Rauscher, F.M., Sanders, R.A. & Watkins III, J.B. 2000. Effects of piperine on antioxidant pathways in tissues from normal and streptozotocin-induced diabetic rats. *Journal of Biochemical and Molecular Toxicology*, 14(6): 329–334.

Ravindran, P.N. (ed.) 2000. *Black Pepper: Piper nigrum*. CRC Press, London. http://doi.org/10.1201/9780203303870

Romeilah, R.M., Fayed, S.A. & Mahmoud, G.I. 2010. Chemical compositions, antiviral and antioxidant activities of seven essential oils. *Journal of Applied Sciences Research*, 6: 50–62.

Sachan, A.K., Kumar, S., Kumari, K. & Singh, D. 2018. Medicinal uses of spices used in our traditional culture: Worldwide. *Journal of Medicinal Plants Studies*, 6: 116–122.

Salehi, B., Calina, D., Docea, A.O., Koirala, N., Aryal, S., Lombardo, D., Pasqua, L., Taheri, Y., Marina Salgado Castillo, C., Martorell, M. & Martins, N. 2020. Curcumin's nanomedicine formulations for therapeutic application in neurological diseases. *Journal of Clinical Medicine*, 9(2): 430.

Sharifi-Rad, J., Rayess, Y.E., Rizk, A.A., Sadaka, C., Zgheib, R., Zam, W., Sestito, S., Rapposelli, S., Neffe-Skocińska, K., Zielińska, D. & Salehi, B. 2020. Turmeric and its major compound curcumin on health: Bioactive effects and safety profiles for food, pharmaceutical, biotechnological and medicinal applications. *Frontiers in Pharmacology*, 11: 01021.

Sharma, K., Kaur, R., Kumar, S., Saini, R.K., Sharma, S., Pawde, S.V. & Kumar, V. 2023. Saponins: A concise review on food related aspects, applications and health implications. *Food Chemistry Advances*, 2: 100191.

Sharma, R.A., Euden, S.A., Platton, S.L., Cooke, D.N., Shafayat, A., Hewitt, H.R., Marczylo, T.H., Morgan, B., Hemingway, D., Plummer, S.M. & Pirmohamed, M. 2004. Phase I clinical trial of oral curcumin: Biomarkers of systemic activity and compliance. *Clinical Cancer Research*, 10(20): 6847–6854.

Shoba, G., Joy, D., Joseph, T., Majeed, M., Rajendran, R. & Srinivas, P.S.S.R. 1998. Influence of piperine on the pharmacokinetics of curcumin in animals and human volunteers. *Planta Medica*, 64(4): 353–356.

Singh, B. & Sharma, R.A. 2015. Plant terpenes: Defense responses, phylogenetic analysis, regulation and clinical applications. *3 Biotech*, 5(2): 129–151.

Singh, N. & Yadav, S.S. 2022. A review on health benefits of phenolics derived from dietary spices. *Current Research in Food Science*, 5: 1508–1523.

Sivarajan, M., Lalithapriya, U., Mariajenita, P., Vajiha, B.A., Harini, K., Madhushalini, D. & Sukumar, M. 2017. Synergistic effect of spice extracts and modified atmospheric packaging towards non-thermal preservation of chicken meat under refrigerated storage. *Poultry Science*, 96(8): 2839–2844.

Sriti, J., Wannes, W.A., Talou, T., Vilarem, G. & Marzouk, B. 2011. Chemical composition and antioxidant activities of Tunisian and Canadian coriander (*Coriandrum sativum* L.) fruit. *Journal of Essential Oil Research*, 23(4): 7–15.

Stojanović-Radić, Z., Pejčić, M., Dimitrijević, M., Aleksić, A., V Anil Kumar, N., Salehi, B., C Cho, W. & Sharifi-Rad, J. 2019. Piperine-a major principle of black pepper: A review of its bioactivity and studies. *Applied Sciences*, 9(20): 4270.

Tavares, W.S., Cruz, I., Petacci, F., Freitas, S.S., Serrão, J.E. & Zanuncio, J.C. 2011. Insecticide activity of piperine: Toxicity to eggs of *Spodoptera frugiperda* (Lepidoptera: Noctuidae) and *Diatraea saccharalis* (Lepidoptera: Pyralidae) and phytotoxicity on several vegetables. *Journal of Medicinal Plants Research*, 5: 5301–5306.

Telci, I., Toncer, O.G. & Sahbaz, N. 2006. Yield, essential oil content and composition of *Coriandrum sativum* varieties (var. vulgare Alef and var. microcarpum DC.) grown in two different locations. *Journal of Essential Oil Research*, 18(2): 189–193.

Tiwari, A., Mahadik, K.R. & Gabhe, S.Y. 2020. Piperine: A comprehensive review of methods of isolation, purification, and biological properties. *Medicine in Drug Discovery*, 7: 100027.

Tudu, C.K., Dutta, T., Ghorai, M., Biswas, P., Samanta, D., Oleksak, P., Jha, N.K., Kumar, M., Proćków, J., Pérez de Lastra, J.M. & Dey, A. 2022. Traditional uses, phytochemistry, pharmacology and toxicology of garlic (*Allium sativum*), a storehouse of diverse phytochemicals: A review of research from the last decade focusing on health and nutritional implications. *Frontiers in Nutrition*, 9: 949554.

US-FDA; Food and Drug Administration, US Department of Health and Human Servises 2017.

Wang, X., Shen, Y., Thakur, K., Han, J., Zhang, J.G., Hu, F. & Wei, Z.J. 2020. Antibacterial activity and mechanism of ginger essential oil against *Escherichia coli* and *Staphylococcus aureus*. *Molecules*, 25(17): 3955.

Wright, J.S. 2002. Predicting the antioxidant activity of curcumin and curcuminoids. *Journal of Molecular Structure: Theochem*, 591(1–3): 207–217.

Xue, Q., Xiang, Z., Wang, S., Cong, Z., Gao, P. & Liu, X. 2022. Recent advances in nutritional composition, phytochemistry, bioactive, and potential applications of *Syzygium aromaticum* L.(Myrtaceae). *Frontiers in Nutrition*, 9. http://doi.org/10.3389/fnut.2022.1002147

Yang, H., Huang, S., Wei, Y., Cao, S., Pi, C., Feng, T., Liang, J., Zhao, L. & Ren, G. 2017. Curcumin enhances the anticancer effect of 5-fluorouracil against gastric cancer through down-regulation of COX-2 and NF-κB signaling pathways. *Journal of Cancer*, 8(18): 3697–3706.

Yashin, A., Yashin, Y., Xia, X. & Nemzer, B. 2017. Antioxidant activity of spices and their impact on human health: A review. *Antioxidants*, 6(3): 70.

Yusuf, A., Fagbuaro, S.S. & Fajemilehin, S.O.K. 2018. Chemical composition, phytochemical and mineral profile of garlic (*Allium sativum*). *Journal of Bioscience and Biotechnology Discovery*, 3(5): 105–109.

Zhang, H.A. & Kitts, D.D. 2021. Turmeric and its bioactive constituents trigger cell signaling mechanisms that protect against diabetes and cardiovascular diseases. *Molecular and Cellular Biochemistry*, 476(10): 3785–3814.

7 Antioxidants from Spice and Condiment Plant Species and Their Applications

Iram Iqbal, Fatima Saqib, Muqeet Wahid, Muhammad Farhaj Latif, Rabia Naz, and Mohammad S. Mubarak

7.1 BACKGROUND

Today's generation shares the problems of a demanding lifestyle and an imbalanced diet (de Araújo et al., 2021). These elements cause harm to human health and are responsible for many fatal metabolic diseases and epigenetic changes (Gupta et al., 2020). Thus, customers always seek food products that satisfy nutritional needs and positively affect people's health and emotional well-being, and decrease illness occurrence (Oliviero et al., 2018). According to Lutz et al. (2009), diet adjustment can save 30% of lives, mainly when it includes more plant-based biomass. Food becomes unsafe to eat because of the harmful chemical process known as lipid oxidation. The meal becomes intolerable and unsatisfying due to the flavor and the odor turning rotten. Oxidative rancidity may result in toxic chemicals, unfavorable off-flavors, unfavorable aromas, and a significant loss in food quality (Decker et al., 2010; Embuscado, 2015). In this respect, antioxidants prevent other molecules from being oxidized. One of the original definitions of oxidation, which is how the term 'oxidation' came to be used, is the reaction of an element or molecule with oxygen (Embuscado, 2015).

For food items vulnerable to this sort of chemical change, using antioxidants has become essential to avoid food going bad because of oxidative deterioration. Within this context, natural antioxidants included in food are polyphenolic compounds, carotenoids, vitamin C, certain protein-based compounds, phospholipids, and sterols (Choe & Min, 2009). Research findings showed that bioactive plant-based polyphenols work as food preservatives. Similar to a typical synthetic antioxidant compound (BHT, 100 ppm), 1.5% lychee pod extract in sheep meat nuggets prevents lipid oxidation (Das et al., 2016). Additionally, the antibacterial effect of polyphenol additions suppresses the development of molds, yeast, and some bacteria, including *mesophilic*, *Staphylococcus aureus*, and coliforms (Martillanes et al., 2020).

The antioxidant potency of polyphenolic compounds is mostly attributed to the presence of hydroxyl groups (Piccolella et al., 2019). Hence, dietary phenolics primarily function as reducing agents, hydrogen donors, and oxygen quenchers, which is how they provide antioxidant activity. Some phenolics act as binders of metal ions that function as oxidation reaction catalysts. In this regard, flavonoids are naturally occurring, polyhydroxylated aromatic substances found in many plant species, such as vegetables and fruits, and also in herbs and spices. Flavonoids can neutralize free radicals like hydroxyl, peroxyl, and superoxide radicals by forming metal ion complexes, rendering them inactive. Additionally, it has been shown that flavonoids can block the enzymes lipoxygenase and cyclooxygenase, responsible for oxidative rancidity in food production. Culinary spices and herbs have a long history of safe use and are good sources of antioxidants. Ancient Egyptians utilized a variety of spices, including cinnamon, cumin, and onion, among others, in their diet, for

DOI: 10.1201/9781003387046-7

health reasons, and for mummification more than 5,000 years ago (Embuscado, 2015). Recent years have witnessed a significant increase in the interest in plant-derived antioxidants among consumers, manufacturers, and food scientists. Familiar sources of phenolic compounds include several culinary herbs and spices, which exhibit greater antioxidant power than fruits, grains, and nuts (Pellegrini et al., 2006; Carlsen et al., 2010; Husen, 2020, 2021, 2022a, 2022b; 2023; Bachheti et al., 2023; Rahman & Husen, 2021, 2022, 2023a, 2023b). These substances have recently attracted much attention in diet-based therapy because of their affordability, accessibility, acceptability, extended administration, and omnipresence qualities (Usman et al., 2022).

It is well acknowledged that a plant-based diet, heavy in nutrient-rich vegetables, fruits, and other plant foods, may lower the risk of illnesses linked to oxidative stress. Since a regular diet contains more than 25,000 bioactive food components, it might be challenging to comprehend nutrition's complicated impact on many chronic illnesses (Carlsen et al., 2010). Many plant-based food items are available that contain numerous phenolic chemicals. In food goods, more than 500 different polyphenols have been identified. Outstanding sources of polyphenolic compounds with a protective impact on the health of humans include tea, coffee, chocolate, vegetables, fruits, spices, and culinary herbs (Issaoui et al., 2020). Spices are dried plant components that enhance the flavor, color, fragrance, and other culinary properties of food products. They are one of the richest sources of polyphenols (Singh et al., 2021b; Kaundal et al., 2022). Moreover, the most prevalent secondary metabolites in plants are phenolics, made principally from L-tyrosine and L-phenylalanine synthesized through the shikimate pathway and containing either one or more hydroxyl groups attached directly to aromatic rings. These compounds include phenolic acids, simple phenols, flavonoids, stilbenes, coumarins, condensed and hydrolyzable tannins, lignans, and lignins (Chirinos et al., 2009; Kumar & Pruthi, 2014). The examination and investigation of more chemical components from plants have become more practical due to advances in plant biochemistry research. Additionally, it is possible to investigate the bioprocesses involved at the molecular level of synthesis of these phytoconstituents (Bostan et al., 2013).

In addition, polyphenols are well-known natural medicinal substances used to treat various chronic or non-communicable diseases, including nephrotoxicity, cardiovascular disease, neurodegenerative diseases, and cancer, according to epidemiological data (Wahle et al., 2010; Mark et al., 2019; Piccolella et al., 2019). Polyphenols also have well-documented pharmacological properties in the literature, including antioxidant, antimicrobial, anti-inflammatory, antidepressant, antilipidaemic, and anticancer activities (Wahle et al., 2010; Kumar & Goel, 2019; Wang et al., 2020; Usman et al., 2022). According to the 'US National Health and Nutrition Examination Survey', 5–10% of individuals in the US use botanical-based supplements, such as spices, for health improvement reasons. Since spices have fewer adverse effects than traditional drugs, are more readily available, and have health advantages, their use may have increased (Su & Li, 2011; Clarke et al., 2015). In this respect, almost half of the participants in a cross-sectional survey study of 703 individuals in the midwestern US, who were knowledgeable about the health benefits of spices (48%) and were keen for the use of spices as 'complementary and alternative medicine' therapies (CAM) therapies (9%) were found to be interested in learning about the health benefits of spices and their use as well as predictors of spice use (51%). Eight of the ten listed spices were known to or used by the majority (>50%) of the participants. Some 54% of participants reported using one or more spices daily, and 64, 58, and 56% of individuals thought ginger, garlic, and cinnamon could boost wellness, respectively. Furthermore, the majority of participants also said that seven out of those ten spices helped avert a particular disease, with more than two-thirds of them citing ginger (72 %), garlic (68 %), and cinnamon (67 %) as examples. Spice use has been studied in pediatric populations in addition to adult populations. For instance, more than one in ten infants and kids received spices, primarily to treat minor illnesses like fussiness or gastrointestinal issues, coughs, and colds (Isbill et al., 2018). On the basis of the preceding section, this chapter summarizes the literature related to antioxidants derived from spices and condiments and evaluates their potential in treating diseases. It also focuses on food spices and condiments as good sources of natural antioxidants.

7.2 ANTIOXIDANTS

7.2.1 Classification

Antioxidants are those substances that prevent or postpone the initiation of oxidation and can be either natural or artificial (Shahidi & Zhong, 2010). Natural antioxidants are in greater demand over synthetic antioxidants due to safety concerns, rising customer preferences for natural products, clear labels, and decreased use of different food additives. Spices and culinary herbs are excellent sources of antioxidants for food preservation due to their naturally occurring antioxidant components. Utilizing antioxidants of natural origin from herbs and spices has extra benefits: they are healthy, and the body can quickly metabolize them. They may also be referred to as natural flavours or spices. On the other hand, synthetic antioxidants might not offer any additional nutritional benefits and might even negatively impact people (Embuscado, 2015). The classification of antioxidants is shown in Table 7.1.

7.2.2 Phytochemicals and Antioxidants from Spices

Since ancient times, people have used spices and herbs for therapeutic purposes, in addition to them being natural sources of flavour and colour. Because they contain potent chemicals, which have an antioxidative effect, herbs and spices are among the best sources of natural antioxidants (Embuscado, 2015). In this respect, a sizeable class of bioactive compounds that are generated from plants, called phytochemicals, may have potential anti-disease protective actions. Carotenoids, plant sterols, glucosinolates, flavonoids, other phenolic chemicals, and other sulphur-containing substances make up this group; more than 6,000 flavonoids have been identified (Jaganath & Crozier, 2010). Phytochemicals are abundant in spices and other plants (Embuscado, 2015). The ability of spices to act as antioxidants, due to their phenolic components, has been the subject of much research work (Lu et al., 2011). The chemical makeup of spices, namely the amount of polyphenolic and other biologically active chemicals, is related to their antioxidant potential. The active substances present in culinary herbs and spices, such as phenolic acids, flavonoids, alkaloids, lignans, essential oils, and the principal antioxidants, are listed in Table 7.2.

TABLE 7.1
Classification of Antioxidants (Embuscado, 2015)

Sr.no	Class	Examples	Functions
1	Chelating agents	• Phosphates • Citric acid • Ethylenediaminetetraacetic acid (EDTA)	Bind or sequester metallic ions that can catalyze oxidation
2	Scavengers of free radicals	• BHAs (Butylated hydroxyanisole) • TBHQ (*tert*-Butylhydroquinone) • BHTs (Butylated hydroxytoluene) • Tocopherols • Propyl gallate • Herb and spice extracts	Donate a hydrogen atom to stop free radicals.
3	Scavengers of oxygen	• Vitamin C (ascorbic acid) • Ascorbates • Erythorbic acid • Ascorbic palmitate • Sulphates and bisulphates	Oxygen-based reactions

TABLE 7.2
Valuable Chemicals in Spices and Herbs

Spices and Herbs	Phytocompounds
Allspice	Pimentol, gallic acid, eugenol, quercetin
Anise	Linalool, pinene, camphene, *trans*- and *cis*-anetholes, eugenol, acetanisole, luteolin-7-glucoside, isoorientin, rutin, apigenin-7-glucoside
Basil	Apigenin, quercetin, catechins, anthocyanins, rutin, kaempferol, limonene, eugenol, terpinene, carvacrol, menthol, geraniol, tannins, safrole, ursolic, rosmarinic acids, *p*-coumaric acid
Bay leaf	1,8-cineole, cinnamtannin
Black pepper	Camphene, piperine, pinene, terpenes, limonene, isoquercetin, sarmentine, piperidine
Caraway	Aromatic aldehydes, monoterpenes, sesquiterpenes, terpene esters, terpenol, terpenon, safranal, terpenal, limonene, kaempferol, tannins, quercetin, caffeic, *p*-coumaric, ferulic, chlorogenic acids
Cardamon	Terpinolene, limonene,1,8-cineole, myrcene, quercetin, caffeic acid, luteolin, kaempferol, pelargonidin
Cinnamon	Terpineol, eugenol, limonene, proanthocyanidins, catechins, tannins, safrole, pinene, linalool, methyl eugenol, benzaldehyde
Cloves	Eugenol, acetyl eugenol, isoeugenol, pinene, sesquiterpenes, vanillin, flavonoids, gallic acid, phenolic acids
Coriander	Borneol, linalool, terpineol, geraniol, cumene, terpinene, pinene, quercetin, caffeic, kaempferol, ferulic, *n*-coumaric, vanillic acids, tocopherols, rutin, pyrogallol
Dill	Myricetin, kaempferol, limonene, catechins, carvone, isorhamnetin, quercetin
Fennel flower	Thymoquinone, thymol, *p*-cymene, thymohydroquinone, carvacrol, nigellicine, hederin, nigellidine, pinene
Fenugreek	Aromatic aldehydes, sesquiterpenes, terpenes
Garlic	Diallyl disulphide, diallyl sulphide, diallyl trisulphide, allicin, *S*-allyl cysteine, allyl isothiocyanate
Ginger	Paradol, gingerol, turmeric, geraniol, borneol, geranial, linalool, zingerol, camphene, zingiberon
Horseradish	Sinigrin, allyl isothiocyanate, phenyl methyl isothiocyanate, asparagine
Lavender	Quercetin, apigenin, limonene, kaempferol, ferulic, glucosides, rosmarinic, caffeic, *p*-coumaric acid
Marjoram	Limonene, terpinene, pinene, *p*-cymene, ferulic, apigenin, sinapinic, syringic, caffeic, vanillic, rosmarinic, 4-hydroxybenzoic acids
Mustard	Kaempferol glucoside, carotene, allyl isothiocyanate, isorhamnetin, isorhamnetin-7-*O*-glucoside
Myrtle	Anthocyanins, limonene, pinene, ellagic acid, gallic acid, myrtocommulone, myricetin-3-*O*-rhamnoside, myricetin-3-*O*-galactoside
Nutmeg	Myricetin, lignans, caffeic acid, orgentin, catechins
Onion	Dipyridyl disulphide, apigenin, quercetin-4-glucoside, rutin, quercetin
Oregano	Luteolin, quercetin, apigenin, myricetin, eriodictyol, diosmetin, caffeic, carvacrol, rosmarinic, thymol, *p*-coumaric, protocatechuic acids
Parsley	Kaempferol, luteolin, myricetin, quercetin, caffeic acid, apigenin
Peppermint	Limonene, menthone, isomenthone, eriocitrin, hesperidin, apigenin, rutin, luteolin, carotenes, caffeic acid, tocopherols, chlorogenic acid, rosmarinic acid, menthol
Red pepper	Lutein, tocopherol, carotene, quercetin, capsanthin, ascorbic acid, capsaicin
Rosemary	Geraniol, rosmanol, apigenin, limonene, pinene, naringin, rosmarinic, vanillic, luteolin, caffeic acids, ursolic acid, carnosol
Saffron	Salicylic acid, flavonoids, crocins, safranal, caffeic, gallic, ferulic, syringic, *n*-catechuic, vanillic acids
Sage	Limonene, carnosol, pinene, catechins, apigenin, rosmarinic acid, luteolin, saponin, carnosine, caffeic acids, vanillic, geraniol
Turmeric	Eugenol, essential oils, curcumins, carotene, caffeic, ascorbic acids, *p*-coumaric, protocatechuic, syringic, vanillic acids

Source: Kaefer & Milner, 2008

7.2.3 Mining of Polyphenols from Spices

Free phenolics (FPs) are soluble and unattached to other biological molecules. Because they are so straightforward, free phenolics can effectively be mined using aqueous or organic solvents. However, bound phenolics (BPs) cannot be extracted using the same methods because they are insoluble and bonded substances. To ease the effective and exact mining of BPs, specific chemical, physical, and biological treatments have been devised, as detailed below.

7.2.3.1 Chemical Methods

In the chemical approach, phenolic molecules are effectively extracted using chemicals. Alkali and acidic mining methods are the two additional divisions of the procedure, depending on the chemicals employed to extract phenolic components. Alkali and acidic methods of phenolic extraction are depicted in Figure 7.1.

7.2.3.2 Alkali Method

The alkali method's demand for a pretreatment procedure makes it slightly more challenging (Wang et al., 2020). The alkali technique uses alkalis, such as sodium hydroxide (NaOH), for mining. The polyphenols bound to the cell wall components are released as a result of the ester bond being broken by the alkali treatment. Although NaOH is predominantly employed in this method's mining procedures, potassium hydroxide (KOH) is also used to break down the structure of cell walls. This process can extract bound phenolics, including vanillic, caffeic, ferulic, chlorogenic, *p*-coumaric, and syringic acids, among others. To stop BPs from oxidizing, the procedure also needs ascorbic acid (vitamin C) and EDTA. In addition, intense alkali environments make phenolic hydroxyl groups vulnerable to deprotonation. In contrast, strongly acidic environments prevent the synthesis of quinones, necessitating the addition of hydrochloric acid (HCl) to correct the pH. Additionally, because the precipitation of protein occurs under fundamental conditions in this approach, a sizeable percentage of the protein becomes insoluble and is washed off. It is also advised that this procedure be used at night in an inert gas atmosphere. Such restrictions on this method prevent its widespread application.

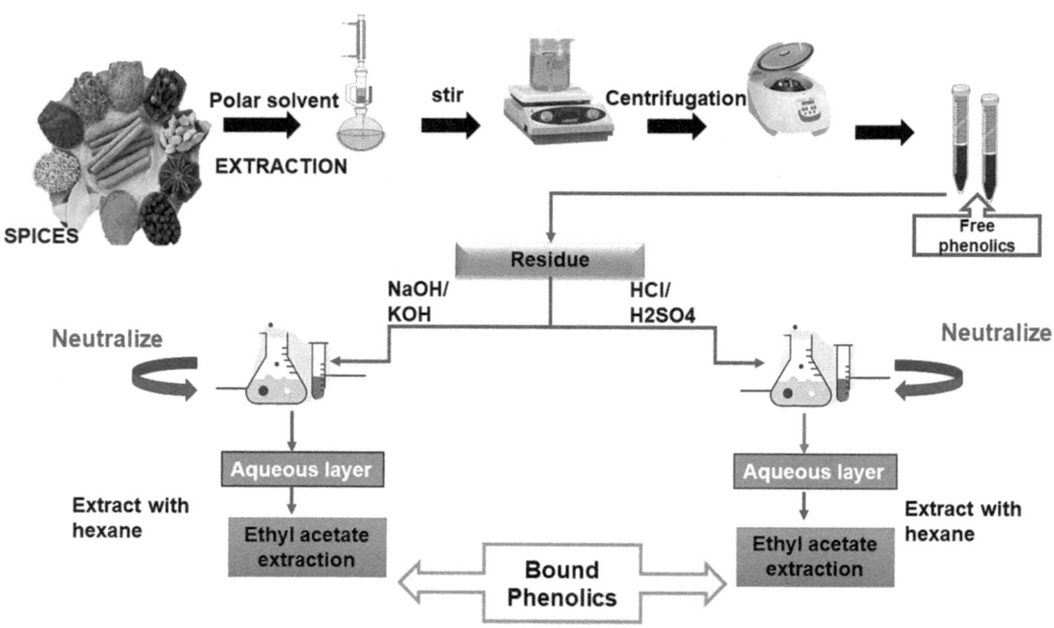

FIGURE 7.1 Alkaline and acidic methods of phenolic extraction.

7.2.3.3 Acidic Method

This is another widely used chemical mining technique. Although hydrochloric acid (HCl) and sulphuric acid (H_2SO_4) are used in this procedure, solid and mild acids such as oxalic acids can also be used for mining. The ester and glycosidic linkages between the components of the cell wall and the polyphenols are broken down by acid hydrolysis (Wang et al., 2020).

7.2.3.4 Biological Methods

Using biological components like enzymes as the foundation of the mining process is known as the biological method. Pectinase, amylase, xylanase, and cellulose are a few examples of enzymes that can hydrolyze carbohydrates and be employed for mining BPs. For example, the enzyme ferulic acid esterase interferes with the interaction between oligosaccharides and ferulic acid, which are part of the plant cell wall. The enzymatic methods are more precise, efficient, and dependable than chemical approaches. Limitations of the chemical method are likewise overcome by the biological method (Panja, 2018; Wang et al., 2020).

7.2.3.5 Physical Methods

Compared to other mining techniques, this process is significantly more affordable. The hydrogen, covalent, and hydrophobic connections between the macromolecules and phenolics are broken down using an intense microwave, ultrahigh-pressure heat treatment, and ultrasound, which release the phenolics from the microstructural forces. At the industrial level, this technique is heavily utilized (Panja, 2018).

7.2.4 Antioxidants as Food Preservatives

Oxidation of foods containing lipids or fats causes oxidative rancidity. Many foods develop off-flavours and off-odours due to this chemical reaction, turning them rancid and giving them harsh, unpleasant tastes and odours. These items might not even be safe to eat. Peroxides and hydroperoxides are produced when the oil or fats in food combine with ambient oxygen molecules, and carbonyl compounds are obtained as a by-product of secondary oxidation. Similarly, aldehydes, ketones, acids, and alcohols are the reaction products from oxidative rancidity that give the unpleasant flavours and odours that make food unpalatable and turn off consumers. Numerous studies have shown that herbs and spices, including oregano, sage, and rosemary, have high antioxidant activity (Cuvelier et al., 1994; Zheng & Wang, 2001; Pizzale et al., 2002). Reduced nutritional and physiological content of lipids results from degradation of lipid-soluble vitamins and important fatty acids by lipid oxidation. Additionally, it has been discovered that the by-products of lipid oxidation can activate the atherosclerotic process by increasing the levels of peroxides and cholesterol in blood serum, inhibiting the activity of enzymes, and causing pathophysiological changes in the mucosa of the digestive tract. One of the theoretically most accessible methods for lowering fat oxidation is using some antioxidants (Karpińska et al., 2001). According to the literature, the use of either synthetic or natural food additives can limit or minimize lipid oxidation and microbial growth in meat products (Renerre et al., 1996; Lee et al., 1997). Various polyphenols have different mechanisms of antioxidant action, but catechin, a flavonoid, was found to penetrate the lipid bilayer of bacterial cells to a great extent, causing the release of intracellular contents to the outside of the cell and, ultimately, cellular death. The synergistic effects of antibiotics were also linked to increased effectiveness and decreased dosage absorption. Creating organic food additives and preservatives can benefit from polyphenols (Takó et al., 2020).

7.2.4.1 Applications of Antioxidants from Spices as Food Preservatives

Foods can be preserved with the help of whole or ground spices and herbs. Inhibiting the oxidation of lipids and microbiological development in fish and meat is another application for spice extracts and essential oils. Thus, it is advantageous to use herbs and spices or products made from them, such as extracts and essential oils, as they are effective at reducing or inhibiting oxidative rancidity and offer additional advantages like clean labels, all-natural ingredients, and lack of genetically modified organisms (GMOs), all of which appeal to consumers (Embuscado, 2015). Table 7.3 summarizes different studies to evaluate the food preservation activity of different spices.

7.2.5 Oxidative Stress and the Role of Antioxidants

The main factor causing many human diseases is oxidative stress (Martins et al., 2016). All aerobic life forms are continually under the oxidative pressure of reactive oxygen species (ROS), formed during the metabolic use of oxygen (Guetens et al., 2002). ROS are produced *via* regular aerophilic metabolism as a by-product, but a rise in its level under stress and can be an elementary health risk (Rahal et al., 2014). The primary cell organelle involved in the formation of ROS is the mitochondrion. Through several oxidative phosphorylation procedures, adenosine triphosphate (ATP) is produced. Instead of four oxygen electron reductions, one or two occur during this process, forming H_2O_2 or O_2, which can then transform into additional ROS (Inoue et al., 2003). Different oxygen-free radicals are influential originators of the peroxidation of lipids, whose function in the aetiology of illnesses is well-established. Once lipid peroxidation is commenced, chain reactions will propagate until the formation of termination products. Consequently, biological systems accumulate fat peroxidation products such as 4-hydroxy-2-nonenol (4-HNE), F2-isoprostanes, and malondialdehyde (MDA). In this respect, DNA bases are highly vulnerable to oxidation by ROS, and 8-hydroxy-2-deoxyguanosine is the most apparent oxidation product from DNA bases.

Mutations and deletions can be caused by the oxidation of DNA bases in both mitochondrial and nuclear DNA. DNA originating from mitochondria is more susceptible to oxidative injury than nuclear DNA due to its proximity to ROS generation sites and insufficient restoration capacity. These oxidative alterations result in changes in enzymatic and structural proteins, which may have significant physiological consequences. Moreover, redox modulations of transcription factors affect (increasing or decreasing) their specific DNA-binding capabilities, thus affecting gene expression (Rahal et al., 2014). Numerous studies have reported the relationship between oxidative stress and disease occurrence and the protective effects of polyphenols as antioxidants. In this context, Hussain et al. (2016) reported that inflammation induced by oxidative stress drives numerous chronic diseases. There is an inversely proportionate link between oxidative stress and inflammation. It was claimed that the antioxidant capacity of polyphenolic compounds is advantageous for combating inflammation and carcinogenesis caused by inflammation (Hussain et al., 2016). In addition, Oliviero et al. (2018) reported that inflammation significantly contributes to the development of osteoarthritis. Therefore, preventing oxidative stress may avoid most diseases, including ageing, inflammation, obesity, osteoporosis, cancer, neurodegenerative diseases, diabetes, and cardiovascular diseases, among others (de Araújo et al., 2021; Singh et al., 2021a).

7.2.5.1 Health Benefits of Polyphenols from Spices

Spices contain phenolic chemicals, which are thought to be naturally occurring pharmaceuticals. Human tissues do not directly absorb the BPs that are consumed. To make them easy to use, they should instead enter the gastrointestinal tract through the stomach and continue to the large intestinal colon in their unbroken state, where they will undergo enzymatic digestion and colonic fermentation (Wang et al., 2020). The body tissues easily absorb and circulate the released simple form of BPs, which then perform numerous therapeutic actions like preventing cancer, Parkinson's disease, Alzheimer's disease , neurodegeneration, oxidative stress, microbial infection, hypertension, obesity, and hyperglycaemia (Hussain et al., 2019). The most popular seasoning ingredients used to enhance the flavour and appearance of food include cloves, chili, dried form of peppermint, star anise, turmeric, dried oregano seeds, celery seeds, rosemary, spearmint, capers, sweet basil, thyme (both dried and fresh), ginger, curry, lemon verbena, cinnamon, cumin, parsley, caraway, and

TABLE 7.3
Uses of Spices and Herbs to Inhibit Oxidation in Foods

Sr no	Spice / Herb	Usage Form	Antioxidant Activity	Oxidation Method	Results	Reference
1	Chinese five-spice ingredients: cloves, cinnamon, pepper, fennel, star anise	Powder		Sensory evaluation thiobarbituric acid reactive substance (TBARS)	All spices and mixtures decreased the cooked ground beef's rancid flavour and odour. The decreased TBA values of the spices compared with the control indicate they have antioxidant effects rather than masking rotten off-flavours.	Dwivedi et al., 2006
2	Coriander and annatto	Seeds and leaves		Conjugated dienes, FFA, TBARS, cholesterol oxides.	TBARS and conjugated dienes were observed in white hake fish meatballs during frozen storage. Anatto and coriander were effective in reducing lipid oxidation. When combined, these spices' increased antioxidant activity was presumably caused by a synergistic effect. The spices' antioxidant activity was reduced during cooking.	Sancho et al., 2011
3	Garam masala spices like caraway, black pepper, cardamom, cinnamon, chili powder, cloves, cumin, coriander, fennel, star anise, ginger, and nutmeg.	Powder		Sensory evaluation TBARS	Garam masala's unique spices worked well to keep cooked beef's TBA levels low during refrigeration and significantly lessened the sense of rancid flavour and odour.	Vasavada et al., 2006
4	Garlic and onion	Powder		Volatile chemicals, free fatty acids (FFA), peroxide value (POV), and thiobarbituric acid reactive substance (TBARS)	With the reduction of oxidative products and hexanal, fresh pork belly and loin showed antioxidant activity from garlic and onion comparable to sodium ascorbate.	Park et al., 2008

(Continued)

TABLE 7.3 (CONTINUED)
Uses of Spices and Herbs to Inhibit Oxidation in Foods

Sr no	Spice / Herb	Usage Form	Antioxidant Activity	Oxidation Method	Results	Reference
5	Lamiaceae (mint) family spices	Marinades –Southwest, Caribbean, herb	Polyphenols determination by HPLC	Increase in heterocyclic amines	Marinade was applied to steaks for an hour before grilling. Compared to controls and liquid blanks, all three marinades considerably reduced imidazo-azarene heterocyclic amines (HCAs). The most significant reduction in total HCAs was shown in the Caribbean blend, followed by the herb and Southwest marinades.	Smith et al., 2008
6	Licorice and rosemary extract	Extract	DPPH, ABTS, phenolic compounds	TBARS, sensory evaluation	In pre-cooked pig patties, licorice extract was a powerful antioxidant to prevent fat/lipid oxidation during chilled and frozen storage, regardless of salt level. The higher potential of this substance to scavenge radicals, especially the hydroxyl radical (•OH), has been linked to its antioxidant action. When administered at the optimal dosages, the effect of licorice extract was more than rosemary extract and nearly as effective as BHA.	Jiang et al., 2013
7	Oregano	Essential oils	Essential oil (EO) analysis	Sensory evaluation	By reducing lipid oxidation, essential oil from oregano increased the shelf life of extra virgin oil while retaining its sensory quality.	Asensio et al., 2012
8	Oregano and Rosemary	Extracts	Total phenolic content (TPC), DPPH assa'	Reduced levels of eicosapentaenoic acid (EPA) and docosahexaenoic acid (DHA)	Long-chain n-3 fatty acid oxidation was prevented by the rosemary and oregano extracts. Oregano extract's antioxidant activity was inferior to that of rosemary extract. The rosemary extract has a three times greater capacity to scavenge free radicals than DPPH.	Bhale et al., 2007
9	Rosemary and rosmarinic acid	Powder form		Heterocyclic amines (HCAs) and mutagenicity of HCA.	Rosmarinic acid and rosemary powder can reduce the formation of heterocyclic amines (HCAs) in beef patties when fried at 400 °C and the mutagenicity of these HCAs.	Balogh et al., 2000
10	Rosemary	Extract		Rancimat test	Rosemary extracts showed marked antioxidant and antibacterial potential when added to chicken frankfurters packed under a vacuum.	Rižnar et al., 2006

(Continued)

TABLE 7.3 (CONTINUED)
Uses of Spices and Herbs to Inhibit Oxidation in Foods

Sr no	Spice / Herb	Usage Form	Antioxidant Activity	Oxidation Method	Results	Reference
11	Sage and garlic	Fresh and flakes		FFA, cholesterol oxides, secondary products of lipid oxidation, degradation of vitamin E	During thermal processing and storage of chicken breast meat at 18 °C, the oxidation of lipids and cholesterol can be reduced and delayed for 90 days by adding 0.1 g of sage/100 g. On the other hand, garlic did not prevent oxidation.	Mariutti et al., 2011
12	Turmeric and shallot	Extract		POV, sensory evaluation TBARS,	Whole gutted rainbow trout fish dipped in either 1.5% turmeric extract or 1.5% shallot extract, or both turmeric and shallot extracts combined (1.5 % + 1.5 %) can delay chemical changes, slow microbial growth, preserve sensory qualities, and lengthen the shelf life of rainbow trout when kept refrigerated.	Pezeshk et al., 2011
13	Thyme	Dried leaves	Total phenolic content TPC, ORAC assay	Rancimat test, sensory evaluation	The phenol-enriched mix of olive oil obtained with a 60: 40 ratio of olive and thyme resulted in a significantly more stable product to oxidation than the control.	Rubió et al., 2012
14	Thyme and lemon balm	Essential oil (EO.)	DPPH assay and. GC of EO,	Lipid peroxide measurement	When kept at 4 °C, thyme and balm essential oils can effectively slow down the degradation of chicken meat and increase its shelf life.	Fratianni et al., 2010

marjoram. Additionally, these spices are an abundant source of numerous biologically active chemicals, such as phenolic compounds. There are tremendous medicinal effects of spices because of these substances (Saraswathi et al., 2020). For example, clove is top of the list and has a total phenol content of approximately 15,188 mg/100 g, related to their potent antioxidant properties. Oregano is high in flavonoids and may be used to treat cancer, oxidative stress, respiratory, inflammatory, and digestive diseases, headaches, rheumatism, and diabetes (Issaoui et al., 2020). Most polyphenols, measured in mg gallic acid equivalents (GAE)/g dry matter, are found in oregano, thyme, and rosemary (Slimestad et al., 2020). Figure 7.2 is a summary of spice-based health benefits of polyphenols and antioxidants. Below are some health benefits of polyphenols,

7.2.5.2 Antioxidant Activity

Over the past 40 years, oxidative stress has gained more attention as a contributing component to aging and numerous pathophysiological aging processes (Hybertson et al., 2011). Antioxidants are substances that counteract the harmful effects of free radicals and the oxidative stress they cause (Yashin et al., 2017). Though living systems also include endogenous antioxidant substances to combat the harmful effects of free radicals, these antioxidants can occasionally be insufficient, necessitating exogenous antioxidants. The use of antioxidants present in food products is rising quickly because of the adverse effects linked to the use of synthetic antioxidants (Sokamte et al., 2019). Research findings showed that polyphenols can change the redox status of a cell under oxidative stress by inhibiting the formation of superoxide ions by the enzyme xanthine oxidase (Basli et al., 2017). Therefore, nutritional antioxidants have received much attention as curative and preventative measures. Moreover, numerous studies have shown that some naturally occurring antioxidants can reduce oxidative stress (Osawa, 1999).

Published research indicated that the quantity and the position of the hydroxyl (–OH) group in a specific phenolic compound determines its antioxidant capacity (Kumar & Goel, 2019). Spices have inherent radical-scavenging properties (RSA) because they have an abundance of phenolics and polyphenols (Yashin et al., 2017). Various studies have shown the potent antioxidant activity of polyphenols from spices. The polyphenols in spices have also been shown to have substantial antioxidant action (Škrovánková et al., 2017). Using the Folin-Ciocaulteu (FC) reagent assay and the 1,1-diphenyl-2-picrylhydrazyl (DPPH) assay, the total phenolic content (TPC) and antioxidant capacity of both aqueous and ethanolic extracts of paprika (*Capsicum annuum*) and pepper (*Piper nigrum*) were evaluated. The TPC of paprika was greater than that of pepper and varied from 14.67 to 28.78 mg GAE/g, whereas that of pepper ranged from 12.03 to 22.88 mg GAE/g. Compared

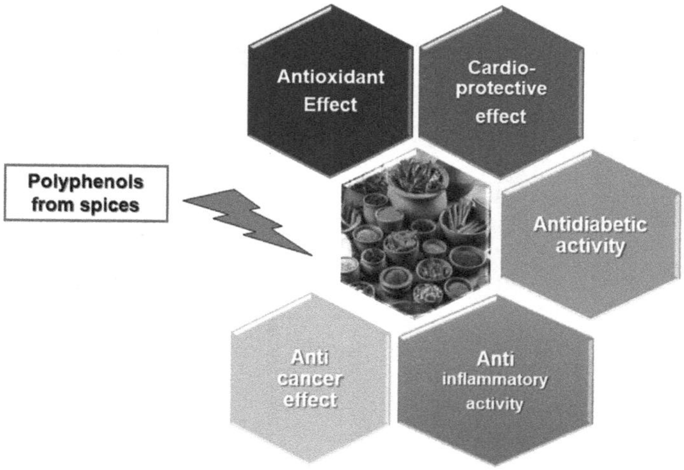

FIGURE 7.2 Health benefits of polyphenols from spices.

with the aqueous extract (4.45–16.24 mg ascorbic acid equivalents (AAE)/g), the ethanolic paprika extract had a greater overall antioxidant activity (8.73–16.17 mg AAE/g). The total antioxidant activity of pepper was 7.07–15.81 mg AAE/g for the aqueous extract and 8.25–15.93 mg AAE/g for the ethanol extract (Škrovánková et al., 2017). In another study (Basli et al., 2017), 12 regularly consumed culinary spices (clove, cumin, black cumin, cinnamon, cardamom, allspice, black cardamom, turmeric, nutmeg, fennel, star anise, and black pepper) were investigated for their TPC and antioxidant efficacy. The results showed that allspice and clove had the highest total phenolic concentration. Clove demonstrated the highest antioxidant capacity in the antioxidant assay, leading to the hypothesis that phenolic content and antioxidant efficacy are positively correlated.

Similarly, researchers conducted a thorough investigation of 39 spices to determine their TPC and antioxidant potential using the 2,2′-azino-bis(3-ethylbenzothiazoline-6-sulphonic acid) (ABTS), DPPH, and 'cupric reducing antioxidant capacity' (CUPRAC) assays (Assefa et al., 2018). According to the ABTS, CUPRAC, and DPPH assays, the total antioxidant capacity ranged from 1.42 to 112.94 mg AAE/g, 0.52–54.47 mg TE/g, and 1.14 to 91.09 mg Trolox equivalents (TE). The FC assay was used to calculate the TPC, which ranged from 2.93 to 160.55 mg of GAE/g dry weight (DW). The most effective antioxidant agent was discovered to be *Syzygium aromaticum* (clove flower buds), then *Pimenta dioica* (allspice fruits), and thirdly, *Cinnamomum verum* (cinnamon bark). The findings also indicated a clear relationship between total phenolic content and antioxidant capability. Sepahpour et al. (2018) evaluated the TPC and antioxidant potential of several solvent extracts of lemongrass, turmeric, curry leaf, and torch ginger. High-performance liquid chromatography (HPLC) was used to determine the total phenolic compounds, while the DPPH and FRAP tests were used to evaluate the antioxidant activity. Results revealed that curry leaf, lemongrass, and torch ginger had the next highest antioxidant activity, with turmeric coming in first. Turmeric (221.68 mg GA/g freeze-dried crude extract), torch ginger (98.10 mg GA/g freeze-dried crude extract), and lemongrass (28.19 mg GA/g freeze-dried crude extract) all yielded the highest TPC when 80 percent solvent was used. According to these findings, the quantity of all phenolic compounds and their antioxidant power is directly correlated (Sepahpour et al., 2018).

7.2.5.3 Anti-inflammatory Activity

Various nutritionally induced oxidative stress causes can bring on inflammation. According to published data, a high intake of macronutrients *via* nuclear factor-kappa B (NF-κB) mediated cell signaling pathways might encourage oxidative stress and contribute to inflammation (Tan et al., 2018), which is accompanied by the stimulation of expression of many inflammatory genes and transcription factors (Hussain et al., 2016). Additionally, it happens due to increased nitric oxide (NO) generation. Pro-inflammatory cytokines such as interleukin (IL)-1, tumor necrosis factor (TNF), and IL-6, IFN, IL-8, and others boost NO generation. Therefore, a fundamental component of any strategy for treating inflammation is to decrease NO generation. Reduced NO generation and, ultimately, prevention of inflammation will result from the downregulation of pro-inflammatory mediators (Singh et al., 2021a). The activity of phenolic chemicals contributes to the decrease in NO generation (Zhang & Tsao, 2016). In both animal models and *in vitro*, capsaicin from red peppers has sufficient antioxidant capacity to reduce oxidative stress in diverse organs or tissues. An animal study found that capsaicin slowed the migration of neutrophils (inflammatory cells) towards the inflammatory focus, which decreased proinflammatory cytokine production and vascular permeability. By regulating messenger molecules generated by obese mouse fat cells and inhibiting macrophage from releasing pro-inflammatory mediators *in vitro*, capsaicin may also reduce inflammation brought on by obesity (Kang et al., 2007; Spiller et al., 2008). Through the control of pro-inflammatory and anti-inflammatory gene expression *in vitro*, cinnamon polyphenol extract reduced the inflammatory processes (Cao et al., 2008).

7.2.5.4 Cardiovascular Activity

Cardiovascular disease (CVD) is the leading cause of mortality worldwide. Age, low-density lipoprotein (LDL) oxidation, high blood pressure (HBP), high cholesterol (HC), inflammation, urine isoprostanes, and hyperglycaemia, are the most significant cardiometabolic health impairment risk factors. Additionally, platelet accumulation and hyperactivation facilitate thrombus formation. Furthermore, dysregulated endothelium leads to the onset of atherosclerotic thrombosis and cardiovascular disorders (Lutz et al., 2019). Anti-atherogenic and anti-inflammatory effects of dietary polyphenols minimize cardiovascular disease risk factors and morbidity rates (Tressera-Rimbau et al., 2017). Therefore, the treatment of one condition will immediately result in the inhibition of all linked ailments. Polyphenols can also intermingle with other biologically active molecules to improve their pharmacological efficacy and confer synergistic effects (Fraga et al., 2019; de Araújo et al., 2021). It has been hypothesized that a diet rich in phenolic compounds might reduce the incidence of coronary heart disorders, such as atherosclerosis. These metabolites can prevent the oxidative degradation of low-density lipoprotein (LDL) and execute their anti-platelet aggregation action, a critical mechanism in the proliferation of atherosclerosis (Lutz et al., 2019). Hence, foods rich in polyphenols from spices might be employed as amending or preventive agents for healing more than one sickness at a time and for general well-being (Yashin et al., 2017).

7.2.5.5 Anti-diabetic Activity

Diabetes is characterized by an imbalance in the hormone insulin and an incorrect blood glucose level (Singh et al., 2021a). It is the top metabolic condition globally, and 592 million may be impacted through 2035 (Ahangarpour et al., 2019). Due to a disparity between the creation of free radicals and a person's capacity to oxidize them, diabetes is recognized as an oxidative stress condition. The destruction of body organs by ROS, which are poorly neutralized by antioxidants and causes inflammation and a range of metabolic problems, is strongly linked to oxidative stress (Furukawa et al., 2004). In preventing oxidative stress and the related dysfunction of insulin and glucose receptors, phenolic acid is a key player. Antioxidants inhibit free radicals in various ways, and phenolic substances, especially phenolic acids, fight oxidative stress and associated drawbacks by blocking the enzymes that produce ROS (Wu et al., 2014). In this respect, chlorogenic and ferulic acids have been shown to stimulate transporters and to be effective anti-diabetic drugs. Ferulic and chlorogenic acids showed the exact mechanism of transporter stimulation and work as anti-diabetic agents (Vetrani et al., 2020).

Various studies have evaluated the role of spice-originating polyphenols in controlling diabetes. Bi and colleagues have reviewed the function of spices in regulating hyperglycaemia. The polyphenolic components of spices are responsible for the anti-diabetic potential of spices. Improved insulin resistance and glucose homoeostasis are benefits of the complex polyphenol combination (Bi et al., 2017). Similarly, Haldar et al. (2018) reported that GLP-1, a glucagon-like peptide, is crucial for maintaining a healthy blood glucose level. It was postulated that blended spices with high dietary polyphenol content might promote postprandial GLP-1 release. Therefore, a clinical trial was conducted on twenty young, healthy-looking Chinese males supplemented with test meals enhanced with various spices and vegetables. The study was randomized, controlled, and dose-dependent. Results showed a significant increase in plasma GLP-1 concentrations, which supported the use of spices high in polyphenols to reduce blood sugar levels (Haldar et al., 2018).

Furthermore, it has been claimed that spice-derived phenolics and polyphenols actively interrupt the signaling for the metabolism of glucose and regulate glucose production (intrahepatic glucose output), insulin release (from islet cells), and glucose uptake (intestinal glucose uptake) with increased activity of insulin. All these actions of polyphenolic compounds will eventually result in a level of blood sugar that is balanced, helping to avoid hyperglycemia (Ge et al., 2017).

7.2.5.6 Anti-cancer Activity

Cancer is characterized by unchecked cell growth and spread in a multifactorial, terrible and dynamic condition. According to a report by the World Health Organization (WHO), it is among the foremost causes of death. More than twice as many people die from cancer yearly than from AIDS, malaria, and tuberculosis combined. By 2030, it is predicted that over 22.2 million people will have passed away from the disease (Reddy et al., 2003; Basli et al., 2017). Polyphenols are thought to be able to stop several important events in cancer development, including angiogenesis, inflammation, differentiation, apoptosis, and metastasis. Because oxidative stress plays a crucial role in carcinogenesis, polyphenols' anti-cancer impact is primarily related to their antioxidant capacity. Over half of all anti-cancer prescription medications licensed globally between the 1940s and 2006 were natural compounds or their derivatives (Efferth et al., 2007). Xu et al. (2020) examined the *in vitro* anti-cancer properties of 15 spice extracts in methanol. Six cancer cell lines were used in the investigation, including those from the colon (HT29), liver (HepG2), pancreas (MIA PaCa2), breast (MCF7), blood (Raji), and lung (A549). Results showed that all spice-originated extracts exhibited notable anti-cancer efficacy. The anti-cancer effects of spices with high polyphenol content were greater (Xu et al., 2020). Polyphenols' ability to act as antioxidants prevents cancer development and cell growth. To counteract the effects of free radicals, the polyphenolic chemicals provide one hydrogen atom. To reduce oxidative stress, these substances can induce the activation of phase I enzymes, which is directly related to the reduction of carcinogenesis. By using both intrinsic and extrinsic mechanisms, the polyphenolic chemicals also aid in the death of cancer cells (Bi et al., 2017). Listed in Table 7.4 are the pharmacological effects of phenolic compounds obtained from spices.

7.3 NOVEL ANTIOXIDANT FORMULATIONS FROM SPICES

Nano-delivery systems have gained popularity in the past several decades due to their high bioavailability and potential for precise targeting. Recently, plant extracts and other phytochemicals have been mixed with nano-formulations. Plants and spices are the most abundant sources of phytochemicals, such as polyphenols, alkaloids, and flavonoids, with powerful antioxidant capabilities. Therefore, dietary supplementation with bioactive phytochemicals with antioxidant and anti-inflammatory effects is a viable therapeutic approach to deal with the body's natural processes that are out of balance. Nano-delivery systems, including nanoparticles, solid lipid nanoparticles, nano-emulsions, micelles, liposomes, and carbon nanotubes (CNTs), can improve the stability and solubility of drugs with superior permeability to improve the ability to target cells, as well as negligible toxicity (Prajapati et al., 2022).

7.4 CONCLUSION AND FUTURE TRENDS

In conclusion, results obtained from this chapter indicate that when compared with other spices, thyme, clove, rosemary, and oregano are substantial sources of polyphenolic content. As a result of their high polyphenol content, these spices can significantly lower the risk of illnesses. Despite scientific evidence of the antioxidative action of several spices, herbs, and teas, rosemary and sage leaf extracts are the natural antioxidant sources used most often. They primarily counteract oxidative stress, which is associated with the development of several illnesses, including diabetes, inflammation, and cancer, among others, by scavenging ROS. This is the primary justification for polyphenols' wide variety of pharmacological effects. Furthermore, including these ingredients in healthy meals helps achieve sustainability objectives. Therefore, these secondary metabolites can be developed into beneficial dietary supplements and an alternative to synthetic antioxidants. Although it is justified to have a diet rich in polyphenols, large amounts must be used with caution. A clinical study should also be conducted to validate the true efficacy of polyphenols and spices. Due to rising consumer demand, researchers will continue to focus on developing more effective

TABLE 7.4
Summary of Phenolic Compounds from Spices and Their Pharmacological Effects

Sr. No	Spice	Biological Origin	Polyphenolic Compound	PubChem ID	Pharmacological Effect	References
1	Ajwain 	*Trachyspermum ammi* L.	Thymol	(Pubchem CID: 21105998)	• Prevents cardiovascular disorders • Antidiabetic and prevent hyperglycaemia • Antimicrobial effect • Nematicidal effect	Chahal et al., 2017; Mughal, 2019; Feng et al., 2020
2	Aniseed 	*Pimpinella anisum* L.	Chlorogenic acid	(CID:1405788)	• Potent antioxidant activity	Bettaieb Rebey et al., 2018
3	Black cardamom 	*Amomum subulatum*	Diosmin	(Pubchem CID: 4444932)	• Latent antioxidant agent	Feng et al., 2020
4	Black cumin 	*Nigella sativa* L.	Thymol	(Pubchem CID: 21105998)	• Prevents cardiovascular disorders • Antidiabetic and prevent hyperglycaemia • Antimicrobial effect • Nematicidal effect	Chahal et al., 2017; Mughal, 2019; Feng et al., 2020
5	Black pepper 	*Piper nigrum* L.	*p*-Hydroxybenzoic acid	(Pubchem CID: 132)	• Strong antioxidant	Feng et al., 2020

(Continued)

TABLE 7.4 (CONTINUED)
Summary of Phenolic Compounds from Spices and Their Pharmacological Effects

Sr. No	Spice	Biological Origin	Polyphenolic Compound	PubChem ID	Pharmacological Effect	References
6	Caraway 	*Carum carvi* L.	Carvacrol	(CID:21105867)	• Excellent antimicrobial • Inhibits oxidative stress • Prevents carcinogenesis	Sachan et al., 2016
7	Celery 	*Apium graveolens* L.	Luteolin	(CID: 444102)	• Potential antioxidant effect	Yashin et al., 2017
8	Cinnamon 	*Cinnamomum verum* J. Presl	Hydroxycinnamic acids	(CID: NR.)	• Antimicrobial and food preservative • Antidiabetic effect • Prevents oxidative stress • Inhibits tumour formation	Rashid et al., 2020; Takó, M. et al., 2020; Singh et al., 2021a
9	Clove 	*Syzygium aromaticum* (L.) Merr.& Perry	Eugenol	(CID: 3314)	• Reduces inflammation • Acts as an antioxidant and antiviral • Antimicrobial effect • Promotes digestion	Issaoui et al., 2020; Saraswathi et al., 2020
10	Coriander 	*Coriandrum sativum* L.	Stigmasterol, sitosterol	(CID:4444352) (CID: 192962)	• Exhibits antioxidant effects. • Protective effect from oxidative damage • Prevents high levels of glucose	Bi et al., 2017; Ge et al., 2017; Issaoui et al., 2020
11	Dill 	*Anethum graveolens* L.	Quercetin	(CID: 4444051	• Potential antioxidant activity • Prevention of cardiovascular diseases • Anticancer effect • Diminishes renal disorders	Yashin et al., 2017

(*Continued*)

TABLE 7.4 (CONTINUED)
Summary of Phenolic Compounds from Spices and Their Pharmacological Effects

Sr. No	Spice	Biological Origin	Polyphenolic Compound	PubChem ID	Pharmacological Effect	References
12	Fennel 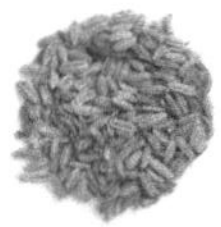	*Foeniculum vulgare* Mill.	Quercetin	(CID: 4444051	• Potential antioxidant activity • Prevention of cardiovascular diseases • Anticancer effect • Diminishes renal disorders	Yashin et al., 2017
13	Fenugreek 	*Trigonella foenum-graecum* L.	Rhaponticin	(CID: 552853)	• Reactive oxygen species (ROS) scavenger • Improves insulin sensitivity and functions of mitochondria	Li et al., 2018
14	Garlic 	*Allium sativum* L.	Quercetin	(CID: 4444051	• Potential antioxidant activity • Prevention of cardiovascular diseases • Anticancer effect • Diminishes renal disorders	Yashin et al., 2017
15	Ginger 	*Zingiber officinale* Rosc.	(+)-Catechin, gallic acid, gingerol	(CID: 8711) (CID: 361) (CID: 391126)	• Acts as an antioxidant • Antimicrobial effect	Yashin et al., 2017; Ghafoor et al., 2020; Takó et al., 2020
16	Mint 	*Mentha piperita* L.	Rosmarinic acid	(CID: 4445104	• Reduces oxidative stress • Anticancer and cytotoxic effects • Antimicrobial effect • Neuroprotective	Elansary et al., 2020; Issaoui et al., 2020

(Continued)

TABLE 7.4 (CONTINUED)
Summary of Phenolic Compounds from Spices and Their Pharmacological Effects

Sr. No	Spice	Biological Origin	Polyphenolic Compound	PubChem ID	Pharmacological Effect	References
17	Mustard 	*Brassica juncea* L. Czern	Kaempferol	(CID: 444395)	• Potential antioxidant effect • Reduces hyperglycaemia • Acts as an anticancer agent	Yashin et al., 2017
18	Oregano 	*Origanum vulgare* L.	Naringenin	(CID: NR.)	• Anti-inflammatory • Anticancer effect • Anti-microbial effect • Acts as immunomodulator • Prevents oxidative stress • Acts as an antioxidant	Issaoui et al., 2020
19	Parsley 	*Petroselinum crispum* Mill.	Apigenin	(CID: 444100)	• Latent antioxidant agent	Yashin et al., 2017
20	Rosemary 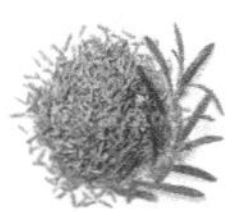	*Rosmarinus officinalis* L.	Rosmarinic acid	(Pubchem CID: 4445104)	• Reduces oxidative stress • Antimicrobial effect • Neuroprotective • Cytotoxic effects and anticancer effects	Elansary et al., 2020; Issaoui et al., 2020
21	Saffron 	*Crocus sativus* L.	Safranal	(CID: 55000)	• Acts as anticonvulsant • Free radical scavenger	Issaoui et al., 2020
22	Sage 	*Salvia officinalis* L.	Carnosol	(CID: 390568)	• Anti-inflammatory activity • Augments formation of cartilage matrix	Oliviero et al., 2018

(Continued)

TABLE 7.4 (CONTINUED)
Summary of Phenolic Compounds from Spices and Their Pharmacological Effects

Sr. No	Spice	Biological Origin	Polyphenolic Compound	PubChem ID	Pharmacological Effect	References
23	Thyme 	*Thymus vulgaris* L.	Rosmarinic acid, caffeic acid	(CID:4445104) (CID: 2423)	• Reduce oxidative stress • Anti-atherosclerotic • Anticancer effect • Anti-inflammatory • Antimicrobial effect	Issaoui et al., 2020
24	Turmeric 	*Curcuma longa.*	Curcumin	(CID: 64)	• Suspension of pro-inflammatory cytokine increase • Acts as an antioxidant • Antidiabetic effect • Apoptogenic activities • Anti-proliferative effect • Chondroprotective effects	Oliviero et al., 2018; Ahangarpour et al., 2019; Mughal, 2019

methods of isolating active compounds, using chemometrics to create a potent blend of natural food antioxidants, and using assorted designs to maximize the potency of herbs, blended spices, or antioxidant compounds. Several research organizations are now isolating antioxidant chemicals from vegetables, berries, fruits, spices, herbs, and teas. As a result, more information on the active compounds of plants will soon be accessible. In addition, technological advancements and research will lead to better methods of growing spices and herbs with more significant amounts of antioxidants and an increase in interest in the food antioxidants found in spices and herbs.

LIST OF ABBREVIATIONS

4-HNE: 4-Hydroxy-2-nonenol
AAE: Ascorbic acid equivalents
ABTS: 2,2′-Azino-bis(3-ethylbenzothiazoline-6-sulphonic acid)
ATP: Adenosine triphosphate
BHT: Butylated hydroxytoluene
BHAs: Butylated hydroxyanisole
BPs: Bound phenolics
CAM: Complementary and alternative medicine
CNTs: Carbon nanotubes
CUPRAC: Cupric reducing antioxidant capacity
CVD: Cardiovascular disease
DHA: Docosahexaenoic acid
DPPH: 1,1-Diphenyl-2-picrylhydrazyl
DW: Dry weight
EDTA: Ethylenedinitrilotetraacetic acid
EO: Essential oil
EPA: Eicosapentaenoic acid
FC: Folin-Ciocaulteu
FPs: Free phenolics
FFA: Free fatty acids
GAE: Gallic acid equivalents
GLP-1: Glucagon-like peptide
GMOs: Genetically modified organisms
HBP: High blood pressure
HCAs: Heterocyclic amines
HC: High cholesterol
HPLC: High-performance liquid chromatography
IL: Interleukin
LDL: Low-density lipoproteins
MDA: Malondialdehyde
NF-B: Nuclear factor-kappa B
NO: Nitric oxide
–OH: Hydroxyl group
POV: Peroxide value
ROS: Reactive oxygen species
RSA: Radical-scavenging properties
TBARS: Thiobarbituric acid reactive substance
TBHQ: *Tert*-butylhydroquinone
TE: Trolox equivalents
TPC: Total phenolic content
WHO: World Health Organization

REFERENCES

Ahangarpour, A., Sayahi, M., & Sayahi, M., 2019. The antidiabetic and antioxidant properties of some phenolic phytochemicals: A review study. *Diabetes & Metabolic Syndrome: Clinical Research & Reviews* 13(1): 854–857.

Asensio, C.M., Nepote, V., & Grosso, N.R., 2012. Sensory attribute preservation in extra virgin olive oil with the addition of oregano essential oil as a natural antioxidant. *Journal of Food Science* 77(9): S294–S301.

Assefa, A.D., Keum, Y.-S., & Saini, R.K., 2018. A comprehensive study of polyphenols contents and antioxidant potential of 39 widely used spices and food condiments. *Journal of Food Measurement and Characterization* 12: 1548–1555.

Bachheti, A., Deepti, B.R.K., Singh, A., Zebeaman, M., Hunde, Y., & Husen, A., 2023. Bioactive constituents and health promoting compounds of underutilized fruits of the northern Himalayas of India: A review. *Food Production, Processing and Nutrition*. https://doi.org/10.1186/s43014-023-00140-5

Balogh, Z., Gray, J.I., Gomaa, E.A., & Booren, A.M., 2000. Formation and inhibition of heterocyclic aromatic amines in fried ground beef patties. *Food and Chemical Toxicology* 38(5): 395–401.

Basli, A., Belkacem, N., & Amrani, I., 2017. Health benefits of phenolic compounds against cancers. In: Marcos Soto-Hernandez, Mariana Palma-Tenango and Maria del Rosario Garcia-Mateos (Eds.), *Phenolic Compounds-Biological Activity*, vol. 8. InTechOpen, London, 193–210.

Bettaieb Rebey, I., Bourgou, S., Aidi Wannes, W., Hamrouni Selami, I., Saidani Tounsi, M., Marzouk, B., Fauconnier, M.-L., & Ksouri, R., 2018. Comparative assessment of phytochemical profiles and antioxidant properties of Tunisian and Egyptian anise (*Pimpinella anisum* L.) seeds. *Plant Biosystems-An International Journal Dealing with all Aspects of Plant Biology* 152(5): 971–978.

Bhale, S.D., Xu, Z., Prinyawiwatkul, W., King, J.M., & Godber, J.S., 2007. Oregano and rosemary extracts inhibit oxidation of long-chain n-3 fatty acids in menhaden oil. *Journal of Food Science* 72(9): C504–508.

Bi, X., Lim, J., & Henry, C.J., 2017. Spices in the management of diabetes mellitus. *Food Chemistry* 217: 281–293.

Bostan, C., Butnariu, M., Butu, M., Ortan, A., Butu, A., Rodino, S., & Parvu, C., 2013. Allelopathic effect of *Festuca rubra* on perennial grasses. *Romanian Biotechnological Letters* 18(2): 8190–8196.

Cao, H., Urban, J.F., Jr., & Anderson, R.A., 2008. Cinnamon polyphenol extract affects immune responses by regulating anti- and proinflammatory and glucose transporter gene expression in mouse macrophages. *The Journal Of Nutrition* 138(5): 833–840.

Carlsen, M.H., Halvorsen, B.L., Holte, K., Bøhn, S.K., Dragland, S., Sampson, L., Willey, C., Senoo, H., Umezono, Y., Sanada, C., Barikmo, I., Berhe, N., Willett, W.C., Phillips, K.M., Jacobs, D.R., Jr., & Blomhoff, R., 2010. The total antioxidant content of more than 3100 foods, beverages, spices, herbs and supplements used worldwide. *Nutrition Journal* 9: 3.

Chahal, K., Dhaiwal, K., Kumar, A., Kataria, D., & Singla, N., 2017. Chemical composition of *Trachyspermum ammi* L. and its biological properties: A review. *Journal of Pharmacognosy and Phytochemistry* 6(3): 131–140.

Chirinos, R., Betalleluz-Pallardel, I., Huamán, A., Arbizu, C., Pedreschi, R., & Campos, D., 2009. HPLC-DAD characterisation of phenolic compounds from Andean oca (*Oxalis tuberosa* Mol.) tubers and their contribution to the antioxidant capacity. *Food Chemistry* 113(4): 1243–1251.

Choe, E., & Min, D.B., 2009. Mechanisms of antioxidants in the oxidation of foods. *Comprehensive Reviews in Food Science and Food Safety* 8(4): 345–358.

Clarke, T.C., Black, L.I., Stussman, B.J., Barnes, P.M., & Nahin, R.L., 2015. Trends in the use of complementary health approaches among adults: United States, 2002–2012. *Natl Health Stat Report* (79): 1–16.

Cuvelier, M.E., Berset, C., & Richard, H., 1994. Antioxidant constituents in sage (*Salvia officinalis*). *Journal of Agricultural and Food Chemistry* 42(3): 665–669.

Das, A.K., Rajkumar, V., Nanda, P.K., Chauhan, P., Pradhan, S.R., & Biswas, S., 2016. Antioxidant efficacy of litchi (*Litchi chinensis* Sonn.) pericarp extract in sheep meat nuggets. *Antioxidants* 5(2): 16.

de Araújo, F.F., de Paulo Farias, D., Neri-Numa, I.A., & Pastore, G.M., 2021. Polyphenols and their applications: An approach in food chemistry and innovation potential. *Food Chemistry* 338: 127535.

Decker, E.A., Elias, R.J., & McClements, D.J. (Eds.)., 2010. *Oxidation in Foods and Beverages and Antioxidant Applications: Management in Different Industry Sectors*. University of Massachusetts, USA: Elsevier.

Dwivedi, S., Vasavada, M.N., & Cornforth, D., 2006. Evaluation of antioxidant effects and sensory attributes of Chinese five-spice ingredients in cooked ground beef. *Journal of Food Science* 71(1): C12–C17.

Efferth, T., Li, P.C., Konkimalla, V.S., & Kaina, B., 2007. From traditional Chinese medicine to rational cancer therapy. *Trends Molecular Medicine* 13(8): 353–361.

Elansary, H.O., Szopa, A., Kubica, P., Ekiert, H., Klimek-Szczykutowicz, M., El-Ansary, D.O., & Mahmoud, E.A., 2020. Polyphenol profile and antimicrobial and cytotoxic activities of natural *Mentha piperita* and *Mentha longifolia* populations in Northern Saudi Arabia. *Processes* 8(4): 479.

Embuscado, M.E., 2015. Spices and herbs: Natural sources of antioxidants–a mini-review. *Journal of Functional Foods* 18(B): 811–819.

Feng, Y., Dunshea, F.R., & Suleria, H.A., 2020. LC-ESI-QTOF/MS characterization of bioactive compounds from black spices and their potential antioxidant activities. *Journal of Food Science and Technology* 57: 4671–4687.

Fraga, C.G., Croft, K.D., Kennedy, D.O., & Tomás-Barberán, F.A., 2019. The effects of polyphenols and other bioactives on human health. *Food & Function* 10(2): 514–528.

Fratianni, F., De Martino, L., Melone, A., De Feo, V., Coppola, R., & Nazzaro, F., 2010. Preservation of chicken breast meat treated with thyme and balm essential oils. *Journal of Food Science* 75(8): M528–M535.

Furukawa, S., Fujita, T., Shimabukuro, M., Iwaki, M., Yamada, Y., Nakajima, Y., Nakayama, O., Makishima, M., Matsuda, M., & Shimomura, I., 2004. Increased oxidative stress in obesity and its impact on metabolic syndrome. *The Journal of Clinical Investigation* 114(12): 1752–1761.

Ge, Q., Chen, L., & Chen, K., 2017. Treatment of diabetes mellitus using iPS cells and spice polyphenols. *Journal of Diabetes Research* 2017: 5837804.

Ghafoor, K., Al Juhaimi, F., Özcan, M.M., Uslu, N., Babiker, E.E., & Ahmed, I.A.M., 2020. Total phenolics, total carotenoids, individual phenolics and antioxidant activity of ginger (*Zingiber officinale*) rhizome as affected by drying methods. *Lwt, 126*, 109354.

Guetens, G., Boeck, G.D., Highley, M., van Oosterom, A.T., & de Bruijn, E.A., 2002. Oxidative DNA damage: Biological significance and methods of analysis. *Critical Reviews in Clinical Laboratory Sciences* 39(4–5): 331–457.

Gupta, J., Sharma, S., Sharma, N.R., & Kabra, D., 2020. Phytochemicals enriched in spices: A source of natural epigenetic therapy. *Archives of Pharmacal Research* 43(2): 171–186.

Haldar, S., Chia, S.C., & Henry, C.J., 2018. Polyphenol-rich curry made with mixed spices and vegetables increases postprandial plasma GLP-1 concentration in a dose-dependent manner. *European Journal of Clinical Nutrition* 72(2): 297–300.

Hussain, M.B., Hassan, S., Waheed, M., Javed, A., Farooq, M. A., & Tahir, A., 2019. Bioavailability and metabolic pathway of phenolic compounds. In: *Plant Physiological Aspects of Phenolic Compounds*. IntechOpen.

Hussain, T., Tan, B., Yin, Y., Blachier, F., Tossou, M.C., & Rahu, N., 2016. Oxidative stress and inflammation: What polyphenols can do for us? *Oxidative Medicine and Cellular Longevity* 2016: 7432797.

Husen, A., 2020. *Exploring Medicinal Plants* (Book Series). CRC Press, 6000 Broken Sound Parkway NW, Suite 300, Boca Raton, FL, 33487–2742.

Husen, A., 2021. *Traditional Herbal Therapy for The Human Immune System*. CRC Press, 6000 Broken Sound Parkway NW, Suite 300, Boca Raton, FL, 33487–2742. https://doi.org/10.1201/9781003137955

Husen, A., 2022a. *Herbs, Shrubs and Trees of Potential Medicinal Benefits*. CRC Press, 6000 Broken Sound Parkway NW, Suite 300, Boca Raton, FL, 33487–2742. https://doi.org/10.1201/9781003205067

Husen, A., 2022b. *Exploring Poisonous Plants: Medicinal Values, Toxicity Response and Other Potential Uses*. CRC Press, 6000 Broken Sound Parkway NW, Suite 300, Boca Raton, FL, 33487–2742. https://doi.org/10.1201/b23017

Husen, A., 2023. *Antidiabetic Medicinal Plants and Herbal Treatments*. CRC Press, 6000 Broken Sound Parkway NW, Suite 300, Boca Raton, FL, 33487–2742.

Hybertson, B.M., Gao, B., Bose, S.K., & McCord, J.M., 2011. Oxidative stress in health and disease: The therapeutic potential of Nrf2 activation. *Molecular Aspects of Medicine* 32(4): 234–246.

Inoue, M., Sato, E.F., Nishikawa, M., Park, A.M., Kira, Y., Imada, I., & Utsumi, K., 2003. Mitochondrial generation of reactive oxygen species and its role in aerobic life. *Current Medicinal Chemistry* 10(23): 2495–2505.

Isbill, J., Kandiah, J., & Khubchandani, J., 2018. Use of ethnic spices by adults in the United States: An exploratory study. *Health Promotion Perspectives* 8(1): 33–40.

Issaoui, M., Delgado, A.M., Caruso, G., Micali, M., Barbera, M., Atrous, H., Ouslati, A., & Chammem, N., 2020. Phenols, flavors, and the Mediterranean diet. *Journal of AOAC International* 103(4): 915–924.

Jaganath, I. B., & Crozier, A., 2010. Dietary flavonoids and phenolic compounds. *Plant Phenolics and Human Health: Biochemistry, Nutrition, and Pharmacology* 1: 1–50.

Jiang, J., Zhang, X., True, A.D., Zhou, L., & Xiong, Y.L., 2013. Inhibition of lipid oxidation and rancidity in precooked pork patties by radical-scavenging licorice (*Glycyrrhiza glabra*) extract. *Journal of Food Science* 78(11): C1686–C1694.

Kaefer, C.M., & Milner, J.A., 2008. The role of herbs and spices in cancer prevention. *Journal of Nutrtional Biochemistry* 19(6): 347–361.

Kang, J.H., Kim, C.S., Han, I.S., Kawada, T., & Yu, R., 2007. Capsaicin, a spicy component of hot peppers, modulates adipokine gene expression and protein release from obese-mouse adipose tissues and isolated adipocytes and suppresses the inflammatory responses of adipose tissue macrophages. *FEBS Letters* 581(23): 4389–4396.

Karpińska, M., Borowski, J., & Danowska-Oziewicz, M., 2001. The use of natural antioxidants in ready-to-serve food. *Food Chemistry* 72(1): 5–9.

Kaundal, R., Kumar, M., Kumar, S., Singh, D., & Kumar, D., 2022. Polyphenolic profiling, antioxidant, and antimicrobial activities revealed the quality and adaptive behavior of viola species, a dietary spice in the Himalayas. *Molecules* 27(12): 3867.

Kumar, N., & Goel, N., 2019. Phenolic acids: Natural versatile molecules with promising therapeutic applications. *Biotechnology Reports* 24: e00370.

Kumar, N., & Pruthi, V., 2014. Potential applications of ferulic acid from natural sources. *Biotechnology Reports* 4: 86–93.

Lee, T.G., Williams, S.K., Sloan, D., & Littell, R., 1997. Development and evaluation of a chicken breakfast sausage manufactured with mechanically deboned chicken meat. *Poultry Science* 76(2): 415–421.

Li, G., Luan, G., He, Y., Tie, F., Wang, Z., Suo, Y., Ma, C., &Wang, H., 2018. Polyphenol stilbenes from fenugreek (*Trigonella foenum-graecum* L.) seeds improve insulin sensitivity and mitochondrial function in 3T3-L1 adipocytes. *Oxidative Medicine and Cellular Longevity* 2018: 7634362.

Lu, M., Yuan, B., Zeng, M., & Chen, J., 2011. Antioxidant capacity and major phenolic compounds of spices commonly consumed in China. *Food Research International* 44(2): 530–536.

Lutz, M., Fuentes, E., Ávila, F., Alarcón, M., & Palomo, I., 2019. Roles of phenolic compounds in the reduction of risk factors of cardiovascular diseases. *Molecules* 24(2):366.

Mariutti, L.R., Nogueira, G.C., & Bragagnolo, N., 2011. Lipid and cholesterol oxidation in chicken meat are inhibited by sage but not by garlic. *Journal of Food Science* 76(6): C909–915.

Mark, R., Lyu, X., Lee, J.J., & Parra-Saldívar, R., Chen, W.N., 2019. Sustainable production of natural phenolics for functional food applications. *Journal of Functional Foods* 57: 233–254.

Martillanes, S., Rocha-Pimienta, J., Gil, M.V., Ayuso-Yuste, M.C., & Delgado-Adámez, J., 2020. Antioxidant and antimicrobial evaluation of rice bran (*Oryza sativa* L.) extracts in a mayonnaise-type emulsion. *Food Chemistry* 308: 125633.

Martins, N., Barros, L., & Ferreira, I.C., 2016. *In vivo* antioxidant activity of phenolic compounds: Facts and gaps. *Trends in Food Science & Technology* 48: 1–12.

Mughal, M.H., 2019. Spices: A review on diabetes mellitus. *Biomedical Journal of Scientific & Technological Research* 15: 11651–11657.

Oliviero, F., Scanu, A., Zamudio-Cuevas, Y., Punzi, L., & Spinella, P., 2018. Anti-inflammatory effects of polyphenols in arthritis. *Journal of the Science of Food and Agriculture* 98(5): 1653–1659.

Osawa, T., 1999. Protective role of dietary polyphenols in oxidative stress. *Mechanisms of Ageing and Development* 111(2–3): 133–139.

Panja, P., 2018. Green extraction methods of food polyphenols from vegetable materials. *Current Opinion in Food Science* 23: 173–182.

Park, S.Y., Yoo, S.S., Shim, J.H., & Chin, K.B., 2008. Physicochemical properties, and antioxidant and antimicrobial effects of garlic and onion powder in fresh pork belly and loin during refrigerated storage. *Journal of Food Science* 73(8): C577–C584.

Pellegrini, N., Serafini, M., Salvatore, S., Del Rio, D., Bianchi, M., & Brighenti, F., 2006. Total antioxidant capacity of spices, dried fruits, nuts, pulses, cereals and sweets consumed in Italy assessed by three different in vitro assays. *Molecular Nutrition & Food Research* 50(11): 1030–1038.

Pezeshk, S., Rezaei, M., & Hosseini, H., 2011. Effects of turmeric, shallot extracts, and their combination on quality characteristics of vacuum-packaged rainbow trout stored at 4 ± 1 °C. *Journal of Food Science* 76(6): M387–M391.

Piccolella, S., Crescente, G., Candela, L., & Pacifico, S., 2019. Nutraceutical polyphenols: New analytical challenges and opportunities. *Journal of Pharmaceutical and Biomedical Analysis* 175: 112774.

Pizzale, L., Bortolomeazzi, R., Vichi, S., Überegger, E., & Conte, L.S., 2002. Antioxidant activity of sage (*Salvia officinalis* and *Salvia fruticosa*) and oregano (*Origanum onites* and *Origanum indercedens*) extracts related to their phenolic compound content. *Journal of the Science of Food and Agriculture* 82(14): 1645–1651.

Prajapati, S.K., Jain, D., Parveen, S., Maji, S., & Deb, P.K., 2022. Nanodelivery of antioxidant herbal extracts, spices, and dietary constituents. *Phytoantioxidants and Nanotherapeutics* 1: 145–171.

Rahal, A., Kumar, A., Singh, V., Yadav, B., Tiwari, R., Chakraborty, S., & Dhama, K., 2014. Oxidative stress, prooxidants, and antioxidants: The interplay. *Biomed Research International* 2014: 761264.

Rahman, S., & Husen, A. 2021. Potential Role of Medicinal Plants in the Cure of Liver and Kidney Diseases. In: Husen, A., Bachheti, R.K., & Bachheti, A. (Eds.), *Non-Timber Forest Products*. Springer International Publishing AG, Gewerbestrasse 11, 6330, Cham, 229–325. https://doi.org/10.1007/978-3-030-73077-2_10

Rahman, S., & Husen, A. 2022. *Asparagus racemosus* (Shatavari) and *Dioscorea villosa* (Wild yams). In: Husen, A. (Ed.), *Herbs, Shrubs, and Trees of Potential Medicinal Benefits*. CRC Press, 6000 Broken Sound Parkway NW, Suite 300, Boca Raton, FL, 33487–2742.

Rahman, S, Husen, A., 2023a. *Trigonella foenum-graecum* L.: A Useful Medicinal Plant for Diabetes and Other Diseases, In: Husen, A. (Ed.), *Antidiabetic Medicinal Plants and Herbal Treatments*. CRC Press, 6000 Broken Sound Parkway NW, Suite 300, Boca Raton, FL, 33487–2742.

Rahman, S, Husen, A., 2023b. Role of Green Chiretta [*Andaphis paniculata* (Burm. F.)] Nees in Diabetes and Other Ailments, In: Husen, A. (Ed.), *Antidiabetic Medicinal Plants and Herbal Treatments*. CRC Press, 6000 Broken Sound Parkway NW, Suite 300, Boca Raton, FL, 33487–2742.

Rashid, Z., Khan, M.R., Mubeen, R., Hassan, A., Saeed, F., & Afzaal, M., 2020. Exploring the effect of cinnamon essential oil to enhance the stability and safety of fresh apples. *Journal of Food Processing and Preservation* 44(12): e14926.

Reddy, L., Odhav, B., & Bhoola, K.D., 2003. Natural products for cancer prevention: A global perspective. *Pharmacology & Therapeutics* 99(1): 1–13.

Renerre, M., Dumont, F., & Gatellier, P., 1996. Antioxidant enzyme activities in beef in relation to oxidation of lipid and myoglobin. *Meat Science* 43(2): 111–121.

Rižnar, K., Čelan, Š., Knez, Ž., ŠKerget, M., Bauman, D., & Glaser, R., 2006. Antioxidant and antimicrobial activity of rosemary extract in chicken frankfurters. *Journal of Food Science* 71(7): C425–C429.

Rubió, L., Motilva, M.J., Macià, A., Ramo, T., & Romero, M.P., 2012. Development of a phenol-enriched olive oil with both its own phenolic compounds and complementary phenols from thyme. *Journal of Agriculture and Food Chemistry* 60(12): 3105–3112.

Sachan, A.K., Das, D.R., & Kumar, M., 2016. Carum carvi - An important medicinal plant. *Journal of Chemical and Pharmaceutical Research* 8(3): 529–533.

Sancho, R.A., de Lima, F.A., Costa, G.G., Mariutti, L.R., & Bragagnolo, N., 2011. Effect of annatto seed and coriander leaves as natural antioxidants in fish meatballs during frozen storage. *Journal of Food Science* 76(6): C838–C845.

Saraswathi, K., Arumugam, P., & Sivaraj, C., 2020. Pharmacological activities of differential parts of selected essential Indian spices. *Journal of Pharmacognosy and Phytochemistry* 9(2): 2024–2033.

Sepahpour, S., Selamat, J., Abdul Manap, M.Y., Khatib, A., & Abdull Razis, A.F., 2018. Comparative analysis of chemical composition, antioxidant activity and quantitative characterization of some phenolic compounds in selected herbs and spices in different solvent extraction systems. *Molecules*, 23(2): 402.

Shahidi, F., & Zhong, Y., 2010. Novel antioxidants in food quality preservation and health promotion. *European Journal of Lipid Science and Technology* 112(9): 930–940.

Singh, N., Rao, A.S., Nandal, A., Kumar, S., Yadav, S.S., Ganaie, S.A., & Narasimhan, B., 2021. Phytochemical and pharmacological review of *Cinnamomum verum* J. Presl-a versatile spice used in food and nutrition. *Food Chemistry* 338: 127773.

Singh, N., Yadav, S.S., Kumar, S., & Narashiman, B., 2021. A review on traditional uses, phytochemistry, pharmacology, and clinical research of dietary spice *Cuminum cyminum* L. *Phytotherapy Research* 35(9): 5007–5030.

Škrovánková, S., Mlček, J., Orsavová, J., Juríková, T., & Dřímalová, P., 2017. Polyphenols content and antioxidant activity of paprika and pepper spices. *Potravinarstvo Slovak Journal of Food Sciences* 11(1): 52–57.

Slimestad, R., Fossen, T., & Brede, C., 2020. Flavonoids and other phenolics in herbs commonly used in Norwegian commercial kitchens. *Food Chemistry* 309: 125678.

Smith, J.S., Ameri, F., & Gadgil, P., 2008. Effect of marinades on the formation of heterocyclic amines in grilled beef steaks. *Journal of Food Science* 73(6): T100–T105.

Sokamte, T.A., Mbougueng, P.D., Tatsadjieu, N.L., & Sachindra, N.M., 2019. Phenolic compounds characterization and antioxidant activities of selected spices from Cameroon. *South African Journal of Botany* 121: 7–15.

Spiller, F., Alves, M.K., Vieira, S.M., Carvalho, T.A., Leite, C.E., Lunardelli, A., Poloni, J.A., Cunha, F.Q., & de Oliveira, J.R., 2008. Anti-inflammatory effects of red pepper (*Capsicum baccatum*) on carrageenan- and antigen-induced inflammation. *The Journal of Pharmacy and Pharmacology* 60(4): 473–478.

Su, D., & Li, L., 2011. Trends in the use of complementary and alternative medicine in the United States: 2002–2007. *Journal of Health Care for Poor and Underserved* 22(1): 296–310.

Takó, M., Kerekes, E.B., Zambrano, C., Kotogán, A., Papp, T., Krisch, J., & Vágvölgyi, C., 2020. Plant phenolics and phenolic-enriched extracts as antimicrobial agents against food-contaminating microorganisms. *Antioxidants* 9(2): 165.

Tan, B.L., Norhaizan, M.E., & Liew, W.P., 2018. Nutrients and oxidative stress: Friend or foe? *Oxidative Medicine and Cellular Longevity* 2018: 9719584.

Tressera-Rimbau, A., Arranz, S., Eder, M., & Vallverdú-Queralt, A., 2017. Dietary polyphenols in the prevention of stroke. *Oxidative Medicine and Cellular Longevity* 2017: 7467962.

Usman, I., Hussain, M., Imran, A., Afzaal, M., Saeed, F., Javed, M., Afzal, A., Ashfaq, I., Al Jbawi, E., & Saewan, A. S., 2022. Traditional and innovative approaches for the extraction of bioactive compounds. *International Journal of Food Properties* 25(1): 1215–1233.

Vasavada, M.N., Dwivedi, S., & Cornforth, D., 2006. Evaluation of garam masala spices and phosphates as antioxidants in cooked ground beef. *Journal of Food Science* 71(5): C292–C297.

Vetrani, C., Costabile, G., Vitale, M., & Giacco, R., 2020. (Poly) phenols and cardiovascular diseases: Looking in to move forward. *Journal of Functional Foods* 71: 104013.

Wahle, K.W.J., Brown, I., Rotondo, D., & Heys, S.D., 2010. Plant phenolics in the prevention and treatment of cancer. In: Giardi, M.T., Rea, G., & Berra, B. (Eds.), *Bio-Farms for Nutraceuticals. Advances in Experimental Medicine and Biology*, vol. 698. Springer, Boston, MA.

Wang, Z., Li, S., Ge, S., & Lin, S., 2020. Review of distribution, extraction methods, and health benefits of bound phenolics in food plants. *Journal of Agriculture and Food Chemistry* 68(11): 3330–3343.

Wu, Y., Ding, Y., Tanaka, Y., & Zhang, W., 2014. Risk factors contributing to type 2 diabetes and recent advances in the treatment and prevention. *International Journal of Medical Science* 11(11): 1185–1200.

Xu, B., Ganesan, K., Mickymaray, S., Alfaiz, F.A., Thatchinamoorthi, R., & Al Aboody, M.S., 2020. Immunomodulatory and antineoplastic efficacy of common spices and their connection with phenolic antioxidants. *Bioactive Compounds in Health and Disease* 3(2): 15–31.

Yashin, A., Yashin, Y., Xia, X., & Nemzer, B., 2017. Antioxidant activity of spices and their impact on human health: A review. *Antioxidants* 6(3): 70.

Zhang, H., & Tsao, R., 2016. Dietary polyphenols, oxidative stress and antioxidant and anti-inflammatory effects. *Current Opinion in Food Science* 8: 33–42.

Zheng, W., & Wang, S.Y., 2001. Antioxidant activity and phenolic compounds in selected herbs. *Journal of Agricultural and Food Chemistry* 49(11): 5165–5170.

8 Allspice (*Pimenta dioica*)

Niccolò Pilla, Paolo Gabrielli, Altero Aguzzi, Massimo Lucarini, and Alessandra Durazzo

8.1 INTRODUCTION

Pimenta dioica is a tree belonging to the Myrtaceae family. It is an evergreen tree with a height of approximately 22 ft. Occasionally the trees grow up to 43 feet tall, with light grey bark and dark green leaves (4- to 8-cm long). It is characterized by a compound inflorescence, axillary, paniculate which is separately branched and composed of many-flowered cymes. The flowers are small, white, produced in the axils of upper leaves. The flowering period in between March and June. The flower develops a fruit or berry, and each one contains two green seeds (Premachandrana & Murthy, 2022). Allspice was originally native to the tropical forests of South and Central America, especially Cuba and Jamaica, southern Mexico, and the West Indies. Commercial Allspice is also obtained from the fruits of another related species, *Pimenta racemosa* (Myrtaceae), although the fruits of the latter are larger and are known to be less aromatic. Jamaica is the world's largest producer of allspice and Jamaican allspice is renowned for being of exceptional quality because it contains a higher level of essential oils, which give it more flavor than allspice grown in different countries (Premachandrana & Murthy, 2022).

Christopher Columbus was the first European to discover this spice during his voyages to the New World at the beginning of the sixteenth century. Originally, Columbus was searching for common spices, such as cinnamon, cloves, and pepper. The pepper (*Piper nigrum* and related species), originally from central Asia, was not found in the new lands, however, so *P. dioica* was called Jamaican pepper. The name allspice was given later by the British, due to its strong aromatic flavour that recalls a mixed aromas of cloves, pepper, cinnamon, and nutmeg (Vaughan & Geissler, 2009, Bown, 1995). *P. dioica* is used extensively in Caribbean and Middle Eastern cuisines as well as in European ones. In particular, it is among the most important spices used in Jamaica, if not the most important. It is used for traditional recipes, such as a rum produced together with the berries, or as a flavouring in Jamaican jerk. It can also be used as a condiment in stews and sauces or to flavour vegetables, and it is used in the preparation of bread and cakes. Allspice has been used in folk medicine for centuries. In each country of the Central Americas, allspice had a different use. For example, in Jamaica, it was drunk together with hot tea to treat colds, dysmenorrhea, and dyspepsia, and the same uses were applied in Costa Rica. In Guatemala, allspice was applied for treating bruises and myalgia. In Cuban medicine, a mix of herbs together with allspice was used to improve digestion. Along with the expansion of this spice, allspice was adopted by the Indian medicine, Ayurveda.

8.2 BIOACTIVE COMPOUNDS AND POTENTIAL BENEFICIAL EFFECTS

Several bioactive molecules are present in the extracts of *P. dioica* (Jirovetz et al., 2007; Ashokkumar et al., 2022). The prevalent classes of compounds found in *P. dioica* are phenolics, sesquiterpenes, monoterpenes, oxygenated esters, steroids, and fatty acids (Premachandran et al., 2022). Polyphenols and hydrolyzable tannins are found in abundance in the essential oil (Yoshimura et al., 2011, Merida-Reyes et al., 2020). One of the major constituents of the oil is eugenol (Ashokkumar et al., 2022). Supercritical fluid extraction has confirmed that eugenol is the major component (77.9%), followed by β-caryophyllene (5.1%), squalene (4.1%), and α-humulene (2.3%) (Marongiu et al.,

DOI: 10.1201/9781003387046-8

2005). Due to the abundance of bioactive compounds, *P. dioica* has been investigated for potential beneficial effects on human health. Many interesting effects have been reported over the years and many more have still to be investigated. The main effects reported for *P. dioica* are antioxidant, antibacterial, antifungal, and anticancer properties (Premachandran et al., 2022).

The antioxidant activity could be mainly attributed to eugenol (Jirovetz et al., 2007). Vhin et al. (2000) have studied the antioxidant activity of the various extracts of the plant, reporting that the ethanol extract exhibited a remarkable antioxidant effect (Vhin et al., 2000). Furthermore, Sánchez-Zarate et al. (2020) concluded how *P. dioica* leaves extracts are a good source of phenolic compounds and have significative antioxidant activity (Sánchez-Zarate et al., 2020).

An *in vivo* study suggested that an aqueous suspension of allspice was able to significantly suppress carrageenan-induced rat paw oedema at a high dose, and also reduced tissue granulation formation, moreover inhibit acetic acid-mediated action, tail-flick reactive time and dose- and time-dependent decrease in yeast-induced hyperpyrexia in mice (Al-Rehailyet al., 2002). A more recent study indicated that allspice can increase the level of the anti-inflammatory cytokine IL-10 (Lorenzo-Leal et al., 2019a).

Like other spices *P. dioica* has shown antibacterial effect, in particular related to the presence of eugenol and phenolic compounds (Burth, 2004).

Lorenzo-Leal et al. (2019b) tested essential oil from *P. dioica* against different bacterial strains, and it was effective against five microbial strains, including *Acinetobacter baumannii*, methicillin-resistant *Staphylococcus aureus, Pseudomonas aeruginosa, S. aureus,* and the fungus *Candida albicans,* with MICs (Minimum Inhibitory Concentrations) ranging between 500 and 2000 μg/mL (Lorenzo-Leal et al., 2019b). Moreover, the allspice essential oil inhibited the growth of *Listeria monocytogenes* and *Salmonella typhimurium*, pathogens associated with foodborne illness (Lorenzo-Leal et al., 2019b).

Allspice essential oil was tested against *A. baumannii*, revealing a good antibacterial effect, and suggesting that the abundance of oxygenated monoterpenes and phenolic compounds contribute to the antibacterial effect (Ismail et al., 2020). Another study found that allspice significantly reduced the amount of verocytotoxin type 1 and 2 produced by *Escherichia coli* O157:H7 strain, indicating again that eugenol, as the main compound, was responsible for this effect (Takemasa et al., 2009).

A high antibacterial activity against *Bacillus subtilis* was observed by Merida-Réyes et al. (2020). Essential oil from allspice showed a MIC (minimum inhibitory concentrations) of 0.31 to 1.25 μL/mL against different species of fungi (Kamble & Patil, 2008). A more recent study indicated that allspice leaf essential oil was an inhibitor of the fungus *Aspergillus flavus* (Sarathambal et al., 2021). In different concentrations, pimento (allspice) oil completely inhibited the growth and the production of aflatoxin B_1 by *A. flavus* (Chaudhari et al., 2020). *In vivo* and *in vitro* studies evaluated the efficacy of an aqueous extract against human breast cancer cells. The extract inhibited the viability and clonogenic development of various types of breast cancer cells, although its oral administration to athymic mice implanted with MDA-MB231 tumours did not show a significative decrease in tumour growth (Zhang et al., 2015). Gallic acid, another compound present in *P. dioica*, has been studied for its potential anti-tumour effects (Hsu et al., 2011; Garcìa-Rivera et al., 2011).

A methanolic extract of *P. dioica* revealed anti-tumour activity against two breast cancer cell lines MCF-7 and SK-BR3, suggesting a potential effect against gastric cancer (Doyle et al., 2018). In the same study, Doyle at al. (2018) investigated the potential oestrogenic activity of allspice extracts; the oestrogen agonist effect could explain the traditional use of this spice in Costa Rica where it was used to manage menopausal symptoms. An *in vivo* study on normotensive and hypotensive rats caused a depression of the central nervous system and a reduction in mean arterial blood pressure (Ciccio et al., 1997). Suarez et al. (2000) observed a dose-dependent reduction of blood pressure in spontaneously hypertensive rats.

8.3 ALLSPICE LITERATURE QUANTITATIVE RESEARCH ANALYSIS ANALYSIS

On March 7, 2023, a bibliographic search for pimenta dioica publications was carried out based on Scopus database content. The Scopus online database (http://www.scopus.com/home.uri, accessed on March 7, 2023) was used to extract bibliometric data using the search string: TITLE-ABS-KEY ("*Pimenta dioica*" OR "Allspice" OR "Jamaica pepper" OR "Myrle pepper"). Publications mentioning the relevant words or their derivates in the title, abstract, or keywords were identified throughout the strategy applied. The evaluated parameters were trends of publications and citations, document type, authorship, and country/region.

For the basic analyses, the functions of the Scopus web online platform named 'Analyze' and as 'Create Citation Report' were used. The 'full records and cited references' were exported to VOSviewer software (version 1.6.19, www.vosviewer.com, accessed on March 7, 2023) (Waltman et al., 2010, Van Eck et al., 2010, 2011, 2014) for additional processing procedures/operations. The terms utilized in the title abstracts of publications and keywords of publications were analyzed by the above-mentioned VOSviewer software. In the map of terms, the bubble size indicates how frequently a term is mentioned in the articles; two bubbles that are close to each other reflect that the terms co-appeared more often in the selected publications. The average citations per publication (CPP) are given by the color of a bubble.

Five was set as the minimum number of occurrences of a keyword. Out of 5013 keywords, 288 met the threshold, and two of them were manually excluded.

Three hundred and eighteen documents were returned by the search, covering the time period from 1918 to 2023 and were cited 11011 times in total. The most cited article is about the bactericidal activities of different spices, including allspice (Friedman et al., 2002). In recent years, there was a steady increase in the number of documents published per year, indicating a growing interest in the potentialities of the plant. Out of 318 documents, 85.7% are articles follow by 6.1 % of reviews and the remainder are conference papers, book chapters, notes, editorials, books, and letters (Figure 8.1). Among the oldest papers, the article published in 1935 by Strong in the *American Journal of Cancer* is an investigation of the effect of allspice oil on spontaneous cancer in mice,

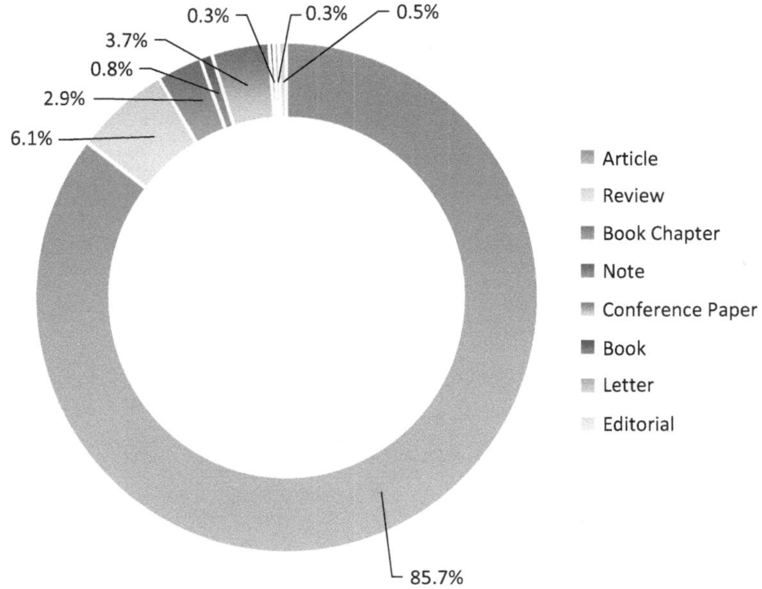

FIGURE 8.1 Distribution of the type documents (Bibliometric data were extracted from the Scopus online database).

demonstrating that a daily administration of allspice oil was effective in the control of carcinoma of the mammary glands.

Figure 8.2 lists the most productive authors according to the bibliographic search. The author with the most publications is Cruz-Oliveira, J., with nine documents. In his most cited article, he investigated the residue of allspice as a biosorbent, indicating another potential use of allspice (Cruz-Oliveira et al., 2013). Zieliński, H., the second most-published author, has based his work on the antioxidant activity of *P. dioica* (Przygodzka et al., 2014, 2015). By contrast, Muzolf-Panek, M. investigated the extracts of different plants, including allspice, as preservative agents during the storage of raw foods (Muzolf-Panek et al., 2019, 2020).

Figure 8.3 shows the most productive countries in terms of papers on allspice research, with the United States leading with 66 documents, followed by Mexico, remembering that *P. dioica* is a traditional spice of Central and South America.

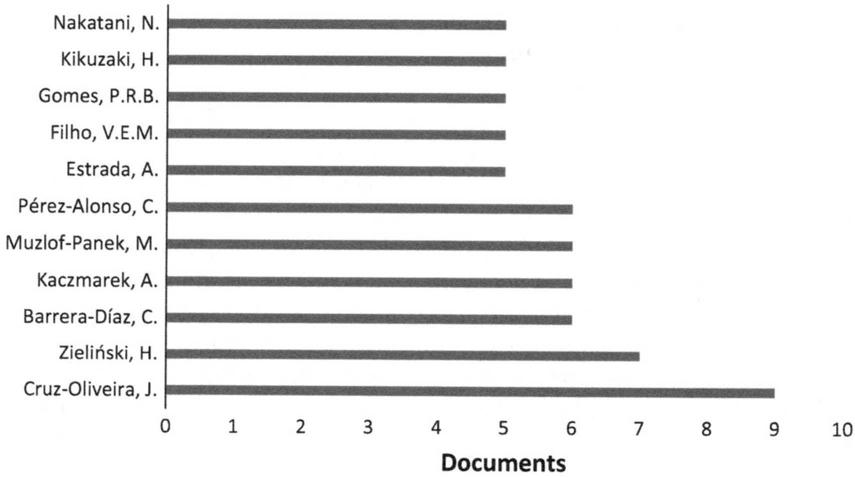

FIGURE 8.2 Most productive authors (Bibliometric data were extracted from the Scopus online database).

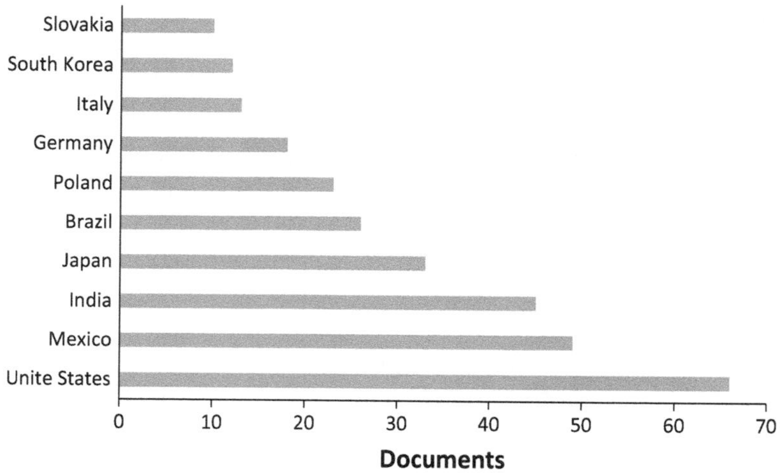

FIGURE 8.3 Most productive countries/territories (Bibliometric data were extracted from Scopus online database).

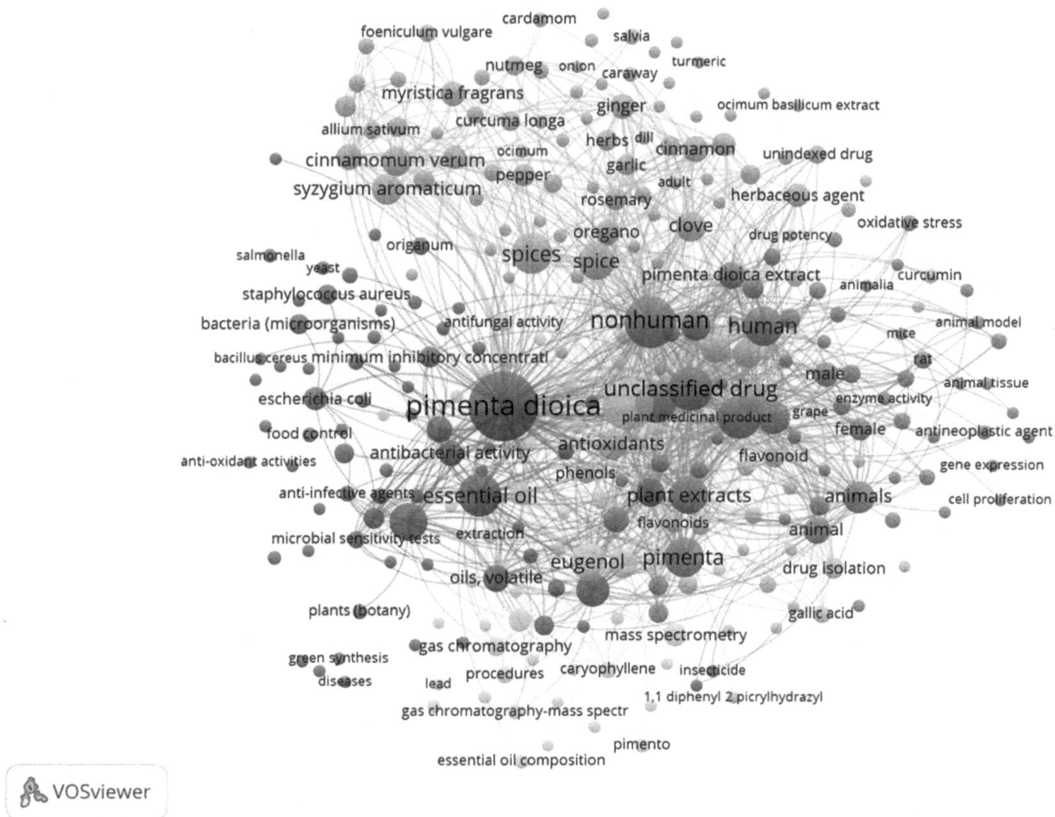

FIGURE 8.4 The term map: the bubble size indicates how frequently a term is mentioned in the articles, then two bubbles that are close to each other reflect that the terms co-appeared more often in the selected publications. The average citations per publications (CPP) are given by the colour of a bubble (Based on data from Scopus and elaborated by VOSviewer software).

Two hundred and eighty-six terms in total were obtained from the literature quantitative analysis on publications, and they are visualized as a term map in Figure 8.4. Terms such as plant extract, pimento dioica, eugenol, essential oil, antioxidant activity, spices are the top recurring keywords.

8.4 FOOD AND NON-FOOD APPLICATIONS

In addition to the potential beneficial health effects, *P. dioica* also has several other applications. The nematicidal activity (Park et al., 2007; Leela et al., 2000) of the essential oil could be used as a valid alternative to common pesticides. Also, the extracted oil from allspice berries has shown anticaricidal effects against the cattle tick *Rhipicephalus microplus* (Martinez-Velazquez et al., 2011). In the food industry, allspice's primary function is to flavour food, or as a preservative in meat production (Parthasarathy et al., 2008; Rao et al.,2012; Rema et al., 2012). Essential oil from allspice has been utilized in cosmetic production, for example in the preparation of soaps, perfumes, after-shave lotions, waxes, candle making, etc. (Premachandran & Murty, 2022). In the perspective of nanonutraceutical applications (Yeung et al., 2020), *P. dioica* has been used as mediator for gold nanoparticles, suggested as a green alternative to conventionally used *in-vivo* theranostic agents (Karhey et al., 2020). Moreover, aqueous and alcoholic extracts of allspice have been investigated as non-toxic corrosion inhibitors for mild steel, exhibiting an excellent inhibitory effect (Anupama et al., 2015).

8.5 CONCLUSION AND FUTURE ASPECTS

Thanks to its bioactive components, *P. dioica* has numerous potentials, not only as a spice. The abundance of bioactive compounds in allspice gives it an important role in the pharmaceutical and nutraceutical fields, especially for eugenol-based products. The possible effects on human health are promising and its traditional use for decades makes it safe and recognized as such worldwide. There are also numerous possible uses beyond the food industry, starting, for example, from its use as a green pesticide to protect the environment. Further studies are still needed to exploit this and other potential uses.

REFERENCES

Anupama, K. K., Ramya, K., Shainy, K. M., & Joseph, A. 2015. Adsorption and electrochemical studies of *Pimenta dioica* leaf extracts as corrosion inhibitor for mild steel in hydrochloric acid. *Materials Chemistry and Physics*, *167*, 28–41.

Ashokkumar, K., Murugan, M., Vellaikumar, S., & Dhanya, M. K. 2022. Essential oil composition of allspice [*Pimenta dioica* (L.) Merill.] leaf from the Western Ghats, India. *Journal of Current Opinion in Crop Science*, *3*(3), 168–172.

Bown, D. 1995. *The Royal Horticultural Society Encyclopedia of Herbs & Their Uses.* Dorling Kindersley Limited, London.

Burt, S. 2004. Essential oils: Their antibacterial properties and potential applications in foods—A review. *International Journal of Food Microbiology*, *94*(3), 223–253.

Chaudhari, A. K., Singh, V. K., Dwivedy, A. K., Das, S., Upadhyay, N., Singh, A. ... Dubey, N. K. 2020. Chemically characterised *Pimenta dioica* (L.) Merr. essential oil as a novel plant based antimicrobial against fungal and aflatoxin B1 contamination of stored maize and its possible mode of action. *Natural Product Research*, *34*(5), 745–749.

Ciccio, J. F. 1997. Effects of acute and subacute administration of *Pimenta dioica* (Myrtaceae) extracts on normal and hypertensive albino rats. *Revista de Biologia Tropical*, *44*, 39–45.

Cruz-Olivares, J., Pérez-Alonso, C., Barrera-Díaz, C., Ureña-Nuñez, F., Chaparro Mercado, M. C., & Bilyeu, B. 2013. Modeling of lead (II) biosorption by residue of allspice in a fixed-bed column. *Chemical Engineering Journal*, *228*, 21–27.

Doyle, B. J., Lawal, T. O., Locklear, T. D., Hernandez, L., Perez, A. L., Patel, U. ... Mahady, G. B. 2018. Isolation and identification of three new chromones from the leaves of *Pimenta dioica* with cytotoxic, oestrogenic and anti-oestrogenic effects. *Pharmaceutical Biology*, *56*(1), 235–244.

El-gizawy, H. A., Mosaad, Y. O., Gobba, N. A. E. K., & Hussein, M. A. 2018. Chemical composition of the essential oil of the leaves of *Pimenta dioica* (L.) Merr. and *Pimenta racemosa* (Mill.) cultivated in Egypt and evaluation of their in-vitro antioxidant and antidiabetic activities. *Int. J. Phytomed*, *10*, 226–234.

Friedman, M., Henika, P. R., & Mandrell, R. E. 2002. Bactericidal activities of plant essential oils and some of their isolated constituents against *Campylobacter jejuni*, *Escherichia coli*, *Listeria monocytogenes*, and *Salmonella enterica*. *Journal of Food Protection*, *65*(10), 1545–1560.

García-Rivera, D., Delgado, R., Bougarne, N., Haegeman, G., & Berghe, W. V. 2011. Gallic acid indanone and mangiferin xanthone are strong determinants of immunosuppressive anti-tumour effects of *Mangifera indica* L. bark in MDA-MB231 breast cancer cells. *Cancer Letters*, *305*(1), 21– .

Hsu, J. D., Kao, S. H., Ou, T. T., Chen, Y. J., Li, Y. J., & Wang, C. J. 2011. Gallic acid induces G2/M phase arrest of breast cancer cell MCF-7 through stabilization of p27Kip1 attributed to disruption of p27Kip1/Skp2 complex. *Journal of Agricultural and Food Chemistry*, *59*(5), 1996–2003.

Ismail, M. M., Samir, R., Saber, F. R., Ahmed, S. R., & Farag, M. A. 2020. Pimenta oil as a potential treatment for *Acinetobacter baumannii* wound infection: *In vitro* and *in vivo* bioassays in relation to its chemical composition. *Antibiotics*, *9*(10), 679.

Rema, J., & Krishnamoorthy, B. 2012. Pimenta. In: K. V. Peter (Eds.) *Handbook of Herbs and Spices*, pp. 238–240. Cambridge: Woodhead Publishing.

Jirovetz, L., Buchbauer, G., Stoilova, I., Krastanov, A., Stoyanova, A., & Schmidt, E. 2007. Spice plants: Chemical composition and antioxidant properties of *Pimenta* Lindl. essential oils, part 2: *Pimenta racemosa* (Mill.) JW Moore leaf oil from Jamaica. *Ernährung*, *31*(7/8), 293–300.

Kamble, V. A., & Patil, S. D. 2008. Spice-derived essential oils: Effective antifungal and possible therapeutic agents. *Journal of Herbs, Spices, and Medicinal Plants*, *14*(3–4), 129–143.

Kharey, P., Dutta, S. B., Gorey, A., Manikandan, M., Kumari, A., Vasudevan, S. ... Gupta, S. 2020. Pimenta dioica mediated biosynthesis of gold nanoparticles and evaluation of its potential for theranostic applications. *ChemistrySelect*, *5*(26), 7901–7908.

Leela, N. K., & Ramana, K. V. 2000. Nematicidal activity of the essential oil of allspice (*Pimenta dioica* L. Merr.*). Journal of Plant Biology*, *27*(1), 75–76.

Lorenzo-Leal, A. C., Palou, E., López-Malo, A., & Bach, H. 2019a. Antimicrobial, cytotoxic, and anti-inflammatory activities of *Pimenta dioica* and *Rosmarinus officinalis* essential oils. *BioMed Research International*, 2019.

Lorenzo-Leal, A. C., Palou, E., & López-Malo, A. 2019b. Evaluation of the efficiency of allspice, thyme and rosemary essential oils on two foodborne pathogens *in-vitro* and on alfalfa seeds, and their effect on sensory characteristics of the sprouts. *International Journal of Food Microbiology*, *295*, 19–24.

Marongiu, B., Piras, A., Porcedda, S., Casu, R., & Pierucci, P. 2005. Comparative analysis of supercritical CO2 extract and oil of *Pimenta dioica* leaves. *Journal of Essential Oil Research*, *17*(5), 530–532.

Martinez-Velazquez, M., Castillo-Herrera, G. A., Rosario-Cruz, R., Flores-Fernandez, J. M., Lopez-Ramirez, J., Hernandez-Gutierrez, R., & del Carmen Lugo-Cervantes, E. 2011. Acaricidal effect and chemical composition of essential oils extracted from *Cuminum cyminum*, *Pimenta dioica* and *Ocimum basilicum* against the cattle tick *Rhipicephalus* (*Boophilus*) *microplus* (Acari: Ixodidae). *Parasitology Research*, *108*(2), 481–487.

Mérida-Reyes, M. S., Muñoz-Wug, M. A., Oliva-Hernández, B. E., Gaitán-Fernández, I. C., Simas, D. L. R., Ribeiro da Silva, A. J., & Pérez-Sabino, J. F. 2020. Composition and antibacterial activity of the essential oil from *Pimenta dioica* (L.) Merr. from Guatemala. *Medicines*, *7*(10), 59.

Muzolf-Panek, M., Kaczmarek, A., Tomaszewska-Gras, J., Cegielska-Radziejewska, R., & Majcher, M. 2019. Oxidative and microbiological stability of raw ground pork during chilled storage as affected by Plant extracts. *International Journal of Food Properties*, *22*(1), 111–129.

Muzolf-Panek, M., Kaczmarek, A., Tomaszewska-Gras, J., Cegielska-Radziejewska, R., Szablewski, T., Majcher, M., & Stuper-Szablewska, K. 2020. A chemometric approach to oxidative stability and physicochemical quality of raw ground chicken meat affected by black seed and other spice extracts. *Antioxidants*, *9*(9), 903.

Parthasarathy, V. A., & Kandiannan, K. 2008. *Spices and Condiments*. New Delhi: New India Publishing Agency. https://www.researchgate.net/publication/ 264286797_Organic_Spices.

Park, I. K., Kim, J., Lee, S. G., & Shin, S. C. 2007. Nematicidal activity of plant essential oils and components from ajowan (*Trachyspermum ammi*), allspice (*Pimenta dioica*) and litsea (*Litsea cubeba*) essential oils against pine wood nematode (*Bursaphelenchus xylophilus*). *Journal of Nematology*, *39*(3), 275.

Premachandran, M. S., & Murthy, P. S. 2022. Ethnobotanical, phytochemical, pharmacological properties and applications of *Pimenta dioica* L. *Journal of Essential Oil Research*, *34*(3), 216–232.

Przygodzka, M., Zielińska, D., Ciesarová, Z., Kukurová, K., & Zieliński, H. 2014. Comparison of methods for evaluation of the antioxidant capacity and phenolic compounds in common spices. *LWT – Food Science and Technology*, *58*(2), 321–326.

Przygodzka, M., Zieliński, H., Ciesarová, Z., Kukurová, K., & Lamparski, G. 2015. Study on Sensory Quality, Antioxidant Properties, and Maillard Reaction Products Formation in Rye-Buckwheat Cakes Enhanced with Selected Spices. *Journal of Chemistry*, vol. 2015, Article ID 418639, 9 pages.

Rao, P. S., Navinchandra, S., & Jayaveera, K. N. 2012. An important spice, *Pimenta dioica* (Linn.) Merill: A review. *International Current Pharmaceutical Journal*, *1*(8), 221–225.

Rehaily, A. J., Al-Said, M. S., Al-Yahya, M. A., Mossa, J. S., & Rafatullah, S. 2002. Ethnopharmacological studies on allspice (*Pimenta dioica*) in laboratory animals. *Pharmaceutical Biology*, *40*(3), 200–205.

Sánchez-Zarate, A., Hernández-Gallegos, M. A., Carrera-Lanestosa, A., López Martínez, S., Chay-Canul, A. J., Esparza-Rivera, J. R., & Velázquez-Martínez, J. R. 2020. Antioxidant and antibacterial activity of aqueous, ethanolic and acetonic extracts of *Pimenta dioica* L. leaves. *International Food Research Journal*, *27*(5), 825-834.

Sarathambal, C., Rajagopal, S., & Viswanathan, R. 2021. Mechanism of antioxidant and antifungal properties of *Pimenta dioica* (L.) leaf essential oil on *Aspergillus flavus*. *Journal of Food Science and Technology*, *58*(7), 2497–2506.

Suárez, A., Ulate, G., & Ciccio, J. F. 2000. *Hypotensive action of an aqueous extract of* Pimenta dioica (Myrtaceae) in rats. *Revista de Biología Tropical*, *48*(1), 53–58.

Takemasa, N., Ohnishi, S., Tsuji, M., Shikata, T., & Yokoigawa, K. 2009. Screening and analysis of spices with ability to suppress verocytotoxin production by *Escherichia coli* O157. *Journal of Food Science*, *74*(8), M461–M466.

Van Eck, N. J., & Waltman, L. 2014. Visualizing bibliometric networks. In: Y. Ding, R. Rousseau, & D. Wolfram (Eds.) *Measuring Scholarly Impact: Methods and Practice*, pp. 285–320. Springer, Berlin.

Van Eck, N. J., & Waltman, L. 2010. Software survey: VOSviewer, a computer program for bibliometric mapping. *Scientometrics*, *84*(2), 523–538.

Van Eck, N. J., & Waltman, L. 2011. Text mining and visualization using VOSviewer. *ISSI Newslett*, *7*, 50–54.

Vaughan, J., & Geissler, C. (2009). *The new Oxford book of food plants*. OUP Oxford.

Vinh, N. D., Takacsova, M., Nhat, D. M., & Kristiano, K. 2000. Antioxidant activitiesof allspice extracts in rapeseed oil. *Czech Journal of Food Sciences*, *18*(2), 49–51.

Waltman, L., van Eck, N. J., & Noyons, E. C. 2010. A unified approach to mapping and clustering of bibliometric networks. *J. Informetr*, *4*(4), 629–635.

Yeung, A. W. K., Souto, E. B., Durazzo, A., Lucarini, M., Novellino, E., Tewari, D. … Santini, A. 2020. Big impact of nanoparticles: Analysis of the most cited nanopharmaceuticals and nanonutraceuticals research. *Current Research in Biotechnology*, *2*, 53–63.

Yoshimura, M., Amakura, Y., & Yoshida, T. 2011. Polyphenolic compounds in clove and pimento and their antioxidative activities. *Bioscience, Biotechnology, and Biochemistry*, *75*(11), 2207–2212.

Zari, A. T., Zari, T. A., & Hakeem, K. R. 2021. Anticancer properties of eugenol: A review. *Molecules*, *26*(23), 7407.

Zhang, L., Shamaladevi, N., Jayaprakasha, G. K., Patil, B. S., & Lokeshwar, B. L. 2015. Polyphenol-rich extract of *Pimenta dioica* berries (Allspice) kills breast cancer cells by autophagy and delays growth of triple negative breast cancer in athymic mice. *Oncotarget*, *6*(18), 16379.

9 Aniseed (*Pimpinella animus*)

Nitinkumar S. Shetty, M. Lakshmikanthan, B. Anusha, and N. V. Anil Kumar

9.1 INTRODUCTION

Pimpinella animus, commonly called aniseed or anise, belongs to the Umbelliferae or Apiaceae families. It is a flowering annual herb (Figure 9.1). It is native to Egypt and the eastern Mediterranean region. It is cultivated in the Middle East, Europe, and Southwest Asia (Salehi Surmaghi 2010)

The plant (Figure 9.2) bears yellowish-brown or greenish-brown fruits and white flowers, growing up to 2–3 feet. It is an aromatic plant. The fruit has a characteristic sweet odour and a pleasant aromatic taste. The dried ripe fruits are rich in essential oil with about 1.5–5% content. The oil also consists of fatty acids (like oleic and palmitic acids) (about 8–11%), protein (about 18%), and carbohydrates (about 4%) (Reineccius 1995).

The essential oils from aromatic plants are used extensively as food flavourings because of their delightful taste. They are heavily used in pharmaceuticals due to their medicinal properties and as perfumes because of their fragrance. Even individual components of essential oils are isolated and exploited for the unique properties they possess.

The essential oil content in aniseed is about 20 mL/kg. Phenylpropanoids, sesquiterpenes, C12-sesquiterpenes, and monoterpenes are some of the secondary metabolites. According to the analysis of volatile aniseed oil conducted by Anastasopoulou et al. (2020), twenty-eight components fall into the above-mentioned secondary metabolite classes. Among the identified components, most of the compounds were present in trace amounts, and very few were present at a higher percentage. The identified compounds are shown in Table 9.1.

Aniseed oil has a higher percentage of phenylpropanoids than other metabolites, with *trans*-anethole being the most abundant chemical compound, followed by estragole and other compounds representing less than 5% of peak area, as measured in gas chromatography-mass spectrometry (GC-MS) analysis (Anastasopoulou et al. 2020).

In addition to these compounds, aniseed oil contains oxidized *trans*-anethole products, such as anisic alcohol, anisaldehyde, anisic acid, anethole epoxide, and anisyl alcohol (Bruneton et al. 1995). The oxidation product, *p*-anisaldehyde, is identified to be neurotoxic, but it is absent from aniseed oil, and its content increases only on prolonged storage of aniseed oil (Sweet 1997).

Trans-anethole is oestrogenic, acts as a local anaesthetic agent, and possesses spasmolytic and expectorant activity, so its high concentration in aniseed oil, makes it a suitable option for treating ailments (Orav et al.2008) Anethole is employed in the food and pharmaceutical industries. The United States Food and Drug Administration (FDA-US) has assured its safety and experimentally proven that it has no toxicity at low doses and is considered non-genotoxic and non-carcinogenic (Ponte et al. 2012; Smith et al. 2002)

Another major compound identified in *P. anisum* seed oil is estragole (5.5%), an alkenyl-benzene, and it also provides flavour and enhances the oil's medicinal properties. Nevertheless, intake of estragole should be minimal as it is 'genotoxic and carcinogenic', as confirmed by the 'Scientific Committee on Food'. Based on an evaluation of the 'Committee of Experts on Flavouring Substances of the Council of Europe', intake of estragole should be limited to 0.05 mg/kg per day (Anastasopoulou et al. 2020).

DOI: 10.1201/9781003387046-9

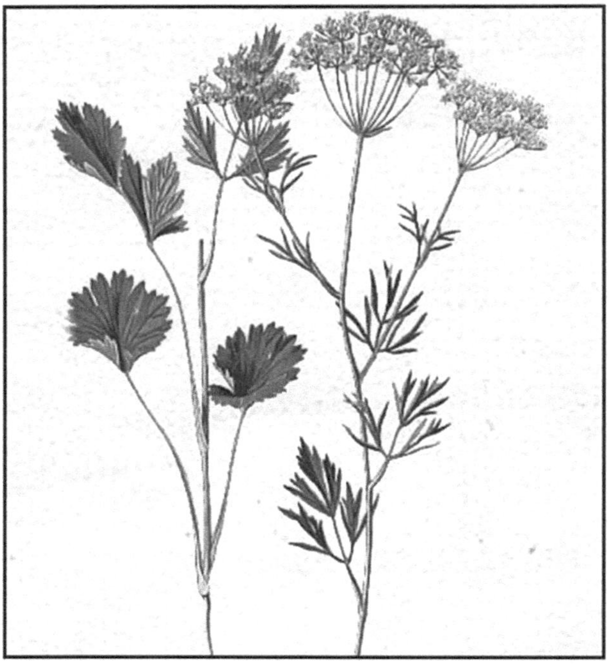

FIGURE 9.1 Pimpinella animus plant.

Plant species from the Apiaceae family are known for their aromatic nature and medicinal properties. *P. anisum* possesses pharmacological and traditional properties, which are discussed below.

FIGURE 9.2 *Aniseed.*

TABLE 9.1
Major Chemical Compounds Identified in Aniseed Oil

Sl No	Class	Compound	Abundance (%)	Structure
1	Phenylpropanoids	*trans*-anethole	89.99	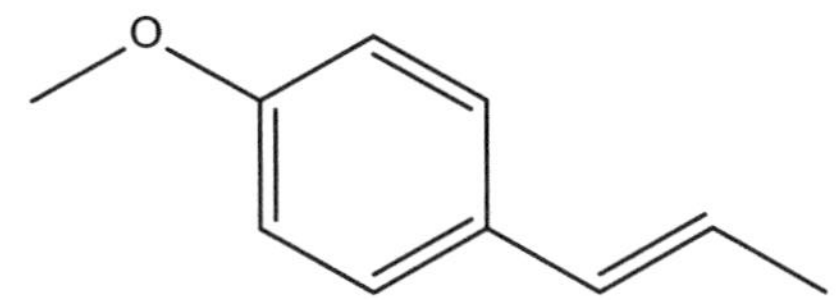
2	Phenylpropanoids	estragole	5.51	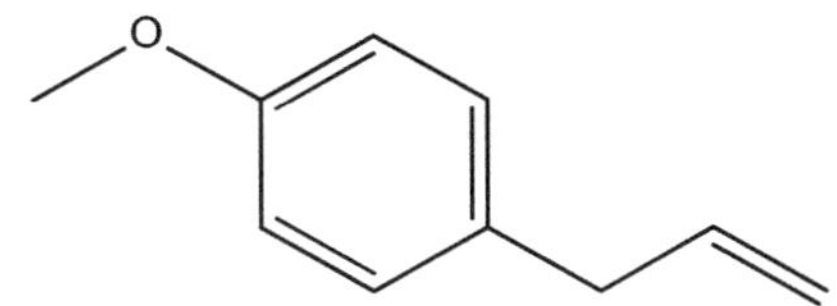
3	Sesquiterpenes	γ-himachalene	1.31	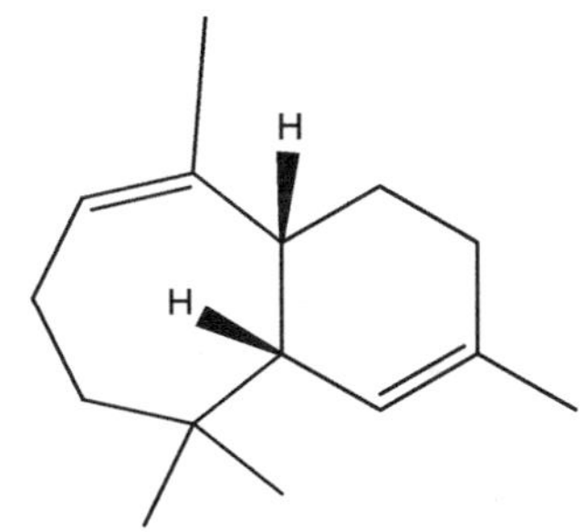
4	Phenylpropanoids	pseudoisoeugenyl-2-methylbutyrate	1.24	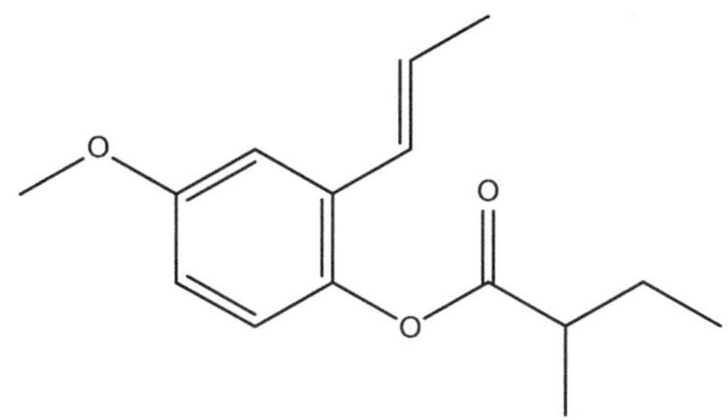

(Continued)

TABLE 9.1 (CONTINUED)
Major Chemical Compounds Identified in Aniseed Oil

Sl No	Class	Compound	Abundance (%)	Structure
5	Phenylpropanoids	*cis*-anethole	0.31	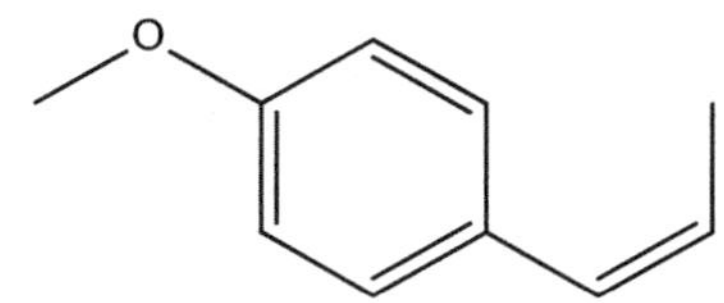
6	Sesquiterpenes	α-himachalene	0.17	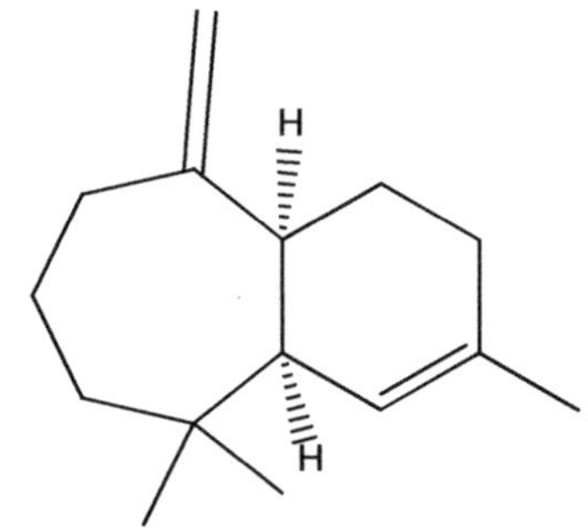
7	Phenylpropanoids	epoxypseudoisoeugenyl-2-methylbutyrate	0.16	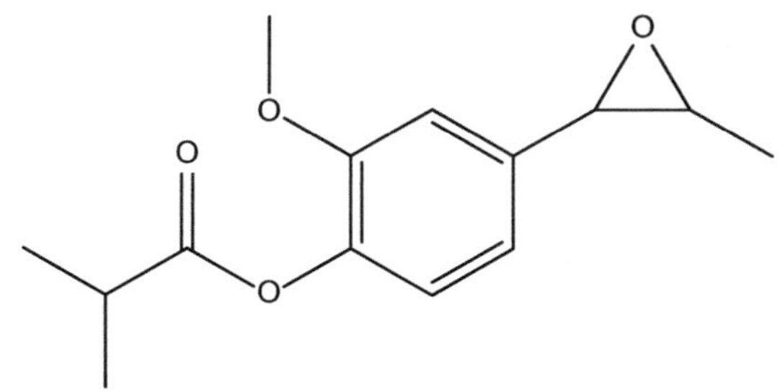

(*Continued*)

TABLE 9.1 (CONTINUED)
Major Chemical Compounds Identified in Aniseed Oil

Sl No	Class	Compound	Abundance (%)	Structure
8	Phenylpropanoids	methyl eugenol	0.14	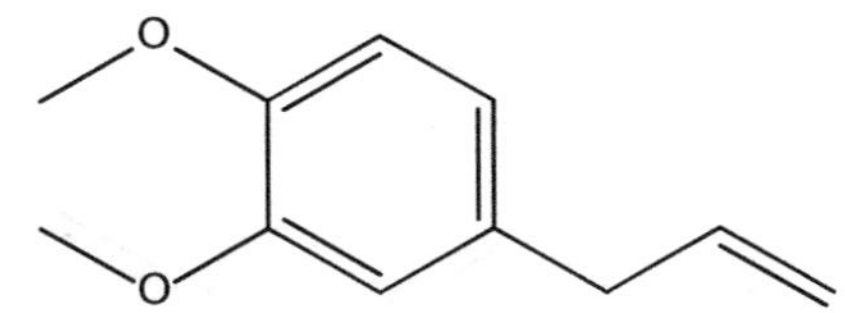
9	Sesquiterpenes	α-zingiberene	0.12	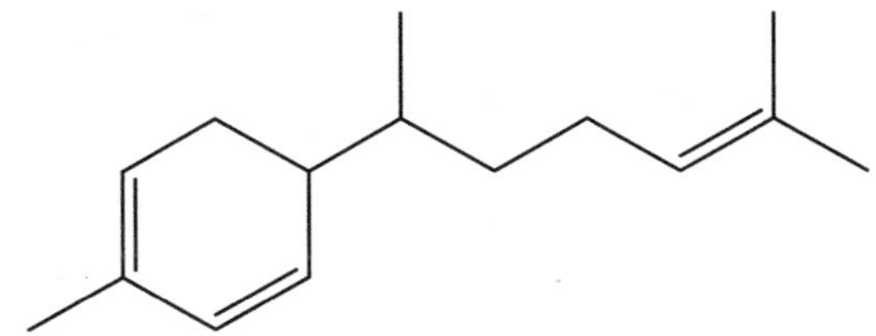
10	Sesquiterpenes	β-himachalene	0.11	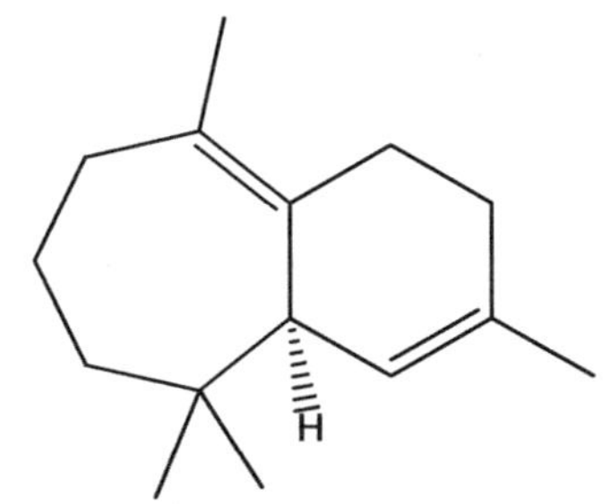
11	Monoterpenes	linalool	0.09	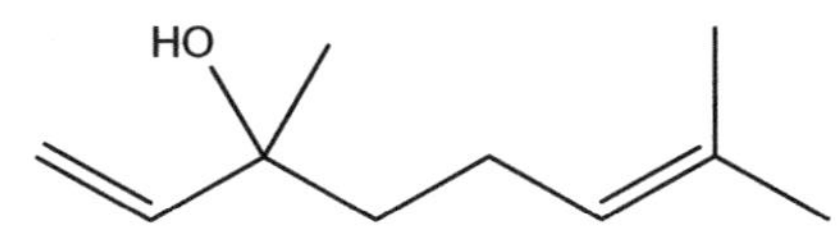

(Continued)

TABLE 9.1 (CONTINUED)
Major Chemical Compounds Identified in Aniseed Oil

Sl No	Class	Compound	Abundance (%)	Structure
12	Sesquiterpenes	β-bisabolene	0.08	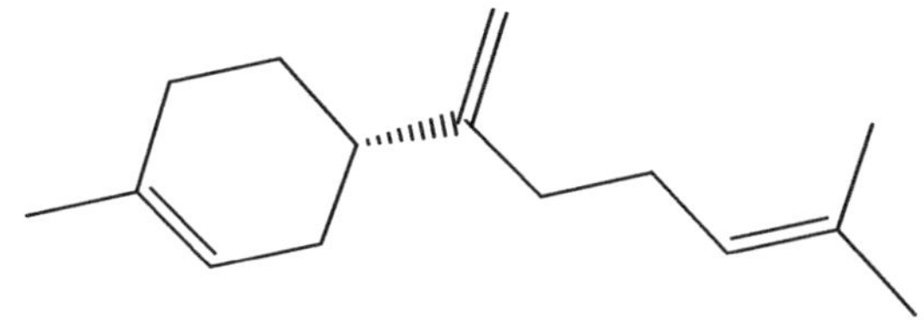
13	C12-sesquiterpenes	geyrene	0.06	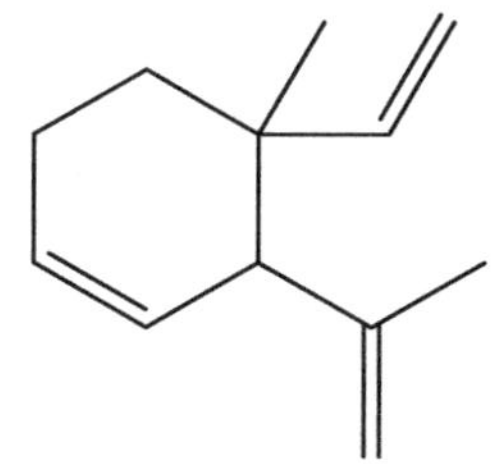
14	Phenylpropanoids	2,5-dimethoxysafrole	0.05	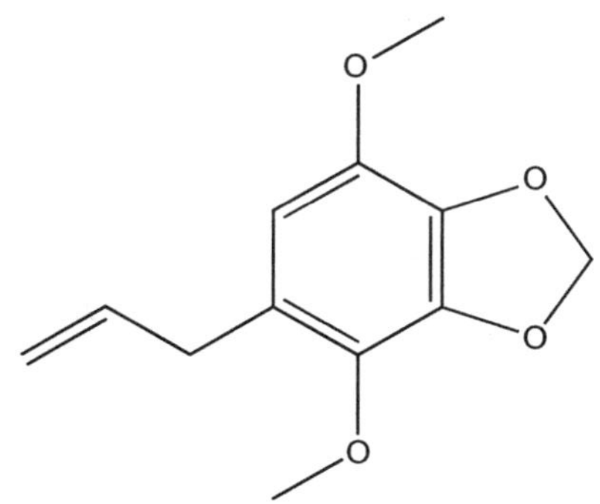

(*Continued*)

TABLE 9.1 (CONTINUED)
Major Chemical Compounds Identified in Aniseed Oil

Sl No	Class	Compound	Abundance (%)	Structure
15	Monoterpenes	α-phellandrene	0.01	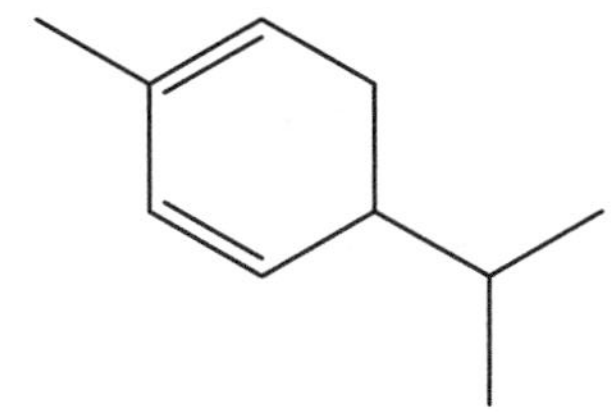
16	Monoterpenes	*p*-cymene	0.01	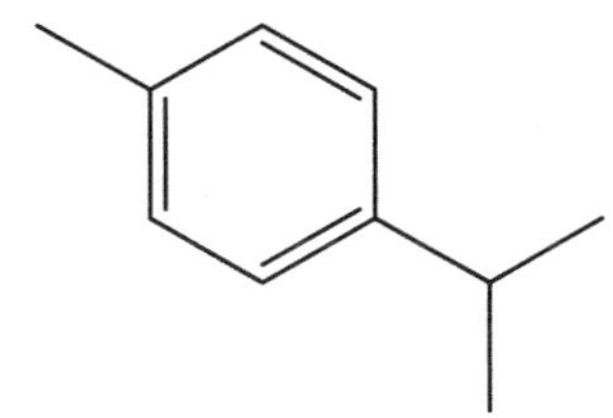
17	Monoterpenes	limonene	0.01	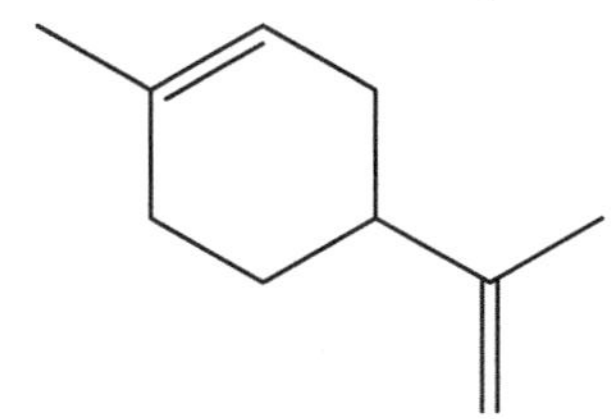
18	Monoterpenes	γ-terpinene	0.01	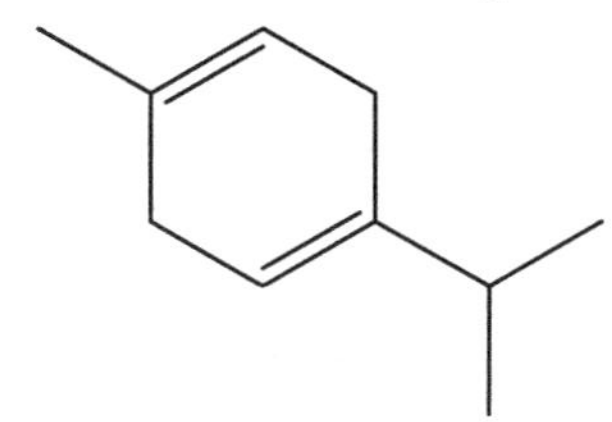

(*Continued*)

TABLE 9.1 (CONTINUED)
Major Chemical Compounds Identified in Aniseed Oil

Sl No	Class	Compound	Abundance (%)	Structure
19	Alkanes	nonacosane	0.01	$CH_3 - (CH_2)_{27} - CH_3$
20	Phenylpropanoids	isoeugenol	-	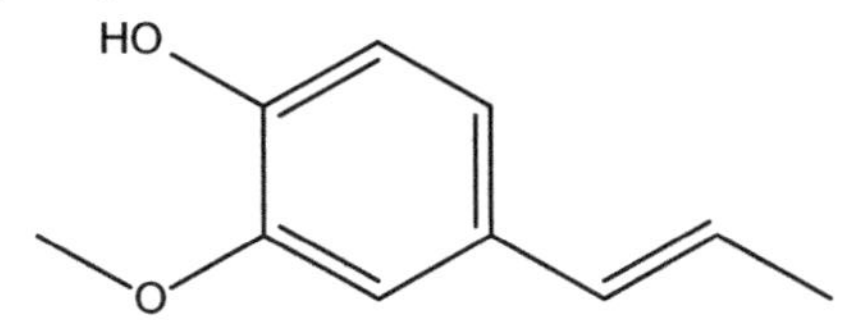
21	Sesquiterpenes	α-cedrene	-	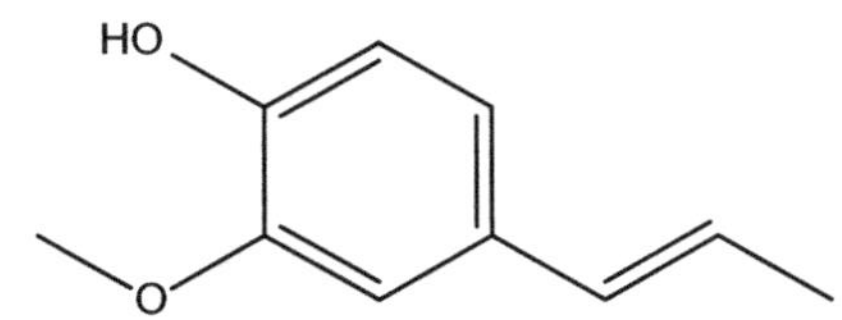
22	Monoterpenes	cycloisolongifolene	-	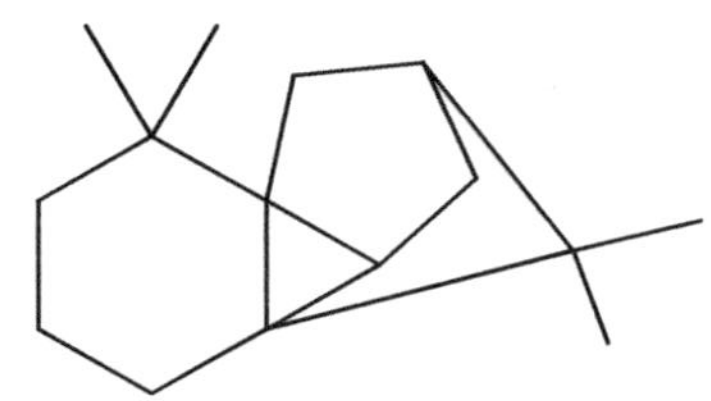
23	Monoterpenes	α-longipinene	-	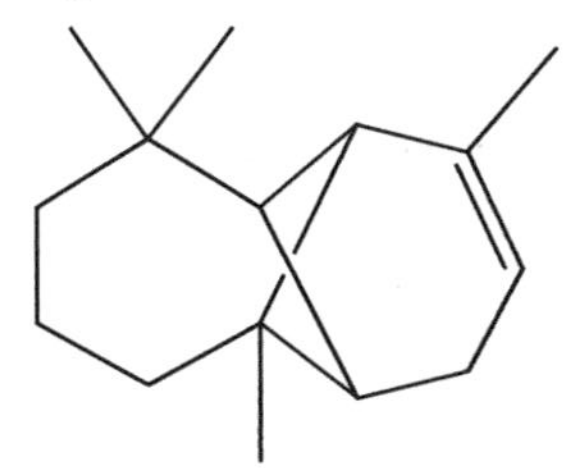

(*Continued*)

TABLE 9.1 (CONTINUED)
Major Chemical Compounds Identified in Aniseed Oil

Sl No	Class	Compound	Abundance (%)	Structure
24	Sesquiterpenes	β-sesquiphellandrene	-	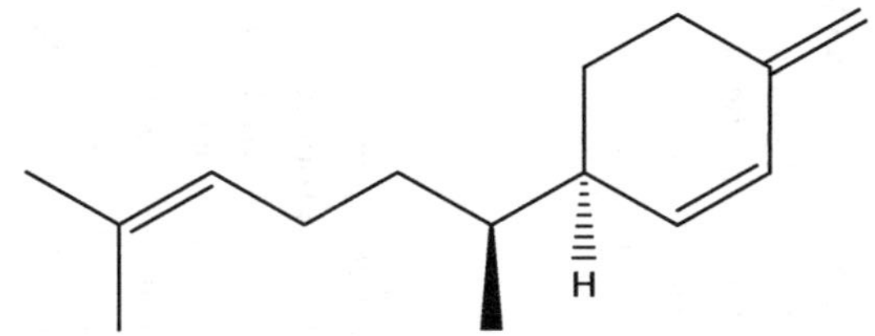
25	Sesquiterpenes	α-copaene-8-ol	-	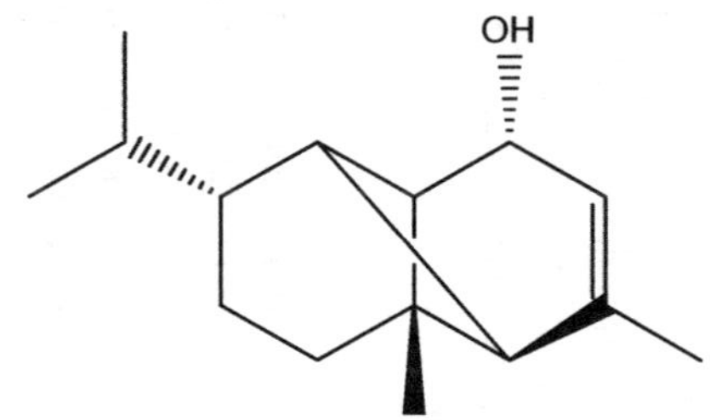
26	Sesquiterpenes	α-bisabolol	-	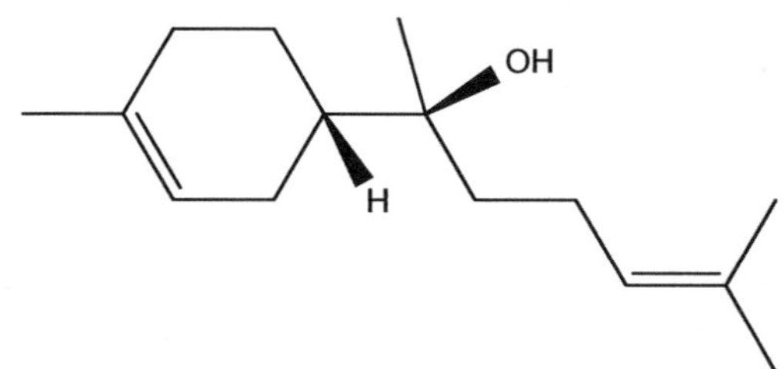
27	Alkanes	docosane	-	$CH_3 - (CH_2)_{20} - CH_3$
28	Alkanes	hexatriacontane	-	$CH_3 - (CH_2)_{34} - CH_3$

9.2 TRADITIONAL USES

Traditional herbal therapy is gaining in popularity and is now regarded as a scientifically valid disease prevention and treatment method. A plant extract is a prospective bioactive agent that can either positively or negatively affect several cellular functions. The most prevalent active components in medicinal plants are essential oils. The plant anise has a high oil content in the seeds. The seeds soaked in hot water are used as a digestive, carminative, antibacterial, diuretic, and folk cure for constipation and insomnia. Anise seeds are utilized in the culinary sector as flavouring ingredients (e.g., Greek ouzo, Lebanese arak, anisette) (Kreydiyyeh et al. 2003).

9.2.1 Food Science Applications

Aniseed is an important plant species in food preparation that has been used as a spice for thousands of years. Before the development of the freezer, many essential oil producers were important in preserving food; today, we may attribute this property to antimicrobial actions. In this context, the complex mixture of secondary metabolites that make up essential oils has a crucial influence on food preservation. Bread, cakes, candies, and beverages can benefit from the flavour of the anise essential oil made from dry, ripe anise grapes. Anise is also significant in food technology because many consumers choose green products free of artificial preservatives. Essential oils also stop food from spoiling and becoming contaminated by microbes, demonstrating their potential for preserving meat, animal products, rice, fruits, and fish meals. A high fat content often inhibits these effects (Hajou et al. 2004).

The essential oil derived from *P. anisum*, with a high anethole concentration (82.8%), showed excellent antioxidant activity in studies employing the DPPH free-radical scavenging assay and the β-carotene/linoleic acid ratio. This study suggests that the essential oil of *P. anisum*, which also contains a considerable amount of anethole, may be helpful in this manner. The food industry needs to know this since antioxidants can prevent food from going bad by preventing the process of lipid peroxidation. Aniseed essential oil offers immense potential for developing new products since, at therapeutic dosages, it has a low toxicity and insignificant hepatotoxicity. Compared to vanillin, another common flavouring compound, anethole, its main component, has a higher LD_{50}. However, 1–5 mL of aniseed essential oil used by humans might cause vomiting and nausea. When large amounts of aniseed-based alcohol are consumed, identical side effects may occur to those seen with absinthe liquors (Albert-Puleo 1980).

Aniseed is a commonly used raw material for teas, infusions, and beverages. This is due to aniseed's sweet flavour and pleasantness, related to its essential oil. *P. anisum*-based traditional alcoholic beverages are produced in several Mediterranean countries, such as Greece, Turkey, France, Spain, and Italy. These beverages include raki, pastis, sambuca, ouzo, arak, and anis. These are prepared by distilling the fruits with ethyl alcohol, generating a flavourful beverage with a high ethanol content. Hence, utilizing essential oils might help preserve the objects for longer by releasing regulated amounts of biocidal chemicals during packaging or in the field. Aniseed, represented mainly by its essential oil, has been adopted in culinary and food production contexts for a long time and is still a never-ending source of supplies for the food industry (Adelakun et al. 2016).

9.2.2 Usage in Agriculture

Numerous research on stored goods, crop pests, cockroaches, ticks, and mosquitoes have demonstrated that the essential oils of members of the Apiaceae family have insecticidal and insect-repellent properties. *Tetranychus urticae*, often known as the red spider mite or two-spotted spider mite, is one of the pests to cause significant crop losses. Recently, these mites have rapidly evolved a tolerance to many chemical insecticides. By hydro-distilling the seeds, anise essential oil was extracted, and the oil's chemical makeup was investigated. The DPPH radical scavenging assay assesses the antioxidant capacities of the volatile oil. Green bean plants are protected from mite damage by anise oil, a natural acaricide, which can also control the gypsy moth, *Lymantria dispar* (Lepidoptera: Erebidae), a significant hardwood pest. (El-Sayed et al. 2022)

9.2.3 Usage as a Growth Enhancer

In the feeding of broiler chickens, anise oil takes the place of antibiotics as a natural growth enhancer. Anise oil was administered to regular meal at various dosages to determine how it affected feed intake, daily liveweight gain, and feed conversion ratio compared with control and antibiotic groups. Two-hundred day-old broiler chicks were divided into five groups of 40 birds each, with each group randomly assigned to a specific treatment diets. There are four copies of each treatment. Groups with inclusion levels of 100 mg/kg, 200 mg/kg, and 400 mg anise oil/kg feed, a positive control group receiving an 0.1% antibiotic (Avilamycin) added to the feed, and a negative control group, with no chemical added to the feed, were created for the experimental study. Feed consumption was equivalent between the five groups ($p > 0.05$). The highest mean daily weight gain was in the 400 mg/kg anise oil group (70.35 g) ($p<0.05$), followed by the antibiotic group (65.84 g), the 100 mg/kg anise oil group (62.57 g), and the 200 mg/kg.kg.kg in the anise oil group (62.47 g), and the control group (61.30 g). Adding 400 mg/kg of anise oil to the diets increased the daily weight compared to the negative control group by almost 15%. In the antibiotic group, this improvement was only 7%. Compared with the antibiotic group, the daily liveweight gain increased by approximately 6.5% when 400 mg/kg of aniseed oil was added to the feed. Compared with the negative control group, the feed conversion ratio increased by almost 12% by adding 400 mg/kg of aniseed oil to the diet. In the antibiotic group, this improvement was 7%. Moreover, the feed conversion ratio

was increased by about 6% compared with the antibiotic group by the addition of 400 mg/kg anise oil to the diets. Anise oil, therefore, proved to be a potential natural growth enhancer for chickens (Ciftci et al. 2005).

9.2.4 Pharmacological Properties

Most species belonging to the family Apiaceae are known for their medicinal properties. The aromatic seeds of the aniseed plant *P. anisum* exhibit known medicinal properties. The secondary metabolites in this plant and particularly in essential oil from the seeds are known to have antibacterial, antifungal, antioxidant, insecticidal, and anti-inflammatory properties. These properties have been employed for treating various medical conditions (Peter 2012).

The inhibitory activity of the aniseed essential oil against various pathogenic bacteria was studied over a period, and positive results were recorded against *Bacillus cereus*, *Staphylococcus aureus*, *Proteus vulgaris*, *Proteus mirabilis*, *Salmonella typhimurium*, and *Streptococcus pyogenes*, whereas *Escherichia coli*, *Pseudomonas aeruginosa*, and *Klebsiella pneumoniae* exhibited resistance to pure anise essential oil (Al-Bayati 2008). Aniseed oil with added benzoic acid or methylparaben exhibited potent activity against *Salmonella enteritidis*, but the growth response of *Listeria monocytogenes* was lowered significantly (Fyfe et al. 1997)

Similarly, anisi aetheroleum oil extracted from aniseed inhibited growth of bacterial species like *Micrococcus luteus* and *Mycobacterium smegmatis*. The effectiveness of this essential oil against ten bacterial species has been observed in a study by Gülçin et al. (2003) As aniseed oil possesses such properties, it has marked applicability to medicine to treat bacterial infections and is used as an antibiotic. The ability to inhibit bacterial growth is observed at enhanced rates in the case of aqueous and methanol extracts of aniseed, compared with the pure essential oil, while acetone extracts did not inhibit bacterial growth (Akhtar et al. 2008).

Trans-anethole and anisaldehyde in aniseed oil are actively involved in exhibiting antifungal activity; they inhibit the growth and toxicity of fungal species belonging to the genera *Alternaria*, *Aspergillus*, *Cladosporium*, *Fusarium*, and *Penicillium* (Shukla and Tripathi 1987). It has been reported that species like *Candida albicans*, *Candida glabrata*, *Candida parapsilosis*, *Alternaria alternata*, *Aspergillus niger*, and *Aspergillus parasiticus* are sensitive to these components of aniseed essential oil. Most of these fungal species are known to produce mycotoxins that cause adverse health effects like acute poisoning or long-term effects in humans and livestock (Koriem 2015).

Hot water extracts of seeds of *P. anisum* contain three lignin-carbohydrate-protein complexes (LC1, LC2, and LC3), which possess antiviral and immunostimulant activities. These complexes have inhibitory effects against HIV type 1 and type 2 viruses, human cytomegalovirus, and measles. The complexes also induce the expression of mRNAs of interleukins (IL-1ß and IL-10) and are also known to activate macrophage cells. The properties of these complexes make aqueous aniseed extracts a preventative measure for infectious diseases (Lee et al. 2011).

Anethole and anisaldehyde are known to have insecticidal properties. *Trans*-anethole, being an active insecticide, is toxic against *Lycoriella ingenua* larvae. (Leung 1980) Anise oil vapours are effective as ovicidal and larvicidal ingredients against mosquitoes like *Culex pipiens* (Erler et al. 2006). These oil vapours are also reported to be lethal to the eggs of two beetle species, *Tribolium confusum* and *Ephestia kuehniella* (Tunç et al. 2000). A study has also been conducted on the control of house dust mites *Dermatophagoides farina* and *Dermatophagoides pteronyssinus*, using the aniseed oil derivative *p*-anisaldehyde, and results showed that this compound is toxic to these mites. Due to its acaricidal activity, *p*-anisaldehyde can be used to control these mites (Lee 2004).

Anise chemical constituents, like ascorbic acid, hydroquinone, isoorientin, myristicin, rutin, and stigmasterol, from plant and fruit parts of *P. anisum* are excellent antioxidants. The antioxidant effect of this plant is better than available synthetic antioxidants. The anisum essential oil is an excellent preservative for spice materials (Andallu et al. 2011). Alcohol and aqueous extracts from Aniseed are potent antioxidants and have good superoxide anion and DPPH radical scavenging, metal chelating, reducing power, and H_2O_2-scavenging activity. The flavonoid concentration in the essential oil correlates positively with the antioxidant activity; *P. anisum* extracts with high flavonoid concentration have significant antioxidant activity (Nickavar and Abolhasani 2009).

Anisi aetheroleum is also known for its anticonvulsant effect. It increases the delay of induced seizures; this has been studied in the case of mice. Various doses of plant oil can enhance the functions of Ca^{2+} channel activity or inhibit Ca^{2+}-activated K^+ channels, increasing the threshold of induced seizures (Heidari 2005).

The inflammatory mediators that are secreted were inhibited by anethole. Studies on mice have shown that the activity of anethole reduces pro-inflammatory mediators, including TNF-α and interleukin-1ß (Mahboubi and Mahboubi 2021).

The analgesic and anti-inflammatory property of essential oils is known to aid and reduce pain in women after childbirth. In women's cases, ingested anise tea stimulates menstruation and reduces menstrual cramps (Albert-Puleo 1980). An experiment on lab rats found that feeding them aniseed methanol extract improved their milk yield compared with rats who were not fed with the extract, inferring that aniseed oil boosts milk production in females. Anethole increases prolactin levels because its competition with dopamine (due to its structural similarity to dopamine) acts antagonistically at dopamine receptor sites, enhancing milk production (Hosseinzadeh et al. 2014).

Higher concentrations of anethole and estragole have relaxing effects mediated *via* opening voltage-dependent Ca^{2+} channels (Soares et al. 2007). Because of its inhibitory effects on muscarinic receptors, anethole has a good relaxant effect. The activation of the NO-cGMP pathway is one potential mechanism for this activity (Nickavar and Abolhasani 2009). Lower doses of anethole exhibit carminative and expectorant or bronchodilator activities. Anethole lubricates the respiratory tract by reducing the viscosity and increasing the volume of respiratory tract mucus. Aqueous extracts like aniseed tea and aniseed decoction relieve respiratory tract congestion and can be a gastrointestinal relaxant (Leung 1980).

In a study of the anti-ulcer property of aniseed oil, the findings showed that anise significantly reduced gastric mucosal damage caused by necrotizing agents and indomethacin. Histological analysis further confirmed the antiulcer efficacy (Shojaii and Abdollahi Fard 2012). Incorporating aniseed into the diet for several days lowers triglycerides, serum cholesterol, and fasting blood sugar. Also, lipid peroxidation in red blood cells (RBC) and plasma and protein oxidation in serum decreases significantly, whereas vitamin A levels increase. Antidiabetic, hypolipidaemic, and antioxidant properties of *P. anisum* seeds account for all of these effects (Rajeshwari et al. 2011).

An antidiuretic effect results from anethole's increased Na^+/K^+-ATPase activity, considerably boosting sodium reabsorption in the kidney. The incidence and occurrence of both invasive and non-invasive adenocarcinomas have decreased due to the anticancer activity of *trans*-anethole (Koriem 2015; Reddy et al. 1993).

9.2.5 Other Potential Uses

As a substitute for current conventional fuels, low-viscosity anise oil can be added to waste cooking oil at a concentration of 80%. It might lead to the discontinuation of using fossil fuels in diesel engines. Additionally, it cuts carbon monoxide and hydrocarbon emissions by 43.8% and 18.6%, respectively (Balasubramanian et al. 2022). Aniseed oil is used to produce packaging film by infusing it with gelatin-alginate. The oil's phenolic content boosts the gelatin-alginate film's antioxidant activity, extending shelf life and maintaining product quality, whereas its antibacterial activity lowers chemical and microbiological spoilage, making the film suitable for food storage (Jafari et al. 2022).

Aniseed oil capsules were reportedly mixed with white metal plain bearings by steam locomotive builders, according to a report in British Railway magazine. The combination's distinctive aroma served as a warning in the event of overheating. Anise is used for fishing, where it is added to fishing lures to attract fish; it can also be converted into a liquid perfume (Gabriel et al. 2008).

9.3 CONCLUSION AND FUTURE REMARKS

The long history of aniseed in traditional medicine in earlier days was not explored scientifically, leading to limited information on the science behind the traditional usage. Researchers explored this natural wonder based on assumptions, leading to the identification of various biological activities. Also, curiosity led to usage in different fields like agriculture, while usage in locomotives is reported in the literature. The biological activity of this herb is mainly due to its constituents like anethole and estragole. Due to its low toxicity, aniseed can be a promising potential material for future drug discovery and development. All the research on this material is at the initial phase of drug development. Very little research has been carried out on lower animal models. More studies are recommended to assess the beneficial effects of *anise* in human models.

LIST OF ABBREVIATIONS

DPPH: 2,2-diphenyl-1-picrylhydrazyl
GC-MS: Gas chromatography-mass spectrometry
HIV: Human Immunodeficiency Virus
LD_{50}: 50% Lethal Dose value
RBC: Red blood cells
TNF-α: Tumor necrosis factor-alpha

REFERENCES

Adelakun, Oluyemisi Elizabeth, Olusegun James Oyelade, and Bosede Folake Olanipekun. 2016. Chapter 7 - use of essential oils in food preservation. In *Essential Oils in Food Preservation, Flavor and Safety*, edited by V. R. Preedy, (pp. 71–84). San Diego: Academic Press.

Akhtar, A, A. A. Deshmukh, A. V. Bhonsle, P. M. Kshirsagar, and M. A. Kolekar. 2008. In vitro antibacterial activity of *Pimpinella anisum* fruit extracts against some pathogenic bacteria. *Veterinary World* 1 (9):272–274.

Al-Bayati, Firas A. 2008. Synergistic antibacterial activity between *Thymus vulgaris* and *Pimpinella anisum* essential oils and methanol extracts. *Journal of Ethnopharmacology* 116 (3):403–406.

Albert-Puleo, Michael. 1980. Fennel and anise as estrogenic agents. *Journal of Ethnopharmacology* 2 (4):337–344.

Anastasopoulou, Eleni, Konstantia Graikou, Christos Ganos, Gioacchino Calapai, and Ioanna Chinou. 2020. *Pimpinella anisum* seeds essential oil from Lesvos island: Effect of hydrodistillation time, comparison of its aromatic profile with other samples of the Greek market. Safe use. *Food and Chemical Toxicology: An International Journal Published for the British Industrial Biological Research Association* 135:110875.

Andallu, Bondada, C. U. Nuts Rajeshwari. 2011. Aniseeds (*Pimpinella anisum* L.) in Health and Disease. *Nuts and Seeds in Health and Disease Prevention*:175–181.

Balasubramanian, Dhinesh, Tanakorn Wongwuttanasatian, Inbanaathan Papla Venugopal, and Amudhan Rajarajan. 2022. Exploration of combustion behavior in a compression ignition engine fuelled with low-viscosity *Pimpinella anisum* and waste cooking oil biodiesel blends. *Journal of Cleaner Production* 331:129999.

Bruneton, Jean, and others. 1995. *Pharmacognosy, Phytochemistry, Medicinal Plants*. Lavoisier Publishing.

Ciftci, Mehmet, Talat Guler, Bestami Dalkiliç, and O Nihat Ertas. 2005. The effect of anise oil (*Pimpinella anisum* L.) on broiler performance. *International Journal of Poultry Science* 4 (11):851–855.

El-Sayed, Salwa M., Nevin Ahmed, Samy Selim, et al. 2022. Acaricidal and antioxidant activities of anise oil (*Pimpinella anisum*) and the oil's effect on protease and acetylcholinesterase in the two-spotted spider mite (*Tetranychus urticae* Koch). *Agriculture* 12 (2):224.

Erler, F., I. Ulug, and B. Yalcinkaya. 2006. Repellent activity of five essential oils against *Culex pipiens*. *Fitoterapia* 77 (7):491–494.

Fyfe, L., F. Armstrong, and J. Stewart. 1997. Inhibition of *Listeria monocytogenes* and *Salmonella enteriditis* by combinations of plant oils and derivatives of benzoic acid: The development of synergistic antimicrobial combinations. *International Journal of Antimicrobial Agents* 9 (3):195–199.

Gabriel, O., K. Lange, E. Dahm, and T. Wendt. 2008. *Von Brandt's Fish Catching Methods of the World*. Wiley.

Gülçın, İlhami, Münir Oktay, Ekrem Kıreçcı, and Ö. İrfan Küfrevıoğlu. 2003. Screening of antioxidant and antimicrobial activities of anise (*Pimpinella anisum* L.) seed extracts. *Food Chemistry* 83 (3):371–382.

Hajou, Reema M. K., Fatma U. Afifi, and Abdelkader H. Battah. 2004. Comparative determination of multi-pesticide residues in *Pimpinella anisum* using two different AOAC methods. *Food Chemistry* 88 (3):469–478.

Heidari, M. R. 2005. Effects of methyl alcoholic extract of *Pimpinella anisum* L. on picrotoxin-induced seizure in mice and its probable mechanism. *HBI_Journals* 10 (3):1–8.

Hosseinzadeh, Hossein, Mohsen Tafaghodi, Shirin Abedzadeh, and Elahe Taghiabadi. 2014. Effect of aqueous and ethanolic extracts of *Pimpinella anisum* L. Seeds on milk production in rats. *Journal of Acupuncture and Meridian Studies* 7 (4):211–216.

Jafari, Roya, Mohsen Zandi, and Ali Ganjloo. 2022. Characterization of alginate-gelatin edible film containing anise (*Pimpinella anisum* L.) essential oil. *Journal of Polymers and the Environment*:1–16.

Koriem, Khaled Mohamed Mohamed. 2015. Approach to pharmacological and clinical applications of Anisi aetheroleum. *Asian Pacific Journal of Tropical Biomedicine* 5 (1):60–67.

Kreydiyyeh, Sawsan Ibrahim, Julnar Usta, Khuzama Knio, Sarine Markossian, and Shawky Dagher. 2003. Aniseed oil increases glucose absorption and reduces urine output in the rat. *Life Sciences* 74 (5):663–673.

Lee, Hoi-Seon. 2004. p-Anisaldehyde: Acaricidal component of *Pimpinella anisum* seed oil against the house dust mites *Dermatophagoides farinae* and *Dermatophagoides pteronyssinus*. *Planta Medica* 70 (03):279–281.

Lee, Jung-Bum, Chihiro Yamagishi, Kyoko Hayashi, and Toshimitsu Hayashi. 2011. Antiviral and immunostimulating effects of lignin-carbohydrate-protein complexes from *Pimpinella anisum*. *Bioscience, Biotechnology, and Biochemistry* 75 (3):459–465.

Leung, Albert Y. 1980. *Encyclopedia of Common Natural Ingredients Used in Food, Drugs, and Cosmetics*. Hoboken, NJ: John Wiley & Sons, Inc.

Mahboubi, Mohaddese, and Mona Mahboubi. 2021. *Pimpinella anisum* and female disorders: A review. *Phytomedicine Plus* 1 (3):100063.

Nickavar, Bahman, and Farideh Al-Sadat Abolhasani. 2009. Screening of antioxidant properties of seven Umbelliferae fruits from Iran. *Pakistan Journal of Pharmaceutical Sciences* 22 (1):30–36.

Orav, Anne, Ain Raal, and Elmar Arak. 2008. Essential oil composition of *Pimpinella anisum* L. fruits from various European countries. *Natural Product Research* 22 (3):227–232.

Peter, K. V. 2012. *Handbook of Herbs and Spices*. Elsevier Science.

Ponte, Edson L., Paloma L. Sousa, Maria V. A. P. Rocha, et al. 2012. Comparative study of the anti-edematogenic effects of anethole and estragole. *Pharmacological Reports* 64 (4):984–990.

Rajeshwari, Ullagaddi, I. Shobha, and B. Andallu. 2011. Comparison of aniseeds and coriander seeds for antidiabetic, hypolipidemic and antioxidant activities. *Spatula DD* 1 (1):9–16.

Reddy, B. S., C. V. Rao, A. Rivenson, and G. Kelloff. 1993. Chemoprevention of colon carcinogenesis by organosulfur compounds. *Cancer Res* 53 (15):3493–3498.

Reineccius, Gary. 1995. Natural flavoring materials. In *Source Book of Flavors*. Springer Science+ Business Media, Dordrecht, the Netherlands.

Salehi Surmaghi, M. H. 2010. *Medicinal Plants and Phytotherapy*. Vol. 1. Tehran, Iran: Donyay Taghziah Press.

Shojaii, Asie, and Mehri Abdollahi Fard. 2012. Review of pharmacological properties and chemical constituents of *Pimpinella anisum*. *ISRN Pharmaceutics* 2012:510795.

Shukla, H. S., and S. C. Tripathi. 1987. Antifungal substance in the essential oil of anise (*Pimpinella anisum* L.). *Agricultural and Biological Chemistry* 51 (7):1991–1993.

Smith, R. L., T. B. Adams, J. Doull, et al. 2002. Safety assessment of allylalkoxybenzene derivatives used as flavouring substances — Methyl eugenol and estragole. *Food and Chemical Toxicology* 40 (7):851–870.

Soares, Pedro Marcos G., Ricardo F. Lima, Alana de Freitas Pires, Emmanuel P. Souza, Ana Maria S. Assreuy, and David N. Criddle. 2007. Effects of anethole and structural analogues on the contractility of rat isolated aorta: Involvement of voltage-dependent Ca^{2+}-channels. *Life Sciences* 81 (13):1085–1093.

Sweet, Doris V. 1997. *Registry of Toxic Effects of Chemical Substances (RTECS): Comprehensive Guide to the RTECS*. US Department of Health and Human Services, Public Health Service, Centers

Tunç, İ., B. M. Berger, F. Erler, and F. Dağlı. 2000. Ovicidal activity of essential oils from five plants against two stored-product insects. *Journal of Stored Products Research* 36 (2):161–168.

10 Asafoetida (*Ferula assa-foetida*)

Niccolò Pilla, Paolo Gabrielli, Altero Aguzzi, Alessandra Durazzo, and Massimo Lucarini

10.1 INTRODUCTION

The Apiaceae family consists of about 455 genera and about 3700 species that belong to this family and they are often used in the form of spices. *Ferula* is one of the largest genera of the Apiaceae family and is categorized in the Peucedaneae tribe and Ferulinae subtribe of the family (Sonigara et al., 2021). Asafoetida (*Ferula assa-foetida* L.) is an oleo-gum-resin obtained from the stems of *Ferula* plants. The *Ferula* genus includes more than 170 species, randomly distributed throughout central Asia, particularly West Afghanistan, Iraq, Turkey, and eastern Iran, Europe, and North Africa. The plant is a monoecious, herbaceous perennial with an unpleasant odor that grows to about 2–3 m in height, with a circular diameter of 30–40 cm. Flowers are pale greenish yellow, produced in large compound umbels, with numerous stem leaves on the inflorescence being densely pubescent. It has whitish-yellow persistent petals (Shahrajabian et al., 2021). The oleo-gum-resin can be obtained by two different methods: an incision of the roots or removal of the stems. Hardened exudates are collected and packaged, ready for export. Asafoetida appears in two main forms, tears and mass. The most common form on the market being asafoetida mass (Evans, 2002). The resin yield, essential oil content, and composition of the essential oils of *F. assafoetida* populations are influenced by genetic, climatic, edaphic factors, etc. The resin and essential oil yield of asafoetida showed significant variability. There is a positive correlation between resin yield and precipitation at the plant-growing site. A progressive increase in essential oil yield was confirmed with increasing temperature; as a result, essential oil content is highest in the sites at the lowest altitude and greatest precipitation rate (Moghaddam & Farhadi, 2015).

10.2 HISTORICAL ASPECTS OF ASAFOETIDA AND TRADITIONAL MEDICINAL USES

Mainly utilized in India, asafoetida has been used by different cultures as a spice and phytochemical-based medicine; this plant is characterized by a bitter taste and a sulphurous flavour. Evidence of the use of asafoetida is reported as starting in ancient Rome, where it was used to replace precious and more expensive spices (Riddle, 1997). In the culture of Nepal, it is believed to have aphrodisiac properties and is consumed in the daily diet. It is used as a treatment for different diseases like gastrointestinal disorders (Rajakumar & Shivanna, 2009), nervous disorders (Ross, 2005), respiratory problems (Srinivasan, 2005), and impotence (Ross, 2005; Kapoor et al., 2000).

10.3 ASAFOETIDA LITERATURE QUANTITATIVE RESEARCH ANALYSIS

On March 1, 2023 a search for asafoetida publications was carried out based on Scopus database content. The Scopus online database (www.scopus.com/home.uri, accessed on March 1, 2023) was used to extract bibliometric data using the search string: TITLE-ABS-KEY ('Assa-foetida" OR 'Asafoetida*' OR 'Asa foetida*' OR 'Stinking gum*' OR 'Devil's dung*'). Publications mentioning the relevant words or their derivates in the title, abstract, or keywords were identified throughout the

DOI: 10.1201/9781003387046-10

applied search strategy. The evaluated parameters were trends of publications and citations, document type, authorship, and country/region.

For the basic analyses, the functions of the Scopus web online platform, named 'Analyze' and as 'Create Citation Report', were used. The 'full records and cited references' were exported to VOSviewer software (version 1.6.19 www.vosviewer.com, accessed on March 1, 2023) (Waltman et al., 2010; Van Eck et al., 2010, 2011, 2014) for additional processing procedures/operations. The terms utilized in the title abstracts of publications and keywords of publications were analyzed by the above-mentioned VOSviewer software. In a term map, the bubble size indicates how frequently a term is mentioned in the articles, then two bubbles that are close to each other reflect that the terms co-appeared more often in the selected publications. The average number of citations per publications (CPP) are given by the colour of a bubble.

Five was set as the minimum number of occurrences of a keyword. Out of 7093 keywords, 533 met the threshold, and three of them were manually excluded.

By the search, 469 publications were returned: the time period covered was from 1828 to 2023 and were cited overall 7497 times. There has been a continuous increase in the number of publications year by year and in the beginning of March in 2023 there were already six publications.

The distribution of the types of documents relative to the 469 publications retrieved is shown in Figure 10.1. 'Article' category represents almost all of the publications, at 82.7%, followed by 'Review' category at 11.9% and 'Book Chapter' category at 1.7%.

One of the most cited documents is an article by Dephour et al. (2009) investigating the antioxidant activity of an aqueous methanolic extract of *F. assa-foetida.*

Figure 10.2 shows the authors who have produced the most documents on the subject. Iranshashi, M. leads this classification with 14 documents. In particular, he focused his investigation on the pharmacology and toxicology of asafoetida's bioactive compounds. Bagheri, S.M. focused his studies on the potential effect of *Ferula assa-foetida* (Bagheri et al., 2016; Bagheri et al., 2017; Bagheri et al., 2014a). His oldest paper investigated the anticonvulsant effect of the plant (Bagheri et al., 2014b), while the newest one is about an effect of asafoetida on ovary polycystic syndrome (Shiei at al., 2022).

Figure 10.3 shows the most productive countries/territory with respect to asafoetida research. The most productive country is Iran with 176 documents, followed by India with 136, whereas, in third place, are the United States with only 25 documents The most-cited document produced in Iran is about the antioxidant activity of asafoetida and the composition of its essential oil (Dehpour et al., 2009), while, in India, the three most-cited articles are published by Platel, K., all of them

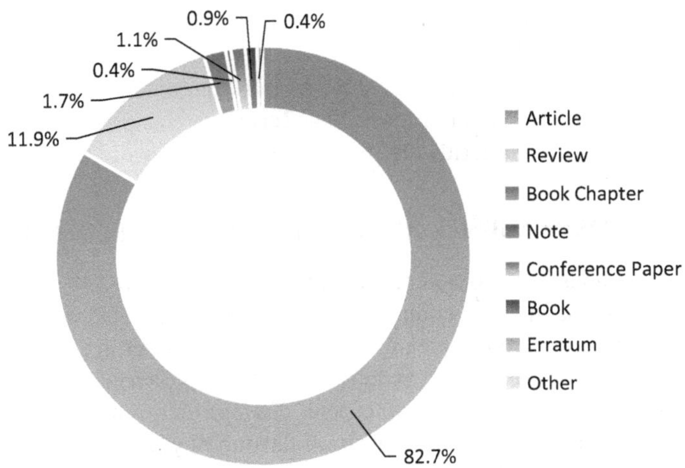

FIGURE 10.1 Distribution of documents by type (Bibliometric data were extracted from the Scopus online database).

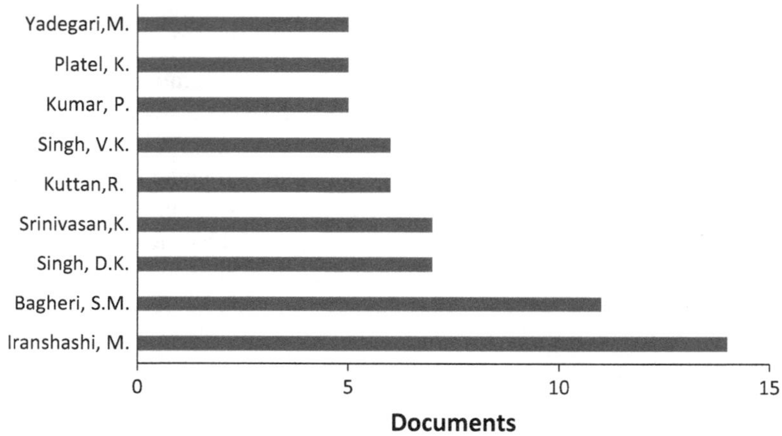

FIGURE 10.2 Most-productive authors (Bibliometric data were extracted from the Scopus online database).

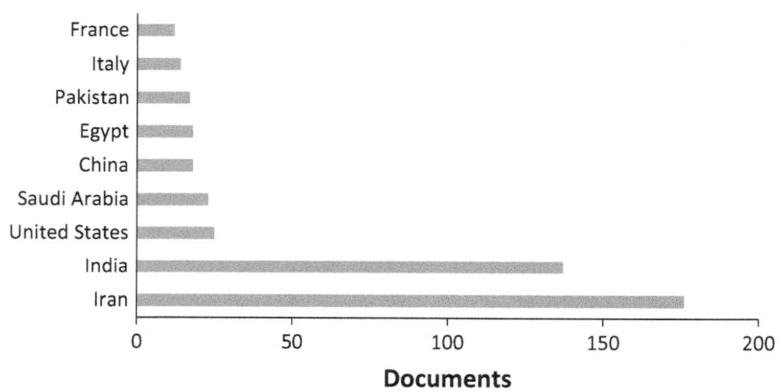

FIGURE 10.3 Most-productive countries/territories (Bibliometric data were extracted from Scopus online database).

describing the effect of asafoetida on digestive activity (Platel et al., 1996, 2000, 2004) A reasonable explanation for these significative trends could be the fact that asafoetida is a spice traditionally largely consumed in Iran and India.

A total of 530 terms were obtained from the literature quantitative analysis on publications, and they are visualized as a term map in Figure 10.4. Terms such as plant extract, medicinal plant, unclassified drug, essential oil, and ferula are the top-recurring keywords.

10.4 BIOACTIVE COMPONENTS AND BENEFICIAL PROPERTIES

Asafoetida composition consists of 68% carbohydrates, 16% moisture, 4% protein, 7% minerals, 4% fibre, and 1% fat. Chemically, it is comprised of four main fractions, including 40–64% resin, 25% gum, 10–17% essential/volatile oils, and 1.5–10% ash (Singh et al., 2023). The main phytochemicals present in the genus *Ferula* are as follows: coumarins, coumarin esters, sesquiterpenes, sesquiterpene lactones, monoterpenes, monoterpene coumarins, prenylated coumarins, sulphur-containing compounds responsible for the unpleasant flavour, phytoestrogens, flavonoids and so on (Sonigra et al., 2021). Due to the abundance of bioactive compounds, various beneficial properties are attributed to asafoetida (Figure 10.5). Generally, the combined action of bioactive components gives the potential beneficial effect of a plant (Durazzo et al., 2018; Durazzo & Lucarini, 2019;

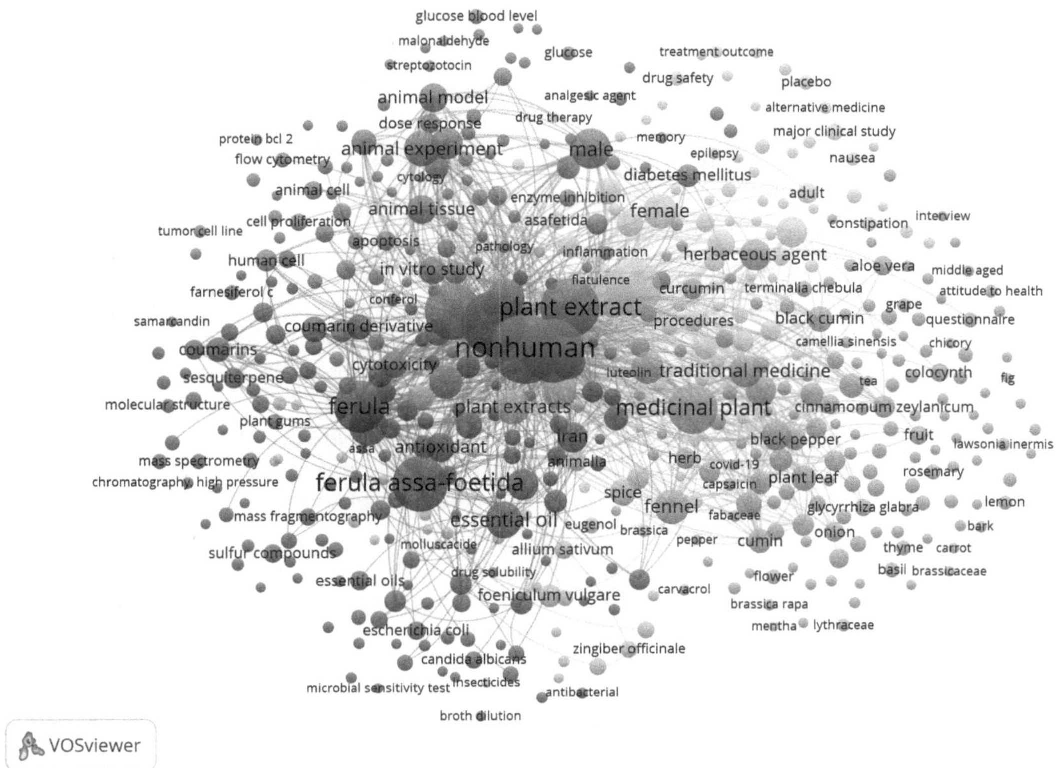

FIGURE 10.4 The term map. The bubble size indicates how frequently a term is mentioned in the articles, then two bubbles that are close to each other reflect that the terms co-appeared more often in the selected publications. The average number of citations per publication (CPP) are given by the the colour of a bubble (Based on data from Scopus and elaborated by VOSviewer software).

Sharifi-Rad et al., 2021; Nazhand et al., 2021). Coumarin derivates are one of the major representative classes, particularly sesquiterpene coumarins (Abd El-Razek et al., 2003), e.g., umbelliprenin ,5-hydroxyumbelliprenin, 8-hydroxyumbelliprenin, tadshiferin, asacoumarin A, 8-acetoxy-5-*S*-hydroxyumbelliprenin, asacoumarin, assafoetidin, franesiferol A, B, and C, galbanic acid, conferol, gummosin, abd assafoetidinol A and B. (Abd El-Razek et al., 2003; Appendino et al., 2006; Lee et

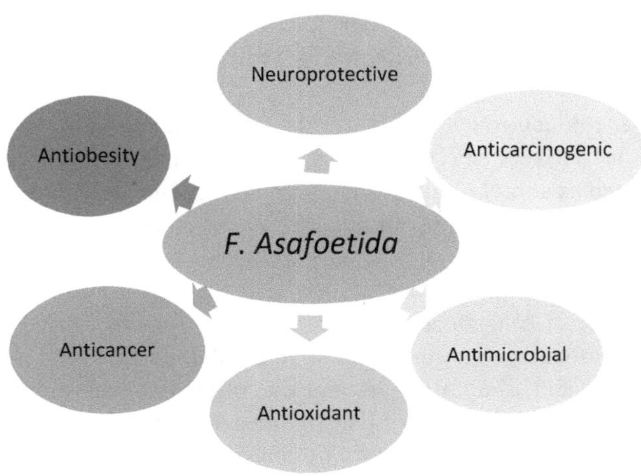

FIGURE 10.5 Potential beneficial effects of *Ferula assa-foetida*

al., 2009; Asghari et al., 2015; Razavi et al., 2016) Related to the presence of bioactive compounds either in the leaf or gum extract, *F. assa-foetida* has shown high antioxidant activity, acting as a potential inhibitor of lipid peroxidation (Dehpour et al., 2009; Niazmand & Razavizadeh, 2021).

An explanation for the traditional use of asafoetida to treat respiratory problems is given by the strong lipoxygenase inhibitory activity (LIA) of umbellinprenin (Iranshahi et al., 2009); currently, LIA is a strategy to treat asthma.

Antinociceptive and anti-inflammatory activity were evaluated in an experimental model of pain and inflammation (Bagheri et al., 2016). This study (Bagheri et al., 2016) reported that a dosage of 10 mg/kg of *F. assa-foetida* has the greatest antinociceptive effect in male albino mice.

Network pharmacology was applied for the identification of the mechanism of the plant *F. assa-foetida* for the treatment of asthma. The study suggested that cynaroside, farnesiferol-B, farnesiferol-C, and assafoetidin can play a significant role in elucidating the mechanism of *F. assa-foetida* for the treatment of asthma (Quasm et al., 2023).

A neuroprotective effect is suggested by different *in vivo* and *in vitro* studies. Tayeboon et al. (2013) demonstrated that F. *assa-foetida* extracts had a protective role on glutamate-induced neurotoxicity in CGN (cerebellar granule neurons) cells, whereas Moghaddam et al. (2014) showed that *F. assa-foetida* extracts in lower concentrations could improve the electrical function and healing process of injured nerves in pyridoxine-induced neuropathy. A recent study has suggested a marked neuroprotective effect of asafoetida, using it as a treatment for mice with Alzheimer's disease, reporting significantly amelioration of cognitive impairment induced by scopolamine, as evidenced by improving learning and memory abilities, and reducing neuronal injury, cholinergic system impairment, oxidative stress, and apoptosis in the hippocampus of mice (Huang et al., 2022).

Another well-known property of *F. assa-foetida* is antimicrobial and antifungal activity. Crude extracts were tested against different fungal and bacterial strains, demonstrating that alcoholic and aqueous extracts have a significant effect against *Bacillus subtilis*, *Staphylococcus aureus*, *Escherichia coli*, *Pseudomonas aeruginosa*, and *Aspergillus niger* (Shrivastava et al., 2012; Bhatnager et al., 2015). Asafoetida essential oil, in association with other essential oils derived from spices, has shown antifungal activity (Sitara et al., 2008; Kamble et al., 2008; Rani et al., 2009;). Daneshkazemi et al. (2019) have revealed that essential oils obtained from seeds and oleo-gum-resin of *Ferula assa-foetida* show antimicrobial effects against oral pathogens.

Another study has shown that oleo-gum-resin obtained from *F. assa-foetida* has an antimicrobial activity against foodborne pathogens and spoilage microorganisms, such as *E. coli*, *S. aureus*, and *L. monocytogenes* (Campana et al., 2022)

Furthermore, *in vivo* and *in vitro* studies have revealed that asafoetida stimulates digestive enzyme activity, in particular with respect to the pancreatic enzymes (Platel et al., 2000; Ramakrishna et al., 2003). The positive influence exerted on the pancreatic digestive enzymes by different spices consumed in a diet could be a further contribution to the well-known action of spices on the stimulation of digestion.

A double-blind, placebo-controlled, randomized study on 43 subjects characterized with functional dyspepsia demonstrated the safety and efficacy of *F. assa-foetida* oleo-gum-resin for the first time, when supplemented as a food-grade formulation with fenugreek dietary fibre (for 30 days at a dosage of 250 mg twice a day, containing around 36% (w/w) of asafoetida gum (Mala et al., 2018). Of the subjects treated, 81% reported significant improvements in overall symptoms (Mala et al., 2018).

Another interesting effect of asafoetida was reported by Azizan et al. (2012). In their study, one administration of asafoetida oleo-gum-resin at doses of 25 or 50 mg in rats resulted in significant decreases in body weight, abdominal fat, and size of epididymal adipocytes, compared with untreated rats. Levels of serum leptin were also significantly decreased in treated rats (Azizan et al., 2012). A more recent study has shown similar effects of an asafoetida-enriched diet in diabetic rats, where two different diets were tested, namely 0.5 g or 2.0 g asafoetida per 100 g animal feed (Esmailidehaj et al., 2022). This study (Esmailidehaj et al., 2022) has reported that fasting blood sugar, HbA1C, oxidative stress and hepatic enzyme activity significantly decreased and plasma level of insulin markedly increased in the rats treated with 0.5 g asafoetida per 100 g feed compared with the diabetic group. The plasma lipid profile was improved in the 0.5 g group. The response of thoracic aorta rings to vasodilator and vasoconstrictor substances was considerably improved in the 0.5 g asafoetida group than in the control group. The group treated with 2.0 g of asafoetida did not have a significant effect on diabetes-induced disorders, suggesting that the effects of asafoetida-enriched diet on biological disorders caused by type 2 diabetes are dependent on the percentage of asafoetida in the diet (Esmailidehaj et al., 2022). Singh (2022) showed that some phytochemicals in *F. assa-foetida* have potential antidiabetic properties, in particular having potential anti-α-amylase and anti-α-glucosidase activity. Thanks to this potential anti-enzyme activity, the phytochemicals could be further used to develop into effective antidiabetic drugs from a natural resource (Singh, 2022). Again, different studies have reported that asafoetida has anti-cancer chemoprevention activity (Saleem et al., 2001; Mallikarjuna et al., 2003; Iranshahi et al., 2008; Pang et al., 2021). *In vitro* studies have observed that daily consumption of asafoetida could decrease tumour multiplicity and reduction in tumour volume (Mallikarjuna et al., 2003; Bagheri et al., 2017; Azani et al., 2021). Umbelliprenin, a squisiterpene coumarin present in *F. assafoetida,* could play a potential therapeutic role by affecting several growth pathways in different cancers (Shazhadi et al., 2020). Ferula oil enriched with organosulphides displayed significant inhibition of cell growth *in vitro* against SKOV3 and A549 cancer cells in a dose-dependent manner (Yatham et al., 2022). A possible effect on the blood glucose level in induced diabetic rats was investigated, reporting that asafoetida extract can reduce the level of blood glucose. This might be due to the presence of tannins and phenolic acids in the extract (Akhlaghiet al., 2012). Mobasheri at al. (2022) have assessed asafoetida's possible anti-inflammatory molecular mechanisms.

10.5 FOOD AND NON-FOOD APPLICATIONS

Due to its antioxidant and antibacterial properties, *F. assa-foetida* could be used as a source of safe and effective natural antioxidants in the food industry to achieve better oxidative stability of fatty foods during storage, while the essential oil obtained from the later stages of *F. assa-foetida* could be used in the health industry as a safe and effective source of antibacterial agents. Furthermore, asafoetida extract could be used as an ecologically protective larvicide; Muturi et al. (2019) have shown that emulsions based on garlic and asafoetida essential oils are highly toxic to mosquito larvae and could be an effective and ecofriendly strategy for controlling vector mosquitoes (Muturi et al., 2019). Another study suggested that silver nanoparticles synthesized using asafoetida resin could be transformed into effective alternative tools for controlling *Adages aegypti* populations (Saranyadevi et al., 2021). Moreover, asafoetida aqueous extract has been studied as a corrosion inhibitor for mild steel (Devikala et al., 2019). Asafoetida safety as flavouring agent was also evaluated by FEEDAP (Additives and Products or Substances used in Animal Feed) (Bampidis et al., 2022), reporting that asafoetida oil from *F. assa-foetida* is safe at the proposed conditions of use of 1.5 mg/kg complete feed for dogs and 0.2 mg/kg complete feed for cats.

10.6 CONCLUSION AND FUTURE ASPECTS

In conclusion various activities have been reported from asafoetida over the years. *F. assa-foetida* has shown interesting potential beneficial effects on human health. Several studies (Iranshashi et al., 2009; Shazhadi et al., 2020; Shing et al., 2022; Mobasheri et al., 2022) have investigated the possible mechanisms by which the various components act, giving scientific demonstration of the traditional uses of *F. assa-foetida* as a treatment for various diseases. Despite that, further research will be needed to better understand the mechanisms of action of these components in order to fully exploit the potential of this plant. Moreover, non-food applications have opened the way for further uses, suggesting a different approach to increasingly important problems (Muturi et al., 2019).

REFERENCES

Abd El-Razek, M. H., Ohta, S., & Hirata, T. 2003. Terpenoid coumarins of the genus *Ferula*. *Heterocycles*, *60*(3), 689–716.

Akhlaghi, F., Rajaei, Z., Hadjzadeh, M., Iranshahi, M., & Alizadeh, M. 2012. Antihyperglycemic effect of asafoetida (*Ferula assa-foetida* oleo-gum-resin) in streptozotocin- induced diabetic rats. *World Applied Sciences Journal*, *17*, 157.e162.

Azani, H., Homayouni Tabrizi, M., Neamati, A., Khadem, F., & Khatamian, N. 2022. The Ferula assa-foetida essential oil nanoemulsion (FAEO-NE) as the selective, apoptotic, and anti-angiogenic anticancer compound in human MCF-7 breast cancer cells and murine mammary tumor models. *Nutrition and Cancer*, *74*(6), 2196–2206.

Azizian, H., Rezvani, M. E., Esmaeli, D. M., & Bagheri, S. M., 2012. Anti-obesity, fat lowering and liver Steatosis protective effects of *Ferula assa-foetida* gum in type 2 diabetic rats: Possible involvement of leptin. Iran J Diabetes Obes 4: 120-126.

Bagheri, S. M., Hedesh, S. T., Mirjalili, A., & Dashti, R. M. H. 2016. Evaluation of anti-inflammatory and some possible mechanisms of antinociceptive effect of *Ferula assa-foetida* oleo-gum-resin. *Journal of Evidence-Based Complementary and Alternative Medicine*, *21*(4), 271–276.

Bagheri, S. M., Mohammadsadeghi, H., Dashti, R. M. H., Mousavian, S. M. M., & Aghaei, Z. A. 2016. Effect of *Ferula assa-foetida* oleo-gum-resin on renal function in normal Wistar rats. *Indian Journal of Nephrology*, *26*(6), 419.

Bagheri, S. M., Rezvani, M. E., Vahidi, A. R., & Esmaili, M. 2014. *Anticonvulsant effect of Ferula assa-foetida* oleo gum resin on chemical and amygdala-kindled rats. *North American Journal of Medical Sciences*, *6*(8), 408.

Bagheri, S. M., Abdian-Asl, A., Moghadam, M. T., Yadegari, M., Mirjalili, A., Zare-Mohazabieh, F., & Momeni, H. 2017. Antitumor effect of *Ferula assa-foetida* oleo gum resin against breast cancer induced by 4T1 cells in BALB/c mice. *Journal of Ayurveda and Integrative Medicine*, *8*(3), 152–158.

Bhatnager, R., Rani, R., & Dang, A. S. 2015. Antibacterial activity of *Ferula assa-foetida*: A comparison of red and white type. *Journal of Applied Biology and Biotechnology*, *3*(2), 018–021.

Campana, R., Tiboni, M., Maggi, F., Cappellacci, L., Cianfaglione, K., Morshedloo, M. R. … Casettari, L. 2022. Comparative analysis of the antimicrobial activity of essential oils and their formulated microemulsions against foodborne pathogens and spoilage bacteria. *Antibiotics*, *11*(4), 447.

Daneshkazemi, A., Zandi, H., Davari, A., Vakili, M., Emtiazi, M., Lotfi, R., & Masoumi, S. M. R. 2019. Antimicrobial activity of the essential oil obtained from the seed and oleo-gum-resin of *Ferula assa-foetida* against oral pathogens. *Frontiers in Dentistry*, *16*(2), 113.

Dehpour, A. A., Ebrahimzadeh, M. A., Fazel, N. S., & Mohammad, N. S. 2009. Antioxidant activity of the methanol extract of *Ferula assa-foetida* and its essential oil composition. *Grasas y Aceites*, *60*(4), 405–412.

Devikala, S., Kamaraj, P., Arthanareeswari, M., & Pavithra, S. 2019. Green Corrosion inhibition of mild steel by asafoetida extract extract in 3.5% NaCl. *Materials Today: Proceedings*, *14*, 590–601.

Durazzo, A., & Lucarini, M. 2019. Editorial: The state of science and innovation of bioactive research and applications, health, and diseases. *Frontiers in Nutrition*, *6*, 178.

Durazzo, A., D'Addezio, L., Camilli, E., Piccinelli, R., Turrini, A., Marletta, L., Marconi, S., Lucarini, M., Lisciani, S., Gabrielli, P., Gambelli, L., Aguzzi, A., & Sette, S. 2018. From plant compounds to botanicals and back: A current snapshot. *Molecules (Basel, Switzerland)*, *23*(8), 1844.

EFSA Panel on Additives and Products or Substances used in Animal Feed (FEEDAP); Bampidis, V., Azimonti, G., Bastos, M. L., Christensen, H., Fašmon Durjava, M., Kouba, M., López-Alonso, M., López Puente, S., Marcon, F., Mayo, B., Pechová, A., Petkova, M., Ramos, F., Sanz, Y., Villa, R. E., Woutersen, R., Brantom, P., Chesson, A., Westendorf, J., Manini, P., Pizzo, F., Dusemund, B. 2022. Safety and efficacy of a feed additive consisting of an essential oil from the gum resin of *Ferula assa-foetida* L. (asafoetida oil) for use in dogs and cats (FEFANA asbl). *EFSA Journal European Food Safety Authority*, *20*(12), e07688.

Esmailidehaj, M., Kahtenaroon, M., Rezvani, M. E., Azizian, H., & Ranjbar, A. 2022. Enriched-asafoetida diet attenuates hyperglycemia, oxidative stress and endothelial dysfunction in type 2 diabetic rats. *Physiology and Pharmacology*, *26*(1), 7–19.

Evans, W. C. 2002. *Volatile Oils and Resins, Trease and Evans Pharmacognosy*, fifteenth ed., p. 286. W.B. Saunders, London.

Huang, Q., Zhang, C., Dong, S., Han, J., Qu, S., Xie, T., Zhao, H., & Shi, Y. 2022. Asafoetida exerts neuroprotective effect on oxidative stress induced apoptosis through PI3K/Akt/GSK3β/Nrf2/HO-1 pathway. *China Medicine, 17*(1), 83.

Iranshahi, M., Askari, M., Sahebkar, A., & Hadjipavlou-Litina, D. 2009. Evaluation of antioxidant, anti-inflammatory and lipoxygenase inhibitory activities of the prenylated coumarin umbelliprenin. *Daru, 17*, 99–103.

Iranshahi, M., Kalategi, F., Rezaee, R., Shahverdi, A. R., Ito, C., Furukawa, H. ... Itoigawa, M. 2008. Cancer chemopreventive activity of terpenoid coumarins from *Ferula* species. *Planta Medica, 74*(02), 147–150.

Kamble, V. A.,, & Patil, S. D. 2008. Spice-derived essential oils: Effective antifungal and possible therapeutic agents. *Journal of Herbs, Spices, and Medicinal Plants, 14*(3–4), 129.e143.

Kapoor, L. D., 2000. *Handbook of Ayurvedic Medicinal Plants*, first ed. CRC Press, New York.

Lee, C. L., Chiang, L. C., Cheng, L. H., Liaw, C. C., Abd El-Razek, M. H., Chang, F. R., & Wu, Y. C. 2009. Influenza A (H1N1) antiviral and cytotoxic agents from *Ferula assa-foetida. Journal of Natural Products, 72*(9), 1568–1572.

Mala, K. N., Thomas, J., Syam, D. S., Maliakel, B., & Krishnakumar, I. M. 2018. Safety and efficacy of *Ferula assa-foetida* in functional dyspepsia: A randomized, double-blind, placebo-controlled study. *Evidence-Based Complementary and Alternative Medicine.*

Mallikarjuna, G. U., Dhanalakshmi, S., Raisuddin, S., & Ramesha Rao, A. 2003. Chemomodulatory influence of *Ferula assa-foetida* on mammary epithelial differentiation, hepatic drug metabolizing enzymes, antioxidant profiles and N-methyl-N-nitrosourea-induced mammary carcinogenesis in rats. *Breast Cancer Research and Treatment, 81*(1), 1–10.

Mobasheri, L., Khorashadizadeh, M., Safarpour, H., Mohammadi, M., Anani Sarab, G., & Askari, V. R. 2022. Anti-inflammatory activity of *Ferula assa-foetida* Oleo-Gum-Resin (asafoetida) against TNF-α-stimulated Human Umbilical Vein Endothelial Cells (HUVECs). *Mediators of Inflammation, Volume 2022, Article ID 5171525.*

Moghadam, F. H., Dehghan, M., Zarepur, E., Dehlavi, R., Ghaseminia, F., Ehsani, S. ... Barzegar, K. 2014. Oleo gum resin of *Ferula assa-foetida* L. ameliorates peripheral neuropathy in mice. *Journal of Ethnopharmacology, 154*(1), 183–189.

Moghaddam, M., & Farhadi, N. 2015. Influence of environmental and genetic factors on resin yield, essential oil content and chemical composition of *Ferula assa-foetida* L. populations. *Journal of Applied Research on Medicinal and Aromatic Plants, 2*(3), 69–76.

Muturi, E. J., Hay, W. T., Behle, R. W., & Selling, G. W. 2019. Amylose inclusion complexes as emulsifiers for garlic and asafoetida essential oils for mosquito control. *Insects, 10*(10), 337.

Niazmand, R., & Razavizadeh, B. M. 2021. *Ferula assa-foetida*: Chemical composition, thermal behavior, antioxidant and antimicrobial activities of leaf and gum hydroalcoholic extracts. *Journal of Food Science and Technology, 58*(6), 2148–2159.

Platel, K., & Srinivasan, K. 2004. Digestive stimulant action of spices: A myth or reality? *Indian Journal of Medical Research, 119*(5), 167.

Platel, P. K., & Srinivasan, K. 1996. Influence of dietary spices or their active principles on digestive enzymes of small intestinal mucosa in rats. *International Journal of Food Sciences and Nutrition, 47*(1), 55–59.

Platel, K., & Srinivasan, K. 2000. Influence of dietary spices and their active principles on pancreatic digestive enzymes in albino rats. *Food/Nahrung, 44*(1), 42–46.

Qasim, M., Abdullah, M., Ashfaq, U. A., Noor, F., Nahid, N., Alzamami, A. ... Khurshid, M. 2023. Molecular mechanism of *Ferula assa-foetida* for the treatment of asthma: Network pharmacology and molecular docking approach. *Saudi Journal of Biological Sciences, 30*(2), 103527.

Rajakumar, N., & Shivanna, M. B. 2009. Ethno-medicinal application of plants in the eastern region of Shimoga district, Karnataka, India. *Journal of Ethnopharmacology, 126*, 64–73.

Ramakrishna Rao, R., Platel, K., & Srinivasan, K. 2003. In vitro influence of spices and spice-active principles on digestive enzymes of rat pancreas and small intestine. *Food/Nahrung, 47*(6), 408–412.

Rani, A., Jain, S., & Dureja, P. 2009. Synergistic fungicidal efficacy of formulations of neem oil, nicotinic acid and *Ferula assa-foetida* with a, b-unsaturated carbonyl com- pounds against Sclerotium rolfsii ITCC 5226 & Macrophomina phaseolina ITCC 0482. *Journal of Pesticide Sciences, 34*, 253.e258.

Razavi, S. M., Nahar, L., Talischi, H., & Sarker, S. D. 2016. Ferulone A and ferulone B: Two new coumarin esters from *Ferula orie*ntalis L. roots. *Natural Product Research, 30*(19), 2183–2189.

Riddle, J. M., 1997. *Eve's Herbs: A History of Contraception and Abortion in the West.* Harvard University Press. Cambridge, United States.

Ross, I. A., 2005. *Chemical Constituents, Traditional and Modern Medicinal Uses, Medicinal Plants of the World*, pp. 223–234. Humana Press Inc., Totowa.

Saleem, M., Alam, A., & Sultana, S. 2001. Asafoetida inhibits early events of carcino-genesis: A chemopreventive study. *Life Sciences*, *68*(16), 1913–1921.

Saranyadevi, S., Suresh, K., Mathiyazhagan, N., Muthusamy, R., & Thirumalaisamy, R. 2021. Silver nanoparticles synthesized using asafoetida resin, characterization of their broad spectrum and larvicidal activity. *Annals of the Romanian Society for Cell Biology 25(4)*, 15035–15049.

Shahrajabian, M. H., Sun, W., Soleymani, A., Khoshkaram, M., & Cheng, Q. 2021. Asafoetida, god's food, a natural medicine. *Pharmacognosy Communications*, *11*(1), 36–39.

Shahzadi, I., Ali, Z., Baek, S. H., Mirza, B., & Ahn, K. S. 2020. Assessment of the antitumor potential of umbelliprenin, a naturally occurring sesquiterpene coumarin. *Biomedicines*, *8*(5), 126.

Sharifi-Rad, J., Quispe, C., Herrera-Bravo, J., Akram, M., Abbaass, W., Semwal, P. ... Calina, D. 2021. Phytochemical constituents, biological activities, and health-promoting effects of *Melissa officinalis*. *Oxidative Medicine and Cellular Longevity*, *2021*, 1–20.

Shieh, A., Bagheri, S. M., Yadegari, M., Javidmehr, D., & Farhadi, Z. 2022. Therapeutic effect of *Ferula assa-foetida* oleo-gum resin in rats with letrozole-induced polycystic ovary syndrome. *Clinical and Experimental Reproductive Medicine*, *49*(4), 239–247.

Shrivastava, V., Bhardwaj, U., Sharma, V., Mahajan, N., Sharma, V., & Shrivastava, G. 2012. Antimicrobial activities of asafoetida resin extracts (a potential Indian spice). *Journal of Pharmacy Research*, *5*, 5022e5024.

Singh, P. P., Joshi, R., Kumar, R., Kumar, A., & Sharma, U. 2022. Comparative phytochemical analysis of *Ferula assa-foetida* with *Ferula jaeschkeana* and commercial oleo-gum resins using GC-MS and UHPLC-PDA-QTOF-IMS. *Food Research International*, *164*, 112434.

Sitara, U., Niaz, I., Naseem, J., & Sultana, N. 2008. Antifungal effect of essential oils on *in vitro* growth of pathogenic fungi. *Pakistan Journal of Botany*, *40*, 409.e414.

Sonigra, P., & Meena, M. 2021. Metabolic profile, bioactivities, and variations in the chemical constituents of essential oils of the *Ferula* genus (Apiaceae). *Frontiers in Pharmacology*, *11*, 608649.

Srinivasan, K. 2005. Spices as influencers of body metabolism: An overview of three decades ofresearch. *Food Research International*, *38*(1), 77–86.

Tayeboon, G. S., Tavakoli, F., Hassani, S., Khanavi, M., Sabzevari, O., & Ostad, S. N. 2013. Effects of *Cymbopogon citratus* and *Ferula assa-foetida* extracts on glutamate-induced neurotoxicity. *In Vitro Cellular and Developmental Biology-Animal*, *49*(9), 706–715.

Van Eck, N. J., & Waltman, L., 2014. Visualizing bibliometric networks. In Y. Ding, R. Rousseau, & D. Wolfram (Eds.), *Measuring Scholarly Impact: Methods and Practice*, pp. 285–320. Springer, Berlin.

Van Eck, N. J., & Waltman, L. 2010. Software survey: VOSviewer, a computer program for bibliometric mapping. *Scientometrics*, *84*(2), 523–538.

Van Eck, N. J., & Waltman, L. 2011. Text mining and visualization using VOSviewer. *ISSI Newsletter*, *7*, 50–54.

Waltman, L., van Eck, N. J., & Noyons, E. C. 2010. A unified approach to mapping and clustering of bibliometric networks. *Journal of Informetrics*, *4*(4), 629–635.

Yatham, P., Shukla, D., Srivastava, A. K., Pragadheesh, V. S., & Kumar, D. 2022. Purification and identification of anticancer organosulfides from *Ferula assa-foetida* gum: Integrative analysis employing GC/GC-MS/RP-HPLC/NMR. *Natural Product Research*, *36*(11), 2869–2874.

11 Traditional Uses, Pharmacological Properties, and other Potential Uses of Black Cumin Seeds

Prairna Balyan, Hina Alim, and Ahmad Ali

11.1 INTRODUCTION

The plant kingdom plays a crucial role in human diets and serves as an unavoidable source of modern medications, in addition to preserving the ecological balance and supplying oxygen for life. Plant-based foods satisfy the body's fundamental nutritional needs, maintain bodily health, and provide an immunity-boosting defence against a variety of diseases (Wirnitzer, 2018; Husen, 2021, 2022, 2023). There is currently a renaissance of interest in the use of plants as a source of food and medicine, which is in line with recent advances in the field of optimal nutrition. Due to its accessibility, low cost, and the perception that natural therapies have fewer side effects than synthetic ones, the use of phytomedicine has recently increased significantly for a wide range of diseases (Shyur & Yang, 2008). It is also encouraged to create new products from natural sources because it is thought that only 15% of the total species of herbs that exist in the world have had their pharmacological potential investigated (De Luca et al., 2012). Minor but essential ingredients, like herbs and spices, which are mostly employed as flavouring additives and preservatives, are among the diversity of food materials that include a wealth of biofunctional molecules (Ali, 2020). Although they are mostly used in cooking, the majority of these culinary herbs and spices are also well-known for their nutraceutical benefits because they have a significant potential to improve health (Bishnoi, 2016).

Nigella sativa L., popularly known as black cumin or black seeds, is a popular culinary plant that has also long been prized for its healing properties (Figure 11.1). Black cumin is native to a large area of the eastern Mediterranean, northern Africa, the Indian subcontinent, and Southwest Asia. It is grown in several nations, including Egypt, Iran, Greece, Syria, Albania, Turkey, Saudi Arabia, India, and Pakistan (Balyan et al., 2022). As a panacea, black cumin has been recommended in traditional medicine for treatment of a wide range of illnesses and disorders, including dermatitis, anorexia, amenorrhoea, arthritis, hypertension, rheumatism, headaches, and back pain. These traditional uses of *N. sativa* seeds are largely attributed to the extensive range of medicinal properties exhibited by these seeds, which include antioxidant, anti-inflammatory, immunomodulatory, anticancer, neuroprotective, antimicrobial, antihypertensive, cardioprotective, antidiabetic, gastroprotective, nephroprotective, and hepatoprotective properties (Yimer et al., 2019).

Several phytoconstituents, primarily alkaloids, saponins, sterols, and essential oils, have been detected in *N. sativa* in phytochemical investigations. However, many of the chemical identities and biological functions of these chemicals remain unknown (Balyan & Ali, 2022). Thymoquinone (TQ), thymohydroquinone, thymol, carvacrol, nigellidine, nigellicine, and α-hederin are primarily responsible for the pharmacological actions and therapeutic advantages of black cumin seed,

DOI: 10.1201/9781003387046-11

FIGURE 11.1 Black cumin seeds.

especially its essential oil (Hannan et al., 2021). TQ, a key component of the essential oil, is the most bioactive molecule and has the broadest range of therapeutic effects among the various active ingredients that have been documented thus far (Fatima Shad et al., 2021). Although receiving less attention in scientific literature, black cumin has a high nutritional value due to its adequate protein and lipid content, as well as its significant amounts of important fatty acids, amino acids, vitamins, and minerals (Ramadan, 2007). Black cumin is a valuable source of nutraceuticals since it contains both active phytochemicals and essential elements that support the immune system and overall health of the human body (Zielińska et al., 2021). In this chapter, we conduct a critical evaluation of the literature on the traditional uses, pharmacological characteristics, and other health advantages of black cumin seeds and their phytoconstituents, as well as the purported underlying molecular mechanisms.

11.2 BLACK CUMIN SEEDS AND THEIR PHYTOCHEMICAL PROFILE

Nigella sativa, a member of the Ranunculaceae family, is an annual flowering plant with linear leaves, bluish flowers, capsule-shaped fruits, and tiny black seeds that resemble cumin and are therefore known as black cumin seeds. It grows to a height of 20 to 30 cm. This miracle herb, or *Haba al-Barakah* (blessed seeds), was referred to as "the herb from heaven" by ancient herbalists (Ahmad et al., 2013). Depending on the growing locations, stage of maturation, processing procedures, and isolation methods, black cumin's phytochemical composition varies. Several chemical classes have been used to categorize the bioactive phytochemicals of black cumin, which include major and minor secondary metabolites (Figure 11.2) (Hannan et al., 2021).

11.2.1 Terpenes and Terpenoids

TQ and its derivatives, such as thymol, carvacrol, α-pinene, thymol, *trans*-anethol, 4-terpineol, thymohydroquinone (THQ), the sesquiterpene longifolene, *p*-cymene, dithymoquinone, and several other chemicals, are members of the terpene and terpenoid families. The existence of quinine components, the most common of which is TQ, contributes to their pharmacological properties.

11.2.2 Alkaloids

Alkaloids of black cumin can be divided into two groups based on their chemical structures: isoquinoline alkaloids, like nigellicimine and nigellicimine-N-oxide, and pyrazole or indazole alkaloids, like nigellidine and nigellicine (Gutiérrez-Grijalva et al., 2020).

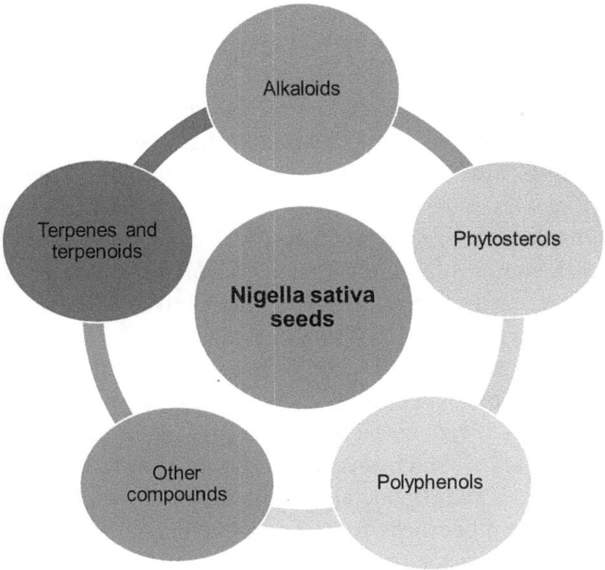

FIGURE 11.2 Phytoconstituents of black cumin seeds.

11.2.3 Phytosterols

Due to their medical and nutraceutical advantages in decreasing total cholesterol and low-density lipoprotein levels, phytosterols – an essential component of the human diet – are attracting more attention (Rocha et al., 2011). The appealing free radical scavenging potential that tocopherols display is anticipated to halt the peroxidation of lipids (Castro et al., 2006). Many sterols are present in the oil obtained from black cumin, with β-sitosterol accounting for the majority (44–54%). The second most important sterol in black cumin oil is stigmasterol, which makes up 16.57 to 20.92% of all sterols. A smaller amount of Δ7-stigmasterol, Δ7-avenasterol, campesterol, and cholesterol are also present in the oil. Black cumin is a powerful natural agent for decreasing blood cholesterol and avoiding cardiovascular illnesses due to the presence of substantial levels of sterols (Dehkordi & Kamkhah, 2008).

11.2.4 Polyphenols

Some of the phenols that have been identified in the seeds of *N. sativa* include apigenin, caffeic acid, caftaric acid, chlorogenic acid, ferulic acid, gentisic acid, quercetin, kaempferol, myricetin, methyl-4-hydroxybenzoate, pyrogallol, *p*-coumaric acid, *p*-hydroxybenzoic acid, ruin, salicylic acid, etc. (Gueffai et al., 2022).

11.2.5 Other Components

Because of the inclusion of various other chemical ingredients, such as specific carbohydrates, essential and non-essential amino acids, glycerolipids, phospholipids, vitamins, minerals, and certain alkane hydrocarbons, black cumin is recognized as an enriched natural product.

11.3 TRADITIONAL USES OF BLACK CUMIN SEEDS

Around the world, there is broad utilization of *N. sativa* as a restorative spice. It is used widely by Unani, Ayurveda, Tibb, and Siddha, which are only a few of the customary clinical frameworks. Oil and seeds have, for quite some time, been utilized in numerous dietary and clinical frameworks in stories. *N. sativa* seeds have generally been reported to fix many circumstances and issues. In Islamic writing, it is viewed as quite possibly having the best healing ability. In Tibb-e-Nabawi, it has been proposed that you use it routinely (prophetic medication). It has a long history of use as liver tonics, diuretics, to treat stomach-related problems, diarrhoea, and as hunger energizers and analgesics, as well as for treating skin issues (Ahmad et al., 2013).

As a zest and seasoning fixative, *N. sativa* has been widely used in a range of food arrangements, including bread, yogurt, pickles, sauces, and mixed greens. Long utilized as a conventional medication in the countries of the Middle Eastern landmass, Africa, Europe, and the Far East, 'black seed' is otherwise called 'black cumin' (the English name), *Habbatul Barakah* (the Arabic name), and *Tikur azmud* (the Amharic name). Early cultivators alluded to *N. sativa* as 'the spice of paradise' and portrayed it as an otherworldly plant. While depicting the black seed's abilities to mend, the Prophet Muhammad (PBUH) expressed, 'Hold tight to utilize this black seed, as it contains a treatment for each sickness save demise'. Avicenna, a famous doctor from the tenth century celebrated for his work, 'The Group of Medication', encouraged using Nigella seeds to help increase the body and to support the body's energy system from exhaustion and downfall. *N. sativa* is likewise alluded to as melanthion by Dioscorides and Hippocrates and is additionally referenced in the Blessed Book of Scriptures for its restorative properties (Yarnell and Abascal, 2011).

Black cumin seeds are used restoratively for different conditions, including different torment conditions, like constant migraine and back torment, diabetes, loss of motion, diseases, irritation, hypertension, and gastrointestinal system-related issues that are treated with different materials. As an effective cure, it has likewise been applied straightforwardly to rashes, nasal abscesses, orchitis, skin inflammation, and sore joints. With the numerous conventional clinical uses of *N. sativa* seeds and the identification of the main active ingredient of *N. sativa*, TQ, this invaluable spice is examined as an effective medication (Yemir et al., 2019).

Exploration in ethnopharmacology uncovered that *Nigella* species are among those most often used in conventional and folk medication. The *Nigella* genus contains a few annual species, with *N. sativa* being the most notable, having a long history of utilization as a medicine. *N. sativa* has been utilized generally since essentially the Tutankhamen period when Pliny the Elder recommended it to treat stomach-related disorders and identified it as solutions for scorpion stings and snake bites. The Pharmacopoeia of India currently prescribes 0.5 to 4 g powder of *N. sativa* seed for use as an energizer to ease inside and for stomach-related issues as well as a carminative. In Bangladesh and India, it has additionally been used to treat diabetes and feminine issues. Like this, *N. sativa* is often utilized in Algerian customary medication to treat both hypertension (high blood pressure) and diabetes (Salehi et al., 2021).

Likewise, as indicated by the Bedouins of Egypt, the seeds are utilized to treat coronary illness, hypertension, and other conditions, while the wooden stem is used to treat jaundice. Comparative uses of the leaves and seeds of *N. sativa* have been archived in Iranian conventional medication, Pakistani customary medication, Moroccan conventional medication, and conventional medication in both of those nations (Iran and Morocco). Additionally, as indicated by the Bedouins of Egypt, the seeds are utilized to treat coronary illness, hypertension, and other circumstances, while the wooden stem is used to treat jaundice. Comparative utilization of *N. sativa* seeds and leaves has been archived in Iranian customary medication, Pakistani conventional medication, Moroccan customary medication, and conventional medication in both of those nations. It has been guaranteed that *N. sativa* blended with honey has antirheumatic and pain-relieving characteristics. In addition, seeds (in fresh, dried, and powdered structures) and leaves are used in Ethiopia as well as Malaysian conventional medication to cure jungle fever. *N. sativa* is not yet utilized as much as possible but is viewed as a panacea in Bangladesh and Qatar for its beneficial effects. In addition, the last uses are boundless in Lebanon, Palestine, Turkey, Mauritius, and Nepal. Furthermore, seeds of *N. sativa* are crushed and utilized as a flavouring additive for food in Bangladesh and Lebanon. Nigella oil can likewise be used topically. Essentially, a few regions of China consume seeds of *Nigella glandulifera* Frey and Sint., which are often added to 'naan' (a dried U-shaped hotcake). In conventional

Ughur medicine, a water decoction of these seeds is used to treat a great many illnesses, particularly diabetes, very much like with different *Nigella* species (Salehi et al., 2021).

N. sativa has generally been utilized to treat a great many illnesses, infections, and conditions that influence the stomach-related framework, respiratory framework, kidney, and liver capability, cardiovascular system, immunological system, and general prosperity (Ahmad et al., 2013). Black seeds are referenced by Avicenna in "The Standard of Medication" since they helped Middle Eastern nations treat different diseases, including stiffness, asthma, bronchitis, and other related inflammatory conditions. Because of its numerous uses, the Arabic name '*Habbatul barakah*' has been given to Nigella, and that implies the gift seed. The seeds can be used to treat worms and skin emissions, as well as heartburn, loss of craving, diarrhoea, dropsy, amenorrhoea, and dysmenorrhoea. The oil is applied topically as an antibacterial and neighborhood sedative.

To prevent vomiting, cooked black seeds are taken internally (Ahmad et al., 2013). They have been utilized to treat provocative lung conditions, such as bronchitis, clogged arteries, and asthma. Rheumatoid arthritis and different types of joint pain are both treated with black seed oil. Numerous logical investigations demonstrate that nigella seeds, in all actuality, do truly have restorative potential (Dabeer et al., 2022). *N. sativa* is regarded as a successful treatment for a huge number of diseases in the Unani Tibb Institute of Medication. The seeds have been utilized for a really long time in the Central, East, and Southeast Asian nations to treat illnesses like asthma, bronchitis, stiffness, and other comparable circumstances including increasing milk creation in nursing mothers, and battle parasitic infestation. Its oil has been utilized to fix side effects of colds, as well as skin problems like dermatitis and blisters.

11.4 PHARMACOLOGICAL PROPERTIES OF BLACK CUMIN SEEDS AND POTENTIAL MECHANISM OF ACTION

Black cumin seeds have been used for centuries in traditional medicine to treat various ailments. In recent years, the pharmacological properties and the mechanism of action of black cumin seeds have been extensively studied, revealing a range of potential therapeutic benefits (Figure 11.3). Some of the main ones are described below.

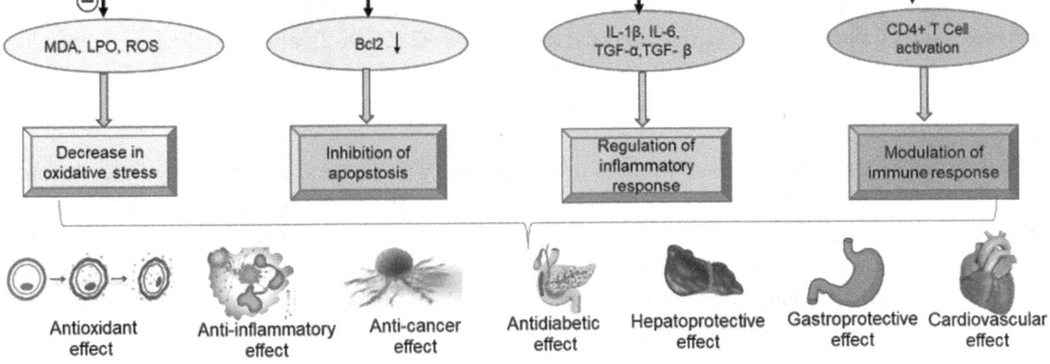

FIGURE 11.3 Pharmacological properties of black cumin seeds and their mechanism of action.

11.4.1 Antioxidant Property

N. sativa seeds have been extensively studied for their potential health benefits, including their antioxidant properties. Recent studies have focused on the antioxidant effects of these seeds in both cell-based *in vitro* models and *in vivo* models (Tavakkoli et al., 2017). In cell-based studies, these seeds have been shown to exert antioxidant effects by reducing oxidative stress and preventing cell damage. For example, a study published in 2019 found that an extract of *N. sativa* seeds was able to reduce oxidative stress in human liver cells exposed to hydrogen peroxide (Yang et al., 2019). In animal studies, these seeds have also been shown to have antioxidant effects. It has been found that *N. sativa* seed oil was able to reduce oxidative stress and improve liver function in rats with liver damage induced by a high-fat diet (Şeker et al., 2021).

N. sativa has been shown to upregulate the activity of antioxidant enzymes such as superoxide dismutase (SOD) and catalase (CAT), which play a crucial role in scavenging reactive oxygen species (ROS) and protecting cells from oxidative damage. It has also been found to increase the level of reduced glutathione (GSH), a powerful antioxidant molecule that helps to neutralize ROS and other harmful substances in the body (Thippeswamy et al., 2005). TQ treatment appears to have protective effects against hydrogen peroxide-induced neurotoxicity in human SH-SY5Y cells. In addition to antioxidant-related genes, TQ treatment also appears to affect various signalling pathways involved in cellular stress response and apoptosis. These include the extracellular signal-regulated protein kinase (ERK)1/2, c-Jun N-terminal kinase (JNK), p53, nuclear factor kappa-light-chain-enhancer of activated B cells (NF-κB), and other pathways (Ismail et al., 2016).

TQ has also been shown to promote the activity of antioxidant enzymes, including superoxide dismutase (SOD) and catalase (CAT), which help to neutralize ROS and protect cells from oxidative damage. Additionally, it has been found to increase levels of reduced glutathione (GSH) (Badary et al., 2021). In addition to promoting antioxidant activity, TQ has been found to downregulate pro-oxidant genes and upregulate antioxidant genes in cells. This means that TQ can help to restore the balance between pro-oxidant and antioxidant processes in cells, which is important for maintaining cellular health and preventing oxidative damage (Mizuno & Fukuhara, 2022).

Further research is needed to understand the underlying mechanisms by which black cumin and its compounds exert their protective effects. This could include investigating the effects of black cumin on specific signalling pathways, gene expression, or metabolic processes. Clinical trials involving human participants are necessary to establish the safety, efficacy, and optimal dosages of *N. sativa* and its compounds. These trials can also help determine which specific disease conditions may benefit from black cumin supplementation, as well as to identify potential adverse effects or drug interactions.

11.4.2 Anti-inflammatory Property

N. sativa seeds exhibit anti-inflammatory properties by inhibiting the production of inflammatory enzymes such as lipoxygenase (LOX) and cyclooxygenase-2 (COX-2), and by reducing the production of pro-inflammatory cytokines (tumor necrosis factor-α (TNF-α) and interleukin-1β (IL-1β), by increasing the production of anti-inflammatory cytokines and by their antioxidant activity (Niu et al., 2021).

Studies have shown that TQ has potent anti-inflammatory properties, which are mediated through the inhibition of pro-inflammatory factors such as TNF-α, IL-1β, IL-6, nitric oxide (NO), cyclooxygenase (COX) 2 and inducible nitric oxide synthase (iNOS). The study by Hossen et al. (2017) reported that TQ was able to inhibit the production of these pro-inflammatory factors in lipopolysaccharide (LPS)-stimulated murine macrophage-like RAW264.7 cells. The mechanism underlying this effect involves the inhibition of IRAK-linked AP-1/NF-κB pathways. IRAK (Interleukin-1 receptor-associated kinase) is a key player in the activation of NF-κB and AP-1 signalling pathways, which are known to play a critical role in the regulation of inflammatory responses (Bennett et al., 2022).

TQ was found to promote the autophosphorylation of TANK-binding kinase 1 (TBK1), which is a key player in the activation of the interferon regulatory factor 3 (IRF-3) signalling pathway. However, it was also found to downregulate the IRF-3 signalling pathways, which ultimately resulted in a reduction in the mRNA abundance of interferons (IFN-α and IFN-β) (Aziz et al., 2018). Although the current evidence for the anti-inflammatory potential of black cumin and TQ is promising, further research is needed to determine their effectiveness in treating human inflammatory conditions

11.4.3 Anti-cancer Effects

Several studies have demonstrated the powerful anticancer properties of *N. sativa* and its chemical components. Researchers indicate that the chemical components of black cumin seeds are chemopreventive and effective at slowing cell growth and inducing apoptosis (Zaky et al., 2021). *In vitro* studies have shown that *N. sativa* seeds and their extracts can inhibit the growth and proliferation of various cancer cell lines, including breast, prostate, lung, and colon cancer cells (Dhaheri et al., 2022). Animal studies have shown that these seeds and their extracts can inhibit the growth of tumours and reduce the incidence of cancer in animals (Korak et al., 2020). Clinical studies have shown that *N. sativa* seeds and their extracts can improve the quality of life and reduce symptoms in cancer patients, such as fatigue, nausea, and pain.

The potential anticancer mechanism of *N. sativa* seeds, its constituents (mainly TQ), and essential oil may involve several pathways, including a) induction of apoptosis cancer cells by activating several signalling pathways, triggering the expression of pro-apoptotic genes, downregulating anti-apoptotic genes and activating caspases (Adinew et al., 2022); b) inhibition of cell proliferation by arresting the cell cycle at various checkpoints, such as the G0/G1, S, and G2/M phases, and suppressing the expression of genes that are involved in cell cycle regulation, leading to the inhibition of cancer cell growth (Khan et al., 2019); c) possession of anti-angiogenic properties, which means that they can inhibit the formation of new blood vessels required for tumour growth and metastasis and also block the expression of pro-angiogenic factors such as VEGF and inhibit the activity of enzymes involved in angiogenesis (Farhangi et al., 2018); d) possession of anti-inflammatory properties by inhibiting the activity of pro-inflammatory cytokines such as TNF-α, IL-6, and COX-2 (Chehl et al., 2009); and e) possession of potent antioxidant properties that can scavenge free radicals and protect cells from oxidative damage (Randhawa & Alghamdi, 2011).

11.4.4 Antidiabetic Effects

Studies have shown that black cumin seeds may have anti-diabetic effects due to their various bioactive compounds, including thymoquinone, thymohydroquinone, and thymol. One study conducted on diabetic rats found that *N. sativa* oil (NSO) was able to significantly reduce blood glucose levels, improve insulin sensitivity, and reduce oxidative stress (Mohebbati et al., 2020). Another study conducted on diabetic human subjects found that consuming black cumin seeds for 12 weeks was able to significantly reduce fasting blood glucose levels, as well as improve other markers of glucose metabolism (Rachman et al., 2017).

N. sativa seed extracts have been demonstrated to have an anti-diabetic effect in either alloxan- or streptozotocin-induced diabetic rats or mice. The mechanism underlying this effect involves a decrease in oxidative stress by increasing the activity of antioxidant enzymes, regulation of blood lipid profiles, improvement of endothelial dysfunction, and enhancement of tissue regeneration and wound healing (Maideen, 2021).

11.4.5 Cardioprotective and Antihypertensive Effects

Several studies have shown that black cumin seeds can help reduce the risk of cardiovascular disease by lowering cholesterol levels, reducing inflammation, and improving the function of the endothelium (the inner lining of blood vessels). Shabana et al. (2013) found that black cumin seeds reduced cholesterol levels in rats and prevented the development of atherosclerosis (hardening of the arteries). In another study, it was shown that black cumin seeds reduced inflammation and improved endothelial function in patients with type 2 diabetes (Farhangi et al., 2018).

Black cumin seeds have also been shown to have antihypertensive effects, meaning that they can help reduce high blood pressure. It has been found that black cumin seeds reduce blood pressure in rats with hypertension, and researchers showed that black cumin seeds reduced blood pressure in patients with mild hypertension (Bush et al., 2020). Black cumin seed probably normalizes hypertension by counteracting the cardiovascular effects of angiotensin II in a model of angiotensin II-induced hypertensive rats (Enayatfard et al., 2018). It is also important to note that black cumin seeds should not be used as a replacement for conventional treatments for high blood pressure or cardiovascular disease without first consulting a healthcare professional.

11.4.6 Hepatoprotective Effects

Several scientific studies have shown that black cumin seeds have hepatoprotective effects. Researchers have found that black cumin seed oil was able to increase levels of antioxidant enzymes in rats with induced liver damage (Miah et al., 2021). Researchers found that these seeds were effective in protecting the liver from damage caused by carbon tetrachloride, a toxic chemical that can cause liver damage (Al-Ghamdi, 2003). Also it has been found that these seeds were effective in protecting the liver from damage caused by high-fat diets (Maideen, 2021).

Supplementation with black cumin extract or its active ingredients increases the levels of GSH, SOD, CAT, and glutathione peroxidase (GPx), which decrease oxidative stress and endoplasmic reticulum stress (reduced activity of ATF4, ATF6, CHOP, GRP78, and XBP1), lowers lipid peroxidation, and lengthens the lifespan of experimental animals (Amin & Hosseinzadeh, 2016). Nitric oxide (NO) is an important signalling molecule in the liver, and its dysregulation has been implicated in various liver diseases, including cirrhosis, hepatitis, and liver cancer (Iwakiri & Kim, 2015). It has been found that *N. sativa* seed extract and its components can reduce NO levels by modulating the activity of nitrite reductase (Ali & Blunden, 2003).

NSO (*Nigella sativa* oil) may have beneficial effects on the liver and could potentially reverse hepatic injury in non-alcoholic fatty liver disease (NAFLD). In a study, it was found that black cumin seed oil was able to improve liver function tests and decrease liver enzymes in patients with NAFLD (Darand et al., 2019). The potential hepatoprotective effects of black cumin seed oil are thought to be due to its antioxidant and anti-inflammatory properties. The oil contains several active compounds, including TQ, which has been shown to have antioxidant and anti-inflammatory effects on the liver.

11.4.7 Gastroprotective Effects

Various factors, such as bacterial and viral infections, affect the gastrointestinal system, including inflammatory conditions and drug-induced side effects (Abrams, 1988). Natural products, such as *N. sativa* seed and TQ, have been shown to have protective effects against these abnormalities. Both NSO and TQ administration can ameliorate cisplastin-induced alterations in the enzymatic and non-enzymatic parameters of the antioxidant defence system in the intestinal mucosa (Shahid et al., 2018). A clinical study demonstrated by researchers in 2019 evaluated the effectiveness of Dosin (6 g/day of black cumin as ground seeds and 12 g/day of honey) in eradicating *Helicobacter pylori* infection, a common cause of peptic and gastric ulcers. The study included 70 patients with confirmed

H. pylori infection who were randomized to receive either standard triple therapy (clarithromycin, amoxicillin, and omeprazole antibiotic therapy) or Dosin for 14 days. The results showed that both treatments were effective in eradicating *H. pylori* infection, with no significant difference between the two groups. However, the Dosin group had a lower incidence of adverse effects compared with the standard triple therapy group (Hashem-Dabaghian et al., 2016). These studies suggest that black cumin seeds may have gastroprotective effects and could potentially be used as a natural remedy for gastric ulcers and other digestive issues. However, further research is needed to fully understand the mechanisms behind these effects and to determine the optimal dosage and duration of treatment.

11.4.8 Other Potential Uses of Black Cumin Seeds

Black cumin seeds have been used for medicinal purposes for centuries in traditional medicine systems such as Ayurveda and Unani. In addition to their use as a spice in cooking, black cumin seeds have several other potential uses, including weight loss, boosting the immune system, improving respiratory health, promoting skin health, relieving pain, treating asthma, and having anti-allergic properties (Harborne, 1994). *N. sativa* seeds may have a beneficial effect on weight loss by reducing appetite, increasing metabolism, and promoting fat burning (Namazi et al., 2018). These seeds have been shown to have immuno-boosting properties and may help fight off infections and diseases (Salem, 2005) and help improve respiratory health by reducing inflammation and opening up airways, making it easier to breathe (Kanter, 2009). *N. sativa* contains antioxidants and anti-inflammatory compounds that may help improve skin health and reduce the appearance of skin ageing (Bordoni et al., 2019). Black cumin seeds may have pain-relieving properties and may be used to relieve pain associated with conditions such as arthritis (Ghannadi et al., 2005). Some studies have suggested that black cumin seed oil may help improve lung function and reduce symptoms of asthma, possibly due to its anti-inflammatory properties (Bordoni et al., 2019). Black cumin seeds may also have anti-allergic properties and have been shown to reduce symptoms of allergic rhinitis (hay fever) in some studies (Ermumcu, 2022). It's important to note that, although black cumin seeds have many potential health benefits, more research is needed to fully understand their effects and determine appropriate dosages.

11.5 CONCLUSION

N. sativa seeds are a rich source of bioactive compounds, including thymoquinone, thymohydroquinone, dithymoquinone, and thymol. These compounds possess various pharmacological properties, including anti-inflammatory, antioxidant, and antimicrobial effects. Studies have shown that *N. sativa* seeds have numerous health benefits, including potential gastroprotective, cardiovascular, liver, neurological, respiratory, and immune system support. At the molecular level, *N. sativa* seeds have been shown to modulate various signalling pathways, including those involved in inflammation, oxidative stress, apoptosis, and cell cycle regulation. *N. sativa* seeds have also been shown to interact with various receptors and enzymes, including those involved in neurotransmission and immune system regulation. Overall, the phytochemistry and molecular pharmacology of *N. sativa* seeds suggest that they have a wide range of potential health benefits, although more research is needed to fully understand their effects and optimal uses.

11.6 FUTURE PERSPECTIVES

The existing evidence on the pharmacological benefits of black cumin and its active ingredient, thymoquinone, suggests that there are promising future perspectives for the use of these natural products in medicine. Research on black cumin and TQ has shown their potential as anti-inflammatory, antioxidant, antimicrobial, anticancer, and neuroprotective agents. Further studies may explore their potential in the treatment of various diseases, including asthma, allergies, diabetes, cancer, and neurodegenerative disorders. In addition, there is a need for further investigation into the pharmacokinetics and pharmacodynamics of black cumin and TQ, including their absorption, metabolism, and excretion in the body. This will help to optimize their dosing and administration optima in clinical settings. Moreover, research may also focus on the potential synergistic effects of black cumin and TQ with other natural products or conventional drugs, as well as their safety and efficacy in combination therapies. Overall, the pharmacological benefits of black cumin and TQ suggest that they have promising future perspectives in the development of new natural products for the treatment and prevention of various diseases.

LIST OF ABBREVIATIONS

CAT: catalase
COX-2: cyclooxygenase-2 (COX-2)
ERF: extracellular signal-regulated protein kinase
IL-1β: interleukin-1β
JNK: c-Jun N-terminal kinase
LOX: lipoxygenase
NAFLD: nonalcoholic fatty liver disease
NF-κB: nuclear factor kappa-light-chain-enhancer of activated B cells
NSO: *Nigella* sativa oil
NO: nitric oxide (NO)
iNOS: nitric oxide synthase
ROS: reactive oxygen species
SOD: superoxide dismutase
TQ: Thymoquinone
TNF-α: tumour necrosis factor-α

REFERENCES

Abrams, G.D. 1988. Pathogenesis of gastrointestinal infections. *The American Journal of Surgical Pathology, 12*(1):76–81.

Adinew, G.M., Messeha, S.S., Taka, E., Badisa, R.B., Antonie, L.M., & Soliman, K.F.A. 2022. Thymoquinone alterations of the apoptotic gene expressions and cell cycle arrest in genetically distinct triple-negative breast cancer cells. *Nutrients, 14*(10):2120. https://doi.org/10.3390/nu14102120.

Ahmad, I., Tripathi, J., Manik, S., Umar, L., & Rabia, J. 2013. Preliminary phytochemical studies of the miracle herb of the century, *Nigella sativa* L. (Black seed). *Indo American Journal of Pharmaceutical Research, 3*:3000–3007.

Ahmad, A., Husain, A., Mujeeb, M., Khan, S.A., Najmi, A.K., Siddique, N.A., Damanhouri, Z.A., & Anwar, F. 2013. A review on therapeutic potential of *Nigella sativa*: A miracle herb. *Asian Pacific Journal of Tropical Biomedicine, 3*(5):337–352. https://doi.org/10.1016/S2221-1691(13)60075-1.

Al-Ghamdi, M.S. 2003. Protective effect of *Nigella sativa* seeds against carbon tetrachloride-induced liver damage. *The American Journal of Chinese Medicine, 31*(5):721–728. https://doi.org/10.1142/S0192415X03001399.

Ali, A. 2020. Herbs that heal: The philanthropic behaviour of nature. *Annals of Phytomedicine, 9*(1):7–17. http://doi.org/10.21276/ap.2020.9.1.2.

Ali, B.H., & Blunden, G. 2003. Pharmacological and toxicological properties of *Nigella sativa. Phytotherapy Research: PTR, 17*(4):299–305. https://doi.org/10.1002/ptr.1309.

Amin, B., & Hosseinzadeh, H. 2016. Black Cumin (*Nigella sativa*) and its active constituent, thymoquinone: An overview on the analgesic and anti-inflammatory effects. *Planta Medica, 82*(1–2):8–16. https://doi.org/10.1055/s-0035-1557838.

Aziz, N., Son, Y.J., & Cho, J.Y. 2018. Thymoquinone Suppresses IRF-3-mediated expression of type I interferons via suppression of TBK1. *International Journal of Molecular Sciences, 19*(5):1355. https://doi.org/10.3390/ijms19051355.

Badary, O.A., Hamza, M.S., & Tikamdas, R. 2021. Thymoquinone: A promising natural compound with potential benefits for COVID-19 prevention and cure. *Drug Design, Development and Therapy, 15*:1819–1833. https://doi.org/10.2147/DDDT.S308863.

Balyan, P., & Ali, A. 2022. Comparative analysis of the biological activities of different extracts of *Nigella sativa* L. seeds. *Annals of Phytomedicine, 11*(1):577–587. http://doi.org/10.54085/ap.2022.11.1.67.

Balyan, P., Akhter, J., Kumar, P., & Ali, A. 2022. Traditional and modern usage of *Nigella sativa* L. (Black cumin). *Annals of Phytomedicine, 11*(2):255–265. http://doi.org/10.54085/ap.2022.11.2.28.

Bennett, J., & Starczynowski, D.T. 2022. IRAK1 and IRAK4 as emerging therapeutic targets in hematologic malignancies. *Current Opinion in Hematology, 29*(1):8–19. https://doi.org/10.1097/MOH.0000000000000693.

Bishnoi, S. 2016. Herbs as functional foods. In: *Functional Foods: Sources and Health Benefits*, D. Mudgill & S. Barak (eds), Scientific Publishers, pp. 141–172.

Bordoni, L., Fedeli, D., Nasuti, C., Maggi, F., Papa, F., Wabitsch, M., De Caterina, R., & Gabbianelli, R. 2019. Antioxidant and anti-inflammatory properties of *Nigella sativa* Oil in human pre-adipocytes. *Antioxidants (Basel, Switzerland), 8*(2):51. https://doi.org/10.3390/antiox8020051.

Bush, B., Peña, T., Bush, R., Zelkha, M., & Lamensdorf, I. 2020. Effects of standardized black seed oil cold press supplement over a six-week period on blood pressure and heart rate in healthy patients: A nonrandomized clinical trial. *Food Science and Nutrition Research, 3*(1):1–5.

Castro, I.A., Rogero, M.M., Junqueira, R.M., & Carrapeiro, M.M. 2006. Free radical scavenger and antioxidant capacity correlation of alpha-tocopherol and trolox measured by three *in vitro* methodologies. *International Journal of Food Sciences and Nutrition, 57*(1–2):75–82. https://doi.org/10.1080/09637480600656199.

Chehl, N., Chipitsyna, G., Gong, Q., Yeo, C.J., & Arafat, H.A. 2009. Anti-inflammatory effects of the *Nigella sativa* seed extract, thymoquinone, in pancreatic cancer cells. *HPB: The Official Journal of the International Hepato Pancreato Biliary Association, 11*(5):373–381. https://doi.org/10.1111/j.1477-2574.2009.00059.x.

Dabeer, S., Rather, M.A., Rasool, S., Rehman, M.U., Alshahrani, S., Jahan, S., Rashid, H., Halawi, M., & Khan, A. 2022. History and traditional uses of black seeds (*Nigella sativa*). *Black Seeds (Nigella Sativa).* https://doi.org/10.1016/b978-0-12-824462-3.00016-0.

Darand, M., Darabi, Z., Yari, Z., Hedayati, M., Shahrbaf, M.A., Khoncheh, A., Hosseini-Ahangar, B., Alavian, S.M., & Hekmatdoost, A. 2019. The effects of black seed supplementation on cardiovascular risk factors in patients with nonalcoholic fatty liver disease: A randomized, double-blind, placebo-controlled clinical trial. *Phytotherapy Research: PTR, 33*(9):2369–2377. https://doi.org/10.1002/ptr.6424.

De Luca, V., Salim, V., Atsumi, S.M., & Yu, F. 2012. Mining the biodiversity of plants: A revolution in the making. *Science (New York, NY), 336*(6089):1658–1661. https://doi.org/10.1126/science.1217410.

Dehkordi, F.R., & Kamkhah, A.F. 2008. Antihypertensive effect of *Nigella sativa* seed extract in patients with mild hypertension. *Fundamental and Clinical Pharmacology, 22*(4):447–452. https://doi.org/10.1111/j.1472-8206.2008.00607.x.

Dhaheri, Y.A., Wali, A.F., Akbar, I., Rasool, S., Razmpoor, M., Jabnoun, S., & Rashid, S. 2022. *Nigella sativa*, a cure for every disease: Phytochemistry, biological activities, and clinical trials. In: *Black Seeds (Nigella sativa)*, A. Khan & M. Rehman (eds), Elsevier, pp. 63–90. ISBN 9780128244623. https://doi.org/10.1016/B978-0-12-824462-3.00011-1.

Enayatfard, L., Mohebbati, R., Niazmand, S., Hosseini, M., & Shafei, M.N. 2018. The standardized extract of *Nigella sativa* and its major ingredient, thymoquinone, ameliorates angiotensin II-induced hypertension in rats. *Journal of Basic and Clinical Physiology and Pharmacology, 30*(1):51–58. https://doi.org/10.1515/jbcpp-2018-0074.

Ermumcu, M.S.K. 2022. Biological activities of black cumin (*Nigella sativa*) seed oil. In: *Multiple Biological Activities of Unconventional Seed Oils*, A. A. Mariod (ed), Academic Press, pp. 43–53. ISBN 9780128241356. https://doi.org/10.1016/B978-0-12-824135-6.00010-6.

Farhangi, M.A., Dehghan, P., & Tajmiri, S. 2018. Powdered black cumin seeds strongly improves serum lipids, atherogenic index of plasma and modulates anthropometric features in patients with Hashimoto's thyroiditis. *Lipids in Health and Disease, 17*(1):59. https://doi.org/10.1186/s12944-018-0704-x.

Fatima Shad, K., Soubra, W., & Cordato, D.J. 2021. The role of thymoquinone, a major constituent of *Nigella sativa*, in the treatment of inflammatory and infectious diseases. *Clinical and Experimental Pharmacology and Physiology, 48*(11):1445–1453. https://doi.org/10.1111/1440-1681.13553.

Ghannadi, A., Hajhashemi, V., & Jafarabadi, H. 2005. An investigation of the analgesic and anti-inflammatory effects of *Nigella sativa* seed polyphenols. *Journal of Medicinal Food, 8*(4):488–493. https://doi.org/10.1089/jmf.2005.8.488.

Gueffai, A., Gonzalez-Serrano, D.J., Christodoulou, M.C., Orellana-Palacios, J.C., Ortega, M.L.S., Ouldmoumna, A., Kiari, F.Z., Ioannou, G.D., Kapnissi-Christodoulou, C.P., Moreno, A., & Hadidi, M. 2022. Phenolics from defatted black cumin seeds (*Nigella sativa* L.): Ultrasound-assisted extraction optimization, comparison, and antioxidant activity. *Biomolecules*, *12*(9):1311. https://doi.org/10.3390/biom12091311.

Gutiérrez-Grijalva, E.P., López-Martínez, L.X., Contreras-Angulo, L.A., Elizalde-Romero, C.A., & Heredia, J.B. 2020. Plant alkaloids: Structures and bioactive properties. In: *Plant-derived Bioactives*, M. Swamy (ed), Springer. https://doi.org/10.1007/978-981-15-2361-8_5.

Hannan, M.A., Rahman, M.A., Sohag, A.A.M., Uddin, M.J., Dash, R., Sikder, M.H., Rahman, M.S., Timalsina, B., Munni, Y.A., Sarker, P.P., Alam, M., Mohibbullah, M., Haque, M.N., Jahan, I., Hossain, M.T., Afrin, T., Rahman, M.M., Tahjib-Ul-Arif, M., Mitra, S., Oktaviani, D.F. … Kim, B. 2021. Black Cumin (*Nigella sativa* L.): A comprehensive review on phytochemistry, health benefits, molecular pharmacology, and safety. *Nutrients*, *13*(6):1784. https://doi.org/10.3390/nu13061784.

Harborne, J.B. 1994. Indian medicinal plants. a compendium of 500 species. *Journal of Pharmacy and Pharmacology*, *46*(11):935. https://doi.org/10.1111/j.2042-7158.1994.tb05722.x.

Hashem-Dabaghian, F., Agah, S., Taghavi-Shirazi, M., & Ghobadi, A. 2016. Combination of *Nigella sativa* and honey in eradication of gastric *Helicobacter pylori* infection. *Iranian Red Crescent Medical Journal*, *18*(11):e23771. https://doi.org/10.5812/ircmj.23771.

Hossen, M.J., Yang, W.S., Kim, D., Aravinthan, A., Kim, J.H., & Cho, J.Y. 2017. Thymoquinone: An IRAK1 inhibitor with *in vivo* and *in vitro* anti-inflammatory activities. *Scientific Reports*, *7*:42995. https://doi.org/10.1038/srep42995.

Husen, A. 2021 *Traditional Herbal Therapy for the Human Immune System*. CRC Press, Boca Raton, FL. https://doi.org/10.1201/9781003137955.

Husen, A. 2022 *Herbs, Shrubs and Trees of Potential Medicinal Benefits*. CRC Press, Boca Raton, FL. https://doi.org/10.1201/9781003205067.

Husen, A. 2023 *Antidiabetic Medicinal Plants and Herbal Treatments*. CRC Press, Boca Raton, FL. https://doi.org/10.1201/b23347.

Ismail, N., Ismail, M., Azmi, N.H., Abu Bakar, M.F., Basri, H., & Abdullah, M.A. 2016. Modulation of hydrogen peroxide-induced oxidative stress in human neuronal cells by thymoquinone-rich fraction and thymoquinone via transcriptomic regulation of antioxidant and apoptotic signaling genes. *Oxidative Medicine and Cellular Longevity*, *2016*:2528935. https://doi.org/10.1155/2016/2528935.

Iwakiri, Y., & Kim, M.Y. 2015. Nitric oxide in liver diseases. *Trends in Pharmacological Sciences*, *36*(8):524–536. https://doi.org/10.1016/j.tips.2015.05.001.

Kanter, M. 2009. Effects of *Nigella sativa* seed extract on ameliorating lung tissue damage in rats after experimental pulmonary aspirations. *Acta Histochemica*, *111*(5):393–403. https://doi.org/10.1016/j.acthis.2008.10.008.

Khan, A., Aldebasy, Y.H., Alsuhaibani, S.A., & Khan, M.A. 2019. thymoquinone augments cyclophosphamide-mediated inhibition of cell proliferation in breast cancer cells. *Asian Pacific Journal of Cancer Prevention*, *20*(4):1153–1160. https://doi.org/10.31557/APJCP.2019.20.4.1153.

Korak, T., Ergül, E., & Sazcı, A. 2020. *Nigella sativa* and cancer: A review focusing on breast cancer, inhibition of metastasis and enhancement of natural killer cell cytotoxicity. *Current Pharmaceutical Biotechnology*.

Maideen, N.M.P., Balasubramanian, R., & Ramanathan, S. 2021. *Nigella sativa* (black seed), A potential herb for the pharmacotherapeutic management of hypertension - A review. *Current Cardiology Reviews*, *17*(4):e230421187786. https://doi.org/10.2174/1573403X16666201110125906.

Miah, P., Mohona, S.B.S., Rahman, M.M., Subhan, N., Khan, F., Hossain, H., Sharker, S.M., & Alam, M.A. 2021. Supplementation of cumin seed powder prevents oxidative stress, hyperlipidemia and non-alcoholic fatty liver in high fat diet fed rats. *Biomedicine and Pharmacotherapy*, *141*:111908. https://doi.org/10.1016/j.biopha.2021.111908.

Mizuno, M., & Fukuhara, K. 2022. Antioxidant and prooxidant effects of thymoquinone and its hydroquinone metabolite. *Biological and Pharmaceutical Bulletin*, *45*(9):1389–1393. https://doi.org/10.1248/bpb.b22-00199.

Mohebbati, R., Abbasnezhad, A., Havakhah, S., & Mousavi, M. 2020. The Effect of *Nigella sativa* on renal oxidative injury in diabetic rats. *Saudi Journal of Kidney Diseases and Transplantation: An Official Publication of the Saudi Center for Organ Transplantation, Saudi Arabia*, *31*(4):775–786. https://doi.org/10.4103/1319-2442.292311.

Namazi, N., Larijani, B., Ayati, M.H., & Abdollahi, M. 2018. The effects of *Nigella sativa* L. on obesity: A systematic review and meta-analysis. *Journal of Ethnopharmacology*, *219*:173–181. https://doi.org/10.1016/j.jep.2018.03.001.

Niu, Y., Wang, B., Zhou, L., Ma, C., Waterhouse, G.I.N., Liu, Z., Ahmed, A.F., Sun-Waterhouse, D., & Kang, W. 2021. *Nigella sativa*: A dietary supplement as an immune-modulator on the basis of bioactive components. *Frontiers in Nutrition*, *8*:722813. https://doi.org/10.3389/fnut.2021.722813.

Rachman, P.N., Akrom, & Darmawan, E. 2017. The efficacy of black cumin seed (*Nigella sativa*) oil and hypoglycemic drug combination to reduce HbA1c level in patients with metabolic syndrome risk. *IOP Conference Series: Materials Science and Engineering*, 259: 012–018.

Ramadan, M.F. 2007. Nutritional value, functional properties and nutraceutical applications of black cumin (*Nigella sativa* L.): An overview. *International Journal of Food Science and Technology*, *42*:10.

Randhawa, M.A., & Alghamdi, M.S. 2011. Anticancer activity of *Nigella sativa* (black seed) - A review. *The American Journal of Chinese Medicine*, *39*(6):1075–1091. https://doi.org/10.1142/S0192415X1100941X.

Rocha, M., Banuls, C., Bellod, L., Jover, A., Victor, V.M., & Hernandez-Mijares, A. 2011. A review on the role of phytosterols: New insights into cardiovascular risk. *Current Pharmaceutical Design*, *17*(36):4061–4075. https://doi.org/10.2174/138161211798764852.

Salehi, B., Quispe, C., Imran, M., Ul-Haq, I., Živković, J., Abu-Reidah, I.M., Sen, S., Taheri, Y., Acharya, K., Azadi, H., Del Mar Contreras, M., Segura-Carretero, A., Mnayer, D., Sethi, G., Martorell, M., Abdull Razis, A.F., Sunusi, U., Kamal, R.M., Rasul Suleria, H.A., & Sharifi-Rad, J. 2021. Nigella plants: Traditional uses, bioactive phytoconstituents, preclinical and clinical studies. *Frontiers in Pharmacology*, *12*:625386. https://doi.org/10.3389/fphar.2021.625386.

Salem, M.L. 2005. Immunomodulatory and therapeutic properties of the *Nigella sativa* L. seed. *International Immunopharmacology*, *5*(13–14):1749–1770. https://doi.org/10.1016/j.intimp.2005.06.008.

Şeker, U., Kaya, S., İrtegun Kandemir, S., Şener, D., Unay Demirel, O., & Nergiz, Y. 2021. Effects of black cumin seed oil on oxidative stress and expression of membrane-cytoskeleton linker proteins, radixin, and moesin in streptozotocin-induced diabetic rat liver. *Hepatology Forum*, *3*(1):21–26.

Shabana, A., El-Menyar, A., Asim, M., Al-Azzeh, H., & Al Thani, H. 2013. Cardiovascular benefits of black cumin (*Nigella sativa*). *Cardiovascular Toxicology*, *13*(1):9–21. https://doi.org/10.1007/s12012-012-9181-z.

Shyur, L.F., & Yang, N.S. 2008. Metabolomics for phytomedicine research and drug development. *Current Opinion in Chemical Biology*, *12*(1):66–71. https://doi.org/10.1016/j.cbpa.2008.01.032.

Tavakkoli, A., Mahdian, V., Razavi, B.M., & Hosseinzadeh, H. 2017. Review on clinical trials of black seed (*Nigella sativa*) and its active constituent, thymoquinone. *Journal of Pharmacopuncture*, *20*(3):179–193. https://doi.org/10.3831/KPI.2017.20.021.

Thippeswamy, N.B., & Naidu, K.A. 2005. Antioxidant potency of cumin varieties—Cumin, black cumin and bitter cumin on antioxidant systems. *European Food and Research Technology*, *220*(5–6):472–476. https://doi.org/10.1007/s00217-004-1087-y.

Wirnitzer, K.C. 2018. Vegan Nutrition: Latest boom in health and exercise. In: *Therapeutic, Probiotic, and Unconventional Foods*, Alexandru Mihai Grumezescu & Alina Maria Holban (eds), Academic Press, pp. 387–453. ISBN 9780128146255. https://doi.org/10.1016/B978-0-12-814625-5.00020-0.

Yang, Q., Noviana, M., Zhao, Y., Chen, D., & Wang, X. 2019. Effect of curcumin extract against oxidative stress on both structure and deformation capability of red blood cells. *Journal of Biomechanics*, *95*:109301. https://doi.org/10.1016/j.jbiomech.2019.07.045.

Yarnell, E., & Abascal, K. 2011. *Nigella sativa*: Holy herb of the middle East. *Alternative Complementary Therapies*, *17*(2):99–105.

Yimer, E.M., Tuem, K.B., Karim, A., Ur-Rehman, N., & Anwar, F. 2019. *Nigella sativa* L. (Black Cumin): A promising natural remedy for wide range of illnesses. *Evidence-Based Complementary and Alternative Medicine: eCAM*, *2019*:1528635. https://doi.org/10.1155/2019/1528635.

Zaky, A.A., Shim, J.H., & Abd El-Aty, A.M. 2021. A review on extraction, characterization, and applications of bioactive peptides from pressed black cumin seed cake. *Frontiers in Nutrition*, *8*:743909. https://doi.org/10.3389/fnut.2021.743909.

Zielińska, M., Dereń, K., Polak-Szczybyło, E., & Stępień, A.E. 2021. The Role of bioactive compounds of *Nigella sativa* in rheumatoid arthritis therapy - Current reports. *Nutrients*, *13*(10):3369. https://doi.org/10.3390/nu13103369.

12 Black Pepper Seeds (*Piper nigrum*)

Rajasekharan Sharika, Kuljira Mongkolpobsin, Sunita Nilkhet, Tewin Tencomnao, and Siriporn Chuchawankul

12.1 INTRODUCTION

Black pepper is often referred to as the 'King of spices' and holds a very important place in the international spice markets. It is one of the most widely consumed spices and is well known for its medicinal properties as well as its purpose in perfumery, as a preservative, and as an insecticide (Goswami and Malviya, 2020; Srinivasan, 2007). Use of black pepper as a medicinal ingredient has been practiced in folk medicine since time immemorial. It has been used as an immune system enhancer and as a treatment for gastric problems, including diarrhoea and indigestion, in addition to which it is also used against insomnia and epilepsy (Butt et al., 2013). Piperine is the major active compound identified from black pepper. Other important compounds include alkaloids, volatile oils, and oleoresins which are responsible for the pungent smell and flavour produced by it (Butt et al., 2013).

12.2 BOTANICAL DESCRIPTION

Piper nigrum belongs to the family Piperaceae and the order Piperales, belonging to the class Magnoliopsida. It is a climbing perennial vine that produces numerous adventitious rootlets that facilitate its attachment to supports for climbing, growing up to 30 ft. The stems are thick with nodes from which the rootlets grow out. The leaves are elliptical and wide with a glossy texture. It produces clusters of berries (peppercorns) that are generally round or oval in shape; based on the varying stages of development at which they are harvested and the processing methods used, they are classified as black, green, or white peppercorns. The peppercorn forms on a slender elongated spike-like structure on the stem which is mostly found opposite each leaf, hence the number of leaves determines the yield of peppercorns (Figure 12.1) (Butt et al., 2013; Nelson et al., 2011; Wulandari et al., 2021; Milenković and Stanojević, 2021).

12.3 DISTRIBUTION AND CULTIVATION PRACTICES

Black pepper has been an integral part of diet due to its flavour and medicinal properties (Butt et al., 2013). It is believed to be a native of India and was known to be domesticated and cultivated in the Western Ghats of the southern Indian peninsular region, being mentioned in ancient Indian medical texts. Black pepper has been cultivated across the globe in Asian, African and Latin American countries. The major producers of black pepper are reported to be Vietnam, India, Indonesia, Malaysia, China, and Brazil (Milenković and Stanojević, 2021; HAO et al., 2012). It is also commercially grown in the tropical regions of Thailand, Madagascar, West Africa, Mexico, and the Pacific Islands (where it is not widely cultivated) (Nelson et al., 2011, Milenković and Stanojević, 2021). Trade in black pepper can be traced as far as 3000 BC by the Babylonians and Assyrians with the people of the Indian subcontinent. Interestingly, black pepper was also found to be present

DOI: 10.1201/9781003387046-12

FIGURE 12.1 Black pepper and its different stages.

with mummified pharaohs, along with gold and silver. Also, its use has been documented in various religious literatures such as the Vedas, the Bible, and the Quran (Milenković and Stanojević, 2021).

Black pepper grows well in wet soil and generally can be harvested after three years of planting and harvesting can be continued for up to 30 years. Uniform rainfall with a spell of dry weather favours ripening and harvest. It is found growing in tropical regions at altitudes from sea level to 600 m above sea level. It thrives over a temperature range of 22° to 30° C and also tolerates a lower temperature range down to 12° C. The vines grow vertically and are generally supported by trees that do not shed their bark, like palms, facilitating climbing of the vines as well as receiving optimal shade. Interplanting and monocropping techniques are often-used methods in the cultivation of black pepper (Nelson et al., 2011). Recently, studies on the introduction of smart farming technology, which integrates agriculture with modern technology to help monitoring temperature and controlling humidity, have been carried out on black pepper. This technique has shown an increased yield in mushroom cultivation (Madato et al., 2022).

12.4 PHYTOCHEMICAL CONSTITUENTS

Black pepper seeds have been used as a spice for a long time. The character of black pepper, including its colour, odour, taste, and medicinal activity, are related to the phytochemical composition of black pepper seeds. Previous studies have reported that black pepper seeds contain starch, fatty acids, fibres, essential oils, and other compounds (Murthy and Bhattacharya, 2008). The harvesting, extraction, and chemical analytical processes affect the composition of phytochemicals and their proportions. Phytochemical composition in black pepper seeds in ethanol, hexane, and methanol extracts reported the abundance of monoterpenoids, sesquiterpenoids, alkaloids, phenols, and phytosterols from dried black pepper seed extracts (Butt et al., 2013). The best-known bioactive compound in black pepper seeds is piperine, an alkaloid compound. The piperine content varies from 1% to 63% and is quantified by GC-MS analysis (Nag et al., 2020; Vieira et al., 2022; Yoon et al., 2015; Zahin et al., 2021; Vadivel et al., 2018).

Piperine is a bioactive alkaloid compound discovered in members of the Piperaceae family and displays antioxidant, anti-inflammation, anti-tumour, and anti-neurodegenerative properties (Tiwari et al., 2020). Other alkaloid derivatives (piperanine, piperettine, piperylin A, piperolein B, and pipericine) in black pepper are found in small amounts (Gorgani et al., 2017).

The other phytochemicals discovered in extracts are sesquiterpenes, including caryophyllene and copaene. Sesquiterpenes belong to the terpenes, which are common in essential oils, representing aroma or volatile compounds (Jeleń and Gracka., 2015). The chemical components in essential oil extracted from black pepper seeds was investigated, and β-caryophyllene was found to be the most significant phytochemical in black pepper seed essential oil at the range of 18.64 to 31.70%, followed by monoterpene compounds, including sabinene, limonene, and β-pinene (Bagheri et al., 2014). β-Caryophyllene concentration was used as the standard compound for quality control of Viphyllin™ black pepper seed extract, which should not be lower than 30% between batches (Sudeep et al., 2021). Furthermore, Piptigrine, an insecticidal amide, was elucidated from the petroleum ether extract of dried seeds at 4.3 mg/5 kg dry weight (Siddiqui et al., 2004). Also, polysaccharides isolated from the water extract of seeds yielded 32.4 g/500 g of seed dry weight (Chun et al., 2002). According to the literature, the abundance of phytochemicals depends on black pepper sources and extraction methods, as the main phytochemical constituents are listed in Table 12.1. Therefore, piperine and β-caryophyllene from black pepper seed, as shown in Figure 12.2, are still the most relevant phytochemicals developed for the food, cosmetics, and pharmaceutical industries.

TABLE 12.1
The Main Phytochemical Constituents Reported in Black Pepper Seeds

Sources	Phytochemicals	Amount	Methods	Key Reference
Ethanol extract	Caryophyllene	2.87%	GC-MS (Gas Chromatography-Mass Spectrometery)	Yousuf et al., 2022
	N-Hexadecanoic acid	1.77%		
	Piperine	1.37%		
	Copaene	1.22%		
	δ-Cadinene	1.22%		
Ethanol extract	Piperine	17.65 to 35.30 mg/g of dry weight	HPLC	Vieira et al., 2022
Ethanol extract	Piperine	133.8 ± 17.6 mg/g of extract	UPLC–MS/MS	Yoon et al., 2015
Essential oil	β-Caryophyllene	18.64 to 25.38%	GC-MS	Bagheri et al., 2014
	Limonene	14.95 to 15.64%		
	β-Pinene	7.27 to 9.71%		
Essential oil	β-Caryophyllene	31.70%	GC-MS	Vieira et al., 2022
	Sabinene	23.13%		
	Limonene	21.06%		
Hexane extract	Piperine	63%	GC-MS	Nag et l., 2020
Methanol extract	Piperine	46.10%	GC-MS	Zahin et al., 2021
	Copaene	12.25%		
Methanol extract	Piperine	49.97%	GC-MS	Vadivel et al., 2018
Petroleum ether	Piptigrine	4.3 mg/5 kg of dry weight	^{1}H-NMR	Sudeep et al., 2021
Viphyllin™	β-Caryophyllene	30%	GC-MS	Bagheri et al., 2014
Water extract	Polysaccharides	32.4 g/500 g of dry weight	Ethanol precipitation	Siddiqui et al., 2004

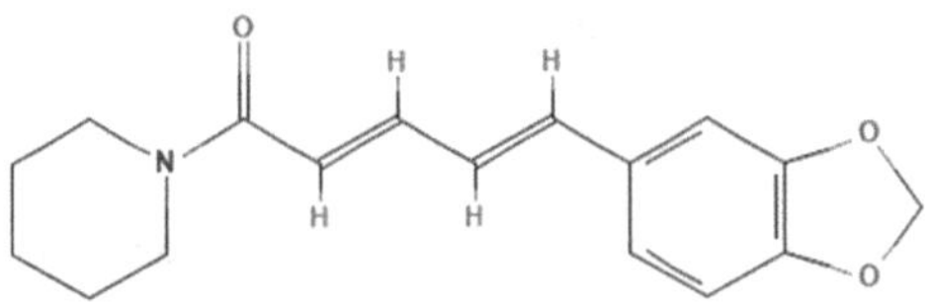

Piperine

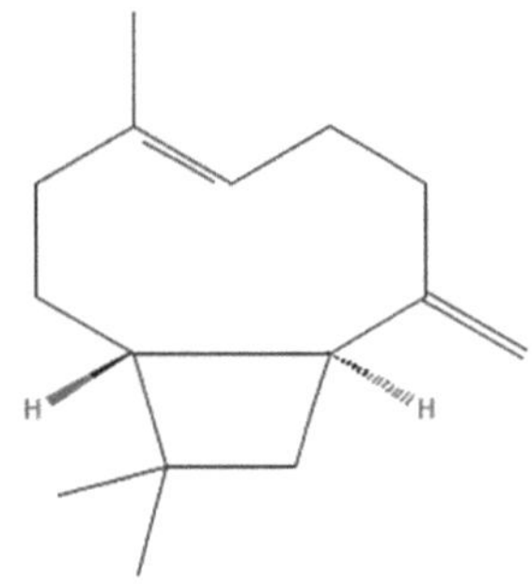

β-caryophyllene

FIGURE 12.2 The structure of piperine and β-caryophyllene.

12.5 PHARMACOLOGICAL STUDIES

Piperine and other constituents isolated from black pepper have several important pharmacological properties against various diseases and metabolic conditions. The important phytochemical nutrients and their pharmacological effects have been listed in Table 12.2.

12.6 ANTI-DIABETIC RESPONSES

Diabetes is a metabolic disorder, which is characterized by an insulin deficiency or insensitivity which can compromise the functioning of multiple organs. Studies have shown that piperine improves insulin levels and the management of hyperglycaemia. Piperine is the most potent component isolated from black pepper. Its lipophilic nature facilitates its absorption through the intestinal wall (Sarfraz et al., 2017). Piperine has also been known to enhance the bioavailability of drugs such as phenobarbitone, rifampicin, pyrazinamide, beta-lactam antibiotics, cefotaxime, amoxicillin, and fluoroquinolones (Atal et al., 2016). In a recent study by Atal et al. (2016), piperine in combination with sub-therapeutic concentrations of metformin was found to significantly reduce the blood glucose level. The essential oil from black pepper is known to have anti-diabetic properties by inhibiting the activities of α-amylase and α-glucosidase, two key enzymes involved in the increase of blood glucose availability through the breakdown of starch (Oboh et al., 2013). The crude extracts were found to manage the diabetic condition induced by streptozotocin (Bandigari et al., 2018).

In different studies, curcumin was found to reduce diabetic cardiomyopathy due to injury to cardiomyocytes, as well as to suppress the apoptosis of neuronal cells by regulating the NFκB and Nrf2/OH-1 (nuclear factor erythroid-related factor2)signalling pathways (Wu et al., 2022, Elsayed et al., 2023). Hyperglycaemia is a known cause of aggravation of preexisting conditions by inducing oxidative stress, inflammation, and apoptosis by enhancing TLR-4/NFκB and blocking Nrf2/OH-1 signalling pathways (Gong et al., 2019). Interestingly, in a different study, the administration of black pepper extract inhibited allergic responses in the respiratory system induced by ovalbumin by enhancing Nrf2/OH-1 activation, leading to anti-inflammatory activity (Bui et al., 2020). This study shows the potential of black pepper extract in regulating one of the key pathways, Nrf2/OH-1, involved in alleviating the effects of diabetes.

PPARγ (peroxisome proliferator-activated receptor-γ) is a nuclear hormone receptor and is known to play a crucial role in both lipid and glucose metabolism. It has been a target for the thiazoledinediones (TZDs) used in the treatment of type 2 diabetes (Larsen et al., 2003). Piperine and its derivatives are well known for their anti-adipogenesis activity by inhibiting the differentiation of fat cells; furthermore, they downregulate PPARγ activity and gene expression (Wang et al., 2020).

A study by Maeda et al. (2018) illustrated the improvement of postprandial hyperglycaemia *in vivo*, through CAMKKβ (Ca^{2+}/calmodulin-dependent protein kinase kinase-beta) -mediated AMPK

TABLE 12.2
The Main Phytochemical Constituents and Their Pharmacological Activities Reported in Black Pepper Seeds

Phytochemical Constituents	Pharmacological Activities	Pathways	Key Reference
Piperine	Antihypertensive activity and cardioprotective effect	-	Wang et al., 2021
	Anti-allergic effect	To inhibit Th2/Th17 responses, mast cell activation, and reduce proinflammatory cytokine levels	Bui et al., 2017
		To improve the epithelial barrier dysfunction via enhancing the activation of Nrf2/HO-1 signalling	Bui et al., 2020
	Anti-cancer activity	To reverse multidrug-resistant cancer cells (MCF-7 and A549 cells) via inhibiting transcription of the correspondıng ABC transporter genes	Li et al., 2011
		To inhibit cell proliferation in PANC-1 human pancreatic cancer cells via inducing G0/G1 arrest and down-regulating cyclin B1, cyclin D1, survivin, and Forkhead box M1 (FoxM1) levels	Jeong et al., 2021
		To inhibit cell proliferation and migration in HCT116, SW480 and DLD-1 colorectal tumour cell lines via impeding canonical Wnt signalling pathway	de Almeida et al.,2020
		To reduce SREBP-1 and FAS expression and suppress EGF-induced MMP-9 expression in human HER2-overexpressing breast cancer cells	Do et al., 2013
		To inhibit the proliferation of prostate cancer DU145, PC-3 and LNCaP cells and induce cell cycle arrest in these cells by downregulating cyclin D1 and cyclin A	Ouyang et al., 2013
	Anti-inflammatory activity	Inhibited expression of IL-2 and IFγ production in PBMC cells stimulated with PHA	Chuchawankul et al., 2012
	Hepatoprotective activity	To reduce CCl_4-induced injury in mice via increasing catalase (CAT), glutathione (GSH) and total superoxide dismutase (T-SOD) activities and reversing CCl_4-elevated total bilirubin (TBIL), glutamate pyruvate transaminase (ALT), aspartate aminotransferase (AST), alkaline phosphatase (AKP), and malondialdehyde (MDA) level	Zhang et al., 2021

(*Continued*)

TABLE 12.2 (CONTINUED)
The Main Phytochemical Constituents and Their Pharmacological Activities Reported in Black Pepper Seeds

Phytochemical Constituents	Pharmacological Activities	Pathways	Key Reference
	Analgesic and anticonvulsant activities	To induce analgesic activity and anti-inflammatory activity in acetic acid induced writhing test and carrageenan-induced paw inflammation in rats, respectively	Tasleem et al., 2014
		To inhibit acetic acid-induced writhing and reverse the analgesic effect and indicate anticonvulsant effect in pentylenetetrazole (PTZ)- and picrotoxin (PIC)-induced seizure models	Bukhari et al., 2013
Piperine-like alkamides (piperettine and piperylin)	Neuroprotective effect	To increase endogenous BDNF protein levels and induce neurite outgrowth in the Neuro-2a cell line	Yun et al., 2018
		To increase PSD-95, p-CaMK II, CaM, and NR2B levels in Permanent Focal Cerebral Ischemia Injury in Rats	Hua et al., 2019
		To activate the AKT/mTOR signalling pathway and suppress autophagy in ischemic stroke mice	Zhang et al., 2020
Piperine, piperettine, piperettyline and feruperine	Neuroprotective effect	Anticholinesterases and antioxidant via inhibiting AChE and BChE activities and enhance DPPH radical-scavenging activity	Tu et al., 2016
A novel piperine derivative (HJ22)	Neuroprotective effect	To induce Keap1-Nrf2-ARE signalling pathway and ameliorate IBO-induced cognitive impairment by alleviating cholinergic damage, oxidative stress, apoptosis and neuroinflammation in Alzheimer's disease	Yang et al., 2020
Piperlonguminine (PE)	Neuroprotective effect	To inhibit NF-κB and MAPK signalling pathways in rats with cerebral ischemia	Yang et al., 2014
		To ameliorate amyloid beta(1–42)-induced spatial memory impairment by attenuation of oxidative stress in the rat hippocampus	Hritcu et al., 2014
	Anti-cancer activity	To up-regulate p53, and down-regulate estrogen receptor (ER), E-cadherin (E-cad), matrix metalloproteinase 9 (MMP-9), matrix metalloproteinase 2 (MMP-2), c-Myc, and vascular endothelial growth factor (VEGF) levels in rats with breast cancer.	Deng et al., 2016

(*Continued*)

TABLE 12.2 (CONTINUED)
The Main Phytochemical Constituents and Their Pharmacological Activities Reported in Black Pepper Seeds

Phytochemical Constituents	Pharmacological Activities	Pathways	Key Reference
		To induce colorectal (HCT116) and lung (A549) cancer cell line apoptosis via increasing ROS production	Tammina et al., 2017
(−)-Kusunokinin, piperloguminine	Anti-cancer activity	To induce cell apoptosis in breast cancer cells (MCF-7 and MDA-MB-468) and colorectal cells (SW-620) via inducing cell arrest at G2/M phase	Sriwiriyajan et al., 2017
		To induce cytotoxic and antiproliferative effect on MCF-7 cells via increasing ROS overproduction and cell cycle arrest at G1/S	de Souza Grinevicius et al., 2016
Alkyl amides	Anti-cancer activity	To exhibit anti-inflammatory, antioxidant and anticancer activities in human tumour cell via suppressing TNF-induced NF-κB activation and inhibiting LPO and COX enzymes	Liu et al., 2010
Piperine and piperic acid	Antioxidant and antimicrobial activities	-	Zarai et al., 2013
Piperine alkaloid		To induce the inhibition of Pseudomonas aeruginosa	Alves et al., 2022
Pipernigramides A–G, pipernigrester A	Anti-inflammatory activity	To inhibit inducible nitric oxide synthase (iNOS)-mediated NO and suppress IL-1β, IL-6, TNF-α, and PGE2 release in RAW 264.7 cells stimulated by lipopolysaccharide	Pei et al., 2020
Caryophyllene and limonene		To reduce acute inflammation in chronic models of inflammation via increasing superoxide dismutase, glutathione and glutathione reductase enzyme levels	Jeena et al., 2014
Piperine alkaloids		To suppress triglycerides (TG), aminotranferases (AST, ALT), alkaline phosphatase (ALP), and bilirubin levels in ethanol-CCl4 induced hepatotoxicity Wistar rats	Nirwane., 2012

(adenosine monophosphate-activated protein kinase)-dependent, GLUT-4 (glucose transporter-4) translocation during treatment with piperine. Piperine was also found to stimulate the expression of glucagon-like protein-1 (GLP-1), which plays an important role in insulin biosynthesis in entero-endocrine cells mediated through the activation of human bitter taste receptors (TAS2Rs) (Huang et al., 2022).

12.7 TRADITIONAL AND OTHER POTENTIAL USES

Black pepper is one of the most widely used food spices across the globe. It has been known to be used in traditional medicine as a remedy against gastric ailments, indigestion, diarrhoea, and fever, as well as to ease pulmonary discomfort (Khan et al., 2021). It was also found that addition of black pepper in food can induce lipase, protease, and pancreatic amylase activity. It has also be shown to stimulate gastrointestinal tract and intestinal absorption (Srinivasan, 2007; Ashokkumar et al., 2021; Feng et al., 2014; Han et al., 2011). The various potential biological activities are contributed to by the chief bioactive molecule, piperine, and essential oils isolated from black pepper. Studies have shown that piperine has high antioxidant potential and tends to protect mice fed a high-fat diet against lipid peroxidation and inflammation (Vijayakumar et al., 2004; Zarai et al., 2013; Jeena et al., 2014; Ashokkumar et al., 2021; Chuchawankul et al., 2012). A potent antibacterial property was shown by piperine, phenolic compounds, and essential oils from black pepper against both Gram-positive and -negative bacteria (Pradhan et al., 1999; Karsha and Lakshmi, Zhang et al., 2017; Morsy et al., 2017). Piperine also exhibited anticancer activity against lung cancer and androgen-dependent and -independent prostate cancer, as well as osteosarcoma (Vijayakumar et al., 2004). In two separate studies, black pepper derivatives were found to show insect-repellent and insecticidal activities (Upadhyay and Jaiswal., 2007; Naseem et al., 2011).

12.8 SAFETY ISSUES

Too much of anything is harmful for the body. A recent study revealed that a bolus intake of piperine in excess of normal human consumption showed undesirable effects on reproductive health, including disturbances in spermatogenesis as well as embryotoxicity (Ziegenhagen et al., 2021). Hence, it is imperative to take any substance, natural or synthetic, with caution.

12.9 CONCLUSION AND FINAL REMARKS

Black pepper seeds possess the key to alleviate several metabolic and other ailments. This chapter describes various studies that have illustrated the bioactive phytochemicals and the underlying mechanism mediating these effects. The chief phytochemical, piperine, and others have been discussed in brief. Regular consumption of black pepper as part of the diet can help in controlling various acute to chronic disorders which have been advocated by traditional medicine and supported by modern research, though more in-depth studies, including animal studies and clinical trials, have to be carried out to analyse the optimum dosage of the bio-actives derived from black pepper for future therapeutic uses.

REFERENCES

Alves, F.S., Cruz, J.N., de Farias Ramos, I.N., do Nascimento Brandão, D.L., Queiroz, R.N., da Silva, G.V., da Silva, G.V., Dolabela, M.F., da Costa, M.L., Khayat, A.S. and de Arimatéia Rodrigues do Rego, J., 2022. Evaluation of antimicrobial activity and cytotoxicity effects of extracts of *Piper nigrum* L. and piperine. *Separations*, 10(1), 21.

Ashokkumar, K., Murugan, M., Dhanya, M.K., Pandian, A. and Warkentin, T.D., 2021. Phytochemistry and therapeutic potential of black pepper [*Piper nigrum* (L.)] essential oil and piperine: A review. *Clinical Phytoscience*, 7(1), 1–11.

Atal, S., Atal, S., Vyas, S. and Phadnis, P., 2016. Bio-enhancing effect of piperine with metformin on lowering blood glucose level in alloxan induced diabetic mice. *Pharmacognosy Research*, 8(1), 56.

Bagheri, H., Manap, M.Y.B.A. and Solati, Z., 2014. Antioxidant activity of *Piper nigrum* L. essential oil extracted by supercritical CO_2 extraction and hydro-distillation. *Talanta*, 121, 220–228.

Bandigari, P., Mohammed, A., Arikilla, S., Chanduri, B., Yema, S. and Chinnala, K., 2018. Evaluation of anti-diabetic activity of seeds of black pepper in streptozotocin induced diabetic rats. *European Journal of Biomedical and Pharmaceutical Sciences, ISSN*, 2349–8870.

Bui, T.T., Fan, Y., Piao, C.H., Van Nguyen, T., Shin, D.U., Jung, S.Y., Hyeon, E., Song, C.H., Lee, S.Y., Shin, H.S. and Chai, O.H., 2020. *Piper nigrum* extract improves OVA-induced nasal epithelial barrier dysfunction via activating Nrf2/HO-1 signaling. *Cellular Immunology*, 351, 104035.

Bui, T.T., Piao, C.H., Song, C.H., Shin, H.S., Shon, D.H. and Chai, O.H., 2017. *Piper nigrum* extract ameliorated allergic inflammation through inhibiting Th2/Th17 responses and mast cells activation. *Cellular Immunology*, 322, 64–73.

Bukhari, I.A., Alhumayyd, M.S., Mahesar, A.L. and Gilani, A.H., 2013. The analgesic and anticonvulsant effects of piperine in mice. *Journal of Physiology and Pharmacology*, 64(6), 789.

Butt, M.S., Pasha, I., Sultan, M.T., Randhawa, M.A., Saeed, F. and Ahmed, W., 2013. Black pepper and health claims: A comprehensive treatise. *Critical Reviews in Food Science and Nutrition*, 53(9), 875–886.

Chuchawankul, S., Khorana, N. and Poovorawan, Y., 2012. Piperine inhibits cytokine production by human peripheral blood mononuclear cells. *Genetics and Molecular Research*, 11(1), 617–627.

Chun, H., Shin, D.H., Hong, B.S., Dai Cho, W., Cho, H.Y. and Yang, H.C., 2002. Biochemical properties of polysaccharides from black pepper. *Biological and Pharmaceutical Bulletin*, 25(9), 1203–1208.

De Almeida, G.C., Oliveira, L.F., Predes, D., Fokoue, H.H., Kuster, R.M., Oliveira, F.L., Mendes, F.A. and Abreu, J.G., 2020. Piperine suppresses the Wnt/β-catenin pathway and has anti-cancer effects on colorectal cancer cells. *Scientific Reports*, 10(1), 11681.

De Souza Grinevicius, V.M.A., Kviecinski, M.R., Mota, N.S.R.S., Ourique, F., Castro, L.S.E.P.W., Andreguetti, R.R., Correia, J.F.G., Wilhem Filho, D., Pich, C.T. and Pedrosa, R.C., 2016. *Piper nigrum* ethanolic extract rich in piperamides causes ROS overproduction, oxidative damage in DNA leading to cell cycle arrest and apoptosis in cancer cells. *Journal of Ethnopharmacology*, 189, 139–147.

Deng, Y., Sriwiriyajan, S., Tedasen, A., Hiransai, P. and Graidist, P., 2016. Anti-cancer effects of *Piper nigrum* via inducing multiple molecular signaling in vivo and in vitro. *Journal of Ethnopharmacology*, 188, 87–95.

Do, M.T., Kim, H.G., Choi, J.H., Khanal, T., Park, B.H., Tran, T.P., Jeong, T.C. and Jeong, H.G., 2013. Antitumor efficacy of piperine in the treatment of human HER2-overexpressing breast cancer cells. *Food Chemistry*, 141(3), 2591–2599.

Elsayed, H.R.H., Rabei, M.R., Elshaer, M.M.A., El Nashar, E.M., Alghamdi, M.A., Al-Qahtani, Z. and Nabawy, A., 2023. Suppression of neuronal apoptosis and glial activation with modulation of Nrf2/HO-1 and NF-kB signaling by curcumin in streptozotocin-induced diabetic spinal cord central neuropathy. *Frontiers in Neuroanatomy*, 17, 1094301.

Feng, X., Liu, Y., Wang, X. and Di, X., 2014. Effects of piperine on the intestinal permeability and pharmacokinetics of linarin in rats. *Molecules*, 19(5), 5624–5633.

Gong, D.J., Wang, L., Yang, Y.Y., Zhang, J.J. and Liu, X.H., 2019. Diabetes aggravates renal ischemia and reperfusion injury in rats by exacerbating oxidative stress, inflammation, and apoptosis. *Renal Failure*, 41(1), 750–761.

Gorgani, L., Mohammadi, M., Najafpour, G.D. and Nikzad, M., 2017. Piperine—The bioactive compound of black pepper: from isolation to medicinal formulations. *Comprehensive Reviews in Food Science and Food Safety*, 16(1), 124–140.

Goswami, A. and Malviya, N., 2020. Reassessing the restorative prospectives of the king of spices black pepper. *Journal of Drug Delivery and Therapeutics*, 10(3), 312–321.

Han, H.K., 2011. The effects of black pepper on the intestinal absorption and hepatic metabolism of drugs. *Expert Opinion on Drug Metabolism and Toxicology*, 7(6), 721–729.

Hao, C.Y., Rui, F.A.N., Ribeiro, M.C., Tan, L.H., Wu, H.S., Yang, J.F., Zheng, W.Q. and Huan, Y., 2012. Modeling the potential geographic distribution of black pepper (*Piper nigrum*) in Asia using GIS tools. *Journal of Integrative Agriculture*, 11(4), 593–599.

Hritcu, L., Noumedem, J.A., Cioanca, O., Hancianu, M., Kuete, V. and Mihasan, M., 2014. Methanolic extract of *Piper nigrum* fruits improves memory impairment by decreasing brain oxidative stress in amyloid beta (1–42) rat model of Alzheimer's disease. *Cellular and Molecular Neurobiology*, 34(3), 437–449.

Hua, S., Wang, B., Chen, R., Zhang, Y., Zhang, Y., Li, T., Dong, L. and Fu, X., 2019. Neuroprotective effect of dichloromethane extraction from *Piper nigrum* L. and *Piper longum* L. on permanent focal cerebral ischemia injury in rats. *Journal of Stroke and Cerebrovascular Diseases*, 28(3), 751–760.

Huang, T.T., Gu, P.P., Zheng, T., Gou, L.S. and Liu, Y.W., 2022. Piperine, as a TAS2R14 agonist, stimulates the secretion of glucagon-like peptide-1 in the human enteroendocrine cell line Caco-2. *Food and Function*, 13(1), 242–254.

Jeena, K., Liju, V.B., Umadevi, N.P. and Kuttan, R., 2014. Antioxidant, anti-inflammatory and antinociceptive properties of black pepper essential oil (*Piper nigrum* Linn). *Journal of Essential Oil Bearing Plants*, 17(1), 1–12.

Jeleń, H.H. and Gracka, A., 2015. Analysis of black pepper volatiles by solid phase microextraction–gas chromatography: A comparison of terpenes profiles with hydrodistillation. *Journal of Chromatography A*, 1418, 200–209.

Jeong, J.H., Ryu, J.H. and Lee, H.J., 2021. In vitro inhibition of *Piper nigrum* and piperine on growth, migration, and invasion of PANC-1 human pancreatic cancer cells. *Natural Product Communications*, 16(11), 1934578X211057694.

Khan, A.U., Talucder, M.S.A., Das, M., Noreen, S. and Pane, Y.S., 2021. Prospect of the black pepper (*Piper nigrum* L.) as natural product used to an herbal medicine. *Open Access Macedonian Journal of Medical Sciences*, 9(F), 563–573.

Larsen, T.M., Toubro, S. and Astrup, A., 2003. PPARgamma agonists in the treatment of type II diabetes: Is increased fatness commensurate with long-term efficacy? *International Journal of Obesity*, 27(2), 147–161.

Li, S., Lei, Y., Jia, Y., Li, N., Wink, M. and Ma, Y., 2011. Piperine, a piperidine alkaloid from *Piper nigrum* re-sensitizes P-gp, MRP1 and BCRP dependent multidrug resistant cancer cells. *Phytomedicine*, 19(1), 83–87.

Liu, Y., Yadev, V.R., Aggarwal, B.B. and Nair, M.G., 2010. Inhibitory effects of black pepper (*Piper nigrum*) extracts and compounds on human tumor cell proliferation, cyclooxygenase enzymes, lipid peroxidation and nuclear transcription factor-kappa-B. *Natural Product Communications*, 5(8), 1934578X1000500822.

Madato, T., Petlamul, W. and Mahamad, K., 2022. Efficiency of smart farm system on enhancement of pepper production. *Naresuan University Journal: Science and Technology (NUJST)*, 30(4), 53–65.

Maeda, A., Shirao, T., Shirasaya, D., Yoshioka, Y., Yamashita, Y., Akagawa, M. and Ashida, H., 2018. Piperine promotes glucose uptake through ROS-dependent activation of the CAMKK/AMPK signaling pathway in skeletal muscle. *Molecular Nutrition and Food Research*, 62(11), 1800086.

Milenković, A.N. and Stanojević, L.P., 2021. Black pepper: Chemical composition and biological activities. *Advanced Technologies*, 10(2), 40–50.

Morsy, N.F. and Abd El-Salam, E.A., 2017. Antimicrobial and antiproliferative activities of black pepper (*Piper nigrum* L.) essential oil and oleoresin. *Journal of Essential Oil Bearing Plants*, 20(3), 779–790.

Murthy, C.T. and Bhattacharya, S., 2008. Cryogenic grinding of black pepper. *Journal of Food Engineering*, 85(1), 18–28.

Nag, A. and Chowdhury, R.R., 2020. Piperine, an alkaloid of black pepper seeds can effectively inhibit the antiviral enzymes of Dengue and Ebola viruses, an *in-silico* molecular docking study. *VirusDisease*, 31(3), 308–315.

Naseem, M.T. and Khan, R.R., 2011. Comparison of repellency of essential oils against red flour beetle *Tribolium castaneum* Herbst (Coleoptera: Tenebrionidae). *Journal of Stored Products and Postharvest Research*, 2(7), 131–134.

Nelson, S.C. and Cannon-Eger, K.T., 2011. Black pepper. *Farm and Forestry Production*, 45(1), 14.

Nirwane, A.M. and Bapat, A.R., 2012. Effect of methanolic extract of *Piper nigrum* fruits in ethanol-CCl_4-induced hepatotoxicity in Wistar rats. *Der Pharmacia Lettre*, 4(3), 795–802.

Oboh, G., Ademosun, A.O., Odubanjo, O.V. and Akinbola, I.A., 2013. Antioxidative properties and inhibition of key enzymes relevant to type-2 diabetes and hypertension by essential oils from black pepper. *Advances in Pharmacological and Pharmaceutical Sciences*, 926047.

Ouyang, D.Y., Zeng, L.H., Pan, H., Xu, L.H., Wang, Y., Liu, K.P. and He, X.H., 2013. Piperine inhibits the proliferation of human prostate cancer cells via induction of cell cycle arrest and autophagy. *Food and Chemical Toxicology*, 60, 424–430.

Pei, H., Xue, L., Tang, M., Tang, H., Kuang, S., Wang, L., Ma, X., Cai, X., Li, Y., Zhao, M. and Peng, A., 2020. Alkaloids from black pepper (*Piper nigrum* L.) exhibit anti-inflammatory activity in murine macrophages by inhibiting activation of NF-κB pathway. *Journal of Agricultural and Food Chemistry*, 68(8), 2406–2417.

Pradhan, K.J., Variyar, P.S. and Bandekar, J.R., 1999. Antimicrobial activity of novel phenolic compounds from green pepper (*Piper nigrum* L.). *LWT – Food Science and Technology*, 32(2), 121–123.

Sarfraz, M., Khaliq, T., Khan, J.A. and Aslam, B., 2017. Effect of aqueous extract of black pepper and ajwa seed on liver enzymes in alloxan-induced diabetic Wister albino rats. *Saudi Pharmaceutical Journal*, 25(4), 449–452.

Siddiqui, B.S., Gulzar, T., Begum, S. and Afshan, F., 2004. Piptigrine, a new insecticidal amide from Piper nigrum Linn. *Natural Product Research*, 18(5), 473–477.

Srinivasan, K., 2007. Black pepper and its pungent principle-piperine: A review of diverse physiological effects. *Critical Reviews in Food Science and Nutrition*, 47(8), 735–748.

Sriwiriyajan, S., Sukpondma, Y., Srisawat, T., Madla, S. and Graidist, P., 2017. (–)-Kusunokinin and piperloguminine from *Piper nigrum*: An alternative option to treat breast cancer. *Biomedicine and Pharmacotherapy*, 92, 732–743.

Sudeep, H.V., Venkatakrishna, K., Amritharaj, Gouthamchandra, K., Reethi, B., Naveen, P., Lingaraju, H.B. and Shyamprasad, K., 2021. A standardized black pepper seed extract containing β -caryophyllene improves cognitive function in scopolamine-induced amnesia model mice via regulation of brain-derived neurotrophic factor and MAPK proteins. *Journal of Food Biochemistry*, 45(12), e13994.

Tammina, S.K., Mandal, B.K., Ranjan, S. and Dasgupta, N., 2017. Cytotoxicity study of *Piper nigrum* seed-mediated synthesized SnO_2 nanoparticles towards colorectal (HCT116) and lung cancer (A549) cell lines. *Journal of Photochemistry and Photobiology B: Biology*, 166, 158–168.

Tasleem, F., Azhar, I., Ali, S.N., Perveen, S. and Mahmood, Z.A., 2014. Analgesic and anti-inflammatory activities of *Piper nigrum* L. *Asian Pacific Journal of Tropical Medicine*, 7, S461–S468.

Tiwari, A., Mahadik, K.R. and Gabhe, S.Y., 2020. Piperine: A comprehensive review of methods of isolation, purification, and biological properties. *Medicine in Drug Discovery*, 7, 100027.

Tu, Y., Zhong, Y., Du, H., Luo, W., Wen, Y., Li, Q., Zhu, C. and Li, Y., 2016. Anticholinesterases and antioxidant alkamides from *Piper nigrum* fruits. *Natural Product Research*, 30(17), 1945–1949.

Upadhyay, R.K. and Jaiswal, G., 2007. Evaluation of biological activities of *Piper nigrum* oil against Tribolium castaneum. *Bulletin of Insectology*, 60(1), 57.

Vadivel, V., Ravichandran, N., Rajalakshmi, P., Brindha, P., Gopal, A. and Kumaravelu, C., 2018. Microscopic, phytochemical, HPTLC, GC–MS and NIRS methods to differentiate herbal adulterants: Pepper and papaya seeds. *Journal of Herbal Medicine*, 11, 36–45.

Vieira, L.V., Juvenato, M.E.M., Krause, M., Heringer, O.A., Ribeiro, J.S., Brandão, G.P., Kuster, R.M. and Carneiro, M.T.W., 2022. The effects of drying methods and harvest season on piperine, essential oil composition, and multi-elemental composition of black pepper. *Food Chemistry*, 390, 133148.

Vijayakumar, R., Surya, D. and Nalini, N., 2004. *Antioxidant efficacy of black pepper (Piper nigrum* L.) and piperine in rats with high fat diet induced oxidative stress. *Redox Report*, 9(2), 105–110.

Wang, D., Zhang, L., Huang, J., Himabindu, K., Tewari, D., Horbańczuk, J.O., Xu, S., Chen, Z. and Atanasov, A.G., 2021. Cardiovascular protective effect of black pepper (*Piper nigrum* L.) and its major bioactive constituent piperine. *Trends in Food Science and Technology*, 117, 34–45.

Wang, Y., Yao, Y., Liu, J., Wu, L., Liu, T., Cui, J. and Lee, D.Y.W., 2020. Synthesis and biological activity of piperine derivatives as potential PPARγ agonists. *Drug Design, Development and Therapy*, 2069–2078.

Wu, X., Zhou, X., Lai, S., Liu, J. and Qi, J., 2022. Curcumin activates Nrf2/HO-1 signaling to relieve diabetic cardiomyopathy injury by reducing ROS in vitro and in vivo. *The FASEB Journal*, 36(9), e22505.

Wulandari, W., Octavia, M.D., Sari, Y.N. and Rivai, H., 2021. *Black Pepper (Piper nigrum L.) Botanical Aspects, Chemical Content, Pharmacological Activities*, 6(1): 83–91.

Yang, T., Sun, S., Wang, T., Tong, X., Bi, J., Wang, Y. and Sun, Z., 2014. Piperlonguminine is neuroprotective in experimental rat stroke. *International Immunopharmacology*, 23(2), 447–451.

Yang, X., Ji, J., Liu, C., Zhou, M., Li, H., Ye, S. and Hu, Q., 2020. HJ22, a novel derivative of piperine, Attenuates ibotenic acid-induced cognitive impairment, oxidativestress, apoptosis and inflammation via inhibiting the protein-protein interaction of Keap1-Nrf2. *International Immunopharmacology*, 83, 106383.

Yoon, Y.C., Kim, S.H., Kim, M.J., Yang, H.J., Rhyu, M.R. and Park, J.H., 2015. Piperine, a component of black pepper, decreases eugenol-induced cAMP and calcium levels in non-chemosensory 3T3-L1 cells. *FEBS Open Bio*, 5(1), 20–25.

Yousuf, M., Khan, H.M.S., Rasool, F., Khan, K.U.R., Usman, F., Ghalloo, B.A., Umair, M., Babalghith, A.O., Kamran, M., Aadil, R.M. and Al Jaouni, S.K., 2022. Chemical profiling, formulation development, in vitro evaluation and molecular docking of *Piper nigrum* seed extract-loaded Emulgel for anti-aging. *Molecules*, 27(18), 5990.

Yun, Y.S., Noda, S., Takahashi, S., Takahashi, Y. and Inoue, H., 2018. Piperine-like alkamides from *Piper nigrum* induce BDNF promoter and promote neurite outgrowth in Neuro-2a cells. *Journal of Natural Medicines*, 72(1), 238–245.

Zahin, M., Bokhari, N.A., Ahmad, I., Husain, F.M., Althubiani, A.S., Alruways, M.W., Perveen, K. and Shalawi, M., 2021. Antioxidant, antibacterial, and antimutagenic activity of *Piper nigrum* seeds extracts. *Saudi Journal of Biological Sciences*, 28(9), 5094–5105.

Zarai, Z., Boujelbene, E., Salem, N.B., Gargouri, Y. and Sayari, A., 2013. Antioxidant and antimicrobial activities of various solvent extracts, piperine and piperic acid from *Piper nigrum*. *LWT – Food Science and Technology*, 50(2), 634–641.

Zhang, C., Zhao, J., Famous, E., Pan, S., Peng, X. and Tian, J., 2021. Antioxidant, hepatoprotective and anti-fungal activities of black pepper (*Piper nigrum* L.) essential oil. *Food Chemistry*, 346, 128845.

Zhang, J., Ye, K.P., Zhang, X., Pan, D.D., Sun, Y.Y. and Cao, J.X., 2017. Antibacterial activity and mechanism of action of black pepper essential oil on meat-borne *Escherichia coli*. *Frontiers in Microbiology*, 7, 2094.

Zhang, Y., He, Q., Yang, M., Hua, S., Ma, Q., Guo, L., Wu, X., Zhang, C., Fu, X. and Liu, J., 2020. Dichloromethane extraction from *Piper nigrum* L. and *P. longum* L. to mitigate ischemic stroke by activating the AKT/mTOR signaling pathway to suppress autophagy. *Brain Research*, 1749, 147047.

Ziegenhagen, R., Heimberg, K., Lampen, A. and Hirsch-Ernst, K.I., 2021. Safety aspects of the use of isolated piperine ingested as a bolus. *Foods*, 10(9), 2121.

13 Cinnamon Bark (*Cinnamomum* species)

Nayomi Deshani De Silva, Keddagoda Gamage Piyumi Wasana, and Anoja Priyadarshani Attanayake

13.1 INTRODUCTION

Cinnamon is among the oldest and best-known spices, which is used in several cultures (Bandara et al., 2012). It is a plant of the laurel family (Lauraceae) of the genus *Cinnamomum* (Hajimonfarednejad et al., 2019). The genus name *Cinnamomum* comes from the Greek word '*kinnamomon*' which refers to the meaning of 'spice' and 'sweet wood'. Cinnamon has been reported in ancient records from early as 2800 BC. Cinnamon was first introduced by Arabs to Western countries and dominated trade in past eras.

Various common names have been used for cinnamon, such as *kurundu* and *davoul kurundu* in Sinhala, *karruwa* and *lavanga pattai* in Tamil, *sanalinga* in Telugu, and *lavanga patte* in Malayalam. In Chinese, it is known as *yuhljuh-kwei*, in Latin as *Cinnamomum,* French as *canella*, and, in Spanish, it is called *Ceilan* (Charles, 2012; Dasanayake, 2016). Overall, *Cinnamomum* contains around 250 species and is spread throughout the world (Bandara et al., 2012; Kumar et al., 2019). Among the *Cinnamomum* species available, *Cinnamomum zeylanicum* Blume., *Cinnamomum cassia* L. J. Presl., *Cinnamomum loureiroi* Nees., and *Cinnamomum burmanni* (Nees & T. Nees) Blume. have great economic importance due to their multiple culinary uses as common spices throughout the world (Bandara et al., 2012; Nabavi et al., 2015; Hajimonfarednejad et al., 2019; Kumar et al., 2019). *Cinnamomum verum* J. Presl. is a synonym of *C. zeylanicum* and is known variously as true cinnamon, and Ceylon cinnamon or Sri Lankan cinnamon. *Cinnamomum aromaticum* Nees. is another synonym for *C. cassia,* and it is commonly known as Chinese cinnamon. However, these *Cinnamomum* species are generally called cinnamon.

Interestingly, a wide range of active phytoconstituents has been identified from various parts of the cinnamon plant and its essential oils. Accordingly, they are continuously being used in pharmaceutical and food-related industries (Ranasinghe et al., 2013; Nabavi et al., 2015). It is well known that, among the parts of the cinnamon plant, cinnamon bark has been widely utilized as a spice and a flavouring agent for centuries. From this usage, the traditional pharmacological properties and other potential uses of cinnamon bark have developed.

13.2 BOTANICAL DESCRIPTION

The cinnamon species are tropical evergreen trees and shrubs 10–15 m in height. The size of the trees varies from small to medium. They consist of thick scabrous bark and strong branches. High-quality cinnamon bark is obtained from three years of age onwards (Thomas & Kuruvilla, 2012). The bark has a pleasant fragrance and a sweet and warm aromatic taste (Jantan et al., 2008; Thomas & Kuruvilla, 2012). The leaves are oblong and are 7–18 cm long. The thick simple leaves have smooth margins, and rough veins parallel to each other. The leaves are red when they are young and turn deep green when they are mature. The plants produce small greenish-to-yellow flowers in panicles. The fruit is a dark-coloured drupe (Thomas & Kuruvilla, 2012) (Figure 13.1).

DOI: 10.1201/9781003387046-13

FIGURE 13.1 Cinnamon plant: (a) Cinnamon in natural habitat (b) dried cinnamon bark (c) cinnamon fruit.

13.3 DISTRIBUTION

The species of *Cinnamomum* are found in tropical and subtropical regions. They are widely distributed in Asia, with some in Australia and South and Central America. Two variants and about 50 species are found in Yunnan, Sichuan, Hubei, Guangxi, and several other provinces in China (Wang et al., 2020), where they grow usually in tropical climates. However, cinnamon species become very rare at latitudes with seasonal climatic conditions. Warm areas, including high humidity, where the average temperature reaches about 27°C are best suited to cinnamon. They are able to grow in a wide range of soils, although high-quality cinnamon can be obtained when trees are grown in sandy soils rich in organic matter. Well-drained soils are good for the growth of cinnamon. Cinnamon requires an average annual rainfall of between 1250 and 2500 mm.

C. zeylanicum is native to Sri Lanka, as well as tropical Asia, and is exotic to several African countries such as Madagascar, Nigeria, Comoros, Mauritius, Ghana, Uganda, Sierra Leone, and Tanzania. The sources of cassia cinnamon are Chinese and are distributed in Vietnam and China. *C. burmanii* is known as Indonesian cinnamon and is distributed in Java and the Sumatra region. *Cinnamomum tamala* (Buch. Ham) Th. G. G. Nees., which is known as Indian cassia, is found in the Northeastern region of Myanmar and India (Thomas & Kuruvilla, 2012; Mbaveng & Kuete, 2017).

13.4 PHYTOCHEMICAL CONSTITUENTS

Cinnamon is made up of many active phytoconstituents, upon which its properties depend. Various extracts/fractions/essential oil taken from various parts of the plant, particularly the bark, exhibit a remarkably different chemical composition with different properties (Błaszczyk et al., 2021). The chemical composition of cinnamon could vary based on several factors such as plant species, growing season, extraction method, tree part, and the stage of maturity. For example, cinnamon oil can be extracted from the middle and upper segments of the cinnamon bark more effectively, and mature cinnamon trees generate the highest yield of cinnamon *trans*-aldehyde (Błaszczyk et al., 2021; Sharifi-Rad et al., 2021). The available quantities and phytoconstituents are identified by different methods including gas chromatography-mass spectrometry (GC-MS) and high-performance liquid chromatography (HPLC). More than 80 compounds have been identified using GC-MS and are categorized under ketones, monoterpene hydrocarbons, sesquiterpene hydrocarbons, oxidized monoterpenes, and oxidized sesquiterpene hydrocarbons (Jayaprakasha & Rao, 2011; Yanakiev, 2020; Błaszczyk et al., 2021; Silva et al., 2022).

A single phytoconstituent or the synergistic effect of different phytoconstituents offers numerous pharmacological properties (Vijayakumar et al., 2020). Cytokines play a vital role as regulators of synovial inflammation. Some cytokines, including tumour necrosis factors (TNF-α) and interleukins (IL)-1, and IL-8 contribute to the promotion of synovial inflammatory responses and lead to cartilage degradation. Polyphenols can achieve anti-inflammatory properties against such cytokines. Furthermore, the antioxidant activity of cinnamon bark is due to the presence of polyphenols (Silva et al., 2022). According to Vijayakumar et al. (2020), various phytoconstituents like alkaloids, flavonoids, phenols, and terpenoids, for the most part activate the α-1-adrenergic receptors to induce β-endorphin secretion which potentiates the secretion of insulin. Furthermore, phytoconstituents in cinnamon bark favour β-cell regeneration (Vijayakumar et al., 2020).

Cinnamon offers various essential oils with multiple characteristics, each of which determines its effectiveness for various potential traditional pharmacological and other potential uses. Cinnamon has a distinctive aroma and smell from the volatile oils present in the bark or leaves. The composition of the extracted oil of various species of *Cinnamomum* has been widely investigated (Jantan & Goh, 1992; Agarwal et al., 2012; Cardoso-Ugarte et al., 2016; Kazemi & Mokhtariniya, 2016; Singh et al., 2023). In general, the essential oil extracted from cinnamon bark consists of cinnamaldehyde (Figure 13.2) as the main compound (60–80%). Other minor constituents are *o*-methoxycinnamaldehyde, *trans*-cinnamic acid, monoterpenoids, and eugenol. In addition, procyanidins, diterpenes, phenylpropanoids, polysaccharides, and mucilage are also present in the bark (Yang et al., 2005; Mohamed et al., 2011; Charles, 2012). Kazemi and Mokhtariniya (2016) reported that *C. zeylanicum* bark oil consists of cinnamic aldehyde (52.3%) and α-copaene (11.42%). In addition, δ-cadinene, styrene benzene, *cis*-calamenene, ethenyl-(CAS), *trans*-cinnamaldehyde, benzaldehyde, α-pinene, naphthalene, α-calacorene, α-curcumene, sativene, α-longipinene, and , cyclohexane, 1-methyl-3-(1-methylethyl) are found at lower concentrations (Kazemi & Mokhtariniya, 2016; Kallel et al., 2019).

These phytoconstituents in cinnamon are fascinating from the point of view of human health. Much research has reported numerous health properties that affect various diseases including diabetes mellitus, Alzheimer's disease, neurological disorders, cancers, and cardiovascular diseases. Cinnamaldehyde has been reported to aid in neuroprotection due to its ability to inhibit tau protein aggregation. Furthermore, eugenol and cinnamaldehyde have the potential to protect against gut injury under conditions of oxidative stress, inflammation, and infections (Sharma et al., 2020; Rachid et al., 2022; Silva et al., 2022).

13.5 USES AS A SPICE, CONDIMENT, AND INVOLVEMENT IN OTHER FOOD PREPARATIONS

A wide variety of spices, especially cinnamon, is widely used as food flavouring agents. Furthermore, cinnamon has anti-inflammatory and antioxidant properties because of the availability of specific bioactive compounds. As a result of its unique properties, cinnamon became a main component

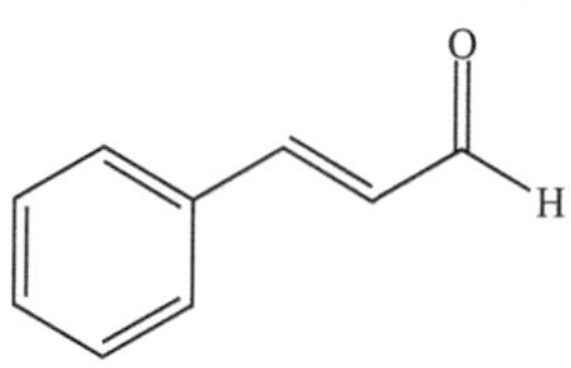

FIGURE 13.2 Chemical structure of cinnamaldehyde.

in cooking, as well as in other food preparations (Kammath et al., 2021). From the ancient world, cinnamon has been used as a common spice by different cultures. During the fourteenth century, Europeans initiated improving and maintaining meat's good quality using cinnamon. It was the quest for spices that led Europeans to explore the world in the fifteenth century. Sri Lanka had been a hub and a leading cradle of the spice trade in the prehistoric era, in which *C. zeylanicum* occupied an exalted position. Cinnamon was the key spice that attracted Europeans to Sri Lanka (Senaratne & Pathirana, 2020).

Cinnamon spice is acquired mainly by drying the central part of the bark, which is marketed as powder or quills. It is reddish brown and has a warm, sweet, spicy, woody aroma and flavour (Charles, 2012; Thomas & Kuruvilla, 2012; Ribeiro-Santos et al., 2017). Cinnamon has been used, particularly in India and China, as a condiment, seasoning, flavouring material, colouring agent, and sometimes as a preservative (Sachan et al., 2018; Moosavi-Movahedi et al., 2021). Cinnamon is the essential spice in curries, garam masala, pickles, teas, and biriyani in India and Sri Lanka (Charles, 2012). In addition, it is extensively used as a flavouring ingredient in alcoholic and non-alcoholic beverages and dessert recipes. Interesting cinnamon-based dessert recipes include apple pie, custard pie, spicy candies, caramel apples, cinnamon buns, milk, and rice pudding. Although cinnamon is used in sweet dishes in Western countries, it is primarily used in spicy dishes of beef, chicken, and lamb in the East (Kowalska et al., 2021). Cinnamon is a main spice in Persian cuisine. In addition, cinnamon is used in powder form when preparing thick soups, sweets, and drinks. In Mexico, it is used as a drink with coffee and chocolate. A stick of cinnamon is used when the Greeks prepare a beef stew, whereas European and Mediterranean regions are known to use cinnamon in a ground form. Latin Americans and Asians use cinnamon in both whole and ground forms. Chinese used ground form in a five-spice blend for soups and sauces (Bandara et al., 2012; Charles, 2012). In addition, it is added to cereals, fruits, and bread dishes in the United States. A sweet taste and a sweet sensation are given as a result of the existence of cinnamaldehyde and could be improved together with sweet foods. This is due to the synergistic effect between the sweet aroma of cinnamon and the sweet taste of sugar (Suriyagoda et al., 2021).

Cinnamon sticks are used in pickles, several beverages, and boiled beef (Thomas & Kuruvilla, 2012). Cinnamon could be mixed with rosewater or other spices to prepare curry powder to obtain a pleasant taste for stews. As a result of the antifungal and antibacterial properties of cinnamon oil, it is added to meat dishes to slow the spoilage of meat. It has considerable potential to act as a preservative for raw foods (Thomas & Kuruvilla, 2012).

While giving colour and flavour to the food by acting as a spice and a condiment, cinnamon has long been considered to possess medicinal value and has been effectively used in indigenous systems.

13.6 TRADITIONAL USES

Cinnamon has been considered a traditional herbal therapy or alternative complementary medicine in the past era. Cinnamon has traditionally been used as a remedy for diarrhoea, allergies, ulcers, arthritis, flatulence, and gastrointestinal spasms. Cinnamon has also been used for treating sore throats, coughs, colds, nausea, inflammation, rheumatism, vomiting, and menstrual disorders. Additionally, it is considered a traditional medicine for the indurations of liver, spleen, breast, stomach, uterus, and abdomen tumours (Bajpai et al., 2005; Abdel-Aziz et al., 2016). As indicated, the aim of the herbal treatment is usually to produce persisting improvements in well-being. Cinnamon bark extracts or its essential oil comprise numerous active compounds which can be used to perform important therapeutic effects and effectively treat various human diseases including diabetes mellitus, Alzheimer's diseases, heart diseases, certain types of cancers, etc. (Koppikar et al., 2010; Rafehi et al., 2012; Da Silva et al., 2019).

Bioactive compounds or extracts in species belonging to the genus *Cinnamomum* have been used in traditional Chinese therapeutics as one of the main materials. Cinnamon bark was prescribed far in the past in China. It is documented that cinnamon ameliorates nephritis, hypertension, and purulent dermatitis as well as potentiating wound healing. Furthermore, it is a remedy for even snake bites (Nagai et al., 1982; Shen et al., 2012). The combination of barks of *C. burmanii* and *Cinnamomum politum* Miq. with hot drinks are recommended to cure diarrhoea and malaria, and also give strength and relieve muscle pain. Cinnamon bark paste is used to treat headaches. In addition, cinnamon bark has been used as a remedy for toothache and to prevent bad breath. More than 600 cinnamon formulations that can be used to manage specific conditions, including piles, amoebiasis, and amenorrhoea, are mentioned in Indian traditional medicine literature (Kumar et al., 2019). In addition, cinnamon is known to treat acne and melisma. The property of the coagulant effect leads to action against haemorrhage. Tissue regeneration and blood flow in the uterus could also improve with cinnamon as a therapeutic activity (Reichling et al., 2009; Rao & Gan, 2014; Nabavi et al., 2015).

13.7 PHARMACOLOGICAL STUDIES ON CINNAMON

Numerous uses of *Cinnamomum* species in traditional medicine encouraged scientific investigations to determine the pharmacological properties associated with cinnamon bark. Scientific experimental approaches attempted to validate the potential uses of species in the *Cinnamomum* genus as sources of therapeutic remedies (Wang et al., 2020). To date, several scientific investigations have been documented on various extracts/fractions and compounds in *Cinnamomum* species, evaluating their role in antidiabetic, antibacterial, antioxidant, anti-inflammatory, antitumour, gastroprotective properties, and cholesterol-lowering effects. A wide variety of pharmacological effects have been proven using numerous preclinical and clinical studies (Kim et al., 2006; Lu et al., 2012; Hamidpour et al., 2015; Zare et al., 2019; Abeysekera et al., 2022). To contextualize the pharmacological activities of cinnamon bark, three well-documented pharmacological activities have been elucidated with cellular and molecular mechanisms. Other evaluated pharmacological properties are shown in Table 13.1. The pharmacological properties of isolated compounds are described in Table 13.2.

13.7.1 Antidiabetic Activities

As already mentioned, cinnamon bark is a promising therapeutic agent for the management of diabetes mellitus for which it has a long track record. Alterations occurred in carbohydrate, lipids, and lipoprotein metabolism, production of insulin deficiency, β-cell dysfunction, and impaired glycaemic index control caused by diabetes mellitus and leading to severe complications (Wild et al., 2004; Sharifi-Rad et al., 2021). Antidiabetic mechanisms involving phytoconstituents derived from *Cinnamomum* species have been described as stimulation of insulin secretion by pancreatic β-cells of the islets of Langerhans, α-amylase and α-glucosidase enzyme inhibition, and increase of the muscle and liver glycogen content. Furthermore, they can stimulate glycogen synthesis, stimulate insulin release, stimulate glucose metabolism, and enhance insulin receptors (Babu et al., 2007; Zhang et al., 2008).

The investigation carried out by Vijayakumar et al. (2020) revealed that the ethanol extract of *C. cassia* bark had a potent antidiabetic activity based on *in vitro*, *in vivo*, and *in silico* studies. An *in vitro* study of the antidiabetic activity of the ethanolic bark extract of *C. cassia* was carried out in terms of α-amylase (IC_{50} = 0.62 mg/mL) and α-glucosidase (IC_{50} = 0.57 mg/mL) inhibitory activity (Vijayakumar et al., 2020). The high activity of α-glucosidase inhibition is supposed to be due to the presence of total polyphenol content. Correlation analysis between the total polyphenol content and the IC_{50} value of α-glucosidase inhibition of the aqueous extract of the bark of *C. burrmanii* exhibited a negative correlation (−0.808) at the significance level of $p < 0.01$ (Ervina et al., 2019). It was also found the contributions of triterpenoids, flavonoids, luteolin, and quercetin inhibit the activity of the α-glucosidase enzyme (Lai et al., 2012; Ervina et al., 2019). *In vitro,* assessment done by Ervina et al. (2019) concluded that the bark extract of *C. burmanni* consists of polymers of

TABLE 13.1
Pharmacological Activities of Various *Cinnamomum* Species Bark Reported from 2018 to 2023

Plant Species	Extract/Fraction	Pharmacological Activity	References
Cinnamomum burmanni (Nees & T. Nees)	Ethyl acetate	Anticancer	Dina et al., 2022
	Aqueous	Antioxidant, gastroprotective, and hepatoprotective	Ervina et al., 2019; Tjandrawinata & Nailufar, 2020; Muhammad et al., 2021
	Ethanol	Antidepressant	Parisa et al., 2020
	Powder	Lowering total cholesterol level	Pulungan & Pane, 2020
Cinnamomum cassia L. J. Presl	Aqueous	Antioxidant, anticancer, anti-arthritic, and gastroprotective	Sharma et al., 2018; Sebai et al., 2019; Eweys et al., 2022
	Silver nanoparticles using aqueous extract	Antioxidant, and anticancer	El-Baz et al., 2023
	Nanoemulsion of *C. cassia* essential oil	Antioxidant, antibacterial, and anticancer	Alam et al., 2023
	Aqueous	Antitumour	Aminzadeh et al., 2022
	Chitosan nanoparticles using essential oil	Anticancer	Xu et al., 2023
	50% ethanol extract	Anticancer	Yu et al., 2019
	95% ethanol extract	Antiproliferative	Joshy et al., 2023
	Ethanol extract	Anticancer	Lin et al., 2022
	Acetone, hexane, ethyl acetate, and chloroform	Antioxidant and antiproliferative	Rad & Movafagh, 2021
Cinnamomum zeylanicum Blume.	Essential oil	Anti-inflammatory, anticancer, antiprotozoal, antibacterial antifungal, antioxidant, acaricidal, anti-Alzheimer, and antipyretic	Shahina et al., 2018; Wang et al., 2018; Alizadeh Behbahani et al., 2020; Kubatka et al., 2020; Tepe & Ozaslan, 2020; Nwanade et al., 2021; Srivastav et al., 2022; Iwiński et al., 2023
	Silver nanoparticles using aqueous or ethanol extracts	Antioxidant	Anjum et al., 2019
	Ethanol extract	Antioxidant, protective effect against ischemia-induced arrhythmias, antilipidaemic activity, and anticholinergic	Sedighi et al., 2018; Gulcin et al., 2019
	Aqueous extract	Antioxidant, antiatherosclerosis, and anticholinergic	Gulcin et al., 2019; Khalisyaseen & Mohammed, 2021
	Powder	Reducing polycystic ovary syndrome risk factors, antihyperlipidaemic, and antioxidant	Beji et al., 2018; Borzoei et al., 2018
	Titanium dioxide nanoparticles, using powder	Antibacterial	Maheswari et al., 2020

TABLE 13.2
Pharmacological Activities of Various Isolated Compounds Reported from 2018 to 2021

Plant Species	Extraction/Fraction	Isolated Compounds	Pharmacological Activity	References
Cinnamomum burmanni (Nees & T. Nees)	Methanol	Catechin, procyanidin B2, cinnamic acid, 3,4 dihydoxybenzaldehyde, quercitrin, protocatechuic acid, epicatechin, and *p*-hydroxybenzaldehyde	Antioxidant	Muhammad et al., 2021
	Natural deep eutectic solvent	*Trans*-cinnamaldehyde, and coumarin	Antioxidant	Ahmad et al., 2020
Cinnamomum cassia L. J. Presl.	Methanol	Anhydrocinnzeylanol, anhydrocinnzeylanine, cinncassiol A, cinncassiol B, cinnzeylanol, cinnzeylanone, cinnzeylanine, cinnacasside A, cinnacasside C, cinnacasside E, 18-hydroxycinnzeylanine, 2,3-anhydroanhydrocinnzeylanine, and 1-acetylcinnacassiol A	Anticancer	Pham et al., 2019
	n-Butanol	(E)-cinnamaldehyde, (Z)-cinnamaldehyde, borneol, styrene, camphene, α-pinene, β-pinene, linalool, terpinen-4-ol, α-terpineol, α-muurolene, geraniol, 3-methoxy-1,2-propanediol, coumarin, and eugenol	Antibacterial	Liang et al., 2019
	Ethanol extract	Catechin, procyanidin B1, procyanidin B2, and (E)-cinnamaldehyde, (Z)-cinnamaldehyde, cinnamyl alcohol, and coumarin	Antibacterial	Liang et al., 2019
Cinnamomum zeylanicum Blume.	Essential oil	Cinnamaldehyde, 1,8-cineole, α-pinene, 9-octadecenoic acid, *trans*–cinnamyl acetate, coumaric acid, 1,4-benzenedicarboxylic acid, phthalic acid, borneol, linalool, and eugenol	Antioxidant and antiproliferative	Kallel et al., 2019
		Trans-cinnamaldehyde and *p*-cymene	Anti-inflammatory	Schink et al., 2018

proanthocyanidin A-type, coumarins, and protonated heterodimers of the flavan-3-ol group, which can assist in complementary therapy of diabetes mellitus (Ervina et al., 2019).

Throughout the process of exploring novel herbal drug leads, the *in vitro* studies are the basis of *in vivo* and clinical trials. The *in vitro* results of Kaskoos (2019) revealed that extracted oil from the bark of *C. zeylanicum* was more effective at inhibiting α-glucosidase enzyme activity, ranging from 25.82 ± 1.51% for the concentration range 31.25 to 1000 μg/mL (Kaskoos, 2019). In addition, investigations on *in vitro* antidiabetic potential of *C. cassia*, *C. zeylanicum*, and *C. loureiroi* bark also reported α-amylase and α-glucosidase inhibitory potentials (Adisakwattana et al., 2011; Nair et al., 2013; Kang et al., 2014; Gulcin et al., 2019). Based on the results of *in vitro* studies on antidiabetic activity, various species of cinnamon markedly inhibit post-prandial hyperglycaemia by inhibiting α-amylase and α-glucosidase enzymes. The above-said enzymes play a vital role in metabolizing polysaccharides and oligosaccharides into glucose that is amenable to intestinal absorption.

Awareness of the antidiabetic activity of cinnamon bark via results from *in vitro* studies consequently leads to pursuit of *in vivo* studies. The antidiabetic and insulin-potentiating activities of the presence of soluble polyphenols in the bark of *C. verum* were proven using streptozotocin (STZ)-induced diabetes rats with an active dose of 200 mg/body weight (Krishnakumar et al., 2014).

Antidiabetic activity of the ethanolic bark extract of *C. cassia* was evaluated using STZ-induced rats by Vijaykumar et al. (2020). Adult albino Wistar rats were assigned to seven groups. Drug administration was done with 300, 400, or 500 mg/kg per day. The level of insulin was reduced in the group of STZ-induced rats due to the destruction of β-cells whereas it returned to normal after the administration of *C. cassia*. Administration of *C. cassia* facilitated the reduction in glucose formation from non-carbohydrate sources, suggesting that this might be one of the reasons for decreasing blood glucose concentration (Vijayakumar et al., 2020). Damage to the pancreatic cells of mice was ameliorated with cinnamon polyphenols at different doses of 0.3, 0.6, or 1.2 g/kg per day. The cinnamon polyphenol intervention group caused a marked reduction in blood glucose in diabetic mice (Li et al., 2013). A study by Babu et al. (2007) showed that cinnamaldehyde isolated from the volatile oil of *C. zeylanicum* bark was capable of lowering the high levels of plasma glucose in STZ-induced diabetes rats (Babu et al., 2007).

Lu et al. (2011) revealed that the intake of cinnamon extracts markedly increased the consumption of extracellular glucose in insulin-resistant HepG2 cells. The results suggested that A- and B-type procyanidin oligomers in various cinnamon extracts/fractions and essential oils have hypoglycaemic activities (Lu et al., 2011). The therapeutic efficacy of several target compounds was assessed by computational methodologies. The experimental results showed that 9-octadecenoic acid has a favourable GlideScore and greater affinity to interact with PPARα/γ, thereby enhancing insulin secretion and reducing the risk of diabetes mellitus (Vijayakumar et al., 2020).

In addition to preclinical and *in silico* studies on the antidiabetic activity of cinnamon bark, extensive clinical studies have brought it to the fore recently. Several clinical studies on cinnamon in the treatment of diabetes mellitus have been conducted due to its high remedial potential (Kirkham et al., 2009; Akilen et al., 2010; Hasanzade et al., 2013; Beejmohun et al., 2014; Sahib, 2016; Zare et al., 2019). The efficacy of cinnamon in patients with type 2 diabetes mellitus was demonstrated through a triple-blind, placebo-controlled, randomized clinical trial. In that study, patients (n=140) received cinnamon bark in 500 mg capsules for three months daily. Glycaemic outcomes, including fasting plasma glucose, fasting insulin, insulin resistance, and glycated heemoglobin, were measured before and after the intervention. The results revealed that cinnamon administration improved glycaemic outcomes (Zare et al., 2019). They proposed that cinnamon could inhibit glycogen synthase kinase-3 and lead to improved glucose uptake (Imparl-Radosevich et al., 1998; Hajimonfarednejad et al., 2018; Zare et al., 2019). The results obtained were supported by the studies of Lu et al. (2012) and Parham et al. (2020) that highlighted that cinnamon is an effective supplement to increase blood glucose control in patients with type 2 diabetes mellitus (Lu et al., 2012; Parham et al., 2020).

13.7.2 Antimicrobial Activities

Due to the antimicrobial properties of cinnamon bark and its oil, several antimicrobial components have been reported to date (Bansod & Rai, 2008; Varalakshmi et al., 2014; Ahmed et al., 2020; Cava-Roda et al., 2021). The mechanisms underlying the antimicrobial effects of the phytoconstituents in cinnamon are quite complex. Cinnamaldehyde and eugenol could inhibit β-lactamase's production by the bacterium and destroy its cell wall (Di Pasqua et al., 2007). Phytoconstituents rich in polyphenols could pass through the phospholipid bilayer of bacterial cell walls. They attach to proteins and thereby restrict their functions. Alteration of membrane permeability is caused by the destruction of the cell wall. Subsequently, this leads to coagulation of the cytoplasm and denaturation of several enzymes (Juven et al., 1994; Burt & Reinders, 2003).

Matan et al. (2006) elucidated the activity of cinnamon oil on several bacteria, namely *Pediococcus halophilus* and *Staphyloccus aureus*, the yeast species *Candida lipolytica*, *Zygosaccharomyces rouxii* and *Debaryomyces hansenii*, and the fungal species *Aspergillus flavus*, *Mucor plumbeus* and *Penicillium roqueforti* (Matan et al., 2006). The results of the above study exhibited the presence of antimicrobial properties in cinnamon oil in terms of antibacterial and antifungal agents. *In vitro* evaluation of *C. zeylanicum* bark extract and its essential oil was performed against *Escherichia coli* O157:H7, *Yersinia enterolitica* O9, and *Klebsiella pneumoniae*, using the disc diffusion method. The essential oil showed potent antibacterial activity against Gram-negative bacteria at minimum inhibitory concentration (MIC) values of 12.5 µL/mL for *E. coli* O157:H7, 6.25 µL/mL for *Y. enterolitica* O9, 1.5 µL/mL for *K. pneumoniae* (Al-Mariri & Safi, 2014).

The exploitation of *Cinnamomum* species for the biosynthesis of nanoparticles as antimicrobial agents has attracted increasing attention due to their remarkable advantages. The study on the formulation of silver nanoparticles using the bark extract of *C. zeylanicum* has proven that it is a good material for the synthesis of nano-based antimicrobial agents. The EC_{50} of formulated silver nanoparticles against *E. coli* BL-21 strain was reported to be 11 ± 1.72 mg/mL, indicating its potency as a bactericide (Sathishkumar et al., 2009). Poly- (D, L- lactide-co-glycolide) nanoparticles encapsulated with cinnamon bark extract were produced to test for activity against *Salmonella enterica* and *Listeria monocytogenes*. The nanoparticles successfully inhibited bacterial growth at one day and three days at concentrations ranging from 224.42 to 549.12 µg/mL (Hill et al., 2013).

A study by Li et al. (2020) assessed how cinnamon oil acts against the gut microbiota throughout the development of inflammatory bowel disease using a mouse model of dextran sodium sulphate-induced colitis. In that study, female pathogen-free mice were used to explore the regulatory effects of cinnamon essential oil on intestinal flora by sequencing the 16S rRNA region. The results suggested that the administration of cinnamon essential oil was effective in modifying gut microbial dysbiosis, particularly in promoting the enhancement of probiotic bacteria (Li et al., 2020). Narayanan et al. (2017) evaluated the efficacy of cinnamon *trans*-cinnamaldehyde as an inhibitory agent for uropathogenic *E. coli* in lower urinary tract infections (Narayanan et al., 2017). Female C57BL/6 mice were fed *ad libitum* with different concentrations (0.1, 0.2, or 0.4%) of *trans*-cinnamaldehyde-containing mouse diet for ten days. The results indicated that the 0.4% concentration of *trans*-cinnamaldehyde effectively reduced uropathogenic *E. coli* with around 3–3.5 log colony forming units (CFU) per gram compared with the control group (Narayanan et al., 2017).

13.7.3 Anti-inflammatory Activity

Many researchers have confirmed the advantages and efficacies of herbs in complementary medicine in the treatment of inflammatory diseases based on scientific evidence (Sharifi-Rad et al., 2021)., Herbal medicine has been used for a prolonged period to prevent or suppress inflammatory diseases. Phytochemical studies have shown that polyphenols, flavonoids, steroids, alkaloids, glycosides, resins, terpenoids, phenolic diterpenes, and curcumin are common cinnamon phytoconstituents with anti-inflammatory activity which act by various mechanisms (Gunawardena et al., 2014). Multiple studies suggested that polyphenols and their monomeric precursors, including flavan-3-ol and gallic acid derivatives, possess anti-inflammatory activities to prevent or suppress chronic inflammatory diseases (Singh et al., 2008). Polyphenols extracted from *C. zeylanicum* have been found to affect immune responses by regulating pro-inflammatory and *GLUT* gene expression. Water-soluble extracts of *C. zeylanicum* reverse TNF-α induced overproduction of intestinal apoB48 by regulating expression of genes involving inflammatory, insulin, and lipoprotein signalling pathways (Cao et al., 2008; Qin et al., 2009). Cinnamaldehyde is particularly effective at inhibiting proinflammatory mediators, including chemokines, interleukins and eicosanoids, and reactive oxygen species involved in Alzheimer's disease.

The *Cinnamomum osmophloeum* Kaneh. bark extracts prepared using solvents such as hexane, ethyl acetate, and methanol proved to be promising anti-inflammatory agents. The study by Rao et al. (2007) demonstrated via *in vitro* studies that the extracts could inhibit increased production of inflammatory mediators such as nitric oxide and cytokines (Rao et al., 2007). *C. zeylanicum* and *C. cassia* were subjected to sequential extraction using water, dichloromethane, ethanol, ethyl acetate, and methanol, and their anti-inflammatory potential was evaluated in terms of the inhibition of nitric oxide production and inhibition of TNF-α production. Among them, the ethyl acetate fraction showed higher inhibitory activity of both *C. zeylanicum* (14.0 ± 4.0 μg/mL) and *C. cassia* (19.7 ± 6.0 μg/mL) extracts for inhibition of nitric oxide production (Gunawardena et al., 2015). Furthermore, E-cinnamaldehyde and *ortho*-methoxycinnamaldehyde were identified as the most abundant compounds in *C. zeylanicum* and *C. cassia*. Indeed, it is proven that they are the most potent compounds for anti-inflammatory activity, exhibiting IC_{50} values for nitric oxide with RAW 264.7 cells of 7.3 ± 1.2 μg/mL and 5.7 ± 1.5 μg/mL respectively (Gunawardena et al., 2015).

The anti-inflammatory activity of various extracts and compounds in cinnamon has been evaluated through different animal models. Type A procyanidin polyphenols are isolated from the *C. zeylanicum* bark and its anti-inflammatory potency was determined using the carrageenan-induced rat paw oedema model by Vetal et al. (2013). At dose levels of 4, 8, or 25 mg/kg, type A procyanidin polyphenols showed significant anti-inflammatory effects (Vetal et al., 2013). The aqueous extract of *C. cassia* inhibits immunological haemolysis at a high concentration of 200 mg/kg. Gamma-irradiated rats were used to explore the defensive role of cinnamon extract in managing inflammatory injuries. In that study, the cinnamon extract was administered daily before and after radiation exposure. The results obtained revealed that cinnamon extracts significantly improved fluctuations in the liver antioxidant system and changes induced to decrease the xanthine oxidoreductase system (Azab et al., 2011).

The impact of clinical trials extends not only to individual patients by establishing a broader selection of effective therapies but also to society by enhancing the value of health care provided. Cinnamon as a food supplement could be considered a safe supplement to relieve pain and other complications of migraine. This claim was supported by Zareie et al. (2020) using a double-blind, placebo-controlled, randomized trial (Zareie et al., 2020). The prescribed dose consisted of 600 mg of dried *C. zeylanicum* bark powder. Serum levels of interleukin-6, calcitonin gene-related peptide, and nitric oxide were measured at baseline and the end of the study. A questionnaire was introduced to record the severity, frequency, and duration of pain attacks. Based on the results, cinnamon consumption significantly decreased serum levels of interleukin-6 and nitric oxide, while it also reduced the duration of attacks and frequency of migraine (Zareie et al., 2020). Furthermore, Shishehbor et al. (2018) conducted a randomized, double-blind clinical trial to investigate the impact of *C. burmannii* bark on inflammatory markers in women with rheumatoid arthritis (Shishehbor et al., 2018). The results revealed that, at the end of the eight-week study period, serum levels of C-reactive protein and TNF-α were significantly decreased in the group of patients who were administered cinnamon at a dose of 500 mg/kg as compared with the placebo group. Compared with the placebo, cinnamon intake significantly reduced the disease activity score (DAS-28, $p < 0.001$) and the number of swollen joints ($p < 0.001$) (Shishehbor et al., 2018).

13.8 SYNERGISTIC EFFECT OF CINNAMON BARK ON PHARMACOLOGICAL ACTIVITIES

The synergistic effect of cinnamon bark with other phytochemicals has been scientifically supported for various pharmacological properties, including antimicrobial, anticancer, and antidiabetic activity (El Atki et al., 2019;Cava-Roda et al., 2021; Khaing et al., 2022). Cinnamon (0.025–0.4 µg/mL), together with thyme (0.05–0.8 µg/mL) or clove (0.05–0.8 µg/mL) resulted in an additive effect against *Bacillus subtilis*, *Staphylococcus aureus*, and *Bacillus cereus*, highlighting the enhanced antibacterial effect (Fei et al., 2011). A profound synergistic effect of five different concentrations of honey (95%–99% w/w) and cinnamon (1%–5% w/w) was recorded against *Streptococcus mutans*. Interestingly, all five concentrations of the combination achieved a higher zone of inhibition at around 200 mg/mL from honey/cinnamon mixtures (Rezvani et al., 2017).

More recently, Nurinda et al. (2022) revealed that the administration of a combination of cinnamon and ginger could show a better antihyperglycaemic effect than the administration of either single extract to STZ-induced diabetic rats. The combination could improve insulin levels, making it possible to manipulate blood glucose levels during hyperglycaemia (Nurinda et al., 2022). Furthermore, control of blood glucose and weight loss improved in diabetic rats at a dose of 200 mg/kg of cinnamon with green tea. Combinations of essential oils including cinnamon, cumin, oregano, and fenugreek have been exerted to intensify insulin sensitivity. The synergistic effect of combined essential oils attributes to lower circulating glucose concentration and systolic blood pressure (Gruenwald et al., 2010).

13.9 OTHER POTENTIAL USES OF CINNAMON BARK

Apart from the culinary and therapeutic uses of cinnamon bark, it has several other potential uses. Cinnamon was used to embalming mummies by Egyptians from ancient times. Furniture, plywood manufacture, and decorative materials are made using cinnamon timber. *Cinnamomum javanicum* Blume. consists of tough timber that can be used for construction (Kumar et al., 2019). In addition, mosquito coils and fragrant joss sticks are manufactured using the mucilage of cinnamon. Cinnamon bark oil has acaricidal and mite-repellent effects, hence it controls house dust mites effectively (Ribeiro-Santos et al., 2017). Furthermore, it is used as an insecticide. Cinnamon oil lends great warmth and sweetness to a fragrance and is commonly found in perfumes and soaps. Cinnamon powder is used in skincare treatments due to its astringent and moisturizing properties (Cardoso-Ugarte et al., 2016).

13.10 SAFETY ISSUES

The extracts, powder, and essential oil of cinnamon bark have been consumed by humans for their spice, culinary and therapeutic uses for thousands of years. In most cases, cinnamon was self-prescribed and bought over-the-counter or obtained from a source other than a registered health practitioner (Abdel-Aziz et al., 2016). Therefore, the safe intake of cinnamon bark has been addressed through the performance of several toxicity studies. Cell viability of the aqueous extract of *C. cassia* was assessed by the MTT assay. Peritoneal macrophages were isolated from male BALM/c mice and were exposed to cinnamon aqueous extracts (20, 100, or 500 mg/kg) for six days. The results of the MTT assay showed that cinnamon water extract was nontoxic up to 400 µg/mL concentration (Hong et al., 2012).

Yun et al. (2018) performed both *in vitro* and *in vivo* safety analyses of long-term general and genetic toxicology for freeze-dried *C. cassia* bark extract. The toxicological evaluation was carried out using Fischer 344 rats and the oral administration of the cinnamon extract (up to 2000 mg/kg) was repeated for 13 weeks continuously. The body weight of the rats remained the same, and this normal condition was reported as safe consumption of cinnamon. High-dose intake (more than

2000 mg/kg) of cinnamon extract resulted in potential nephrotoxicity and hepatotoxicity in rats as evidenced by the obvious increase in kidney or liver weight (Yun et al., 2018). Genetic toxicity indicated that cinnamon extract supplementation is neither mutagenic nor clastogenic (Yun et al., 2018).

Vijayakumar et al. (2022) conducted a study on the acute toxicity and defensive effects of *C. cassia* bark extract in an attempt to investigate dose fixation for therapeutic application. Administration of *C. cassia* ethanolic bark extracts in the dosage range 1000–5000 mg/kg to albino Wistar rats resulted in normal behaviour without any unusual signs and zero death up to 5000 mg/kg after 48 h from the treatment. It is noted that the rats were physically active after cinnamon administration. Based on the results of liver and renal markers, no significant changes were reported in the liver and kidney parameters upon supplementation of 3000 mg/kg and so the bark extract is considered a safe consumption level (Vijayakumar et al., 2022). Furthermore, the toxicity effect of the ethanolic extract of *C. cassia* bark was evaluated in STZ-induced diabetic rats at 300, 400, or 500 mg/kg concentrations. It was evident that increased levels of aspartate aminotransferase (AST), alanine aminotransferase (ALT), and alkaline phosphatase (ALP) which occurred in STZ-induced rats were decreased significantly with the administration of the ethanolic extract of *C. cassia* bark. The results of the histopathology study showed that there was no significant change in the architecture of the cells. The greatest restoration of β-cells of the islet was recorded at a dose of 400 mg/kg (Vijayakumar et al., 2022).

It is imperative to conduct well-planned safety studies in humans before advocating the regular use of plant extracts/compounds for any medicinal benefits. A phase 1 clinical trial was conducted to evaluate the pharmacodynamic properties and safety of the water extract of *C. zeylanicum* (Ranasinghe et al., 2017). Thirty healthy adults were recruited for this study which was conducted for three months with increasing daily doses of 85 mg, 250 mg, and 500 mg as a capsule at monthly intervals. There were no significant changes in the anthropometric parameters during the aforementioned period. Biochemical assessment of full blood count, kidney function tests, liver function tests, fasting blood glucose, and total cholesterol level remained within the normal range without any significant alteration during the study period, confirming the safety of consumption of water extract of *C. zeylanicum* as a herbal capsule (Ranasinghe et al., 2017).

In general, cinnamon bark is considered safe, even during pregnancy. However, the daily intake of cinnamon varies between different investigations. During pregnancy, the dosage did not exceed 0.7 mg/kg cinnamaldehyde. Dried cinnamon herb is prescribed at 0.5 g to 2 g three times per day while the essential oil is prescribed at one to three drops, three times per day. Doses could be changed according to disease conditions (Abascalj & Yarnell, 2010; Hamidpour et al., 2015). Irritation and allergic reactions in the skin or mucus membrane could occur due to excessive use of cinnamon (Hamidpour et al., 2015).

13.11 CULTIVATION PRACTICES

Cinnamon was a prominent commodity in economic life from the thirteenth to the nineteenth centuries. Currently, it has been identified as a crop that has a high export value, and cultivation proceeds on a traditional and commercial level. Cinnamon is produced worldwide at a level of about 27,000–35,000 tons annually. However, the yield depends on the type of species, variety, and tree age. Cinnamon is cultivated largely in Madagascar, the Seychelles, China, and Sri Lanka. It has also been cultivated in India and Vietnam on a small scale (Nabavi et al., 2015; Suriyagoda et al., 2021). Cinnamon can easily grow under tropical conditions on various soil types, such as loam, sandy, and gravelly soils. It is particularly amenable to growing in the range occurring from the west coast to the south coast of Sri Lanka. However, it has been observed that soil quality and climate changes could vary production and quality (Senaratne & Pathirana, 2020). For instance, sandy soils enriched with humus produce high-quality cinnamon. The optimal temperature for cinnamon cultivation is between 20 and 30°C with an annual rainfall range of 1250–2500 mm. Cinnamon is commonly propagated by seedlings and by vegetative propagation. Cinnamon production can be increased by the application of fertilizers containing urea, phosphate, and potash (Ranatunga et al., 2004).

13.12 CONCLUSION AND FUTURE REMARKS

Since ancient times, cinnamon bark has been valued as an excellent spice and condiment due to its strong fragrance. Cinnamon has a unique role in the fields of folk medicine, agriculture, and the food and pharmaceutical industries. It should be noted that the availability of a wide variety of chemical compounds, including cinnamaldehyde, contributes to its high medicinal value. Contemporary investigations have confirmed pharmacological properties including antidiabetic, antimicrobial, anticancer, anti-inflammatory, and antioxidant activities via preclinical and clinical studies. However, those investigations are restricted to very few species among the *Cinnamomum* genus. Therefore, research on different extracts/fractions/isolated compounds in other species through preclinical and clinical studies is warranted. There is a lack of clinical trials on other pharmacological properties other than for antidiabetic activity, and therefore, the clinical efficacy and safety issues of cinnamon drug lead chemicals for diverse disease conditions are unclear. In addition, only a few attempts have so far been made to synthesize cinnamon-encapsulated nanoparticles. Therefore, the formulation of drug lead/food supplements using encapsulated nanoparticles of cinnamon extracts/isolated compounds could be a crucial research area in the near future. Furthermore, it is desirable to assess the synergistic effect of cinnamon with other extracts or phytochemicals as a herbal drug.

LIST OF ABBREVIATIONS

GC-MS: Gas chromatography-mass spectrometry
HPLC: High-performance liquid chromatography
IC_{50}: Concentration (of extracts) causing 50% inhibition
STZ: Streptozotocin

REFERENCES

Abascalj, D.K. & Yarnell, E. 2010. The medicinal uses of cinnamon. *Integrative Medicine* 9(1): 29.

Abdel-Aziz, S.M., Aeron, A. & Kahil, T.A. 2016. *Health benefits and possible risks of herbal medicine: Microbes in food & health*. Springer, Cham. https://doi.org/10.1007/978-3-319-25277-3_6.

Abeysekera, W.P.K.M., Premakumara, G.A.S., Ratnasooriya, W.D. & Abeysekera, W.K.S.M. 2022. Anti-inflammatory, cytotoxicity, and antilipidemic properties: Novel bioactivities of true cinnamon (*Cinnamomum zeylanicum* Blume.) leaf. *BMC Complementary Medicine and Therapies* 22(1): 1–15.

Adisakwattana, S., Lerdsuwankij, O., Poputtachai, U., Minipun, A. & Suparpprom, C. 2011. Inhibitory activity of cinnamon bark species and their combination effect with acarbose against intestinal α-glucosidase and pancreatic α-amylase. *Plant Foods for Human Nutrition* 66(2): 143–148.

Agarwal, R., Pant, A. & Prakash, O. 2012. Chemical composition and biological activities of essential oils of *Cinnamomum tamala*, *Cinnamomum zeylenicum* and *Cinnamomum camphora* growing in Uttarakhand. In: Khemani, L., Srivastava, M., Srivastava, S. (eds) *Chemistry of phytopotentials: Health, energy and environmental perspectives*. Springer, Berlin, Heidelberg. https://doi.org/10.1007/978-3-642-23394-4_18.

Ahmad, I., Arifianti, A.E., Sakti, A.S., Saputri, F.C. & Mun'im, A. 2020. Simultaneous natural deep eutectic solvent-based ultrasonic-assisted extraction of bioactive compounds of cinnamon bark and sappan wood as a dipeptidyl peptidase IV inhibitor. *Molecules* 25(17): 3832.

Ahmed, H.M., Ramadhani, A.M., Erwa, I.Y., Ishag, O.A.O. & Saeed, M.B. 2020. Phytochemical screening, chemical composition and antimicrobial activity of *Cinnamon verum* bark. *International Research Joyrnal of Pure and Applied Chemistry* 21: 36–43.

Akilen, R., Tsiami, A., Devendra, D. & Robinson, N. 2010. Glycated haemoglobin and blood pressure-lowering effect of cinnamon in multi-ethnic type 2 diabetic patients in the UK: A randomized, placebo-controlled, double-blind clinical trial. *Diabetic Medicine* 27(10): 1159–1167.

Al-Mariri, A. & Safi, M. 2014. *In vitro* antibacterial activity of several plant extracts and oils against some Gram-negative bacteria. *Iranian Journal of Medical Sciences* 39(1): 36.

Alam, A., Ansari, M.J., Alqarni, M.H., Salkini, M.A. & Raish, M. 2023. Antioxidant, antibacterial, and anticancer activity of ultrasonic nanoemulsion of *Cinnamomum cassia* L. essential oil. *Plants* 12(4): 834.

Alizadeh Behbahani, B., Falah, F., Lavi Arab, F.,Vasiee, M. & Tabatabaee Yazdi, F. 2020. Chemical composition and antioxidant, antimicrobial, and antiproliferative activities of *Cinnamomum zeylanicum* bark essential oil. *Evidence-Based Complementary and Alternative Medicine* 2020: 1–8.

Aminzadeh, Z., Ziamajidi, N., Abbasalipourkabir, R., Daei, S., Helbi, S. & Moridnia, A. 2022. Antitumor activities of aqueous cinnamon extract on 5637 cell line of bladder cancer through glycolytic pathway. *International Journal of Inflammation* 2022: 1–9.

Anjum, S., Jacob, G. & Gupta, B. 2019. Investigation of the herbal synthesis of silver nanoparticles using *Cinnamon zeylanicum* extract. *Emergent Materials* 2(1): 113–122.

Azab, K.S., Mostafa, A.H.A., Ali, E.M. & Abdel-Aziz, M.A. 2011. Cinnamon extract ameliorates ionizing radiation-induced cellular injury in rats. *Ecotoxicology and Environmental Safety* 74(8): 2324–2329.

Babu, P.S., Prabuseenivasan, S. & Ignacimuthu, S. 2007. Cinnamaldehyde - A potential antidiabetic agent. *Phytomedicine* 14(1): 15–22.

Bajpai, M., Pande, A., Tewari, S. & Prakash, D. 2005. Phenolic contents and antioxidant activity of some food and medicinal plants. *International Journal of Food Sciences and Nutrition* 56(4): 287–291.

Bandara, T., Uluwaduge, I. & Jansz, E. 2012. Bioactivity of cinnamon with special emphasis on diabetes mellitus: A review. *International Journal of Food Sciences and Nutrition* 63(3): 380–386.

Bansod, S. & Rai, M. 2008. Antifungal activity of essential oils from Indian medicinal plants against human pathogenic *Aspergillus fumigatus* and *Aspergillus niger*. *World Journal of Medical Sciences* 3(2): 81–88.

Beejmohun, V., Peytavy-Izard, M., Mignon, C., Muscente-Paque, D., Deplanque, X., Ripoll, C. & Chapal, N. 2014. Acute effect of Ceylon cinnamon extract on postprandial glycemia: Alpha-amylase inhibition, starch tolerance test in rats, and randomized crossover clinical trial in healthy volunteers. *BMC Complementary and Alternative Medicine* 14: 1–11.

Beji, R.S., Khemir, S., Wannes, W.A., Ayari, K. & Ksouri, R. 2018. Antidiabetic, antihyperlipidemic and antioxidant influences of the spice cinnamon (*Cinnamomum zeylanicumon*) in experimental rats. *Brazilian Journal of Pharmaceutical Sciences* 54(2): 1–8.

Błaszczyk, N., Rosiak, A. & Kałużna-Czaplińska, J. 2021. The potential role of cinnamon in human health. *Forests* 12(5): 648.

Borzoei, A., Rafraf, M., Niromanesh, S., Farzadi, L., Narimani, F. & Doostan, F. 2018. Effects of cinnamon supplementation on antioxidant status and serum lipids in women with polycystic ovary syndrome. *Journal of Traditional and Complementary Medicine* 8(1): 128–133.

Burt, S.A. & Reinders, R.D. 2003. Antibacterial activity of selected plant essential oils against *Escherichia coli* O157: H7. *Letters in Applied Microbiology* 36(3): 162–167.

Cao, H., Urban Jr, J.F. & Anderson, R.A. 2008. Cinnamon polyphenol extract affects immune responses by regulating anti-and proinflammatory and glucose transporter gene expression in mouse macrophages. *The Journal of Nutrition* 138(5): 833–840.

Cardoso-Ugarte, G.A., López-Malo, A. & Sosa-Morales, M.E. 2016. *Cinnamon (Cinnamomum zeylanicum) essential oils. Essential oils in food preservation, flavor and safety*. Academic Press. https://doi.org/10.1016/B978-0-12-416641-7.00038-9.

Cava-Roda, R., Taboada-Rodríguez, A., López-Gómez, A., Martínez-Hernández, G.B. & Marín-Iniesta, F. 2021. Synergistic antimicrobial activities of combinations of vanillin and essential oils of cinnamon bark, cinnamon leaves, and cloves. *Foods* 10(6): 1406.

Charles, D.J. 2012. *Cinnamon. Antioxidant properties of spices, herbs and other sources*. Springer, New York. https://doi.org/10.1007/978-1-4614-4310-0_19.

Da Silva, M.L.T., Bernardo, M.A.S., Singh, J. & De Mesquita, M.F. 2019. *Beneficial uses of cinnamon in health and diseases: An interdisciplinary approach. The role of functional food security in global health*. Academic Press. https://doi.org/10.1016/B978-0-12-813148-0.00033-5.

Dasanayake, R. 2016. Cinnamon: A spice of an indigenous origin-historical study. *Narrations* 1(2): 46–58.

Di Pasqua, R., Betts, G., Hoskins, N., Edwards, M., Ercolini, D. & Mauriello, G. 2007. Membrane toxicity of antimicrobial compounds from essential oils. *Journal of Agricultural and Food Chemistry* 55(12): 4863–4870.

Dina, S., Siregar, M.F.G., Jusuf, N.K., Hasibuan, P.A., Andrijono, A., Bachtiar, A., Lumbanraja, S.N. & Nasution, I.P. 2022. Anticancer potential of cinnamon bark extract (*Cinnamomum burmanii*) with cisplatin combination against p-glycoprotein and apoptotic influx biomarkers. *Open Access Macedonian Journal of Medical Sciences* 10(A): 958–964.

El-Baz, Y., Moustafa, A., Ali, M., El-Desoky, G., Wabaidur, S. & Faisal, M. 2023. An analysis of the toxicity, antioxidant, and anti-cancer activity of cinnamon silver nanoparticles in comparison with extracts and fractions of *Cinnamomum cassia* at normal and cancer cell levels. *Nanomaterials* 13(5): 945.

El Atki, Y., Aouam, I., El Kamari, F., Taroq, A., Nayme, K., Timinouni, M., Lyoussi, B. & Abdellaoui, A. 2019. Antibacterial activity of cinnamon essential oils and their synergistic potential with antibiotics. *Journal of Advanced Pharmaceutical Technology and Research* 10(2): 63.

Ervina, M., Lie, H.S., Diva, J., Tewfik, S. & Tewfik, I. 2019. Optimization of water extract of *Cinnamomum burmannii* bark to ascertain its *in vitro* antidiabetic and antioxidant activities. *Biocatalysis and Agricultural Biotechnology* 19: 101152.

Eweys, A.S., Zhao, Y.S. & Darwesh, O.M. 2022. Improving the antioxidant and anticancer potential of *Cinnamomum cassia* via fermentation with *Lactobacillus plantarum*. *Biotechnology Reports* 36: e00768.

Fei, L., Ding, Y.C., Ye, X. & Ding, Y. 2011. Antibacterial effect of cinnamon oil combined with thyme or clove oil. *Agricultural Sciences in China* 10(9): 1482–1487.

Gruenwald, J., Freder, J. & Armbruester, N. 2010. Cinnamon and health. *Critical Reviews in Food Science and Nutrition* 50(9): 822–834.

Gulcin, I., Kaya, R., Goren, A.C., Akincioglu, H., Topal, M., Bingol, Z., Cetin Çakmak, K., Ozturk Sarikaya, S.B., Durmaz, L. & Alwasel, S. 2019. Anticholinergic, antidiabetic and antioxidant activities of cinnamon (*Cinnamomum verum*) bark extracts: Polyphenol contents analysis by LC-MS/MS. *International Journal of Food Properties* 22(1): 1511–1526.

Gunawardena, D., Govindaraghavan, S. & Münch, G. 2014. *Anti-inflammatory properties of cinnamon polyphenols and their monomeric precursors. Polyphenols in human health and disease.* Academic Press. https://doi.org/10.1016/B978-0-12-398456-2.00030-X.

Gunawardena, D., Karunaweera, N., Lee, S., Van Der Kooy, F., Harman, D.G., Raju, R., Bennett, L., Gyengesi, E., Sucher, N.J. & Münch, G. 2015. Anti-inflammatory activity of cinnamon (*C. zeylanicum* and *C. cassia*) extracts-identification of E-cinnamaldehyde and *o*-methoxycinnamaldehyde as the most potent bioactive compounds. *Food and Function* 6(3): 910–919.

Hajimonfarednejad, M., Nimrouzi, M., Heydari, M., Zarshenas, M.M., Raee, M.J. & Jahromi, B.N. 2018. Insulin resistance improvement by cinnamon powder in polycystic ovary syndrome: A randomized double-blind placebo controlled clinical trial. *Phytotherapy Research* 32(2): 276–283.

Hajimonfarednejad, M., Ostovar, M., Raee, M.J., Hashempur, M.H., Mayer, J.G. & Heydari, M. 2019. Cinnamon: A systematic review of adverse events. *Clinical Nutrition* 38(2): 594–602.

Hamidpour, R., Hamidpour, M., Hamidpour, S. & Shahlari, M. 2015. Cinnamon from the selection of traditional applications to its novel effects on the inhibition of angiogenesis in cancer cells and prevention of Alzheimer's disease, and a series of functions such as antioxidant, anticholesterol, antidiabetes, antibacterial, antifungal, nematicidal, acaracidal, and repellent activities. *Journal of Traditional and Complementary Medicine* 5(2): 66–70.

Hasanzade, F., Toliat, M., Emami, S.A. & Emamimoghaadam, Z. 2013. The effect of cinnamon on glucose of type II diabetes patients. *Journal of Traditional and Complementary Medicine* 3(3): 171–174.

Hill, L.E., Taylor, T.M. & Gomes, C. 2013. Antimicrobial efficacy of poly (DL-lactide-co-glycolide)(PLGA) nanoparticles with entrapped cinnamon bark extract against *Listeria monocytogenes* and *Salmonella typhimurium*. *Journal of Food Science* 78(4): N626–N632.

Hong, J.W., Yang, G.E., Kim, Y.B., Eom, S.H., Lew, J.H. & Kang, H. 2012. Anti-inflammatory activity of cinnamon water extract *in vivo* and *in vitro* LPS-induced models. *BMC Complementary and Alternative Medicine* 12(1): 1–8.

Imparl-Radosevich, J., Deas, S., Polansky, M.M., Baedke, D.A., Ingebritsen, T.S., Anderson, R.A. & Graves, D.J. 1998. Regulation of PTP-1 and insulin receptor kinase by fractions from cinnamon: Implications for cinnamon regulation of insulin signalling. *Hormone Research in Paediatrics* 50(3): 177–182.

Iwiński, H., Różański, H., Pachura, N., Wojciechowska, A., Gębarowski, T. & Szumny, A. 2023. *In vitro* evaluation of antiprotozoal properties, cytotoxicity effect and anticancer activity of new essential-oil based phytoncide mixtures. *Molecules* 28(3): 1395.

Jantan, I.B. & Goh, S.H. 1992. Essential oils of *Cinnamomum* species from Peninsular Malaysia. *Journal of Essential Oil Research* 4(2): 161–171.

Jantan, I.B., Moharam, Karim, B.A., Santhanam, J. & Jamal, J.A. 2008. Correlation between chemical composition and antifungal activity of the essential oils of eight *cinnamomum*. species. *Pharmaceutical Biology* 46(6): 406–412.

Jayaprakasha, G. & Rao, L.J.M. 2011. Chemistry, biogenesis, and biological activities of *Cinnamomum zeylanicum*. *Critical Reviews in Food Science and Nutrition* 51(6): 547–562.

Joshy, M.R., Lakshmi Thangavelu, M., Jeevitha, R. & Perumal, E. 2023. *In vitro* anti-proliferative and pro-apoptotic activities of *Cinnamomum cassia* bark extract on osteosarcoma cells. *Journal of Survey in Fisheries Sciences* 10(1S): 233–242.

Juven, B., Kanner, J., Schved, F. & Weisslowicz, H. 1994. Factors that interact with the antibacterial action of thyme essential oil and its active constituents. *Journal of Applied Bacteriology* 76(6): 626–631.

Kallel, I., Hadrich, B., Gargouri, B., Chaabane, A., Lassoued, S., Gdoura, R., Bayoudh, A. & Ben Messaoud, E. 2019. Optimization of cinnamon (*Cinnamomum zeylanicum* Blume.) essential oil extraction: Evaluation of antioxidant and antiproliferative effects. *Evidence-Based Complementary and Alternative Medicine* 2019: 1–11.

Kammath, A.J., Nair, B. & Nath, L.R. 2021. Curry versus cancer: Potential of some selected culinary spices against cancer with *in vitro, in vivo,* and human trials evidences. *Journal of Food Biochemistry* 45(3): e13285.

Kang, B.H., Racicot, K., Pilkenton, S. & Apostolidis, E. 2014. Evaluation of the *in vitro* anti-hyperglycemic effect of *Cinnamomum cassia*-derived phenolic phytochemicals, via carbohydrate hydrolyzing enzyme inhibition. *Plant Foods for Human Nutrition* 69(2): 155–160.

Kaskoos, R.A. 2019. GC/MS profile and *in-vitro* antidiabetic activity of *Cinnamomum zeylanicum* Blume, bark and *Trachyspermum ammi* (L.) Sprague, seeds. *Journal of Essential Oil Bearing Plants* 22(2): 535–544.

Kazemi, M. & Mokhtariniya, S. 2016. Essential oil composition of bark of *Cinnamomum zeylanicum. Journal of Essential Oil-Bearing Plants* 19(3): 786–789.

Khaing, E.M., Phaechamud, T. & Intaraphairot, T. 2022. Synergistic anticancer activity of cinnamon bark oil and imatinib mesylate combination on colorectal cancer cell lines. *Key Engineering Materials.* Trans Tech Publications. https://doi.org/10.4028/p-71k0os.

Khalisyaseen, O. & Mohammed, M.T. 2021. Identification of some antioxidant active compounds in true cinnamon (*Cinnamomum zeylanicum*) bark extract. *NVEO-Natural Volatiles and Essential Oils Journal* 8(4): 7565–7577.

Kim, S.H., Hyun, S.H. & Choung, S.Y. 2006. Anti-diabetic effect of cinnamon extract on blood glucose in db/db mice. *Journal of Ethnopharmacology* 104(1–2): 119–123.

Kirkham, S., Akilen, R., Sharma, S. & Tsiami, A. 2009. The potential of cinnamon to reduce blood glucose levels in patients with type 2 diabetes and insulin resistance. *Diabetes, Obesity and Metabolism* 11(12): 1100–1113.

Koppikar, S.J., Choudhari, A.S., Suryavanshi, S.A., Kumari, S., Chattopadhyay, S. & Kaul-Ghanekar, R. 2010. Aqueous cinnamon extract (ACE-c) from the bark of *Cinnamomum cassia* causes apoptosis in human cervical cancer cell line (SiHa) through loss of mitochondrial membrane potential. *BMC Cancer* 10: 1–12.

Kowalska, J., Tyburski, J., Matysiak, K., Jakubowska, M., Łukaszyk, J. & Krzymińska, J. 2021. Cinnamon as a useful preventive substance for the care of human and plant health. *Molecules* 26(17): 5299.

Krishnakumar, I., Issac, A., Johannah, N., Ninan, E., Maliakel, B. & Kuttan, R. 2014. Effects of the polyphenol content on the anti-diabetic activity of *Cinnamomum zeylanicum* extracts. *Food and Function* 5(9): 2208–2220.

Kubatka, P., Kello, M., Kajo, K., Samec, M., Jasek, K., Vybohova, D., Uramova, S., Liskova, A., Sadlonova, V. & Koklesova, L. 2020. Chemopreventive and therapeutic efficacy of *Cinnamomum zeylanicum* L. bark in experimental breast carcinoma: Mechanistic *in vivo* and *in vitro* analyses. *Molecules* 25(6): 1399.

Kumar, S., Kumari, R. & Mishra, S. 2019. Pharmacological properties and their medicinal uses of *Cinnamomum*: A review. *Journal of Pharmacy and Pharmacology* 71(12): 1735–1761.

Lai, Y.C., Chen, C.K., Tsai, S.F. & Lee, S.S. 2012. Triterpenes as α-glucosidase inhibitors from *Fagus hayatae. Phytochemistry* 74: 206–211.

Li, A.L., Ni, W.W., Zhang, Q.M., Li, Y., Zhang, X.,Wu, H.Y., Du, P., Hou, J.C. & Zhang, Y. 2020. Effect of cinnamon essential oil on gut microbiota in the mouse model of dextran sodium sulfate-induced colitis. *Microbiology and Immunology* 64(1): 23–32.

Li, R., Liang, T., Xu, L., Li, Y., Zhang, S. & Duan, X. 2013. Protective effect of cinnamon polyphenols against STZ-diabetic mice fed high-sugar, high-fat diet and its underlying mechanism. *Food and Chemical Toxicology* 51: 419–425.

Liang, Y., Li, Y., Sun, A. & Liu, X. 2019. Chemical compound identification and antibacterial activity evaluation of cinnamon extracts obtained by subcritical n-butane and ethanol extraction. *Food Science and Nutrition* 7(6): 2186–2193.

Lin, C.Y., Hsieh, Y.S., Chu, S.C., Hsu, L.S., Huang, S.C. & Chen, P.N. 2022. Reduction of invasion and cell stemness and induction of apoptotic cell death by *Cinnamomum cassia* extracts on human osteosarcoma cells. *Environmental Toxicology* 37(6): 1261–1274.

Lu, T., Sheng, H., Wu, J., Cheng, Y., Zhu, J. & Chen, Y. 2012. Cinnamon extract improves fasting blood glucose and glycosylated hemoglobin level in Chinese patients with type 2 diabetes. *Nutrition Research* 32(6): 408–412.

Lu, Z., Jia, Q., Wang, R., Wu, X., Wu, Y., Huang, C. & Li, Y. 2011. Hypoglycemic activities of A-and B-type procyanidin oligomer-rich extracts from different cinnamon barks. *Phytomedicine* 18(4): 298–302.

Maheswari, P., Ponnusamy, S., Harish, S., Muthamizhchelvan, C. & Hayakawa, Y. 2020. Syntheses and characterization of *Syzygium aromaticum, Elettaria cardamomum* and *Cinnamomum verum* modified TiO_2 and their biological applications. *Materials Science in Semiconductor Processing* 105: 104724.

Matan, N., Rimkeeree, H., Mawson, A., Chompreeda, P., Haruthaithanasan, V. & Parker, M. 2006. Antimicrobial activity of cinnamon and clove oils under modified atmosphere conditions. *International Journal of Food Microbiology* 107(2): 180–185.

Mbaveng, A. & Kuete, V. 2017. *Cinnamon species. Medicinal spices and vegetables from Africa.* Academic Press. https://doi.org/10.1016/B978-0-12-809286-6.00017-0.

Mohamed, S.S.H., Hansi Priscilla, D. & Thirumurugan, K. 2011. Cinnamon extract inhibits α-glucosidase activity and dampens postprandial glucose excursion in diabetic rats. *Nutrition and Metabolism* 8: 1–11.

Moosavi-Movahedi, A.A., Miran, M., Salami, M. & Emam-Djomeh, Z. 2021. *Spices as traditional remedies: Scientifically proven benefits. Rationality & scientific lifestyle for health.* University of Tehran Science and Humanities Series. Springer, Cham. https://doi.org/10.1007/978-3-030-74326-0_6.

Muhammad, D.R.A., Tuenter, E., Patria, G.D., Foubert, K., Pieters, L. & Dewettinck, K. 2021. Phytochemical composition and antioxidant activity of *Cinnamomum burmannii* Blume. extracts and their potential application in white chocolate. *Food Chemistry* 340: 127983.

Nabavi, S.F., Di Lorenzo, A., Izadi, M., Sobarzo-Sánchez, E., Daglia, M. & Nabavi, S.M. 2015. Antibacterial effects of cinnamon: From farm to food, cosmetic and pharmaceutical industries. *Nutrients* 7(9): 7729–7748.

Nagai, H., Takizawa, T., Nishiyori, T. & Koda, A. 1982. Experimental glomerulonephritis in mice as a model for immunopharmacological studies. *The Japanese Journal of Pharmacology* 32(6): 1117–1124.

Nair, S.S., Kavrekar, V. & Mishra, A. 2013. *In vitro* studies on alpha amylase and alpha glucosidase inhibitory activities of selected plant extracts. *European Journal of Experimental Biology* 3(1): 128–132.

Narayanan, A., Muyyarikkandy, M., Mooyottu, S., Venkitanarayanan, K. & Amalaradjou, M. 2017. Oral supplementation of *trans*-cinnamaldehyde reduces uropathogenic *Escherichia coli* colonization in a mouse model. *Letters in Applied Microbiology* 64(3): 192–197.

Nurinda, E., Kusumawardani, N., Wulandari, A., Fatmawati, A., Emelda, E., Nisa, H., Hasan, N., Irianti, W., Rohmah, M. & Lestari, P. 2022. Pharmacological study: Synergistic antidiabetic activity of cinnamon Bark and zingiber extract in streptozotocin-induced diabetic rats. *Open Access Macedonian Journal of Medical Sciences* 10(T8): 1–6.

Nwanade, C.F., Wang, M., Wang, T., Zhang, X., Zhai, Y., Zhang, S., Yu, Z. & Liu, J. 2021. The acaricidal activity of cinnamon essential oil: Current knowledge and future perspectives. *International Journal of Acarology* 47(5): 446–450.

Parham, M., Bagherzadeh, M., Asghari, M., Akbari, H., Hosseini, Z., Rafiee, M. & Vafaeimanesh, J. 2020. Evaluating the effect of a herb on the control of blood glucose and insulin-resistance in patients with advanced type 2 diabetes (a double-blind clinical trial). *Caspian Journal of Internal Medicine* 11(1): 12.

Parisa, N., Hidayat, R., Maritska, Z. & Prananjaya, B.A. 2020. Antidepressant effect of cinnamon (*Cinnamomum burmannii*) bark extract in chronic stress-induced rats. *Open Access Macedonian Journal of Medical Sciences* 8(A): 273–277.

Pham, V.C., Nguyen, T.T.A., Vu, T.O., Cao, T.Q., Min, B.S. & Kim, J.A. 2019. Five new diterpenoids from the barks of *Cinnamomum cassia* (L.) J. Presl. *Phytochemistry Letters* 32: 23–28.

Pulungan, A. & Pane, Y.S. 2020. The benefit of cinnamon (*Cinnamomum burmannii*) in lowering total cholesterol levels after consumption of high-fat containing foods in white mice (*Mus musculus*) models. *F1000 Research* 9: 168.

Qin, B., Dawson, H., Polansky, M. & Anderson, R. 2009. Cinnamon extract attenuates TNF-α-induced intestinal lipoprotein ApoB48 overproduction by regulating inflammatory, insulin, and lipoprotein pathways in enterocytes. *Hormone and Metabolic Research* 41(07): 516–522.

Rachid, A.P., Moncada, M., Mesquita, M.F.D., Brito, J., Bernardo, M.A. & Silva, M.L. 2022. Effect of aqueous cinnamon extract on the postprandial glycemia levels in patients with type 2 diabetes mellitus: A randomized controlled trial. *Nutrients* 14(8): 1576.

Rad, S.K. & Movafagh, A. 2021. Study of antioxidant, antiproliferative and DNA damage protecting activities of *Cinnamomum cassia* extracts obtained by sequential extraction. *Recent Patents on Food, Nutrition and Agriculture* 12(1): 45–57.

Rafehi, H.,Ververis, K. & Karagiannis, T. 2012. Controversies surrounding the clinical potential of cinnamon for the management of diabetes. *Diabetes, Obesity and Metabolism* 14(6): 493–499.

Ranasinghe, P., Galappaththy, P., Constantine, G.R., Jayawardena, R., Weeratunga, H.D., Premakumara, S. & Katulanda, P. 2017. *Cinnamomum zeylanicum* (Ceylon cinnamon) as a potential pharmaceutical agent for type-2 diabetes mellitus: Study protocol for a randomized controlled trial. *Trials* 18(1): 1–8.

Ranasinghe, P., Pigera, S., Premakumara, G., Galappaththy, P., Constantine, G.R. & Katulanda, P. 2013. Medicinal properties of 'true'cinnamon (*Cinnamomum zeylanicum*): A systematic review. *BMC Complementary and Alternative Medicine* 13(1): 1–10.

Ranatunga, J., Senanayake, U. & Wijesekera, R. 2004. *Cinnamon and cassia: The genus Cinnamonum*. CRC Press.

Rao, P.V. & Gan, S.H. 2014. Cinnamon: A multifaceted medicinal plant. *Evidence-Based Complementary and Alternative Medicine* 2014: 1–11.

Rao, Y.K., Fang, S.H. & Tzeng, Y.M. 2007. Evaluation of the anti-inflammatory and anti-proliferation tumoral cells activities of *Antrodia camphorata*, *Cordyceps sinensis*, and *Cinnamomum osmophloeum* bark extracts. *Journal of Ethnopharmacology* 114(1): 78–85.

Reichling, J., Schnitzler, P., Suschke, U. & Saller, R. 2009. Essential oils of aromatic plants with antibacterial, antifungal, antiviral, and cytotoxic properties-an overview. *Complementary Medicine Research* 16(2): 79–90.

Rezvani, M.B., Niakan, M., Kamalinejad, M., Ahmadi, F.S. & Hamze, F. 2017. The synergistic effect of honey and cinnamon against *Streptococcus mutans* bacteria. *Asian Pacific Journal of Tropical Biomedicine* 7(4): 314–320.

Ribeiro-Santos, R., Andrade, M., Madella, D., Martinazzo, A.P., Moura, L.D.A.G., De Melo, N.R. & Sanches-Silva, A. 2017. Revisiting an ancient spice with medicinal purposes: Cinnamon. *Trends in Food Science and Technology* 62: 154–169.

Sachan, A., Kumar, S., Kumari, K. & Singh, D. 2018. Medicinal uses of spices used in our traditional culture: Worldwide. *Journal of Medicinal Plants Studies* 6(3): 116–122.

Sahib, A.S. 2016. Anti-diabetic and antioxidant effect of cinnamon in poorly controlled type-2 diabetic Iraqi patients: A randomized, placebo-controlled clinical trial. *Journal of Intercultural Ethnopharmacology* 5(2): 108.

Sathishkumar, M., Sneha, K., Won, S., Cho, C.W., Kim, S. & Yun, Y.S. 2009. *Cinnamon zeylanicum* bark extract and powder mediated green synthesis of nano-crystalline silver particles and its bactericidal activity. *Colloids and Surfaces B: Biointerfaces* 73(2): 332–338.

Schink, A., Naumoska, K., Kitanovski, Z., Kampf, C.J., Fröhlich-Nowoisky, J., Thines, E., Pöschl, U., Schuppan, D. & Lucas, K. 2018. Anti-inflammatory effects of cinnamon extract and identification of active compounds influencing the TLR2 and TLR4 signaling pathways. *Food and Function* 9(11): 5950–5964.

Sebai, H., Rtibi, K., Selmi, S., Jridi, M., Balti, R. & Marzouki, L. 2019. Modulating and opposite actions of two aqueous extracts prepared from *Cinnamomum cassia* L. bark and *Quercus ilex* L. on the gastrointestinal tract in rats. *RSC Advances* 9(38): 21695–21706.

Sedighi, M., Nazari, A., Faghihi, M., Rafieian-Kopaei, M., Karimi, A., Moghimian, M., Mozaffarpur, S.A., Rashidipour, M., Namdari, M. & Cheraghi, M. 2018. Protective effects of cinnamon bark extract against ischemia-reperfusion injury and arrhythmias in rat. *Phytotherapy Research* 32(10): 1983–1991.

Senaratne, R. & Pathirana, R. 2020. *Cinnamon - Botany, agronomy, chemistry and industrial applications*. Springer Nature Switzerland AG. https://doi.org/10.1007/978-3-030-54426-3.

Shahina, Z., El-Ganiny, A.M., Minion, J., Whiteway, M., Sultana, T. & Dahms, T.E. 2018. *Cinnamomum zeylanicum* bark essential oil induces cell wall remodelling and spindle defects in *Candida albicans*. *Fungal Biology and Biotechnology* 5: 1–16.

Sharifi-Rad, J., Dey, A., Koirala, N., Shaheen, S., El Omari, N., Salehi, B., Goloshvili, T., Cirone Silva, N.C., Bouyahya, A. & Vitalini, S. 2021. *Cinnamomum* species: Bridging phytochemistry knowledge, pharmacological properties and toxicological safety for health benefits. *Frontiers in Pharmacology* 12: 600139.

Sharma, H., Chauhan, P. & Singh, S. 2018. Evaluation of the anti-arthritic activity of *Cinnamomum cassia* bark extract in experimental models. *Integrative Medicine Research* 7(4): 366–373.

Sharma, S., Mandal, A., Kant, R., Jachak, S. & Jagzape, M. 2020. Is cinnamon efficacious for glycaemic control in type-2 diabetes mellitus. *Journal of Pakistan Medical School* 30: 32.

Shen, Y., Jia, L.N., Honma, N., Hosono, T., Ariga, T. & Seki, T. 2012. Beneficial effects of cinnamon on the metabolic syndrome, inflammation, and pain, and mechanisms underlying these effects-a review. *Journal of Traditional and Complementary Medicine* 2(1): 27–32.

Shishehbor, F., Rezaeyan Safar, M., Rajaei, E. & Haghighizadeh, M.H., 2018. Cinnamon consumption improves clinical symptoms and inflammatory markers in women with rheumatoid arthritis. *Journal of the American College of Nutrition*, *37*(8): 685–690.

Silva, M.L., Bernardo, M.A., Singh, J. & De Mesquita, M.F. 2022. Cinnamon as a complementary therapeutic approach for dysglycemia and dyslipidemia control in type 2 diabetes mellitus and its molecular mechanism of action: A review. *Nutrients* 14(13): 2773.

Singh, B., Nathawat, S. & Sharma, R.A. 2023. Antimicrobial potential of Indian *Cinnamomum* species. *Saudi Journal of Biological Sciences* 30(2): 103549.

Singh, M., Arseneault, M., Sanderson, T., Murthy, V. & Ramassamy, C. 2008. Challenges for research on polyphenols from foods in Alzheimer's disease: Bioavailability, metabolism, and cellular and molecular mechanisms. *Journal of Agricultural and Food Chemistry* 56(13): 4855–4873.

Srivastav, G., Gupta, D., Dubey, A. & Kumar, N. 2022. Investigation of anti-pyretic activity of cinnamon oil in Wistar rat. *Journal for Research in Applied Sciences and Biotechnology* 1(3): 51–56.

Suriyagoda, L., Mohotti, A.J., Vidanarachchi, J.K., Kodithuwakku, S.P., Chathurika, M., Bandaranayake, P.C., Hetherington, A.M. & Beneragama, C.K. 2021. "Ceylon cinnamon": Much more than just a spice. *Plants, People, Planet* 3(4): 319–336.

Tepe, A.S. & Ozaslan, M. 2020. Anti-Alzheimer, anti-diabetic, skin-whitening, and antioxidant activities of the essential oil of *Cinnamomum zeylanicum*. *Industrial Crops and Products* 145: 112069.

Thomas, J. & Kuruvilla, K.M. 2012. *Cinnamon. Handbook of herbs and spices*. Woodhead Publishing. https://doi.org/10.1533/9780857095671.182.

Tjandrawinata, R.R. & Nailufar, F. 2020. Gastroprotective effect of DLBS2411 bioactive fraction from *Cinnamomum burmannii* against ethanol-induced gastric damage in rats. *Journal of Experimental Pharmacology* 12: 87–95.

Varalakshmi, B., Anand, A.V., Karpagam, T., Bai, J.S. & Manikandan, R. 2014. *In vitro* antimicrobial and anticancer activity of *Cinnamomum zeylanicum* Linn. bark extracts. *International Journal of Pharmacy and Pharmaceutical Sciences* 6(1): 12–18.

Vetal, S., Bodhankar, S.L., Mohan, V. & Thakurdesai, P.A. 2013. Anti-inflammatory and anti-arthritic activity of type-A procyanidine polyphenols from bark of *Cinnamomum zeylanicum* in rats. *Food Science and Human Wellness* 2(2): 59–67.

Vijayakumar, K., Prasanna, B., Rengarajan, R., Rathinam, A., Velayuthaprabhu, S. & Vijaya Anand, A. 2020. Anti-diabetic and hypolipidemic effects of *Cinnamon cassia* bark extracts: An *in vitro*, *in vivo*, and *in silico* approach. *Archives of Physiology and Biochemistry* 129(2): 1–11.

Vijayakumar, K., Rengarajan, R., Suganthi, N., Prasanna, B., Velayuthaprabhu, S., Shenbagam, M. & Vijaya Anand, A. 2022. Acute toxicity studies and protective effects of *Cinnamon cassia* bark extract in streptozotocin-induced diabetic rats. *Drug and Chemical Toxicology* 45(5): 2086–2096.

Wang, J., Su, B., Jiang, H., Cui, N., Yu, Z., Yang, Y. & Sun, Y. 2020. Traditional uses, phytochemistry and pharmacological activities of the genus *Cinnamomum* (Lauraceae): A review. *Fitoterapia* 146: 104675.

Wang, Y., Zhang, Y., Shi, Y.Q., Pan, X.H., Lu, Y.H. & Cao, P. 2018. Antibacterial effects of cinnamon (*Cinnamomum zeylanicum*) bark essential oil on *Porphyromonas gingivalis*. *Microbial Pathogenesis* 116: 26–32.

Wild, S., Roglic, G., Green, A., Sicree, R. & King, H. 2004. Global prevalence of diabetes: Estimates for the year 2000 and projections for 2030. *Diabetes Care* 27(5): 1047–1053.

Xu, X., Li, Q., Dong, W., Zhao, G., Lu, Y., Huang, X. & Liang, X. 2023. *Cinnamon cassia* oil chitosan nanoparticles: Physicochemical properties and anti-breast cancer activity. *International Journal of Biological Macromolecules* 224: 1065–1078.

Yanakiev, S. 2020. Effects of cinnamon (*Cinnamomum* spp.) in dentistry: A review. *Molecules* 25(18): 4184.

Yang, Y.C., Lee, H.S., Lee, S.H., Clark, J.M. & Ahn, Y.J. 2005. Ovicidal and adulticidal activities of *Cinnamomum zeylanicum* bark essential oil compounds and related compounds against *Pediculus humanus capitis* (Anoplura: Pediculicidae). *International Journal for Parasitology* 35(14): 1595–1600.

Yu, C.H., Chu, S.C., Yang, S.F., Hsieh, Y.S., Lee, C.Y. & Chen, P.N. 2019. Induction of apoptotic but not autophagic cell death by *Cinnamomum cassia* extracts on human oral cancer cells. *Journal of Cellular Physiology* 234(4): 5289–5303.

Yun, J.W., You, J.R., Kim, Y.S., Kim, S.H., Cho, E.Y., Yoon, J.H., Kwon, E., Jang, J.J., Park, J.S. & Kim, H.C. 2018. *In vitro* and *in vivo* safety studies of cinnamon extract (*Cinnamomum cassia*) on general and genetic toxicology. *Regulatory Toxicology and Pharmacology* 95: 115–123.

Zare, R., Nadjarzadeh, A., Zarshenas, M.M., Shams, M. & Heydari, M. 2019. Efficacy of cinnamon in patients with type II diabetes mellitus: A randomized controlled clinical trial. *Clinical Nutrition* 38(2): 549–556.

Zareie, A., Sahebkar, A., Khorvash, F., Bagherniya, M., Hasanzadeh, A. & Askari, G. 2020. Effect of cinnamon on migraine attacks and inflammatory markers: A randomized double-blind placebo-controlled trial. *Phytotherapy Research* 34(11): 2945–2952.

Zhang, W., XU, Y.C., Guo, F.J., Meng, Y. & Li, M.L. 2008. Anti-diabetic effects of cinnamaldehyde and berberine and their impacts on retinol-binding protein 4 expression in rats with type 2 diabetes mellitus. *Chinese Medical Journal* 121(21): 2124–2128.

14 Clove Buds (*Syzygium aromaticum*)

Kanchanlata Tungare, Mustansir Bhori, Swaroopa Amit Patil, Pooja Iyer, Namdev Jagtap, Aayushi Kadam, and Sekar Vijayakuma

14.1 INTRODUCTION

As a spice, clove has been employed as a food and medicinal plant preservative since the olden days. Clove trees are found at altitudes of 200 m above sea level, and flowers are produced after four years of planting in fields in coastal areas. The binomial of the clove tree is *Syzygium aromaticum* (L.) Merr. & Perry (syn. *Eugenia caryophyllata*) from the Myrtaceae family (Cortés-Rojas et al., 2014). Stretching back in time to 300 BC, Indonesia was in the limelight of European, Indian, Chinese, and Indian traders for cloves as well as nutmeg, popular Indonesian spices at that time, and cloves were a major trading commodity (Sihotang et al., 2018).

14.1.1 Role of Clove Buds in Diet

Clove buds are a well-known source of bioactive compounds that offer potential health benefits. They contain eugenol, a compound that has shown antimicrobial and anti-inflammatory properties. In addition, they are rich in antioxidants, which have been associated with the reduction of oxidative stress and inflammation. A study found that clove buds had the highest antioxidant capacity of 24 different herbs and spices tested (Jaganathan et al., 2014). Research has also suggested that clove buds may have a positive effect on blood sugar levels and could be beneficial for people with diabetes. A randomized controlled trial found that consuming 2–3 g of clove powder per day for 30 days led to significant improvements in fasting blood sugar levels and insulin resistance in people diagnosed with type 2 diabetes (Khan et al., 2006). These findings suggest that incorporating clove buds into the diet may be a helpful addition to a diabetes management plan. In addition to the potential health benefits mentioned earlier, some studies have suggested that clove buds may have pain-relieving properties. A study found that clove oil was effective in reducing pain during dental procedures (Chohan et al., 2010). Clove buds are also used in traditional medicine practices, such as Ayurveda and Traditional Chinese Medicine, where they are believed to have a warming effect on the body and to promote digestion (Batiha et al., 2019). Additionally, clove buds are a common ingredient in culinary applications, particularly in dishes from Indian, Middle Eastern, and African cuisines. As with any food or supplement, it is important to use clove buds in moderation and to be aware of any potential side effects or interactions with medications. Some individuals may be allergic to cloves or suffer from gastrointestinal distress after consuming this spice (Zheng et al., 1992).

In recent years, the United States Department of Agriculture has collaborated with private companies and universities from various states to create a database of concentrations of polyphenol compounds and their antioxidant activity. Clove had the highest score with regard to polyphenol concentration and the highest antioxidant capacity of any spice tested. This was determined using the database's classification of the top 100 dietary sources (Pérez-Jiménez et al., 2010). Clove buds have been known to play a multifaceted role in the diet. In a recent review article (Haro-González et

DOI: 10.1201/9781003387046-14

al., 2021), CEO is high in biological activity and, at the same time, possesses a wide range of uses in animal and human health. It is used as a flavouring in our daily diet and the food industry, and it has medicinal properties, such as anticancer, antimicrobial, antidiabetic, antiviral, anti-inflammatory, and many others (Salama et al., 2018; Kaur & Kaushal, 2019; Hussain et al., 2017). Culinary applications for clove buds include studding for *Solanum lycopersicum* L. (tomato), salads, soups, *Allium cepa* L. (onions), and as a flavouring substance for meat products, biscuits, pickles, sandwiches, candies, chewing gum, chocolates, and many other dishes found in various cuisines around the world (Hussain et al., 2017). Consuming and chewing clove buds is beneficial to individuals' oral hygiene and teeth care. According to folklore traditional medicine, clove oil is a potent antiseptic and analgesic (Pulikottil & Nath, 2015).

14.1.2 Medicinal Uses of Clove Buds

For centuries, clove buds have traditionally been used in medicine due to their various medicinal properties. The main active compound in clove buds, eugenol, possesses antibacterial, antifungal, and anti-inflammatory properties (Meena et al., 2013). Studies have shown that clove buds may be effective in pain relief, particularly in the treatment of dental pain. Clove oil has been found to reduce pain and inflammation following dental procedures (Athamneh et al., 2019). Additionally, clove buds have been found to exhibit anti-cancer properties, with research indicating that they can prevent or sometimes arrest the proliferation of cancer cells (Nirmala et al., 2019). Some studies have also demonstrated the power of clove extract to trigger cell destruction in breast cancer cells (Singh et al., 2012). Furthermore, research suggests that clove buds might be fruitful in the treatment of digestive abnormalities, such as ulcerative colitis, because of their prominent anti-inflammatory properties (Khare et al., 2021). These findings highlight the potential of clove buds as a natural remedy with various medicinal applications.

CEO is widely used for its medicinal properties. CEO is an essential oil obtained from raw vegetable material, which has a complex composition and a distinct odour (European Pharmacopoeia, 2019). CEO contains a plethora of phenolic molecules that possess biological activities including antibacterial, antifungal, and insecticidal effects (El-Saber Batiha et al., 2020). Nyachoti et al. (2006) observed that the intestinal villus height is reduced by the effects of various bacteria in the pig intestines, which can lead to less surface area for absorption in the intestine and hence indigestion. However, clove buds have been found to increase villus height and improve the villus-height-to-crypt-depth ratio. This phenomenon was indicated in a study by Gandomani et al. (2014) in egg-laying hens. A study conducted on the B16-F1 cell line of melanoma cells indicated that eugenol, a major compound derived from clove buds, upregulates the expression of major enzymes involved in melanogenesis in these cells, such as tyrosinases, TRP-2, and TRP-1, without showing cytotoxicity. This finding suggests that eugenol in clove buds can be a safe and potent agent for the treatment of hypopigmentary disorders (Uto et al., 2022).

Clove buds have been utilized for their potential medicinal properties. In one study, clove oil was found to be as potent as benzocaine in making pain milder during needle insertion in individuals with dental anxiety, indicating its potential in treating dental pain (Alqareer et al., 2006). Clove buds have also been demonstrated to possess antifungal activity, with CEO inhibiting the growth of multiple types of fungi (Taheri et al., 2023). Furthermore, clove buds have antiparasitic properties and may be effective against specific types of parasites (Nassan et al., 2015). Lastly, clove buds have been traditionally used to alleviate digestive issues, and some studies have suggested their potential protective effects on the digestive system (Dzoyem et al., 2014). These findings suggest that clove buds may have a diverse range of medicinal uses and could be a useful addition to traditional medical treatments (El-Saber Batiha et al., 2020).

14.2 DESCRIPTION, SCIENTIFIC CLASSIFICATION, NUTRITIONAL CONTENT AND CHEMICAL COMPOUNDS ISOLATED FROM CLOVE BUDS

Syzygium aromaticum is an intermediate-sized, evergreen tree that grows up to 8–12 m tall. It hails from Maluku Islands in the eastern part of Indonesia. This plant species has its roots in the Myrtaceae family, which is a taxon of the Dicotyledonae and is classified under the order Myrtales. The tree possesses a woody stem and produces essential oil, with flowers appearing in multiples of either five or four. The flowers of this species generally have five petals, with numerous brightly coloured stamens. The leaves of *S. aromaticum* are evergreen and mostly opposite, displaying a simple structure with an entire margin in most cases. The phloem is located on both sides of the xylem, as opposed to just the outside, which is commonly observed in other plant species. The trade in this spice has been a valuable stimulator of economic growth in the Maluku Islands, contributing to the region's development over time (El-Saber Batiha et al., 2020).

Clove buds are a rich source of several essential nutrients, including vitamins, minerals, and antioxidants. Information from the United States Department of Agriculture (USDA, 2021) states that 100 g of clove buds contain 65% of the recommended daily intake of vitamin C, 20% of vitamin K, 20% of calcium, and 30% of iron. Furthermore, clove buds are rich in dietary fibre, protein, and healthy fats. Additionally, they contain a range of phenolic molecules, like eugenol and gallic acid, which exhibit strong antioxidant properties, providing protection against oxidative stress and inflammation (Jayaprakasha et al., 2007; Cai et al., 2004; Sharma et al., 2017). Recent studies have suggested that incorporating clove buds into the diet can provide many positive health benefits, such as better blood sugar control (Kumar et al., 2013), reduced cholesterol levels (Al-Hashimi et al., 2019), as well as reduced risk of some types of cancer (Li et al., 2019). Therefore, the consumption of clove buds is highly recommended to maintain a healthy lifestyle.

The main chemical compound extracted and isolated from clove buds, which imparts the aroma to it, is eugenol, which has been studied for its analgesic, anti-inflammatory, and antioxidant effects. The other compounds extracted from clove buds are essential oils, β-caryophyllene, acetyl eugenol, vanillin, crategolic acid, gallotannic acid, tannins, sesquiterpenes, and methyl salicylate, the latter being a painkiller. The flavonoids include kaempferol, rhamnetin, eugenin, and eugenitin, whereas the triterpenoids include stigmasterol, oleanolic acid, and campesterol (Bensky et al., 2004). Clove buds also contain compounds such as -caryophyllene, acetyl eugenol, and vanillin, which have been observed to have antimicrobial, antifungal, and anti-inflammatory activities (El-Saber Batiha et al., 2020). Research studies have demonstrated the potential of clove bud compounds to improve dental health, treat respiratory infections, and protect against oxidative stress. Previous literature supports that eugenol-rich clove extract mediated improvement in dental health by reducing plaque formation and gingival inflammation (Pulikottil and Nath, 2015). Another study demonstrated the antimicrobial activity of clove bud essential oil against a range of pathogenic bacteria (Nuñez and Aquino, 2012). These findings suggest that clove bud compounds have promising potential as natural remedies for various health conditions.

14.3 TOXICITY AND PHARMACOKINETICS

The United States Food and Drug Administration has given buds of clove, oleoresins, and CEOs the status of 'Generally Regarded as Safe' (GRAS) for application as food additives (FDA-US, Washington, DC, 1978). CEOs and its most significant extracted component, eugenol, are put into the group as low risk pesticides by the EPA, and food or other commercial substances containing these ingredients are exempt from FIFRA requirements (EPA, 2011). Several chronic and acute toxicity studies of CEOs have revealed an oral dose LD_{50} of 3597.5 mg/kg with no adverse effects when demonstrated in accordance with sub-chronic toxicity tests, with NAOEL levels ranging from 900 to 2000 mg/kg/day (Shalaby et al., 2011; WHO, 2005).

When taken orally, eugenol is rapidly absorbed and metabolized in the liver, with a 24 h period of elimination from the body (Cortés-Rojas, 2014). Eugenol, the major component of CEO in its isomorphic form, exhibits a much higher rate of cytotoxicity (Atsumi et al., 2005). Although clove buds have many potential health benefits, it is important to consider the potential toxicity of these compounds as well. Eugenol has been shown to have toxic effects at high doses (Ndung'u et al., 2017). In animal studies, high doses of eugenol have been associated with liver and kidney damage, as well as deleterious effects on the central nervous system (CNS) (Gunes et al., 2019). In addition, certain clove products, such as clove oil, have been reported to cause allergic reactions and skin irritation (Leung & Foster, 1996). While clove buds are generally considered safe when used in culinary amounts, it is important to exercise caution when using clove-based products in medicinal or therapeutic contexts. A study discovered that high doses of eugenol caused liver toxicity in rats (Pal et al., 2018). Additionally, a case report documented a severe allergic reaction in rats who consumed a large amount of clove oil (Yassin et al., 2010).

Several studies have been demonstrated to evaluate the toxicity profile of clove buds in various animal models. An investigation on studying the acute and subacute toxicity of clove bud essential oil in rats revealed that acute administration of CEO at a dosage of 2000 mg/kg body weight did not produce any significant toxic effects in rats (Kurokawa et al., 1995). However, subacute administration of CEO at a dosage of 500 mg/kg body weight for 28 days led to some haematological and biochemical alterations (Hassan et al., 2014), indicating mild toxicity. Another study evaluated the genotoxic potential of clove bud essential oil and suggested that clove bud essential oil did not induce any significant genotoxic effects in either *in-vitro* or *in-vivo* assays, indicating that the essential oil is not likely to be mutagenic or carcinogenic (Ramesh et al., 2010). Another study conducted on rats found that acute administration of eugenol at a dose of 2000 mg/kg body weight did not produce any significant toxic effects (Suresh et al., 2012). However, subacute administration of eugenol at a dose of 250 mg/kg body weight for 28 days resulted in mild toxicity. Despite the widespread use of clove buds as a natural remedy for various ailments, due to their antimicrobial and anti-inflammatory properties, it is important to note that, like any other natural product, clove buds can exhibit toxicity at higher doses (Jaganathan et al., 2014). Furthermore other researchers presented a study where rats with oral administration of clove buds at doses of up to 2000 mg/kg body weight did not exhibit any significant adverse effects on the general health or behaviour of the rats, but subchronic administration of clove buds at a dose of 1000 mg/kg body weight for 28 days resulted in some haematological and histopathological alterations, indicating mild toxicity (Kim et al., 2015). Another study on mice found that acute administration of clove bud essential oil at doses of up to 2000 mg/kg body weight did not produce any significant toxic effects, but repeated administration of clove bud essential oil at a dose of 200 mg/kg body weight for 28 days led to some histopathological alterations in the liver and kidneys, indicating mild toxicity (Kundu et al., 2016). Also, along the same lines, another work on subchronic administration of eugenol at doses of up to 200 mg/kg body weight for 90 days did not produce any marked toxic effects on the general health or behavior of the rats. However, higher doses of eugenol (400 and 800 mg/kg body weight) lead to some histopathological alterations in the liver and kidneys, indicating mild toxicity (Vijayasteltar et al., 2019). While clove buds and their essential oil are generally considered safe when consumed in moderation, excessive consumption may result in toxicity. It is important to use clove buds and their essential oil in moderation and under the guidance of a health care professional.

Clove buds have been used for centuries for their medicinal properties, and their pharmacokinetic properties have been the subject of several studies. The pharmacokinetics of clove buds refers to the way in which the body absorbs, distributes, metabolizes, and eliminates the active compounds present in clove buds. In a study by Yang et al. (2007), the pharmacokinetics of eugenol, the primary active compound in clove buds, were investigated in rats. The study found that, after oral administration, eugenol was quickly taken up from the gastrointestinal tract (GI) and reached the highest plasma concentrations within 15 minutes. The liver extensively metabolized eugenol, and the metabolites were eliminated primarily in the urine. Similarly, a study by Li et al.

(2015) investigated the pharmacokinetics of β-caryophyllene, another active compound in clove buds, in rats. The study found that β-caryophyllene was also rapidly absorbed from the GI tract and reached peak plasma concentrations within 30 minutes of oral administration. β-caryophyllene was metabolized in the liver and eliminated primarily in the urine. A study by Yoshinari et al. (2006) searched for the pharmacokinetics of the same molecules in humans. The study showed that both compounds were rapidly absorbed from the gastrointestinal tract and reached peak plasma concentrations within 1–2 hours of oral administration. These compounds were metabolized in the liver and eliminated mainly in the urine. Similarly, a study by Yang et al. (2007) found that eugenol and Li et al. (2015) found that β-caryophyllene were rapidly absorbed in rats and metabolized in the liver. These findings indicate the need for frequent dosing of clove buds to maintain therapeutic levels in the body. Moreover, the metabolism of these compounds in the liver could affect drug interactions and toxicity, emphasizing the importance of understanding the pharmacokinetic properties of clove buds. Various studies have investigated the pharmacokinetic properties of clove buds and their active compounds in both humans and animal models. Abdul Manan et al. (2018) found that eugenol, the main active compound in clove buds, was rapidly absorbed and eliminated with a half-life of approximately 4.5 hours in healthy human volunteers. Khan et al. (2019) investigated the pharmacokinetics of eugenol and β-caryophyllene in rats and found that both compounds were rapidly absorbed and distributed to various tissues, with β-caryophyllene having the longer half-lifee. Matsubara et al. (2018) examined the effects of ageing on the pharmacokinetics of eugenol in rats and found that ageing significantly altered the pharmacokinetics of eugenol, with a longer half-life and reduced clearance rate in older rats compared with younger rats. Furthermore, Wu et al. (2019) conducted a study in rats and found that eugenol had a higher absorption rate and reached a higher plasma concentrations than either β-caryophyllene or acetyl eugenol. The metabolism and excretion of these compounds varied, with eugenol being mainly metabolized in the liver and excreted in the urine, whereas β-caryophyllene was mainly excreted in the faeces. Similarly, a study by Schnitzler et al. (2010) in humans found that eugenol was rapidly absorbed and metabolized, with a half-life of approximately 3.3 hours. Eugenol was primarily metabolized in the liver and excreted in the urine, with only a small amount excreted in the faeces. Xiao et al. (2017) conducted a study in rats and found that eugenol was rapidly absorbed and distributed to various tissues, including the liver, kidney, and lung. Eugenol was mainly metabolized in the liver to create conjugates, which were then removed via urine and bile. This finding was supported by another group of researcher who suggested that eugenol was rapidly absorbed in the small intestine and metabolized in the liver, with its metabolites being excreted in the urine and faeces (Hua et al., 2013). Another study in healthy human volunteers found that eugenol was also rapidly absorbed and metabolized, with a short half-life, being mainly metabolized in the liver and excreted in the urine as glucuronide and sulphate conjugates (Dai et al., 2012). Additionally, a study explored the pharmacokinetics of eugenol in rats with carbon tetrachloride-induced liver injury. The study revealed that liver injury caused changes in the metabolism of eugenol, leading to reduced elimination, and higher plasma concentrations of eugenol and its metabolites (Lu et al., 2012). The literature presented highlights the importance of understanding the pharmacokinetic properties and suggests that eugenol is rapidly absorbed and metabolized, with the majority being metabolized in the liver and excreted in the urine. β-caryophyllene, on the other hand, is mainly excreted in the faeces. The pharmacokinetics of these compounds may also be altered in cases of liver injury. Therefore, it is essential to gain a comprehensive understanding of the pharmacokinetics of clove buds and their active compounds to ensure their safe and effective use in traditional medicine. Further studies are necessary to fully elucidate the pharmacokinetics of clove buds and their bioactive compounds in humans.

14.4 PHARMACOLOGICAL PROPERTIES AND ACTIVITIES OF CLOVE BUDS

This next section will explore the pharmacological activities exhibited by clove buds and CEO, as well as the medicinal uses and activities of the compounds extracted and isolated from these plant materials.

14.4.1 Antifungal Activity

Recent research has confirmed the traditional use of clove buds in treating fungal infections due to their antifungal properties. CEO has been found to exhibit potent antifungal effects against various species of fungi, like *Candida albicans*, *Aspergillus niger*, and *Trichophyton mentagrophytes* (Singh et al., 2015). In addition, CEO has also been observed to exhibit a very good antifungal effect against dermatophytes (skin infection-causing fungi) such as *Trichophyton rubrum*, *Trichophyton mentagrophytes*, and *Microsporum gypseum* (Singh et al., 2014). Another study, conducted by Chami and colleagues in 2004, tested the antifungal activity of carvacrol and eugenol, components extracted from clove buds, in a vaginal candidiasis disease model. The study used histological and microbial procedures to distinguish the control group from the samples. The results and observations of the study suggested that carvacrol and eugenol could potentially serve as promising agents for the prophylaxis and employment in the therapeutic treatment for vaginal candidiasis (Chami et al., 2004). In a research study by Furletti et al. (2011), they revealed that eugenol exhibited potent antifungal activity against all tested *Candida* species, including fluconazole-resistant strains. Similarly, a study by Shreaz et al. (2016) investigated the antifungal potential of CEO against several pathogenic fungi and found that CEO showed very high antifungal activity against *Candida albicans*, *Aspergillus fumigatus*, and *Fusarium oxysporum*. Another research study assessed the antifungal activity of CEO against two important foodborne fungal pathogens, *Aspergillus flavus* and *Aspergillus parasiticus*, and found that it significantly inhibited their growth (Mohsenzadeh et al., 2018). Additionally, a study investigated the antifungal activity of clove extracts against various dermatophytes, including *Trichophyton rubrum* and *Microsporum canis*, and found that the extracts exhibited potent antifungal activity against these fungi (Kumar et al., 2015; Kumar et al., 2016). Also, CEO and eugenol have been shown to have potent antifungal activity against fluconazole-resistant strains (Ghannoum et al., 2010). A study by Garg et al. (2011) employed a rat model of oral candidiasis, in which the rats were immunosuppressed. The study found that eugenol, when formulated as solid lipid nanoparticles, showed promising improvement in therapeutic effectiveness.

Clove buds have also shown potential in combating various plant pathogens. Shao et al. (2017) assessed in detail the antifungal activity of CEO against several phytopathogenic fungi, like *Botrytis cinerea*, *Colletotrichum gloeosporioides*, and *Fusarium oxysporum* f. sp. *niveum*. The in-depth analysis found that CEO effectively inhibited the growth of all tested fungi and had strong antifungal activity. Duman and Korukluoglu (2016) investigated the antifungal activity of CEO against postharvest fungal pathogens of peaches. The study found that CEO exhibited significant antifungal activity against *Penicillium digitatum*, *Penicillium italicum*, and *Aspergillus niger*, which are common postharvest pathogens causing fruit decay. Clove buds have also demonstrated potent antifungal activity against a range of food spoilage and mycotoxigenic fungi. One study evaluated the effectiveness of CEO against *Aspergillus flavus*, a common fungal pathogen known for producing carcinogenic mycotoxins called aflatoxins. Results showed that CEO was effective in slowing down or stopping the growth of *A. flavus* and lessening the production of aflatoxins (Singh et al., 2015). Another study (Bajpai et al., 2013) investigated the antifungal activity of CEO against food spoilage fungi, including *Penicillium expansum* and *Alternaria alternata*. The study found that CEO demonstrated potent antifungal activity against these fungi by inhibiting their growth and spore germination.

Clove buds have been reported to have potent antifungal activities against various food spoilage and mycotoxigenic fungi as well as fungal pathogens that cause human infections. The antifungal activity of clove buds is attributed to the role of eugenol. Eugenol is known to disrupt fungal cell membranes and inhibit fungal growth. Numerous studies have investigated the antifungal properties of clove buds and their extracts against a range of fungal pathogens. These findings suggest that clove buds and their extracts may have potential applications in the development of natural antifungal agents for the treatment of fungal infections (Table 14.1).

TABLE 14.1
Studies on Antifungal Properties of Clove Oil and Its Components

Fungal Pathogen	Study	Clove Source	Antifungal Activity	Mechanism of Action
Aspergillus flavus	Singh et al., 2015	Essential oil	Inhibition of growth; reduction of aflatoxin production	Disruption of fungal cell membranes
Aspergillus flavus	Mohsenzadeh et al., 2018	Essential oil	Inhibition of growth	Not specified
Aspergillus niger, *Penicillium digitatum*, *Penicillium italicum*	Duman & Korukluoglu, 2016	Essential oil	Significant antifungal activity	Disruption of fungal cell membranes
Botrytis cinerea, *Colletotrichum gloeosporioides*, *Fusarium oxysporum* f. sp. *niveum*	Shao et al., 2017	Essential oil	Inhibition of growth	Not specified
Candida species	Ghannoum et al., 2010	Essential oil and eugenol	Significant antifungal activity	Disruption of fungal cell membranes
Dermatophytes	Kumar et al., 2015	Essential oil	Significant antifungal activity	Disruption of fungal cell membranes
Penicillium expansum, *Alternaria alternata*	Bajpai et al., 2013	Essential oil	Potent antifungal activity; inhibition of growth and spore germination	Disruption of fungal cell membranes

14.4.2 Antiviral Activity

Research has shown that clove buds have significant antiviral activity against various human viral pathogens. One study found that clove extract and eugenol, its major component, exhibited potent antiviral effect against herpes simplex virus type 1 (HSV-1) and type 2 (HSV-2), with eugenol being more effective than the extract (Lau et al., 2008). Another team found that clove oil demonstrated significant antiviral effect against HSV-1 *in vitro*, with a powerful ability to inhibit the virus at the early steps of infection (Gilling et al., 2014). In addition, studies evaluating the antiviral activity of clove extracts against hepatitis B virus (HBV) found that clove extracts showed potent antiviral activity against HBV *in vitro* by inhibiting virus replication and gene expression (Wang et al., 2019). A study by Siddiqui et al. (2020) on HSV-1 revealed a similar finding of significant CEO-mediated inhibition of HSV-1 replication *in vitro*. With similar objectives, another team of researchers investigated anti-HSV type-1 activity of aqueous extracts of clove buds and observed the activity to be more prevalent in the brain. After testing for toxicity in mice, proven combinations were established (Daniel AN et al., 2009). According to one study, the antiviral activity of clove buds was demonstrated by one of its isolated components, eugenin, which was found to be effective against HSV strains at a concentration of 5 μg/mL. The study revealed that the major targets of eugenin were the viral DNA synthesis machinery, specifically inhibiting the polymerase of the viral DNA (Kurokawa et al., 1998). Another study, by Choi et al. (2013), evaluated the antiviral activity of clove extracts against human immunodeficiency virus type 1 (HIV-1), the agent that causes the onset of acquired immunodeficiency syndrome (AIDS). The results showed that clove extracts displayed powerful antiviral activity against HIV-1, inhibiting its replication *in vitro*. Literature also reports that CEO significantly reduces the cell division of human cytomegalovirus (HCMV) *in vitro*, suggesting its potential as an antiviral agent against this virus (Astani et al., 2014). These findings suggest that clove buds and their extracts have potential applications for the formation of natural antiviral agents for the therapeutic treatment of viral infections. Furthermore, two separate studies investigated the antiviral activity of clove buds against the influenza virus, and reported that they effectively inhibited the virus replication *in vitro*, highlighting its potential as a natural antiviral agent (Yong et al., 2017; Ali et al., 2021). Another study evaluated the antiviral activity of clove extracts against human norovirus, which is responsible for viral gastroenteritis, and discovered that the extracts showed strong antiviral activity against the virus *in vitro* by inhibiting its replication (Yin et al., 2018). The literature also has evidence of study against the severe acute respiratory syndrome coronavirus 2 (SARS-CoV-2), the causative agent of the COVID-19 pandemic, and Zika virus (ZIKV). The findings of these studies showed that the extract displayed promising inhibitory effects against SARS-CoV-2 *in vitro* (Ganguly et al., 2020), whereas eugenol induced inhibition of ZIKV replication *in vitro* as well as *in vivo* (Li et al., 2017). Antiviral activity of clove extracts against the human respiratory syncytial virus (RSV), a common cause of respiratory infections in children, has also been studied and reported to exhibit a potent anti-viral activity against RSV (Yuan et al., 2018). The various studies discussed here suggest that clove and its extracts have significant antiviral activity against a plethora of viruses that include HSV, HBV, HIV type 1, influenza virus, human norovirus, severe acute respiratory syndrome coronavirus 2, Zika virus, influenza A virus, and human respiratory syncytial virus. Thus, the antiviral properties of clove buds and their extracts have been demonstrated in multiple studies on a large number of viruses, which have attributed their effectiveness to the presence of eugenol. Eugenol has been found to inhibit viral entry and replication in host cells, making it a potential natural alternative to synthetic antiviral agents in the treatment of viral infections in humans. These findings highlight the potential of clove buds and their extracts as natural antiviral agents in the treatment of various human viral pathogens (Table 14.2).

14.4.3 Antibacterial Activity

Food is crucial for sustaining human life as it provides energy through metabolism. However, certain pathogens, such as *E.coli* and *Listeria monocytogenes*, can cause foodborne illnesses such as indigestion, diarrhoea, and fever. To combat these microorganisms, studies have been conducted to test the effectiveness of various substances such as CEOs and formulations of carvacrol with eugenol. Promising results have been reported for eugenol as an inhibitor of the growth of these pathogens on surfaces that come in contact with food products (Burt et al., 2003; Pérez-Conesa et al.,

TABLE 14.2
Studies on Antiviral Properties of Clove Oil and Its Components

Virus	Antiviral Activity and Mechanism of Action
HSV-1	Inhibited replication of HSV-1 by inhibiting viral DNA synthesis machinery through the eugenin compound extracted from clove buds (Kurokawa et al., 1998)
HIV-1\	Exhibited potent antiviral activity against HIV-1, inhibiting its replication in vitro, possibly through the eugenol compound found in clove extracts (Choi et al., 2013)
Influenza virus	Effectively inhibited the replication of influenza virus *in vitro*, possibly through clove buds' essential oil (Yong et al., 2017)
Human Norovirus	Exhibited strong antiviral activity against human norovirus, inhibiting its replication *in vitro*, possibly through clove extracts (Yin et al., 2018)
SARS-CoV-2	Effectively inhibited the replication of SARS-CoV-2 *in vitro*, possibly through clove bud extract (Ganguly et al., 2020)
ZIKV	Significantly inhibited the replication of ZIKV *in vitro* and *in vivo*, possibly through the eugenol compound found in clove buds (Li et al., 2017)

2006). Also, a study published by Siddiqui et al. (2013) investigated the antibacterial effects of clove extracts against a range of pathogens that contaminate food and multiply to spread disease, include *Salmonella typhimurium*, *E. coli*, *L. monocytogenes*, and *S. aureus*. The outcomes showed that clove extracts exhibited powerful antibacterial effects against all tested strains, with a minimum inhibitory concentration (MIC) in a range from 0.06 to 0.25 mg/mL. Another research study, published by Alqareer et al. (2012), evaluated the antibacterial effect of clove oil against oral bacteria. The findings revealed that clove oil had potent antibacterial activity against *Streptococcus mutans*, a major pathogenic agent of dental caries, with a minimum bactericidal concentration (MBC) of 0.25% v/v. Similar observation was made by Fadel et al. (2021) with respect to *S. mutans*.

In addition to their potent antibacterial effect against a range of pathogenic bacteria, clove extracts have also been found to have synergistic results when combined with other natural compounds, such as tea tree oil, cinnamon oil, and thyme oil (Siddiqui et al., 2013). Agastian et al. (2006) conducted a study to analyse the antibacterial effects of clove oil and its major component, eugenol, against various strains of Gram-positive and Gram-negative bacteria. The findings revealed that both clove oil and eugenol demonstrated significant antibacterial activity against all bacterial strains tested, including *S. aureus*, *E. coli*, *Pseudomonas aeruginosa*, and *Salmonella typhimurium*. Similarly, a study published by Kalemba and Kunicka (2003) investigated the antibacterial activity of CEO against several pathogenic bacteria, including methicillin-resistant *Staphylococcus aureus* (MRSA). The study found that CEO had potent antibacterial activity against all tested strains, including MRSA, with a MIC ranging from 0.01% to 0.25%. Other studies have also had similar outcomes, describing the antibacterial effects of clove extracts against a plethora of bacterial strains, including those that are resistant to antibiotics. For instance, Ghannoum et al. (2010) conducted a study which revealed that a combination of clove oil and tea tree oil exhibited synergistic antibacterial effects against MRSA and other Gram-positive bacteria. Another team of researchers also suggested antibacterial activity against MRSA with an MIC of 6.25 mg/mL (Kedia et al., 2014). This finding was also supported by another team of researchers with respect to MRSA and vancomycin-resistant enterococci (VRE) (Sharma et al., 2017). Similarly, Oyedemi et al. (2010) found significant antibacterial activity, with MICs ranging from 0.31 to 5 mg/mL against MRSA. Furthermore, eugenol has been demonstrated to enhance the activity of certain antibiotics, such as ciprofloxacin and ampicillin, against bacterial infections (Bajpai et al., 2013).

Eugenol has been discovered to display effective antibacterial properties against Gram-positive as well as Gram-negative bacteria. According to a study performed by Domans and Deans in 2000, CEO demonstrated a broad spectrum of activity against 25 different species of both Gram-negative and Gram-positive bacteria (Domans & Deans, 2000). In a study conducted by Mondal et al. (2016), CEO demonstrated its antibacterial activity against several bacterial species, Gram-negative bacteria (*E. coli*, *P. aeruginosa*, *Proteus vulgaris*, and *K. pneumoniae*) as well as Gram-positive bacteria (*S. aureus*, *Bacillus cereus*, *S. mutans*, and *Streptococcus pyogenes*). The study showed that CEO exhibited significant antibacterial activity against all tested bacterial strains, with MICs ranging from 0.039 to 0.625 mg/mL. Additionally, a study by Al-Bayati and Al-Mola (2008) that quantified the antibacterial activity of CEO against *S. aureus*, *E. coli*, *P. aeruginosa*, and *S. typhimurium* also suggested significant antibacterial activity, with MICs ranging from 0.12 to 0.50 mg/mL. A similar observation was reported by Bajpai et al. (2013) with respect to significant antibacterial activity against *S. aureus*, *E. coli*, *B. subtilis*, and *P. aeruginosa*. Chaieb et al. (2007) also reported that CEO showed antibacterial activity against *E. coli*, *S. aureus*, and *Salmonella enteritidis*, with MICs ranging from 0.1% to 0.6%. Other studies have also documented the antibacterial activity of clove against *Helicobacter pylori*, *P. aeruginosa*, and *Enterococcus faecalis*. The active compounds in clove, namely eugenol, eugenyl acetate, and β-caryophyllene, are believed to be responsible for its antibacterial properties. Thus, the literature suggests that clove has significant antibacterial properties against various bacterial species, including antibiotic-resistant bacteria. The active compounds responsible for its antibacterial activity have been identified as eugenol, β-caryophyllene, and other phenolic compounds. These compounds disrupt the bacterial cell membrane, inhibiting bacterial growth and division. The findings suggest that clove may possess potential applications to the treatment and therapy against bacterial infections.

14.4.4 Antioxidant Activity

Oxidative stress results from an imbalance between reactive oxygen species (ROS) and antioxidants (enzymes and non-enzymes) in the body, which can lead to cellular damage, inflammation, and chronic illnesses, such as cancer, cardiovascular disease, and neurodegenerative disorders. Clove buds contain a high concentration of antioxidants, which have been found to scavenge free radicals and prevent lipid peroxidation *in vitro* and *in vivo*. The high concentration of phenolic acids, flavonoids, and anthocyanins in clove buds provide powerful antioxidant effects that neutralize free radicals and protect cells from damage (Shan et al., 2010; Shreaz et al., 2013). Moreover, clove buds possess anti-inflammatory properties that can further contribute to their antioxidant effects by reducing chronic inflammation, which is linked to many chronic diseases (Ali et al., 2016). Studies have suggested that multi-therapeutic properties of clove buds make them a potentially useful natural remedy for various ailments (Dharmendra et al., 2021). Additionally, clove buds have been shown to have analgesic effects and may help relieve pain (Patil et al., 2012). These properties make clove buds a promising candidate for therapeutic use as a natural antioxidant source.

Numerous studies have demonstrated the antioxidant properties of clove buds. Krishnaswamy et al. (2008) reported that clove bud extract showed significant antioxidant activity *in vitro*, as measured by the DPPH and FRAP assays. Chakrabarti et al. (2011) showed that CEO had potent free radical scavenging activity and caused reduced lipid peroxidation in rats. The antioxidant properties of clove buds have also been associated with potential therapeutic benefits in various diseases. El-Soud et al. (2015) found that clove bud extract reduced oxidative stress and inflammation in rats with ulcerative colitis. Likewise, Lee et al. (2013) demonstrated that clove bud extract protected against neuronal damage and cognitive impairment in a mouse model of Alzheimer's disease by reducing oxidative stress and inflammation.

An *in-vitro* analysis was conducted to determine the antioxidant activity of 26 different spices, with a focus on the phenolic compounds present in them. The study utilized 2,2'-azino-bis (3-ethylbenzothiazoline-6-sulphonic acid) [ABTS] activity and HPLC to identify and quantify the compounds responsible for antioxidant activity. The results showed that clove buds had a high polyphenol content and therefore exhibited high antioxidant activity (Shan et al., 2010). Moreover, clove buds exhibited antioxidant activity comparable to the common food preservative, butylated hydroxytoluene (BHT), as demonstrated by Bamdad et al. (2006), who used various *in-vitro* model systems like ferric thiocyanate and hydroxyl radical to quantify the activity. Aqueous extracts of clove and lavender also exhibited high antioxidant activity, inhibiting up to 95% of free radicals such as superoxide radicals at concentrations of 20, 40, or 60 μg/mL. Authors reported strong hydrogen donating abilities of clove extracts, which could be a reason for the high antioxidant activities observed (Gülçina Ì et al. 2004). Eugenol stabilizes radicals like phenoxy by donating a hydrogen atom and forming stable intermediates that do not propagate oxidation. This is a chromophore phenomenon employed as an antioxidant mechanism, allowing eugenol to reduce multiple DPPH radicals with just one hydrogen atom (Gülçin et al., 2011).

Clove buds have been found to possess anticancer effects, which may be attributed to their high antioxidant properties. Studies have demonstrated that clove buds can inhibit the growth and induce apoptosis in various cancer cell lines, such as breast, lung, liver and colon cancer (Prasad et al., 2010; Wang et al., 2014). Eugenol has been identified as a potential agent for inducing cell cycle arrest and apoptosis in cancer cells (Padhy et al., 2022). Furthermore, the antioxidant properties of clove buds are also beneficial for maintaining skin health. A study by Khan et al. (2017) revealed that clove bud oil can protect against UV-induced skin damage and oxidative stress in mice, while Mukherjee et al. (2017) reported that clove bud extract can protect against oxidative stress and skin ageing in human skin fibroblasts. From an application point of view, while developing antioxidant

TABLE 14.3
Antioxidant Properties of Clove Oil and Its Components

Study	Antioxidant Activity	Health Benefit
Shan et al., 2010	High antioxidant activity in clove buds due to high polyphenol content	Protection against oxidative stress-related diseases
Bamdad et al., 2006	Clove buds exhibit strong antioxidant activity, comparable to BHT	Protection against oxidative stress-related diseases
Gülçina Ì et al., 2011	Clove and lavender extracts show inhibitory activities against free radicals and superoxide	Protection against oxidative stress-related diseases
Chatterjee D et al., 2013	Encapsulated clove powder shows controlled release of antioxidants and prolongs shelf life	Application as functional food ingredient
Krishnaswamy et al., 2008	Clove bud extract exhibits significant antioxidant activity *in vitro*	Protection against oxidative stress-related diseases
Chakrabarti et al., 2011	CEO has strong free radical scavenging activity and inhibits lipid peroxidation	Protection against oxidative stress-related diseases
El-Soud et al., 2015	Clove bud extract reduces oxidative stress and inflammation in rats with ulcerative colitis	Therapeutic effects in ulcerative colitis
Lee et al., 2013	Clove bud extract protects against neuronal damage and cognitive impairment in a mouse model of Alzheimer's disease	Therapeutic effects in Alzheimer's disease
Prasad et al., 2010	Clove buds exhibit antiproliferative and proapoptotic effects against various cancer cell lines	Anticancer effects
Wang et al., 2014	Clove buds exhibit antiproliferative and proapoptotic effects against various cancer cell lines	Anticancer effects
Banerjee et al., 2014	Eugenol found in clove buds can induce cell cycle arrest and apoptosis in cancer cells	Anticancer effects
Khan et al., 2017	Clove bud oil protects against UV-induced skin damage and oxidative stress in mice	Skin-protective effects
Mukherjee et al., 2017	Clove bud extract has a protective effect against oxidative stress and skin ageing in human skin fibroblasts	Skin-protective effects

formulations, controlled release of antioxidants can be achieved by encapsulating clove powder obtained through spray drying with arabic gum, maltodextrin, or other wall materials. The shelf life of these encapsulated and unencapsulated extracts rich in components extracted and isolated from clove were tested with soybean oil, demonstrating the potential for practical application (Chatterjee et al., 2013) (Table 14.3).

14.4.5 Antinociceptive Activity

Clove's therapeutic effects have been attributed to the presence of eugenol, which has been used for centuries to alleviate toothache, joint pain, and muscle spasms (Cortés-Rojas et al., 2014). Eugenol's analgesic activity is due to its ability to inhibit voltage-gated sodium and calcium ion channels, as supported by research studies (Li et al., 1052008). Moreover, eugenol's antinociceptive activity has been linked to the capsaicin receptor present on sensory terminals in the spinal cord (Ohkubo & Shibata, 1997). In one study, significant antinociceptive effects were observed at 50, 75, and 100 mg/kg doses (Daniel et al., 2009). The inhibition of gamma-aminobutyric acid (GABA) $\alpha 1\beta 2\gamma 2$ expressed in the ganglia of the trigeminal nerve is believed to be the mechanism of action underlying eugenol's antinociceptive effects (Beltrán-Villalobos et al., 2017; Khalilzadeh et al., 2016).

Therefore, the analgesic and antinociceptive effects of clove buds can be attributed to eugenol, which can possibly be used as a therapeutic in reducing pain and discomfort.

14.4.6 Anticancer Activity

The essential oil derived from cloves (CEO) contains eugenol, α-humulene, and β-caryophyllene, that possess antitumour properties with potential for cancer prevention and treatment (Haro-González et al., 2021). The anticancer property of CEO is accredited to its antioxidant properties, which can counteract the harmful effects of reactive oxygen species (ROS) on signaling pathways, cell proliferation, angiogenesis, and metastasis (Morsy, 2017; Bayala & Bagora, 2014). CEO has demonstrated efficacy against various types of cancer, including cervical cancer, through mechanisms such as the activation of detoxifying enzymes (Bayala et al. 2014). The use of clove extracts for cancer treatment has been extensively researched by various scientists in recent years.

Taha Yassin et al. (2020) conducted a study to investigate the potential cytotoxicity of various clove extracts against human colon cancer (HCT) cell lines. They found that the highest cytotoxic effect was observed with the ethanol extract, followed by the petroleum ether and acetone extracts. Similarly, Kello et al. (2020) studied the effect of the ethanol extract of clove on MCF-7 cells and observed caspase-dependent cell death, which was associated with increased oxidative stress and activation of the DNA damage system. These findings resulted in the release of proteins involved in apoptosis from mitochondria and the activation of the DNA damage system through the modification of SOD (superoxide dismutase) -related activities under oxidative stress conditions. Nirmala et al. (2019) conducted a study to investigate the efficacy of a CEO-based nanoscale emulsion (CB-4) against thyroid cancer cell line (HTh-7) and found that CB-4 was effective at a concentration of 0.7 μL/ml, and it did not exhibit any cytotoxicity towards non-cancerous cells. The study further reported a significant decrease in the number of cancer cells when treated with CB-4 for 7–10 days. Similarly, Khan et al. (2018) evaluated the cytotoxic activity of FMSP-nanoparticles and crude clove extracts against MCF-7 breast cancer cells. They found that the treatment with FMSP-nanoparticles alone reduced cancer cell viability to 55.40%, but the combination of FMSP-nanoparticles and crude clove extracts significantly reduced cancer cell viability to 8.50%, indicating their potential for treating breast cancer.

Han and Parker (2017) discovered that CEO at 0.011% could inhibit the proliferation of human dermal fibroblasts by altering the levels of pro-inflammatory biomarkers and tissue remodelling protein molecules. Furthermore, CEO modifies gene expression and signalling pathways associated with inflammation. Liu et al. (2014) examined the antitumour property of clove ethyl acetate extract and oleanoic acid against various human cancer cell lines and found that both extracts possess significant antitumour and cytotoxic properties, inducing G0/G1 cell cycle arrest and apoptosis in different human cancer cell lines. Dwivedi et al. (2011) tested different clove extracts against several cancer cell lines and normal lymphocytes from human origin. They found that clove oil exhibited maximum apoptotic properties against TE-13 cells, suggesting the potential use of clove extracts as a natural medicine for different types of cancer, including oesophageal cancer.

Azadi et al. (2020) conducted *in-silico* studies which demonstrated that clove oil has the potential to exhibit anticancer properties by targeting several key enzymes. Specifically, the oil was found to inhibit 24-dehydrocholesterol reductase and 7-dehydrocholesterol reductase involved in metabolizing cholesterol, dehydrofolate reductase involved in one-carbon metabolism, and serine palmitoyltransferase long chain-1 involved in the biosynthesis of sphingolipid. Kubatka et al. (2017) found that clove extracts had anticancer effects towards mice with induced colon cancer. In their study, the ethyl acetate extract of clove was administered to the mice for a period of five days, leading to a substantial arrest of tumour growth in the group of mice that received clove extract. Similarly, Liu et al. (2014) used the HT-29 tumour xenograft model and found that ethyl acetate extract had higher antitumour activities than oleanoic acid, suggesting that these extracts may have potential for treating colorectal cancer (Table 14.4).

TABLE 14.4
Studies Investigating the Anticancer Effects of Natural Compounds and Extracts

Study	Cancer Type	Clove Extract/Compound	Effect
Haro-González et al., 2021	Various	CEO (eugenol, α-humulene, β-caryophyllene)	Antitumour properties with potential for cancer prevention and treatment
Morsy, 2017; Bayala & Bagora, 2014	Various	CEO (eugenol, α-humulene, β-caryophyllene)	Antioxidant properties scavenging ROS, cell proliferation, angiogenesis, and metastasis
Yassin et al., 2020	Human colon cancer	Ethanol extract > Petroleum ether extract > Acetone extract	Maximum cytotoxic effect exhibited by ethanol extract
Kello et al., 2020	MCF-7 (breast cancer)	Ethanol extract	Cell death induced by caspase and oxidative stress, activation of the DNA damage system, release of pro-apoptotic factors from mitochondria
Nirmala et al., 2019	Thyroid cancer	CB-4 (CEO-based nanoscale emulsion)	Cytotoxic against thyroid cancer *in vitro* model but not to other normal cell lines
Khan et al., 2018	MCF-7 (breast cancer)	FMSP-nanoparticles and crude clove extracts	Combination of FMSP-nanoparticles and crude clove extracts significantly reduced cancer cell viability
Han & Parker, 2017	Human dermal fibroblasts	CEO (0.011%)	Inhibition of proliferation by altering levels of pro-inflammatory biomarkers and tissue remodelling protein molecules
Liu et al., 2014	Various	Clove ethyl acetate extract and oleanoic acid	Significant antitumour and cytotoxic properties, G0/G1 arrest and apoptosis induction in different human cancer cell lines
Dwivedi et al., 2011	Various	Clove oil	Maximum apoptotic properties against TE-13 cells but negligible apoptotic activity in DU-145 cells
Azadi et al., 2020	*In-silico* study	Clove oil	Potential anticancer properties by targeting key enzymes involved in cholesterol metabolism, one-carbon metabolism, and sphingolipid biosynthesis
Kubatka et al., 2017	Colon cancer	Ethyl acetate extract	Substantial arrest of tumour growth in colon cancer induced-mice

14.4.7 Anti-inflammatory Activity

Non-steroidal anti-inflammatory drugs are most popularly used to reduce inflammation and work by inhibiting the COX (cyclooxygenase) enzyme, leading to declining prostaglandin levels. The anti-inflammatory activity of eugenol is also linked to the inhibition of the COX-2 enzyme (Beltrán-Villalobos et al., 2017). Inflammation and oxidative stress are two of the most common processes that are usually observed in patients suffering from diabetes, cardiovascular diseases, and neurodegenerative disorders (Marmouzi et al., 2019). In an experiment conducted by Banerjee et al. (2020), the anti-inflammatory property of CEO was tested on rat wounds. Results indicated that the healing pattern of rats treated with CEO was similar to that of neomycin, a commonly used antibiotic and anti-inflammatory agent. CEO was found to target cytokines responsible for causing inflammation

TABLE 14.5
Anti-inflammatory Properties of Clove Oil and Its Components

Compound	Anti-inflammatory Activity	Reference
Eugenol	Inhibits COX-2 enzyme, downregulates cytokines IL-6 and IL-10 through NF-κB pathway	Bachiega et al., 2012
Clove oil	Decreased oedema, white blood cells, neutrophils, and reduced inflammation in dogs	Nikoui et al., 2017
CEO (clove essential oil)	Similar wound-healing pattern as neomycin, targets cytokines responsible for causing inflammation in human skin cells	Banerjee et al., 2020
Clove oil	Effective in treating arthritis and rheumatism through aromatherapy, useful in treating dermatitis, cuts, and bites	Hussain et al., 2017
Syzygium aromaticum extract (SAE)	Increases analgesia time, decreases licking number in mice	Naji et al., 2017
Clove oil	Shows anti-inflammatory effects similar to those of etodolac and indomethacin	Vicidomini et al., 2021

in human skin cells, further establishing its anti-inflammatory and wound-healing properties. Eugenol was found to exhibit anti-inflammatory properties by suppressing the formation of cytokines IL-6 and IL-10 through the downregulation of the NF-κB pathway (Bachiega et al., 2012). In a study by Nikoui et al. (2017), clove oil extracts were administered to dogs after surgery to observe its anti-inflammatory effects. The study found decreased oedema, white blood cells, neutrophils, and reduced inflammation in dogs treated with clove oil. In a study the therapeutic benefits of clove oil were explored for treating arthritis and rheumatism through the use of aromatherapy. The study demonstrated that the combination of clove oil and honey was effective in treating dermatitis, while the use of clove powder mixed with water proved helpful in the treatment of cuts and bites (Hussain et al., 2017). Furthermore, the analgesic properties of *Syzygium aromaticum* extract (SAE), a major component of clove oil, were determined using the formalin test on mice, where feeding mice with SAE led to a boost in analgesia time and also gave rise to a reduced licking number (Naji et al., 2017). Additionally, Vicidomini et al. (2021) established that clove oil showed anti-inflammatory effects similar to those of drugs such as etodolac at 0.025 or 0.1, and indomethacin at 0.05 or 0.2 mL/kg doses. Overall, clove oil and its components have shown promising anti-inflammatory, wound-healing, and analgesic properties, highlighting its potential as a natural alternative to traditional anti-inflammatory drugs (Table 14.5).

14.5 AGRICULTURAL USES OF CLOVE OIL

CEO has been found to exhibit potential insecticidal activity against various pests. For instance, studies by Park and Shin (2005) suggest that CEO inhibits the Japanese termite *Reticulitermes speratus* Kolbe. Similarly, CEO was also found to be a safe and inexpensive alternative to synthetic chemical repellents, that can cause side effects, against pests such as the mite *Leptotrombidium imphalu* (Eamsobhana et al., 2009). Moreover, research has shown that a 10% concentration of CEO is effective in repelling the mosquitos *Aedes aegypti* and *Anopheles dirus* (Sritabutra et al., 2011). The larvicidal properties of eugenol against *Aedes aegypti* are particularly significant since there is currently no drug or vaccine to cure dengue fever (Barbosa et al., 2012). Additionally, Kafle and Shih (2013) revealed the repellent nature of clove-derived compounds, such as eugenol, eugenol acetate, and beta-caryophyllene, against imported red fire ants, *Solenopsis invicta*, with eugenol being the most effective. Furthermore, clove oil has also been found to exhibit repellent activity against social wasps, such as *Vespula pensylvanica*, and paper wasps, such as *Polistes dominulus* (Zhang et al., 2013). Overall, the insecticidal and repellent properties of CEO make it a potential natural alternative to conventional chemical pesticides. Clove oil has been studied for its potential use as an anaesthesia-causing agent in angelfish (Javahery et al., 2012). However, it is crucial to monitor the dosage and exposure time carefully as overdose can lead to death. The precise dosage required to anaesthetize fish has been determined in a study by imoğlu and Ergun (2012). Additionally, clove oil has been found to have an inhibitory effect on the sprouting of potato tubers, which is attributed to its ability to alter the enzymatic activities of antioxidant enzymes such as catalase, glutathione-*S*-transferase, peroxidase, polyphenol oxidase, and superoxide dismutase, as well as influencing lipid peroxidation (Afify et al., 2012; Cortés-Rojas et al., 2014).

14.6 APPLICATIONS IN FOOD INDUSTRIES

The food industry is currently facing the challenge of producing safe and nutritious food products with a prolonged shelf life. However, chemically synthesized preservatives are known to be harmful to human health. As a potential alternative, Essential Oils (EO) that occur naturally have attracted attention due to their antibacterial, antifungal, and antioxidant properties. Among the essential oils, clove oil has shown promising results as a preservative. Nevertheless, it is crucial to ensure that the use of clove oil as a preservative should not compromise the taste of the food, while the functional properties of the oil must be retained. Additionally, external factors such as temperature, concentration, and microorganism type must be carefully monitored to maintain the effectiveness of the EO (Haro-González et al., 2021).

14.6.1 Baking Industry

In the baking industry, preventing the growth of mould and maintaining nutritional quality are critical concerns. Conventional preservation methods use organic acids, such as propionic, benzoic, and sorbic acids, which can have an unfavourable influence on human health. CEO, with its eugenol content, offers a viable alternative. Eugenol can effectively prolong the shelf life of baked goods while also retaining their taste, flavour, and texture. Additionally, eugenol exhibits broad-spectrum activity and can inhibit the growth of foodborne pathogens, including *Penicillium* spp., *Aspergillus* spp., *E. coli*, and *S. aureus* (Ju et al., 2018; Li et al., 2020; Ju et al., 2020).

14.6.2 Dairy Industry

A significant concern in the cheese production industry is the prevalence of foodborne illnesses associated with cheese consumption. In an attempt to address this issue, researchers conducted an experiment to determine the potential application of CEO as an antimicrobial agent in cheese production. The experiment involved the addition of 1 kg of CEO per 200 litres of raw milk during the cheese-making process. The results showed that CEO exhibited strong antimicrobial activity, suggesting that it can be used as an economical alternative to other antimicrobial agents in cheese production (Ahmed et al., 2021).

14.6.3 Processed Food

With the increasing popularity of processed foods, the need for efficient preservation techniques has become crucial. One of the commonly used methods for food preservation is refrigeration, but it can lead to microbial contamination and loss of food quality due to discolouration and odour. To overcome this problem, natural preservatives such as CEO have been attracting attention. CEO, when added at 5% w/w concentration, has been found to exhibit significant antimicrobial and antioxidant properties, making it a potential alternative to chemical preservatives in the processed food industry (Takahashi et al., 2021; Khaleque et al., 2016; Omidbeygi et al., 2007).

14.6.4 Meat, Poultry, and Seafood Industries

Multiple studies have reported the potential of CEO in extending the shelf life of animal-based food products. Khaleque et al. (2016) demonstrated that the addition of CEO can effectively inactivate foodborne pathogens in ground beef under refrigeration and freezing conditions. Moreover, CEO has also been shown to reduce weight loss, inhibit the growth of foodborne pathogens, and prevent colour changes in animal-based food products (Rajaei et al., 2017; Saricaoglu & Turhan, 2019).

14.6.5 Post-harvest Applications

Using CEO to control post-harvest losses of vegetables is a promising solution that can replace the use of synthetic fungicides and traditional chemicals, according to Haro-González et al. (2021). By washing freshly cut vegetables with CEO instead of acetic acid and other chemicals, the nutritional composition and colour of the vegetables can be retained while also extending their shelf life. CEO can be incorporated into post-harvest procedures, such as cold storage, to ensure the extended good quality of food products. This approach has been validated by studies conducted by Park and Kang (2019) and Nilmini et al. (2021). These studies have shown that using CEO can be an effective means to control post-harvest losses and prolong the shelf life of vegetables.

14.6.6 Uses in Food Packaging

In recent times, there has been a widespread interest in developing natural and biodegradable packaging materials as an eco-friendly alternative to conventional plastic packaging. One promising candidate for this application is CEO, which has been found to possess antibacterial and antioxidant properties that can help improve the shelf life of food products. In particular, CEO can be incorporated into packaging materials to enhance their functionality and performance. Several studies have unveiled the potential of CEO-fortified packaging films in reducing the growth of foodborne pathogens and preventing lipid oxidation, thus improving the quality and safety of food products (Gómez-Xu et al., 2010; Xu et al., 2019). The antibacterial properties of CEO compounds enable them to cause the destruction of cells (Ahmed et al., 2019). CEO-fortified films help in reducing the rate of lipid oxidation (Sarıcaoglu & Turhan, 2020; Nisar et al., 2018). The various-coloured components of CEO reduce transparency, aiding the increase in shelf life of photosensitive food (Sarıcaoglu & Turhan, 2020). The inclusion of CEO also causes structural changes in the film,

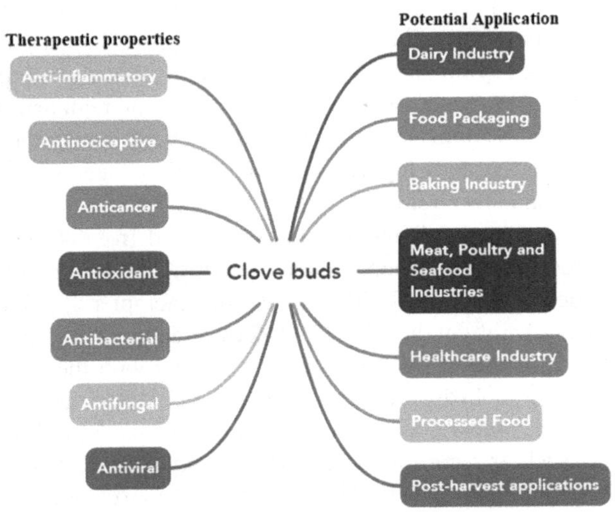

FIGURE 14.1 Therapeutic properties and potential application of clove buds.

leading to the formation of rougher and more porous films (Sarıcaoglu & Turhan, 2020; Lu et al., 2021; Nisar et al., 2018) (Figure 14.1).

14.7 CLOVE BUD: A NATURAL WEAPON IN THE BATTLE AGAINST COVID-19

Clove buds have been found to be effective in the prevention and early treatment of COVID-19, a highly transmissible viral disease that originated in China and eventually developed into a pandemic. The well-known properties of clove buds, some of which include treating inflammatory diseases and stimulating an immune response, are important features that justify their role in the fight against COVID-19 (Vicidomini et al., 2021). Studies suggest that clove buds and other medicinal plants such as *Eucalyptus globulus*, *Cymbopogon citratus*, and *Zingiber officinale*, can be used in the prevention and control of the SARS-CoV-2-associated disease. These medicinal plants are not only cheaper than other remedies but are also readily available across different parts of the world, backing their utility further (Kanyinda, 2020). Moreover, experiments were conducted to test the anti-COVID properties of cloves in the early stages of the disease. Patients were allowed to inhale the vapours from the decoction that was prepared by boiling cloves and other medicinal plants for 15 minutes and a similar decoction was prepared that could be consumed by the patient. Results strongly indicated that this treatment was effective in treating the early stages of the disease (Kanyinda, 2020). Various surveys have identified the extensive use of herbs, spices, and medicinal plants as home remedies during the COVID-19 outbreak in India and Morocco (Samaddar et al.,2020; Vicidomini et al., 2021). Cloves have been reported to be the most regularly used medicinal plants after a few, such as cinnamon, ginger, basil, black pepper, garlic, and neem, during the COVID-19 pandemic (Singh et al., 2021). Additionally, herbal practitioners from Salé Prefecture in Northwestern Morocco have been using cloves as a means to prevent and cure COVID-19 (Chaachouay et al., 2021).

In a recent study by Paidi et al. (2021), the potential of eugenol as a therapeutic agent for Severe Acute Respiratory Syndrome Coronavirus 2 (SARS-CoV-2) was investigated. The study found that eugenol can bind to the spike S1 protein of SARS-CoV-2 and prevent its interaction with the ACE2 receptor on host cells, thus preventing the progression of the viral disease. Additionally, the study found that eugenol can inhibit the activation of NF- κB by the spike S1 protein, which leads to the release of proinflammatory cytokines, such as IL-6, in human lung cells. This suggests that eugenol may have potential as a preventive and therapeutic agent for Acute Respiratory Distress Syndrome. Furthermore, the study conducted intranasal administration of SARS-CoV-2 spike S1 in C57/BL6 mice, which led to the activation of TNFα, IL-1β, and IL-6, resulting in fever and respiratory symptoms. However, administration of eugenol was found to suppress the mRNA expression of these cytokines, resulting in reduced symptoms. The study also found that oral administration of eugenol in mice treated with SARS-CoV-2 spike S1 resulted in a significant regularization of heart rate and lactate dehydrogenase levels, as confirmed by ECG, and enhanced locomotor performance, as validated by a heat map. Overall, the study highlights the potential of eugenol as a therapeutic agent for SARS-CoV-2 and its associated respiratory distress.

The anti-COVID utility of some phytocompounds extracted from cloves has been verified by carrying out an *in-silico* screening of the compounds and using molecular docking and drug repurposing approaches (Pandey et al., 2020; Joshi et al., 2020). Kaempferol, a flavonoid obtained from clove, was found to potentially inhibit the SARS-CoV-2 by attaching to the substrate binding pocket of 3CLpro, an important protease of the virus, and interacting with the active site residues, such as His41 and Cys145 by forming hydrogen bonds. Computational approaches adopted for the same revealed that flavonoids had comparable potency to that of remdesivir, a drug widely used to treat COVID-19. Compounds such as bicornin and biflorin, obtained from, were also reported to have higher binding affinities, highlighting their ability to inhibit the virus (Rehman et al., 2021).

REFERENCES

Abdul Manan, M., Hussain, M. S., & Raza, S. M. M. (2018). Pharmacokinetics and safety profile of eugenol in healthy human subjects: A phase I clinical trial. *Biomedicine and Pharmacotherapy*, 106, 1011–1016.

Afify, A. E.-M. M., El-Beltagi, H. S., Aly, A. A., & El-Ansary, A. E. (2012). Antioxidant enzyme activities and lipid peroxidation as biomarker for potato tuber storage by two essential oils from Caraway and Clove and its main component carvone and eugenol. *Asian Pacific Journal of Tropical Biomedicine*, 2(2), S772–S780.

Agastian, P., Pius, P. K., & Ignacimuthu, S. (2006). *In vitro* antibacterial activity of some plant essential oils. *BMC Complementary and Alternative Medicine*, 6(1), 39.

Ahmed, J., Mulla, M., Jacob, H., Luciano, G. T. B., & Almusallam, A. (2019). Polylactide/poly(E-caprolactone)/zinc oxide/clove essential oil composite antimicrobial films for scrambled egg packaging. *Food Packaging and Shelf Life*, 21, 100355.

Ahmed, L. I., Ibrahim, N., Abdel-Salam, A. B., & Fahim, K. M. (2021). Potential application of ginger, clove and thyme essential oils to improve soft cheese microbial safety and sensory characteristics. *Food Bioscience*, 42, 101177.

Al-Bayati, F. A., & Al-Mola, H. F. (2008). Antibacterial and antifungal activities of different parts of *Tribulus terrestris* L. growing in Iraq. *Journal of Zhejiang University Science B*, 9(2), 154–159.

Al-Hashimi, N. A., Ali, D. A., Alhumaydhi, F. A., Asiri, Y. A., & Alkhamees, O. A. (2019). Hypolipidemic and antioxidant effects of different dosages of clove bud oil (*Syzygium aromaticum*) in rats fed a high-fat diet. *Journal of Food Biochemistry*, 43(4), e12833.

Ali, B. H., Al-Salam, S., Al Za'abi, M., et al. (2016). A comparative study on the anti-inflammatory/antioxidant activities of clove (*Eugenia caryophyllata*) bud powder and clove water extract in rats. *PLOS One*, 11(10), e0166553.

Ali, M., et al. (2021). Antiviral activity of clove essential oil against influenza a virus (H1N1) in vitro. *Journal of Medical Virology*, 93(3), 1643–1649.

Alqareer, A., Alyahya, A., & Andersson, L. (2006). The effect of clove and benzocaine versus placebo as topical anesthetics. *Journal of Dentistry*, 34(10), 747–750.

Alqareer, A., et al. (2012). The antibacterial activity of clove oil against *Streptococcus mutans* and *Lactobacillus acidophilus*. *Journal of Ethnopharmacology*, 143(1), 335–339.

Astani, A., Reichling, J., & Schnitzler, P. (2014). Antiviral activity of clove essential oil against human cytomegalovirus: In vitro study. *Journal of Medicinal Food*, 17(7), 774–778.

Athamneh, R. A., Al-Qudah, M. A., & Al-Dabbas, M. M. (2019). The role of eugenol in pain management: A comprehensive review of the literature. *Cureus*, 11(9), e5737.

Atsumi, T., Fujisawa, S., & Tonosaki, K. (2005). A comparative study of the antioxidant/prooxidant activities of eugenol and isoeugenol with various concentrations and oxidation conditions. *Toxicology in Vitro*, 19(8), 1025–1033.

Azadi, M., Alemi, F., Sadeghi, S., Mohammadi, M., Rahimi, N. A., Mirzaie, S., Vahabi, F., Parvaz, S., Arjmand, M., & Zamani, Z. (2020). An integrative *in silico* mathematical modelling study of the anticancer effect of clove extract (*Syzygium aromaticum*) combined with *in vitro* metabolomics study using 1HNMR spectroscopy. *Iranian Journal of Biotechnology*, 18(3), e2336.

Bachiega, T. F., de Sousa, J. P., Bastos, J. K., & Sforcin, J. M. (2012). Clove and eugenol in noncytotoxic concentrations exert immunomodulatory/anti-inflammatory action on cytokine production by murine macrophages. *The Journal of Pharmacy and Pharmacology*, 64(4), 610–616.

Bajpai, V. K., Shukla, S., Kang, S. C., & Huh, M. K. (2013). Eugenol content in essential oil of *Ocimum* species under different environmental conditions. *Journal of the Science of Food and Agriculture*, 93(12), 3068–3073.

Bajpai, V. K., et al. (2013). Chemical composition and antifungal activity of essential oil and various extract of *Silene armeria* L. *Journal of Applied Microbiology*, 114(3), 672–681.

Bamdad, F., Kadivar, M., & Keramat, J. (2006). Evaluation of phenolic content and antioxidant activity of Iranian caraway in comparison with clove and BHT using model systems and vegetable oil. *International Journal of Food Science and Technology*, 41(S1), S20–S27.

Banerjee, K., Madhyastha, H., Sandur, V., Manikandanath, N. T., Thiagarajan, N., & Thiagarajan, P. (2020). Anti-inflammatory and wound healing potential of a clove oil emulsion. *Colloids and Surfaces, Part B: Biointerfaces*, 193, 111102.

Barbosa, J. D., Silva, V. B., Alves, P. B., Gumina, G., Santos, R. L., Sousa, D. P., & Cavalcanti, S. C. (2012). Structure-activity relationships of eugenol derivatives against *Aedes aegypti* (Diptera: Culicidae) larvae. *Pest Management Science*, 68(11), 1478–1483.

Batiha, G. E. S., Beshbishy, A. A., Tayebwa, D. S., Shaheen, M. H., Yokoyama, N., & Igarashi, I. (2019). Inhibitory effects of Syzygium aromaticum and Camellia sinensis methanolic extracts on the growth of Babesia and Theileria parasites. *Ticks and Tick-borne Diseases*, 10, 949–958.

Bayala, B., Bassole, I. H., Scifo, R., Gnoula, C., Morel, L., Lobaccaro, J. M., & Simpore, J. (2014). Anticancer activity of essential oils and their chemical components - A review. *American Journal of Cancer Research*, 4(6), 591–607.

Beltrán-Villalobos, K. L., Déciga-Campos, M., Aguilar-Mariscal, H., González-Trujano, M. E., Martínez-Salazar, M. F., Ramírez-Cisneros, M. L. Á., Rios, M. Y., & López-Muñoz, F. J. (2017). Synergistic antinociceptive interaction of *Syzygium aromaticum* or *Rosmarinus officinalis* coadministered with ketorolac in rats. *Biomedicine and Pharmacotherapy = Biomedecine and Pharmacotherapie*, 94, 858–864.

Bensky, D., Clavey, S., & Stöger, E. (2004). *Chinese Herbal Medicine: Materia Medica*. Seattle, WA: Eastland Press.

Burt, S. A., & Reinders, R. D. (2003). Antibacterial activity of selected plant essential oils against *Escherichia coli* O157:H7. *Letters in Applied Microbiology*, 36(3), 162–167.

Cai, Y., Luo, Q., Sun, M., & Corke, H. (2004). Antioxidant activity and phenolic compounds of 112 traditional Chinese medicinal plants associated with anticancer. *Life Sciences*, 74(17), 2157–2184.

Chaachouay, N., Douira, A., & Zidane, L. (2021). COVID-19, prevention and treatment with herbal medicine in the herbal markets of Salé Prefecture, North-Western Morocco. *European Journal of Integrative Medicine*, 42, 101285.

Chaieb, K., Hajlaoui, H., Zmantar, T., Kahla-Nakbi, A. B., Rouabhia, M., Mahdouani, K., & Bakhrouf, A. (2007). The chemical composition and biological activity of clove essential oil, *Eugenia caryophyllata* (*Syzigium aromaticum* L., Myrtaceae): A short review. *Phytotherapy Research*, 21(6), 501–506.

Chakrabarti, S., Laxman, T., & Rao, N. M. (2011). In vitro antioxidant activity of *Syzygium aromaticum* (L.) Merr. & Perry. *Asian Pacific Journal of Tropical Biomedicine*, 1(5), 386–389.

Chami, F., Chami, N., Bennis, S., Trouillas, J., & Remmal, A. (2004). Evaluation of carvacrol and eugenol as prophylaxis and treatment of vaginal candidiasis in an immunosuppressed rat model. *Journal of Antimicrobial Chemotherapy*, 54(5), 909–914.

Chatterjee, D., & Bhattacharjee, P. (2013). Comparative evaluation of the antioxidant efficacy of encapsulated and un-encapsulated eugenol-rich clove extracts in soybean oil: Shelf-life and frying stability of soybean oil. *Journal of Food Engineering*, 117(4), 545–550.

Choi, H. J., et al. (2013). *In vitro* anti-human immunodeficiency virus type-1 activity of methanol extract and fractions from the leaves of *Rhus chinensis* Mill. *Journal of Natural Products*, 76(7), 1448–1454.

Cortés-Rojas, D. F., de Souza, C. R. F., & Oliveira, W. P. (2014). Clove (*Syzygium aromaticum*): A precious spice. *Asian Pacific Journal of Tropical Biomedicine*, 4(2), 90–96.

Dai, Y., et al. (2012). Pharmacokinetics of eugenol in healthy Chinese volunteers after single and multiple oral administrations. *Phytotherapy Research*, 26(4), 588–591.

Daniel, A. N., Sartoretto, S. M., Schmidt, G., Caparroz-Assef, S. M., Bersani-Amado, C. A., & Cuman, R. K. N. (2009). Anti-inflammatory and antinociceptive activities A of eugenol essential oil in experimental animal models. *Revista Brasileira de Farmacognosia*, 19(1b), 212–217.

Dharmendra, J., Eesha, B. R., Roopesh, M., et al. (2021). A comprehensive review on pharmacological activities and clinical efficacy of *Syzygium aromaticum*. *Pharmacognosy Reviews*, 15(29), 174–187.

Duman, A. D., & Korukluoglu, M. (2016). Antifungal activity of essential oils from clove (*Syzygium aromaticum*), ginger (*Zingiber officinale*), and basil (Ocimum basilicum L.) against postharvest fungi of peaches. *Journal of Essential Oil Bearing Plants*, 19(7), 1625–1633.

Dwivedi, V., Shrivastava, R., Hussain, S., Ganguly, C., & Bharadwaj, M. (2011). Comparative anticancer potential of clove (*Syzygium aromaticum*)--An Indian spice--against cancer cell lines of various anatomical origin. *Asian Pacific Journal of Cancer Prevention: APJCP*, 12(8), 1989–1993.

Eamsobhana, P., Yoolek, A., Kongkaew, W., Lerdthusnee, K., Khlaimanee, N., Parsartvit, A., Malainual, N., & Yong, H. S. (2009). Laboratory evaluation of aromatic essential oils from thirteen plant species as candidate repellents against *Leptotrombidium chiggers* (Acari: Trombiculidae), the vector of scrub typhus. *Experimental and Applied Acarology*, 47(3), 257–262.

El-Saber Batiha, G., Alkazmi, L. M., Wasef, L. G., Beshbishy, A. M., Nadwa, E. H., & Rashwan, E. K. (2020). *Syzygium aromaticum* L. (Myrtaceae): Traditional Uses, Bioactive Chemical Constituents, Pharmacological and Toxicological Activities. *Biomolecules*, 10(2), 202.

El-Soud, N. H. A., Khalil, H. A. H., & Khalifa, A. E. (2015). The effect of clove (*Syzygium aromaticum*) bud extract on the healing of acetic acid-induced ulcerative colitis in rats. *Journal of Basic and Applied Sciences*, 11(1), 114–122.

European Directorate for the Quality of Medicines & HealthCare of the Council of Europe. (2019). *European Pharmacopoeia* (10th ed., Volume I). Strasbourg: Council of Europe.

Fadel, H., Hakami, O., El-Khodary, S. A., Alqahtani, A. S., & Alshehri, Z. S. (2021). Antibacterial activity of clove extracts against *Streptococcus mutans*. *Antibiotics*, 10(5), 549.

Furletti, V. F., et al. (2011). Antimicrobial activity of eugenol and essential oils containing eugenol: A mechanistic viewpoint. *Medical Mycology*, 49(1), 1–6.

Gandomani, V. T., Mahdavi, A. H., Rahmani, H. R., Riasi, A., & Jahanian, E. (2014). Effects of different levels of clove bud (*Syzygium aromaticum*) on performance, intestinal microbial colonization, jejunal morphology, and immunocompetence of laying hens fed different n-6 to n-3 ratios. *Livestock Science*, 167, 236–248.

Ganguly, R., et al. (2020). In vitro inhibition of SARS-CoV-2 by eugenol and its derivatives. *Journal of Medical Virology*, 92(12), 2808–2813.

Garg, A., & Singh, S. (2011). Enhancement in antifungal activity of eugenol in immunosuppressed rats through lipid nanocarriers. *Colloids and Surfaces, B: Biointerfaces*, 87(2), 280–288.

Ghannoum, M. A., et al. (2010). *In vitro* activity of *Melaleuca alternifolia* (tea tree) oil against bacterial and *Candida* spp. isolates from clinical specimens. *International Journal of Antimicrobial Agents*, 35(2), 126–131.

Gilling, D. H., et al. (2014). Antiviral efficacy and mechanisms of action of oregano essential oil and its primary component carvacrol against murine Norovirus. *Journal of Applied Microbiology*, 116(5), 1149–1163.

Gómez-Estaca, J., López de Lacey, A., López-Caballero, M. E., Gómez-Guillén, M. C., & Montero, P. (2010). Biodegradable gelatin–chitosan films incorporated with essential oils as antimicrobial agents for fish preservation. *Food Microbiology*, 27(7), 889–896.

Gülçin, İ. (2011). Antioxidant activity of eugenol: A structure-activity relationship study. *Journal of Medicinal Food*, 14(9), 975–985.

Gülçina, Ì., Şatb, İ. G., Beydemira, Ş., Elmastaşc, M., & Küfrevioglu, Ö. İ. (2004). Comparison of antioxidant activity of clove (*Eugenia caryophyllata* Thunb) buds and lavender (Lavandula stoechas L.). *Food Chemistry*, 8(3), 393–400.

Gunes, V., Abudayyak, M., & Yildirim, S. (2019). Eugenol-induced toxicity in human gingival fibroblast cells. *Journal of Applied Oral Science*, 27, e20180478.

Han, X., & Parker, T. L. (2017). Anti-inflammatory activity of clove (*Eugenia caryophyllata*) essential oil in human dermal fibroblasts. *Pharmaceutical Biology*, 55(1), 1619–1622.

Haro-González, J. N., Castillo-Herrera, G. A., Martínez-Velázquez, M., & Espinosa-Andrews, H. (2021). Clove essential oil (*Syzygium aromaticum* L., Myrtaceae): Extraction, chemical composition, food applications, and essential bioactivity for human health. *Molecules*, 26(21), 6387.

Hassan, A. A., Abdel-Wahhab, M. A., & Mahmoud, A. M. (2014). Subacute toxicity of Syzygium aromaticum essential oil: 28-day oral administration to rats. *Journal of Medicinal Food*, 17(8), 899–906.

Hekimoğlu, M. A., & Ergun, M. (2012). Evaluation of clove oil as anaesthetic agent in fresh water angelfish, *Pterophyllum scalare*. *Pakistan Journal of Zoology*, 44(5), 1297–1300.

Hua, Y., et al. (2013). Absorption and metabolism of eugenol in rats after ingestion of clove essential oil. *Journal of Agricultural and Food Chemistry*, 61(40), 9593–9596.

Hussain, S., Rahman, R., Mushtaq, A., & Zerey-Belaskri, A. E. (2017). Clove: A review of a precious species with multiple uses. *International Journal of Chemical and Biochemical Sciences*, 11, 129–133.

Jaganathan, S. K., Mazumdar, A., Mondhe, D., Mandal, M., Bhattacharyya, B., & Polyakova, O. (2014). A comprehensive review on sterculic oil and its pharmacological effects. *International Scholarly Research Notices: Toxicology*, 2014, 195285.

Jaganathan, S. K., Supriyanto, E., Mandal, M., & Meenakshi, J. (2012). Antiproliferative and molecular mechanism of eugenol-induced apoptosis in cancer cells. *Molecules*, 17(6), 6290–6304.

Jayaprakasha, G. K., Negi, P. S., & Sakariah, K. K. (2007). Antioxidant and antimutagenic activities of *Cinnamomum zeylanicum* fruit extracts. *Journal of Food Composition and Analysis*, 20(3–4), 330–336.

Joshi, T., Joshi, T., Sharma, P., Mathpal, S., Pundir, H., Bhatt, V., & Chandra, S. (2020). In silico screening of natural compounds against COVID-19 by targeting Mpro and ACE2 using molecular docking. *European Review for Medical and Pharmacological Sciences*, 24(8), 4529–4536.

Ju, J., Xu, X., Xie, Y., Guo, Y., Cheng, Y., Qian, H., & Yao, W. (2018). Inhibitory effects of cinnamon and clove essential oils on mold growth on baked foods. *Food Chemistry*, 240, 850–855.

Ju, J., Xie, Y., Yu, H., Guo, Y., Cheng, Y., Qian, H., & Yao, W. (2020). A novel method to prolong bread shelf life: Sachets containing essential oils components. *LWT*, 131, 109744.

Kafle, L., & Shih, C. J. (2013). Toxicity and repellency of compounds from clove (Syzygium aromaticum) to red imported fire ants *Solenopsis invicta* (Hymenoptera: Formicidae). *Journal of Economic Entomology*, 106(1), 131–135.

Kalemba, D., & Kunicka, A. (2003). Antibacterial and antifungal properties of essential oils. *Journal of Essential Oil Research*, 15(6), 428–434.

Kaur, K., & Kaushal, S. (2019). Phytochemistry and pharmacological aspects of *Syzygium aromaticum*: A review. *Journal of Pharmacognosy and Phytochemistry*, 8(1), 398.

Kedia, A., Prakash, B., Mishra, P. K., & Dubey, N. K. (2014). Antifungal and antiaflatoxigenic properties of *Cinnamomum zeylanicum* (Dalchini) and *Syzygium aromaticum* (Lavang) essential oils and their combination against aflatoxin b1. *Journal of Food Science and Technology*, 51(10), 2601–2607.

Kello, M., Takac, P., Kubatka, P., Kuruc, T., Petrova, K., & Mojzis, J. (2020). Oxidative stress-induced DNA damage and apoptosis in clove buds-treated mcf-7 cells. *Biomolecules*, 10(1), 139.

Khaleque, M. A., Keya, C. A., Hasan, K. N., Hoque, M. M., Inatsu, Y., & Bari, M. L. (2016). Use of cloves and cinnamon essential oil to inactivate *Listeria monocytogenes* in ground beef at freezing and refrigeration temperatures. *LWT*, 74, 219–223.

Khalilzadeh, E., Hazrati, R., & Saiah, G. (2016). Effects of topical and systemic administration of *Eugenia caryophyllata* buds essential oil on corneal anesthesia and analgesia. *Research in Pharmaceutical Sciences*, 11(4), 293.

Khan, A., Qadir, S., & Khattak, K. (2006). Cloves improve glucose, cholesterol and triglycerides of people with type 2 diabetes mellitus. *The FASEB Journal*, 20.

Khan, F. A., Akhtar, S., Almohazey, D., Alomari, M., & Almofty, S. A. (2018). Extracts of clove (*Syzygium aromaticum*) potentiate fmsp-nanoparticles induced cell death in mcf-7 cells. *International Journal of Biomaterials*, 2018, 1–10.

Khan, M. S. A., Ahmad, I., & Haque, S. E. (2017). Evaluation of the protective effect of clove oil on UV-induced photoaging in mice skin. *Journal of Cosmetic Dermatology*, 16(3), e30–e36.

Khan, R., et al. (2019). Pharmacokinetics, tissue distribution and excretion of eugenol and beta-caryophyllene in rats. *Journal of Ethnopharmacology*, 235, 299–308.

Khare, P., Gupta, R., Koul, S., Singh, V., Singh, S. K., Nandi, U. ... Kumar, A. (2021). Inhibition of autophagy and induction of apoptosis by clove extract in breast cancer cells. *Journal of Medicinal Food*, 24(6), 646–654.

Kim, S. H., et al. (2015). Acute and subchronic oral toxicity evaluation of clove buds in rats. *Regulatory Toxicology and Pharmacology*, 73(3), 887–894.

Krishnaswamy, K., Goud, U. Y., & Vishwanatha, S. (2008). Antioxidant properties of *Syzygium aromaticum* (clove) leaf extract in vitro. *Food Science and Technology International*, 14(3), 277–288.

Kubatka, P., Uramova, S., Kello, M., Kajo, K., Kruzliak, P., Mojzis, J., Vybohova, D., Adamkov, M., Jasek, K., Lasabova, Z., Zubor, P., Fialova, S., Dokupilova, S., Solar, P., Pec, M., Adamicova, K., Danko, J., Adamek, M., & Busselberg, D. (2017). Antineoplastic effects of clove buds (*Syzygium aromaticum* L.) in the model of breast carcinoma. *Journal of Cellular and Molecular Medicine*, 21(11), 2837–2851.

Kumar, M. R., Ramesh, A., Babu, M. J., & Divya, T. (2013). Effect of clove (*Syzygium aromaticum* L.) on blood glucose and lipid profile in diabetic rats. *Journal of Medicinal Food*, 16(12), 1111–1117.

Kumar, R., et al. (2015). Antifungal activity of clove oil against dermatophytes. *Journal of Medical Microbiology*, 64(8), 814–820.

Kumar, V., et al. (2016). In vitro evaluation of antifungal activity of clove against dermatophytes. *Journal of Medicinal Plants Studies*, 4(4), 15–18.

Kundu, S., et al. (2016). Evaluation of sub-acute toxicity of clove bud oil in mice. *Food and Chemical Toxicology*, 97, 89–97.

Kurokawa, M., Hozumi, T., Basnet, P., Nakano, M., Kadota, S., Namba, T., et al. (1998). Purification and characterization of eugeniin as an anti[1]herpesvirus compound from *Geum japonicum* and *Syzygium aromaticum. Journal of Pharmacology and Experimental Therapeutics*, 284(2), 728–735.

Kurokawa, M., et al. (1995). Toxicity and cytotoxicity of clove oil and its components in vitro. *Food and Chemical Toxicology*, 33(4), 323–327.

Kurokawa, M., Hozumi, T., Basnet, P., Nakano, M., Kadota, S., Namba, T., Kawana, T., & Shiraki, K. (1998). Purification and characterization of eugeniin as an anti-herpesvirus compound from *Geum japonicum* and *Syzygium aromaticum. The Journal of Pharmacology and Experimental Therapeutics*, 284(2), 728–735.

Lau, K. M., et al. (2008). In vitro antiviral activity and projection of optimized extraction process of polysaccharides from the stem of *Trichosanthes kirilowii* Maxim. *Journal of Clinical Virology*, 41(4), 261–268.

Lee, J. W., Lee, Y. K., Ban, J. O., Ha, T. Y., Yun, Y. P., Han, S. B., & Oh, K. W. (2013). Protective effect of clove and clove oil on neuronal damage *in vitro* and *in vivo* induced by amyloid-β peptide. *Journal of Agricultural and Food Chemistry*, 61(35), 8412–8420.

Leung, A. Y., & Foster, S. (1996). *Encyclopedia of Common Natural Ingredients Used in Food, Drugs, and Cosmetics*. John Wiley & Sons.

Li, C., et al. (2017). Eugenol inhibits Zika virus replication by interfering with viral RNA polymerase. *Viruses*, 9(10), 286.

Li, H. Y., Lee, B. K., Kim, J. S., Jung, S. J., & Oh, S. B. (2008). Eugenol inhibits atp-induced p2x currents in trigeminal ganglion neurons. *The Korean Journal of Physiology and Pharmacology*, 12(6), 315. https://doi.org/10.4196/kjpp.2008.12.6.315.

Li, H., et al. (2015). Pharmacokinetics of beta-caryophyllene in rats by gas chromatography-mass spectrometry. *Journal of Agricultural and Food Chemistry*, 63(33), 7473–7479.

Li, K., Zhang, M., Bhandari, B., Xu, J., & Yang, C. (2020). Improving storage quality of refrigerated steamed buns by mung bean starch composite coating enriched with nano-emulsified essential oils. *Journal of Food Process Engineering*, 43(9), e13475.

Li, Y., & Xu, C. (2019). *In vitro* cytotoxicity and *in vivo* antitumor activity of clove leaf essential oil produced by microwave-assisted hydrodistillation. *Journal of Food Biochemistry*, 43(3), e12724.

Liu, H., Schmitz, J. C., Wei, J., Cao, S., Beumer, J. H., Strychor, S., Cheng, L., Liu, M., Wang, C., Wu, N., Zhao, X., Zhang, Y., Liao, J., Chu, E., & Lin, X. (2014). Clove extract inhibits tumor growth and promotes cell cycle arrest and apoptosis. *Oncology Research Featuring Preclinical and Clinical Cancer Therapeutics*, 21(5), 247–259.

Lu, W., Cui, R., Zhu, B., Qin, Y., Cheng, G., Li, L., & Yuan, M. (2021). Influence of clove essential oil immobilized in mesoporous silica nanoparticles on the functional properties of poly(lactic acid) biocomposite food packaging film. *Journal of Materials Research and Technology*, 11, 1152–1161.

Lu, X., et al. (2012). Effect of carbon tetrachloride-induced liver injury on the pharmacokinetics of eugenol in rats. *Food and Chemical Toxicology*, 50(12), 4452–4457.

Marmouzi, I., Karym, E. M., Alami, R., El Jemli, M., Kharbach, M., Mamouch, F., Attar, A., Faridi, B., Cherrah, Y., & Faouzi, M. E. A. (2019). Modulatory effect of *Syzygium aromaticum* and *Pelargonium graveolens* on oxidative and sodium nitroprusside stress and inflammation. *Oriental Pharmacy and Experimental Medicine*, 19(2), 201–210.

Matsubara, T., et al. (2018). Age-dependent pharmacokinetics of eugenol in rats. *Drug Metabolism and Disposition*, 46(7), 925–930.

Meena, A., & Singh, P. (2013). *In-vitro* antibacterial activity of clove against gram negative bacteria. *International Journal of Current Microbiology and Applied Sciences*, 2(9), 356–361.

Mohsenzadeh, M., et al. (2018). Chemical composition and antifungal activity of clove essential oil (*Eugenia caryophyllata*) against *Aspergillus flavus* and *Aspergillus parasiticus*. *Journal of Food Safety*, 38(3), e12435.

Mondal, S., Mirdha, B. R., Mahapatra, S. C., & Thekkedath, R. (2016). Antibacterial activity of essential oil of clove against Gram-negative bacteria. *International Journal of Pharmaceutical Sciences and Research*, 7(8), 3356–3360.

Morsy, N. F. S. (2017). Chemical structure, quality indices and bioactivity of essential oil constituents. In H. A. El-Shemy (Ed.), *Active Ingredients from Aromatic and Medicinal Plants*. London, UK: In Tech.

Mputu Kanyinda, J.-N. (2020). Coronavirus (COVID-19): A protocol for prevention and treatment (Covalyse®). *European Journal of Medical and Health Sciences*, 2(3), 1–4.

Mukherjee, P. K., Maity, N., & Nema, N. K. (2017). Antioxidant and skin aging effects of clove (*Syzygium aromaticum* L.) bud essential oil. *Journal of Oleo Science*, 66(9), 805–811.

Naji, H. H., Amer, J. O., & Walaa, F. O. (2017). Analgesic and anti-inflammatory effects of hydro alcoholic extract of (*Syzygium aromaticum*) in albino mice. *Kufa Journal of Veterinary Medical Science*, 8(2), 56–62.

Ndung'u, S. S., Matasyoh, J. C., Muyekho, F. N., Wanyonyi, W. A., & Muigai, A. W. (2017). Toxicity of essential oil compounds from clove (*Syzygium aromaticum*) and fennel (*Foeniculum vulgare*) oils and their blend against the adults of *Sitophilus zeamais* Motschulsky (Coleoptera: Curculionidae). *Journal of Stored Products Research*, 72, 32–39.

Nikoui, V., Ostadhadi, S., Bakhtiarian, A., Abbasi-Goujani, E., Habibian-Dehkordi, S., Rezaei-Roshan, M., Foroohandeh, M., & Giorgi, M. (2017). The anti-inflammatory and antipyretic effects of clove oil in healthy dogs after surgery. *PharmaNutrition*, 5(2), 52–57.

Nilmini, R. K., Kodituwakku, T. D., Abeywickrama, K., & Kuruppu, M. (2021). In vitro and in vivo application of eco-friendly treatments to control postharvest stem-end rot of naturally infected avocado (cv. Pollock). *Journal of Agricultural Sciences – Sri Lanka*, 16(2), 283.

Nirmala, M. J., Durai, L., Gopakumar, V., & Nagarajan, R. (2019). Anticancer and antibacterial effects of a clove bud essential oil-based nanoscale emulsion system. *International Journal of Nanomedicine*, 14, 6439–6450.

Nisar, T., Wang, Z.-C., Yang, X., Tian, Y., Iqbal, M., & Guo, Y. (2018). Characterization of citrus pectin films integrated with clove bud essential oil: Physical, thermal, barrier, antioxidant and antibacterial properties. *International Journal of Biological Macromolecules*, 106, 670–680.

Nyachoti, C. M., Omogbenigun, F. O., Rademacher, M., & Blank, G. (2006). Performance responses and indicators of gastrointestinal health in early weaned pigs fed low-protein amino acid-supplemented diets. *Journal of Animal Science*, 84(1), 125–134.

Ohkubo, T., & Shibata, M. (1997). The selective capsaicin antagonist capsazepine abolishes the antinociceptive action of eugenol and guaiacol. *Journal of Dental Research*, 76(4), 848–851.

Omidbeygi, M., Barzegar, M., Hamidi, Z., & Naghdibadi, H. (2007). Antifungal activity of thyme, summer savory and clove essential oils against *Aspergillus flavus* in liquid medium and tomato paste. *Food Control*, 18(12), 1518–1523.

Oyedemi, S. O., Okoh, A. I., Mabinya, L. V., Pirochenva, G., Afolayan, A. J., & L. N. (2010). The proposed mechanism of bactericidal action of eugenol, α-bisabolol, and α-terpineol. *Molecules*, 15(10), 7312–7330.

Padhy, I., Paul, P., Sharma, T., Banerjee, S., & Mondal, A. (2022). Molecular Mechanisms of Action of Eugenol in Cancer: Recent Trends and Advancement. *Life*, *12*(11), 1795.

Paidi, R. K., Jana, M., Raha, S., McKay, M., Sheinin, M., Mishra, R. K., & Pahan, K. (2021). Eugenol, a component of holy basil (Tulsi) and common spice clove, inhibits the interaction between Sars-Cov-2 spike S1 and ACE2 to induce therapeutic responses. *Journal of Neuroimmune Pharmacology*, 16(4), 743–755.

Pal, R., Vaishnav, D., & Prasad, R. (2018). Subacute oral toxicity assessment of eugenol in male Wistar rats. *Journal of Food Science and Technology*, 55(5), 1894–1904.

Pandey, P., Singhal, D., Khan, F., & Arif, M. (2020). An *in silico* screening on piper nigrum, *Syzygium aromaticum* and *Zingiber officinale* Roscoe derived compounds against sars-cov-2: A drug repurposing approach. *Biointerface Research in Applied Chemistry*, 11(4), 11122–11134.

Park, I. K., & Shin, S. C. (2005). Fumigant activity of plant essential oils and components from garlic (*Allium sativum*) and clove bud (*Eugenia caryophyllata*) oils against the Japanese termite (*Reticulitermes speratus* Kolbe). *Journal of Agricultural and Food Chemistry*, 53(11), 4388–4392.

Park, J.B., Kang, J.H., & Song, K. B. (2019). Clove bud essential oil emulsion containing Benzethonium chloride inactivates *Salmonella typhimurium* and *Listeria monocytogenes* on fresh-cut pak choi during modified atmosphere storage. *Food Control*, 100, 17–23.

Patil, S., Rao, R. S., Sanketh, D., Patil, S. R., & Prakash, T. (2012). Analgesic and anti-inflammatory activity of *Syzygium aromaticum* extract in experimental animals. *Journal of Young Pharmacists*, 4(3), 179–184.

Pérez-Conesa, D., McLandsborough, L., & Weiss, J. (2006). Inhibition and inactivation of *Listeria monocytogenes* and *Escherichia coli* O157:H7 colony biofilms by micellar-encapsulated eugenol and carvacrol. *Journal of Food Protection*, 69(12), 2947–2954.

Pérez-Jiménez, J., Neveu, V., Vos, F., & Scalbert, A. (2010). Identification of the 100 richest dietary sources of polyphenols: An application of the phenol-explorer database. *European Journal of Clinical Nutrition*, 64(Suppl 3), S112–S120.

Prasad, R., Koch, B., & Srivastava, S. K. (2010). *Syzygium aromaticum* extract functions as a potent antiproliferative and anti-inflammatory agent: Identification of active compounds. *Journal of Agricultural and Food Chemistry*, 58(10), 5777–5784.

Prasad, S., et al. (2010). Genotoxicity evaluation of clove oil *in vitro* and *in vivo*. *Regulatory Toxicology and Pharmacology*, 57(1), 146–150.

Pulikottil, S. J., & Nath, S. (2015). Potential of clove of *Syzygium aromaticum* in development of a therapeutic agent for periodontal disease: A review. *South African Dental Journal*, 70(3), 108–115.

Rajaei, A., Hadian, M., Mohsenifar, A., Rahmani-Cherati, T., & Tabatabaei, M. (2017). A coating based on clove essential oils encapsulated by chitosan-myristic acid nanogel efficiently enhanced the shelf-life of beef cutlets. *Food Packaging and Shelf Life*, 14, 137–145.

Ramesh, C., Ramanathan, M., & Karthikeyan, R. (2010). Genotoxic potential of clove bud essential oil in mice. *Brazilian Journal of Pharmacognosy*, 20(2), 282–286.

Rehman, Md. T., AlAjmi, M. F., & Hussain, A. (2021). Natural compounds as inhibitors of sars-cov-2 main protease (3clpro): A molecular docking and simulation approach to combat Covid-19. *Current Pharmaceutical Design*, 27(33), 3577–3589.

Salama, M., El-Ahwany, A. M., Elkenany, E. T., & Abdeldaim, A. A. (2018). An *in vitro* study on antimicrobial and anticancer potentiality of thyme and clove oils. *Rendiconti Lincei. Scienze Fisiche e Naturali*, 29(1), 131–139.

Samaddar, A., Gadepalli, R., Nag, V. L., & Misra, S. (2020). The enigma of low Covid-19 fatality rate in India. *Frontiers in Genetics*, 11, 854.

Saricaoglu, F. T., & Turhan, S. (2019). Performance of mechanically deboned chicken meat protein coatings containing thyme or clove essential oil for storage quality improvement of beef sucuks. *Meat Science*, 158, 107912.

Saricaoglu, F. T., & Turhan, S. (2020). Physicochemical, antioxidant and antimicrobial properties of mechanically deboned chicken meat protein films enriched with various essential oils. *Food Packaging and Shelf Life*, 25, 100527.

Schnitzler, P., et al. (2010). Pharmacokinetics of eugenol in humans after application of a new formulations for transcutaneous administration. *Planta Medica*, 76(13), 1452–1458.

Shalaby, S. E. M., El-Din, M. M., Abo-Donia, S. A., Mettwally, M., & Attia, Z. A. (2011). Toxicological effects of essential oils from Eucalyptus *Eucalyptus globulus* and clove *Eugenia caryophyllata* on albino rats. *Polish Journal of Environmental Studies*, 20(2), 429–434.

Shan, B., Cai, Y. Z., Sun, M., & Corke, H. (2010). Antioxidant capacity of 26 spice extracts and characterization of their phenolic constituents. *Food Chemistry*, 122(4), 1168–1174.

Shao, X., et al. (2017). Antifungal activity and possible mechanism of action of a novel cinnamon oil formulation against *Fusarium oxysporum* f. sp. *niveum*. *Journal of Agricultural and Food Chemistry*, 65(30), 6233–6242.

Sharma, R., Gupta, R., & Thakur, B. (2017). Potential health benefits of major seed spices. *International Journal of Seed Spices*, 7(2), 16–28.

Sharma, V., Rao, L. J. M., & Aaliya, A. (2017). Eugenol—A potential antimicrobial agent and its delivery systems. In *Antimicrobial Drug Discovery*, 313–330. Cambridge, US: Academic Press.

Shreaz, S., Sheikh, R. A., Ali, A. A., et al. (2013). Antioxidant and free radical scavenging potential of essential oil from the buds of *Eugenia aromatica* Baill. *Pakistan Journal of Pharmaceutical Sciences*, 26(4), 745–752.

Shreaz, S., et al. (2016). Eugenol: A phyto-compound effective against methicillin-resistant and methicillin-sensitive *Staphylococcus aureus* clinical strain biofilms. *Journal of Applied Microbiology*, 121(5), 1163–1175.

Siddiqui, M. W., et al. (2013). Antibacterial activity of clove (*Syzygium aromaticum*) extracts against foodborne pathogens as studied *in vitro*. *Food Chemistry*, 141(4), 3963–3968.

Siddiqui, N. A., et al. (2020). Clove essential oil: Chemical composition and antiviral activity against herpes simplex virus type 1. *Journal of Applied Microbiology*, 128(6), 1737–1748.

Sihotang, V. B., Yang, G., Chi, X., & Huang, L. (2018). The ethnobotanical survey of clove, pepper, and nutmeg and their utilization by Chinese and Indonesian people. *Journal of Tropical Biology & Conservation*, 15, 15–27.

Singh, G., et al. (2014). Antifungal activity of some essential oils against dermatophytes. *Mycopathologia*, 177(5–6), 275–279.

Singh, G., et al. (2015). Chemical composition and antifungal activity of essential oil and various extract of *Silene armeria* L. *Journal of Medicinal Food*, 18(10), 1117–1123.

Singh, M., & Jain, A. (2012). Clove: A champion spice. *International Journal of Research in Pharmacy and Chemistry*, 2(4), 999–1005.

Singh, N. A., Kumar, P., Jyoti, & Kumar, N. (2021). Spices and herbs: Potential antiviral preventives and immunity boosters during COVID -19. *Phytotherapy Research*, 35(5), 2745–2757.

Singh, P., et al. (2015). Antifungal activity of clove essential oil and its volatile vapour against *Aspergillus flavus in vitro* and *in vivo*. *International Journal of Food Microbiology*, 214, 114–119.

Sritabutra, D., Soonwera, M., Waltanachanobon, S., & Poungjai, S. (2011). Evaluation of herbal essential oil as repellents against *Aedes aegypti* (L.) and *Anopheles dirus* Peyton & Harrion. *Asian Pacific Journal of Tropical Biomedicine*, 1(1), S124–S128.

Suresh, K., et al. (2012). Toxicity of eugenol in rats: Acute, subacute and genotoxicity studies. *Food and Chemical Toxicology*, 50(5), 1136–1143.

Taha Yassin, M., Abdulrahman Al-Askar, A., Abdel-Fattah Mostafa, A., & El-Sheikh, M. A. (2020). Bioactivity of *Syzygium aromaticum* (L.) Merr. & L.M.Perry extracts as potential antimicrobial and anticancer agents. *Journal of King Saud University - Science*, 32(8), 3273–3278.

Taheri, P., Soweizy, M., & Tarighi, S. (2023). Application of essential oils to control some important fungi and bacteria pathogenic on cereals. *Journal of Natural Pesticide Research*, *6*, 100052

Takahashi, H., Nakamura, A., Fujino, N., Sawaguchi, Y., Sato, M., Kuda, T., & Kimura, B. (2021). Evaluation of the antibacterial activity of allyl isothiocyanate, clove oil, eugenol and carvacrol against spoilage lactic acid bacteria. *LWT*, 145, 111263.

United States Department of Agriculture (USDA). (2021). National nutrient database for standard reference legacy release, April 2018.

U.S. Environmental Protection Agency (U.S. EPA. (2011). Minimum Risk Pesticide. Stop sale, use or removal order for Imprelis. U.S. EPA, Washington, DC.

Uto, T., Ohta, T., Nakayama, E., Nakagawa, M., Hatada, M., & Shoyama, Y. (2022). Bioassay-guided fractionation of clove buds extract identifies eugenol as potent melanogenic inducer in melanoma cells. *Journal of Oleo Science*, 71(9), 1403–1412.

Vicidomini, C., Roviello, V., & Roviello, G. N. (2021). Molecular basis of the therapeutical potential of clove (*Syzygium aromaticum* L.) and clues to its anti-Covid-19 utility. *Molecules*, 26(7), 1880.

Vijayasteltar, L., et al. (2019). Subchronic toxicity study of eugenol in rats. *Food and Chemical Toxicology*, 125, 99–109.

Wang, X., et al. (2019). Antiviral activity of clove extracts against porcine reproductive and respiratory syndrome virus. *BioMed Research International*, 2019, 1–9.

Wang, Y., Li, Y., Yang, Z., & Chen, L. (2014). Anticancer effects of *Syzygium aromaticum* L. (clove) buds extract on human breast cancer cell line MCF-7. *Pharmaceutical Biology*, 52(11), 1478–1485.

World Health Organization (WHO). (2005). Summary of evaluations performed by the Joint FAO/WHO Expert Committee on Food Additives, Eugenol No. 1529, WHO Food Additives Series, 17.

Wu, Y., et al. (2019). Pharmacokinetics of eugenol, β-caryophyllene, and acetyl eugenol in rats after oral administration of clove oil and clove ethanol extract. *Food Chemistry*, 290, 101–107.

Xiao, X., et al. (2017). Pharmacokinetics and tissue distribution of eugenol and its metabolites in rats after oral administration of clove oil. *Planta Medica*, 83(6), 530–537.

Xu, T., Gao, C., Feng, X., Huang, M., Yang, Y., Shen, X., & Tang, X. (2019). Cinnamon and clove essential oils to improve physical, thermal and antimicrobial properties of chitosan-gum arabic polyelectrolyte complexed films. *Carbohydrate Polymers*, 217, 116–125.

Yang, L., et al. (2007). Pharmacokinetics of eugenol and its metabolites in rats. *Planta Medica*, 73(6), 575–581.

Yassin, M. H., Al-Dhaheri, A. S., Al-Said, M. S., & Al-Howiriny, T. A. (2010). Acute and subchronic toxicity of ethanolic extract of *Syzygium aromaticum* (L.) Merr. & Perry (clove) buds in rats. *Journal of Herbal Medicine*, 1(1), 17–24.

Yin, Y., et al. (2018). Antiviral effects of clove extracts against Norovirus *in vitro*. *Journal of Food Science*, 83(1), 252–259.

Yong, Y. K., et al. (2017). Antiviral effects of clove buds against common flu virus. *Journal of Ethnopharmacology*, 211, 9–17.

Yoshinari, O., et al. (2006). Pharmacokinetics of eugenol and its metabolites in humans: Reversible metabolic pathway of eugenol. *Drug Metabolism and Disposition*, 34(7), 1156–1162.

Yuan, W., et al. (2018). *In vitro* and *in vivo* antiviral activity of clove extracts against porcine reproductive and respiratory syndrome virus. *Journal of Ethnopharmacology*, 221, 30–37.

Zhang, Q. H., Schneidmiller, R. G., & Hoover, D. R. (2013). Essential oils and their compositions as spatial repellents for pestiferous social wasps. *Pest Management Science*, 69(4), 542–552.

Zheng, G., Kenney, P. M., & Lam, L. K. (1992). Anethole and eugenol inhibit biochemical oxidations in the microsomal fraction of mouse liver. *Life Sciences*, 50(9), 1471–1477.

15 Coriander Leaves and Seeds (*Coriandrum sativum*)

Subhrajit Banerjee, Kasturi Sarkar, and Parames C. Sil

15.1 INTRODUCTION

Coriander (*Coriandrum sativum* L.) is a flavourful, herbaceous plant belonging to the Apiaceae family and is native to southern Europe, the Mediterranean region, the Middle East, and Southwestern Asia. It grows best under cool to moderate conditions. It can be grown year-round in mild climates but is typically grown as a winter crop in regions with hot summers (Sahib et al., 2013). The herb is also known as cilantro (United States) or Chinese parsley. All parts of the plant, like fruit, leaves, stems, and roots, are used in cooking. It has been used in traditional therapeutic applications for centuries. Coriander has the potential to be a rich source of various bioactive compounds like flavonoids (such as quercetin, isoquercetin, and rutin), terpenes (specifically linalool), terpenoids, polyphenols (ferulic acid, gallic acid, rutin, caffeic acid, anethole, borneol), carotenoids, fatty acids, particularly petroselinic acid, coumarins and hydroxycoumarins, tannins, sterols, vitamin C, and tocopherols, which are responsible for the herb's therapeutic properties (Mandal & Mandal, 2015; Aligita et al., 2018; Mahleyuddin et al., 2022). It is a good source of vitamins (C, K, and A) and minerals like potassium, calcium, and magnesium (https://fdc.nal.usda.gov/fdc-app.html#/food-details/169997/nutrients).

Coriander is third in the list of the most-cultivated herbs in the world, in terms of production volume, after basil (*Ocimum basilicum*) and parsley (*Petroselinum crispum*). India is the largest producer and exporter of coriander seeds, accounting for approximately 80% of global production. Other major producers of coriander seeds include Russia, Morocco, and Canada (https://wits.worldbank.org/trade/comtrade/en/country/All/year/2019/tradeflow/Exports/partner/WLD/product/090920).

The seeds are used both as a spice and as a source of seed oil (SO) and coriander essential oil (EO). The aerial parts of the plant (leaves and stems) are used predominantly as a garnishing herb (Khayri et al., 2023), which adds a fresh, citrus/lemony/sweet/spicy flavour, and a bright green freshness to many different dishes. Leaves are chopped or bruised to bring out the flavour and then sprinkled to garnish soups, stews, and curries (Singletary, 2016; Kumar et al., 2022). Leaves are used for dressings and marinades. The seeds have a warm, nutty flavour. To bring out the best flavour, the seeds are toasted in a dry skillet over medium heat and then the seeds are ground into a powder. Whole seeds are extensively used as a spice. and as a common ingredient of spice blends in different countries (*garam masala, khmeli-suneli*). The EO extracted from coriander seeds, leaves, and stems by steam distillation has extensive uses in perfumes, soaps, and other beauty products (Zhang et al., 2006; Khayri et al., 2023).

People exhibiting genetic variation in an olfactory receptor gene dislike the smell of coriander and perceive the flavour to be like soap. This happens due to polymorphism in the gene that detects aldehyde chemicals present in soap and coriander as being alike (Eriksson et al., 2012). Different varieties and cultivars of coriander exist throughout the world, differing in their phytochemical constituents. Distinct varieties differ according to agronomic traits and yield qualities, and are named 'Caribe', 'Confetti', 'Dwarf Lemon', 'Marino', 'Indian coriander', 'Jantar', and 'Thüringer'

DOI: 10.1201/9781003387046-15

FIGURE 15.1 Differently aged coriander plants. The leafy young plants with dense foliage (1), plants at their flowering stage (2), flowers (3) and the fruits (4). The pictures are of plants grown during the winter months in West Bengal, India. The mature plant grows to around 4 and 4.5 feet above the ground with the mature fruit pods harvested after browning.

are available. Variation exists between cultivars in the production of foliage, bolting (slow or fast), aroma (sweet/spicy or lemony), leaf growth, seed size, and EO yields; the characteristics and traits are commonly known to gardeners and cooks (https://plantura.garden/uk/herbs/coriander/coriander-overview). Standard coriander is the most widely cultivated variety, and it is grown for its leaves and seeds. Coriander seeds grown in India have a slightly sweeter flavour than other types refer (Figure 15.1). A type grown in Morocco has a more intense flavour than standard coriander. Some different plant species have coriander-like flavours and are consumed under the name of coriander, including Vietnamese coriander (*Persicaria odorata*) and wild coriander (*Eryngium foetidum*) (Önder, 2018).

15.2 PHYTOCHEMICAL CONSTITUENTS

Hydrodistillation of coriander seeds and aerial parts are used for fractionating the EOs. Dry fruits yield EO of around 0.03 and 2.6%, whereas the yield of seed oil varies between 9.9 and 27.7%. Phytochemical screening of plants showed the presence of EO, tannins, terpenoids, reducing sugars, alkaloids, phenolics, flavonoids, fatty acids, sterols, and glycosides (Al- Snafi et al., 2016). The chemical classes of phytochemicals obtained by different methods of solvent extractions were recently reviewed by Mahleyuddin et al. (2022). They also reviewed the traditional uses as reported from different countries. The difference in age of plant seeds results in variations in the yields of components of essential volatiles (Msaada et al., 2009). Differences in percentage components of major components like linalool, geranyl acetate, camphor, and other monoterpene alcohols, esters, and ketones in harvests from different geographical regions have been reviewed by Sahib et al. (2013) and Mandal and Mandal (2015). Thakur et al. (2021) reviewed the fatty oil constituents of coriander. SOs contain tocopherols (327.47 μg/g), where γ-tocopherol was the major tocopherol (26.40 μg/g), followed by δ-tocopherol (13.5 μg/g) and α-tocopherol (11.7 μg/g). The other types are γ-tocotrienol (238 μg/g), followed by α-tocotrienol and δ-tocotrienols. Sterol concentrations were 36.93 to 51.86 mg/g of oil. Among the sterols, stigmasterol, sitosterol, campestetol, avenasterol, ergosterol, and lanosterol were present in varying amounts in different solvent extracts across different countries. The fatty acid profile of the lipids in SO was dominated by the presence of petroselinic acids (65.7–80.9%), followed by oleic, palmitic, and stearic acids. Unsaturated acids, like linolenic, arachidic, and palmitoleic acids, were found in trace amounts. Major lipids were triacylglycerol (95%), diacylglycerol, and free fatty acids. Coriander SO contains low levels of polar lipids like phospholipids (6.7–8.1%), primarily consisting of phosphatidylcholine, ethanolamine, and inositol.

Seed processing steps, like drying (microwave *versus* sun drying), dehusking of fruit coats, irradiating (for microbial decontamination), blanching of leaves in food industries, and different extraction methods of EO using different solvents (propane *versus* subcritical water extractions), were reviewed by Bhat et al. (2014). Al-Snafi (2016) reviewed extensively the pharmacological actions, including its aphrodisiac/reproductive uses and detoxifying potentials. Al-Snafi (2016), Chahal et al. (2017), Algita et al. (2018), Asfaw (2021) and Sobhani et al. (2022) listed the natural compounds isolated from different parts of leaves and seeds, highlighting the major constituents of leaf volatiles and seed EO as well as the main phytochemicals in extracts. Sahib et al. (2013), in their review, studied the variation of phytochemicals in EOs of fruits from different geographical regions. The EO phytochemical contents of fruits varied across different geographical regions. Among the major constituents, apart from linalool, was geranyl acetate, which was found in Indian fruits (8.12%), Bangladesh fruits (17.6%), and in Italian and Finnish fruits in much lower quantities in the range of 2-4%. Similarly, α-pinene was found in Indian fruits (4.09%), Algerian fruits (3.41%) Italian fruits (1–3%) and Canadian fruits (2.7%) but in lower quantities than this in Tunisian fruits (1.65%). γ-Terpinene was found at the highest concentration in Bangladesh fruits (14.4%), followed by Italian fruits (4–7%), followed by Tunisian fruits (2.15%), whereas, in Iranian fruits, there was a considerable amount of variation, from 0.42 to 7.28%. Cymene was found at the highest level in Italian fruits (6.3%), followed by Iranian fruits (0.84–3.77%) and Bangladesh fruits (1.27%). Limonene was present in Italian fruits (1–2%) and in Finnish fruits (1.9–2.6%). Citronellal/citronellol was found in Bangladesh fruits (1.96%) and camphor was found in Algerian fruits (concentration not specified), Italian fruits (6.3–6.4%), and Finnish fruits (4.8–4.9%). Some of the countries reported that fruits contained cedrene (3.87% in Bangladesh), α-thujene (3.98–7.87% in Iranian fruit samples), thymol (1.85% of Tunisian fruits), and nerol (4.6% in Italian fruits), none of which did the other reports mention (or only traces found, which were not significant). Chahal et al. (2017) described variations in percentage composition of phytochemicals in coriander seeds and leaves. Linalool was found to be the main phytochemical in EO of seeds (but present at much lower concentrations in leaves) while (*E*)-2-decanal was the major compound in leaf extracts. The EOs of leaves were comprised of twenty-seven compounds, where (*E*)-2-decenal, decanal, (*E*)-2-dodecenal and (*E*)-2-tetradecenal were the main components in four cultivars studied by Kumar et al. (2022). The aroma composition of freshly chopped leaves (extracted from the leaves of four cultivars) and EO contents were different among cultivars. The aroma of freshly chopped leaves was characterized by thirteen compounds, including six aldehydes, four alcohols, one ester, and one hydrocarbon. Six-carbon chain length alcohols and acetates gave the aroma a green note. Aroma characterization, such as floral, pleasant, turpentine, green, herbal, cooling, earthy, spicy, sweet, or rose-like, of the volatile oil (EO) constituents by odour profiling of constituents to assign characteristic perceptions of odour notes revealed linalool to be floral/pleasant while geranyl acetate and lemonol were sweet, and pinene, myrcene, and undecanal resembled turpentine, as identified by a trained odourist panel (Ravi et al., 2007). The EO of coriander is a natural product of these several dominant odour components.

Others include γ-terpinene, cuminal, borneol, myrcene, and lemonol (Chahal et al., 2017). Together, these components make up 70–98% of the oil and have distinct floral, grassy, pleasant and citrusy (for linalool), pleasant, floral, rose, herbal (for geranyl acetate), woody and spicy (α-pinene), spicy (for cuminal), sweet and rose-like odour notes (lemonol) and strongly floral (citronellol) odours. Overall, the concentration ranges of these components vary, with linalool being the most abundant (57–75%), followed by geranyl acetate (8.9–24.5%) and α-pinene (2.3–23.2%). Interestingly, oil samples from northern India were found to have higher levels of α-pinene (8.9–23.2%), compared with samples from southern India, although the latter had higher levels of geranyl acetate (5.4–24.5%). These regional differences may contribute to variations in the sensory profile of oil from different sources (Ravi et al., 2007).

The percentages of different components in coriander products may not be representative of other coriander fruits from the same regions or different harvest years or of different varieties and plant species, or from sites with different drainage/soil characteristics, spraying regimes (insecticides, pesticides and hormones), as well as light duration/wavelength, these all being vital factors for the accumulation of the secondary plant metabolites. In this chapter, an attempt has been made to compile all the lists of phytochemicals published in different studies, with deliberate omission of percentage content as the values varied tremendously in the minor constituent fractions. The names of phytochemicals and the plant parts from which they are derived are listed in Table 15.1. The table summarizes the nutrient contents as well. All phytochemicals present in the seed and plant extracts and in SOs and EOs are listed. The table is adapted from reviews of Sahib et al. (2013), Laribi et al. (2015), Mandal and Mandal (2015), Al-Snafi (2016), Chahal et al. (2017), Asfaw (2021), Sobhani et al. (2022), Mahleyuddin et al. (2022), and Al-Khairi et al. (2023), and from articles by Hussain et al. (2018) and Chew et al. (2022);The natural compounds are arranged in alphabetical order.

15.3 TRADITIONAL USES OF CORIANDER AS A MEDICINAL PLANT

Coriander is used as a traditional remedy for several ailments in many cultures (Kumar, 2020). It has been used as a spice in ancient Greek and Roman cultures. Romans brought the herb to Britain. Coriander has been used in traditional medicine in India to treat gastrointestinal, respiratory, and urinary issues. Seeds or an aqueous extract of seeds are often used for rheumatoid arthritis, inflammation, and joint pain (Nair et al., 2013; Mahleyuddin et al., 2022). In *Ayurveda,* it has been used to treat urinary diseases, liver and skin conditions, haemorrhoids, and as a memory booster (Thakur et al., 2021). In *Unani* medicine, seeds or an aqueous infusion of leaves are applied to quench thirst and combat melancholia (Wichtl, 2004; Mahleyuddin et al., 2022). The leaves of the herb find its traditional application as stomachic, carminative, and digestive stimulant in India (Platel & Srinivasan, 2004; Mahleyuddin et al., 2022). In Pakistan, the entire plant is used to treat digestive complaints like flatulence, dysentery, diarrhoea, and vomiting, whereas the aerial parts of the plant are used as an antiviral and neuroenergizer (Khare, 2008; Mahleyuddin et al., 2022). Tamilians in Sri Lanka use coriander to regulate hormones and reduce menstrual cramps (Jesuthasan & Uluwaduge, 2017). In China, the whole plant is used to treat measles, diabetes, aerophagy, and gastroenteritis, while seeds are often used to treat influenza and bad breath (Khare, 2008). In some Asian regions, the EO of the herb is used as gastric secretion stimulant, and for treatment of gastric ulcers or mouth infections (Bashir & Safdar, 2020). The fruits and other parts have been conventionally utilized for treating various ailments, such as fever, burning sensation, loss of appetite, bloating, thirst, headache, abdominal pain, oral issues, gout, rheumatism, inflammation, bronchitis, fever, dysentery, vomiting, and digestive problems. (Ceska et al., 1988; Sharma, 1995). In some cultures, coriander leaves are applied topically to treat chest pain, gall bladder problems, and coughs, and as an aphrodisiac (Varier, 1994).The aqueous extract of the roasted seed is prescribed for liver diseases (Taherian et al., 2012). The leaf decoction is used to treat mouth ulcers and eye redness (Paniagua-Zambrana et al., 2020). Coriander seed extract is used as an antifertility agent and contraceptive in some countries and as a diabetes remedy in Saudi Arabia (Singh, 1986; Al-Said et al., 1987). In

TABLE 15.1
Natural Compounds Identified in Different Extract Preparations of Coriander Listed in Alphabetical Order.

Chemical Class	Phytochemical Components	Plant Part	Reference
Benzoic acid derivative			
	Salicylic acid	Above-ground parts	Oganesyan et al., 2007
Coumarins			
	Esculetin, esculin, scopoletin, 4-hydroxycoumarin, umbelliferone, dicoumarin	Above-ground parts	Oganesyan et al., 2007
Carotenoids			
	β-carotene, total carotenoids, anthocyanins	Leaves at mature and young plant stage, fresh and dry seeds	Divya et al., 2012
EO			
	Decanal, (*E*)-2-decenal, 2-decen-1-ol, 1-decanol, undecanal, (*E*)-2-undecenal, dodecanal, (*E*)-2-dodecenal, (*E*)-2-tridecanal, tetradecanal, (*E*)-2-tetradecenal, (*E*)-2-pentadecenal, phytol, 1-eicosanol, 2-decenoic acid, 2-dedecanal, 2-undecenal, cyclododecane, decamethylene glycol, dodecanoic acid, *E*-11-tetradecenoic acid, capric acid , nonanoic acid, *E*-undecanoic acid, tridecanoic acid, undecanoic acid, undecyl alcohol, 2-dodecenol, tridecanal, 2-dodecenal, tetradecenol, nonane, *E*-2-decen-1-ol, *E*-2-undecen-1-ol, *E*-2-dodecen-1-ol, *trans*-2-undecen-1-ol	Leaves	Potter, 1996; Eyres et al., 2005; Bhuiyan et al., 2009; Matasyoh et al., 2009; Begnami et al., 2010; Freires et al., 2014
	Benzofuran,2,3-dihydro, hexadecanoic acid, methyl ester, 2,4a-epiox y-3,4,5,6,7,8,-hexahydro-2,5,5,8a-tetramethyl-2h-1-benzofuran, 2-methyoxy-4-vinylpheno, 2,3,5,6-tetrafluroanisole, 2,6-dimethyl-3-aminobenzoquinone, dodecanoic acid	Flowers	Dharmalingam and Nazni et al., 2013
Phenylpropenes	Anethole, apiol, estragole, eugenol, eugenyl acetate	Seed and pericarp of fruit	Raal et al., 2004; Msaada et al., 2009; Anwar et al., 2011; Marichali et al., 2014

(Continued)

TABLE 15.1 (CONTINUED)
Natural Compounds Identified in Different Extract Preparations of Coriander Listed in Alphabetical Order.

Chemical Class	Phytochemical Components	Plant Part	Reference
Monoterpenoids	α-pinene, α-thujene, camphene, sabinene, β-pinene, β-myrcene, Δ^3-carene, *p*-cymene, *p*-cymen-8-ol, phellandrene, menthol, carvone, *p*-mentha-1,4-dien-7-ol, *p*-mentha-1,8-diene, enthadien-1-ol, borneol, thymol, carvacrol, α-terpinene, anethol, limonene, D-limonene, γ-terpinene, terpinolene, terpinene-4-ol, α-terpineol, 1,8-cineole, (*Z*)-β-ocimene, citronellal, citronellol, β-citronellol, *cis*-linalool oxide (furanoid), linalool, linalyl acetate, *trans*-linalool oxide, camphor, geraniol, geranial, geranyl acetate, nerol, neral, neryl acetate, lavandulol, cuminaldehyde, *cis*-dihydrocarvone, citral, *m*-cymene, citronellyl acetate, α-cedrene, α-farnesence, β-sesquiphellandrene.		Bhuiyan et al., 2009; Msaada et al., 2009; Zoubiri&Baaliouamer, 2010; Anwar et al., 2011; Neffati et al., 2011; Orav et al., 2011; Marichali et al., 2014
Sesquiterpenoids	*cis*-nerolidol, δ-elemene, α-elemol, β-caryophyllene, α-cubebene, δ-cadinene, α-humulene, germacrene D, santalol		Msaada et al., 2009; Anwar et al., 2011; Neffati et al., 2011; Orav et al., 2011; Marichali et al., 2014
Fatty Acids	Petroselinic acid, linoleic acid, oleic, palmitic acid, myristic acid tridecanoic acid, arachidic acid, heptadecanoic acid, tetradecanoic acid, hexadecanoic acid, stearic acid, palmitoleic acid, α-linolenic acid, gadoleic acid, erucic acid, pentadecanoic acid	Whole fruit, pericarp,seed, leaf, stem and root	Marichali et al., 2014
Flavonoids	Hyperoside, rutin, hesperidin, vicenin, diosmin, luteolin, Apigenin, orientine, dihydroquercetin, catechin, arbutin	Above-ground parts	Oganesyan et al., 2007
Isocoumarins	Coriandrone A, coriandrone B, coriandrone C, coriandrone D, coriandrone E, coriandrin, dihydrocoriandrin	Whole plant	Ceska et al., 1988; Taniguchi et al., 1996
Phenolic acids	Ferulic acid, gallic acid, caffeic acid	Above-ground parts	Oganesyan et al., 2007
Polyphenols			

(Continued)

TABLE 15.1 (CONTINUED)
Natural Compounds Identified in Different Extract Preparations of Coriander Listed in Alphabetical Order.

Chemical Class	Phytochemical Components	Plant Part	Reference
	Kaempherol, quercetin, 3'-*O*-mesquercetin, 4'-*O*-mesquercetin, acacetin, vanillic acid, *p*-coumaric acid, *cis*-ferulic acid, *trans*-ferulic acid	Leaves	Nambiar et al., 2010
Phytosterols			
	Cholesterol, stigmasterol, β-sitosterol, δ-stigmasterol, Δ^5-avenasterol, Δ^5-24-stigmastadienol, Δ^7-stigmasterol, Δ^7-avenasterol	Whole fruit, seed, pericarp	Sriti et al., 2009
Tocopherols			
	α-tocopherol, β-tocopherol, γ-tocopherol, δ-tocopherol, α-tocotrienol, δ-tocotrienol, γ-tocotrienol	Whole fruit, seed, pericarp	Sriti et al., 2010

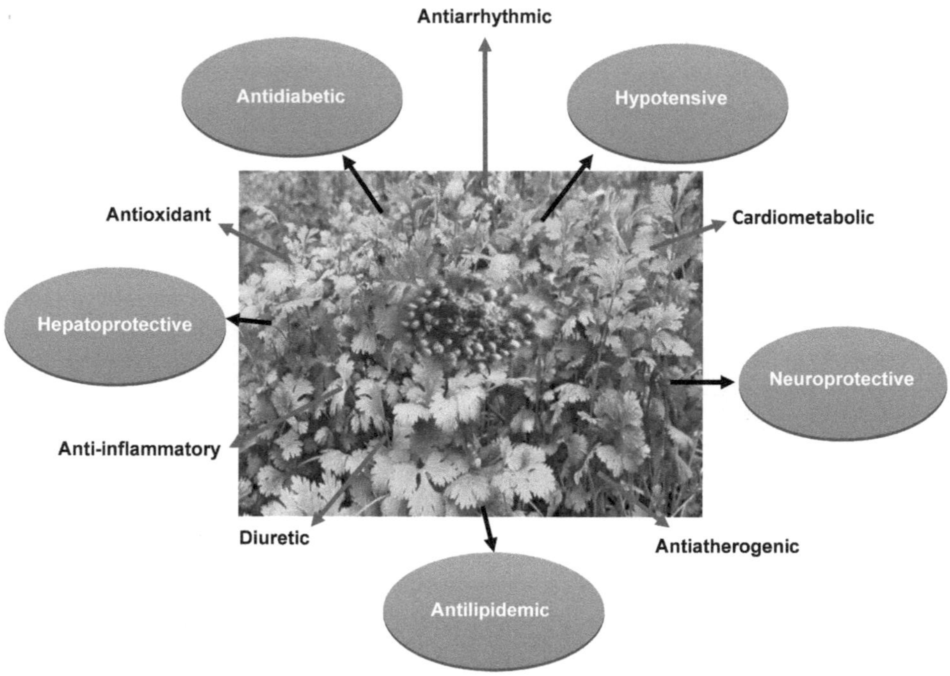

FIGURE 15.2 Pictorial representation of the health benefits of coriander.

Fiji, coriander leaves and fruits are used for skin-related problems, eye diseases, headaches, and bad breath (Sachan et al., 2018). Iranians use the juice of young coriander leaves to treat anxiety and insomnia, while the fruits are used to enhance appetite and treat convulsions (Heider, 1992), whereas powdered seeds or dry extract are used traditionally as carminative, stimulant, and diuretic agent and to ameliorate dyspeptic complaints and loss of appetite (Aissaoui et al., 2008; Taherian et al., 2012; Grieve, 2013). Germans use coriander as a digestive aid (Pieroni & Gray, 2008). In Turkey, coriander fruits and seeds are used as an appetizer, digestive and carminative enhancer (Ugulu et al., 2009), whereas Palestinians use it to boost strength after childbirth and treat urinary tract infections (Abu-Rabia, 2005). In Morocco, oral administration of coriander plant parts are a common practice as a diuretic and to treat other renal diseases (Hassar, 1999). In some parts of Europe, fruits of coriander are used against worm infestations and to treat rheumatism (Bashir & Safdar, 2020). People in the Mediterranean belt commonly utilize this plant to address gastrointestinal problems like anorexia and diarrhoea. It is also used to alleviate spasms, gastric issues, bronchitis, and gout (Sayed-Ahmed et al., 2017). Tunisians use this herb as a flavouring or food supplement and in traditional medicine to treat various conditions, including inadequate lactation, postpartum issues, and eczema (Momin et al., 2012). Coriander is also a popular ingredient in Indian, Chinese, and American cuisine (Bruneton,1995). It is popularly used as a seasoning and preservative agent in the culinary industry due to its distinct flavour and aroma (Singletary, 2016). It can be used in the form of leaves or seeds, in whole or ground form, and the powdered fruit of coriander is often used as a flavouring agent in dishes such as meat, fish, and baked goods in India (Mahendra & Bisht, 2011; Mahleyuddin et al.,2022). A pictorial representation of health benefits of coriander is shown in Figure 15.2.

15.4 PHARMACOKINETIC ACTIVITIES OF CORIANDER

The pharmacokinetic studies of the application and feeding of coriander seeds, aerial parts, or their extracts, SO, hydrodistilled oils (EO), and components of major fractional portions of the oils, like linalool, have been studied by numerous authors. Components of extracts, EO, or volatile oils are used in these studies. Components implicated in the biological and pharmacological effects to account for coriander pharmacokinetic activities are summarized in Table 15.2.

15.4.1 Antidiabetic Activity

Gray and Flatt (1999) found that adding aqueous coriander aqueous to the diet or drinking water (of streptozotocin-induced diabetic mice) resulted in a reduction in hyperglycaemia. An increase in glucose transport, glucose breakdown, and glucose incorporation into glycogen was observed in isolated murine abdominal muscles, which were effects comparable to those of insulin. *In vitro*, the extracts also stimulated insulin secretion from the cell lines used. The study suggested that coriander may have the potential as a natural treatment for diabetes. Aissaoui et al. (2011) evaluated the aqueous extracts of coriander seeds in hyperglycaemic/hyperlipidaemic rats. They found that even a single oral dose could normalize glucose levels within six hours. The extracts lowered serum glucose, lipid profile, atherosclerotic indices, and insulin resistance in their study. The animals maintained healthy body weight with lower serum urea and creatinine levels. Sreelatha and Inbavalli (2012) used the alloxan-induced diabetic model to evaluate the effects of 80% ethanolic extracts of coriander leaves and stems which they found to be more effective than glibenclamide in lowering blood sugar levels and achieving antioxidant protection in diabetes. Simultaneously, these authors observed improvements in lipid profile and hepatic levels of antioxidant enzymes which they attributed to the phenolic contents of the extracts. Aligita et al. (2018) evaluated the ethanolic extracts of leaves for their potential α-glucosidase-inhibiting abilities, which they found to be better than the standard drug acarbose. In *in vivo* studies in alloxan-induced diabetic rats, there was regeneration in pancreatic β-cells (compared to glibenclamide). Kajal and Singh (2019) evaluated seed extracts (petroleum ether) in the antioxidant protection of diabetic renal damages in streptozotocin-induced diabetes and found lower free radical stress (lowered peroxidation of lipids; increased free radical scavenging) in renal tissues. After treatment for 45 days, they found significant improvements in serum biochemical parameters like creatinine, glucose, and lipids and lowered levels of advanced glycation end-products in renal tissues, lowered albumin in urine and better creatinine clearance. They attributed the effects to quercetin, rutin, and kaempferol present in coriander seeds. Hajlaoui et al. (2021) found the antidiabetic actions of EO better (nine times higher) than the standard antidiabetic acarbose. The EO consisted of linalool, γ-terpinene, and α-pinene as the major phytochemicals. Mechchate et al. (2021) identified the polyphenols in coriander seeds as the antidiabetic principle responsible for the effect. Polyphenolic fractions of coriander seed controlled high fasting blood glucose in alloxan-induced diabetic mice. When administered for four weeks, the polyphenols had alleviating effects on fasting sugars, while challenges of glucose were well tolerated. Overall, the animals were healthy, with better insulin activity, synthesized good lipids [correction in ratio between high-density lipoprotein (HDL) and low-density lipoprotein (LDL) affecting the enzyme lipoprotein lipase], and had lowered hepatic marker enzyme levels [aspartate aminotransferase (AST) and alanine aminotransferase (ALT)]. Renal clearances of urea and creatinine were good. Synergistic anti-inflammatory effects in the rat paw model of inflammation were observed. Nine polyphenols, which included various catechins, rutin, chlorogenic acid, vanilic acid, and oleuropein, were synergistically responsible for the effect. In a recent study by Mahmoud et al .(2022), EO was found to have potential for antidiabetic and antioxidant effects. The rats (dexamethasone-treated insulin-resistant animals) were given coriander oil at 1mL/kg/day doses through oral gavage. The results showed that the oil reversed the decrease in the weight of the pancreas. Reduced glutathione (GSH) and BCL2 contents of the pancreas increased with reductions of malondialdehyde (MDA) levels in the pancreas (in comparison to metformin). Serum HDL levels increased and corrections were seen in serum glucose, insulin, triglyceride (TG), and total cholesterol (TC) levels. Amount of glucose utilization in the oral glucose tolerance and insulin resistance tests was evaluated and found to be normalized. Immuno-expression studies revealed a higher level of apoptosis marker BAX in pancreatic tissues with a rise in the BAX/BCL2 ratio. Additionally, histopathological examination showed that the treatment with coriander oil improved congestions, vacuolations, and haemorrhages in pancreatic islets. The study attributed the effects to the major component of EO, linalool.

15.4.2 Active Cardiovascular Activity

Numerous animal studies have demonstrated that water extracts of coriander leaves and stems ('aerial parts') as well as seeds had an effect on reducing blood pressure in rats with drug-induced hypertension, ameliorating arrhythmia, correcting cholesterol levels, inflammation levels, and hyperlipidaemic and atherogenic indices (Mahleyuddin et al., 2022). The extracts contain antioxidants that may help prevent damage to blood vessels and lower the risk of cardiac and vascular damage. The anti-inflammatory properties of the spice also act on the vascular system and arteries. Seed extracts reduced antiarrhythmic activity in potassium- and barium chloride-induced bradycardia and arrhythmia, respectively. They were found to reduce pulse rate and normalize ECG waves and reduce levels of cardiac markers, such as ALT, lactate dehydrogenase (LDH), AST, and creatine kinase-MB (CK-MB) fractions. Coriander seed extract had comparable antitachycardia efficacy to propranolol (a β-blocker drug) but was less effective than atropine at restoring rhythms, and polyphenolic compounds present in coriander are implicated in these actions. It is hypothesized that polyphenols in seed bind to the β-adrenergic receptors and thereby regulate heart rate. Extracts may possibly prevent arrhythmias and improve overall cardiovascular health by slowing down the generation of action potential in myocytes (leading to bradycardia/negative chronotropic actions) (Rehman et al., 2016). Dhyani et al. (2020), in an isoproterenol-induced heart failure model in Wistar rats, found that extracts reversed severe impairment in left ventricular functions, baroreflex responses, lipid peroxidation, and increased expression of endothelin receptors (A and B), thus providing significant protection against heart failure. Wang et al. (2022) observed a lowering of blood pressure in rats induced by a high fructose/salt diet. This was attributed to vasodilation by increased serum levels of prostaglandin, NO (and its synthase), diuresis of sodium (increasing expression of Na^+ proton exchanger), lowered levels of uric acid and glucolipids in blood. Coriander flavonoids, phytosterols, terpenes, and phenolic acids are implicated in its cardiovascular actions.These compounds also have elicited actions like blocking the function of the angiotensin-converting enzyme (ACE), which generates the physiological hypertensive peptide angiotensin. Other cardioprotective effects include the ability to inhibit hyperlipidaemia and cardiometabolic disorders. Additionally, the herb has a diuretic effect which increases the elimination of electrolytes and fluids, reducing the blood volume.

The flavonoids present in coriander leaves, like quercetin, rutin, apigenin, and luteolin, are shown to exhibit blood pressure-lowering effects on their own or in combination (as present in the crude extracts). A potential mechanism for managing hypertension involves the inhibition of ACE, which regulates blood pressure by controlling the production of nitric oxide (NO). Free hydroxyl groups on the flavonoids present in the coriander extracts chelate with the zinc ion on the enzyme's active site and form hydrogen bridges with the nearby amino acids, thereby lowering ACE activity. Flavonoid-rich fractions isolated from the leaves (containing 17 flavonoids) inhibited ACE with IC_{50} values of 28.91 μg/mL which was comparable with other plant flavonoid-fraction-ACE inhibitions. The activity was greater than that of the synthetic drug captopril (ACE inhibitor) at concentrations of greater than 50 μg/mL for both the drug and the dose of the extract (Hussain et al., 2018).

The management of dyslipidaemia is crucial for preventing cardiovascular diseases, particularly in diabetic patients (Miller, 2009). Coriander extracts seem to modulate different enzymes of lipid metabolism pathways and reduce TG and cholesterol levels. For instance, coriander inhibits the activity of 3-hydroxy-3-methyl-glutaryl-coenzyme A reductase (HMG-CoA reductase) which leads to decreased production of cholesterol, enhances lecithin-cholesterol acyltransferase (LCAT) and tissue lipase activities, resulting in more lipid breakdown. Coriander SO has also been shown to decrease TC, TG, and LDL and increase HDL in rats (Ramadan et al., 2008). Seed powder reduces cholesterol concentration in the liver, intestine, proximal and distal colon, cholesterol-to-phospholipid ratio, and the number of tumours in the intestine and colon of rats with colon cancer (Chithra & Leelamma, 2000). In addition, studies by Aissaoui et al. (2011) and Sreelatha and Innavali (2012) both reported improvement of cardioprotective index (HDL/LDL and HDL/TC) and lowering of atherogenic index [(TC–HDL)/HDL] with lower triglyceride levels in blood. A 250 mg coriander/kg bodyweight/day administration to high-fat diet-induced atherosclerotic rabbits significantly reduced serum lipid parameters and increased HDL. It also increased the excretion of cholesterol and phospholipids, reduced lipid peroxidation, and elevated GSH content and CAT activity with an increase in faecal cholesterol and phospholipid output. Sharma et al. (2014) observed that seed extract prevented elevated lipid indices and LDL oxidation-mediated macrophage activation (*in vitro*) and alleviated pathophysiological alterations of high-cholesterol diet-induced atherosclerosis in atherogenic rats. In the *in vitro* experiments conducted by Patel et al. (2013), the seed extracts led to a reversal of copper-mediated cell-free oxidation of LDL, indicating its potent antioxidant potential, which mitigated lipotoxicity in RAW 264.7 cells and prevented cell death. Elevated levels of malondialdehyde (MDA), lipid hydroperoxide, and protein carbonyl, as well as a progressive increase in conjugate diene, were seen which could be reversed by seed extracts. Immunostaining of cell adhesion molecules in plaques validated the anti-atherogenic progressions. Furthermore, extracts lowered cell-mediated LDL oxidation by the lowered MDA production and oxidized LDL-mediated cell death. Enzymatic alterations in the trapped LDL-C and extracellular matrix lead to an increase in the production of adhesion molecules and releases chemokines from endothelial cells, resulting in monocyte recruitment. Coriander extract has also been found to reduce the number of foam cells by inhibiting the expression of SRB1 (Mahleyuddin et al., 2022). Pre-treatment (for 30 days) with methanolic extract of coriander seeds at a dose of 200 or 300 mg/kg protected rats from isoproterenol-induced myocardial necrosis. Necrosis that reversed was a result of lowered lipid peroxidations and levels of endogenous antioxidants and ATPases, lipidaemia and cardiac damage brought on by the extracts. The study concluded that the polyphenolic content of coriander seed extracts is responsible for preventing oxidative damage by effectively scavenging the isoproterenol-generated reactive oxygen species (ROS) (Patel et al., 2012).

Coriander has demonstrated diuretic effects by interacting with the muscarinic receptors of endothelial cells, similar to furosemide, a standard diuretic (Jabeen et al., 2009). Continuous intravenous infusion of an aqueous extract (administered in two doses) was applied to anaesthetized rats and compared with furosemide. The results showed induced diuresis, loss of electrolytes, and increased glomerular filtration rate in a dose-dependent manner, validating its traditional use as a plant with diuretic characteristics in Moroccan traditional medicine (Aissaoui et al., 2008). Studies have shown that methanol and aqueous extracts of coriander leaves can increase the excretion of sodium in urine relative to potassium ('potassium sparing'), indicating a safe profile for the diuretic effect (Thuraisingam et al., 2019). Fractionation of extract contents in the aqueous and organic fractions with simultaneous assays has led to the separation of spasmogenic and spasmolytic components of coriander seeds. The crude extract of coriander induced relaxation of arterial contractions with reductions in blood pressure. The extracts assayed *in vitro* in rabbit aorta and guinea pig atria had both cholinergic and calcium channel-blocking effects (like verapamil) which might be responsible for the smooth muscle actions (Jabeen et al., 2009). Kousar et al. (2011) evaluated that salbutamol-induced injury could be reversed in both pre-treatment and post-treatment in rabbits with fresh leaf extracts. Rectification of lipid profile and antioxidant profile was observed, with increased superoxide dismutase (SOD), catalase (CAT), and peroxidase activities. It had effects on gut microbiota changes with increases in *Bifidobacterium* and *Oscillibacter* (healthy communities), which the researchers could correlate with the other metabolic effects observed. Coriander warrants further clinical trials on humans as an herbal remedy for cardiovascular diseases.

15.4.3 Antioxidant Activity

Hashim et al. (2005) observed pre-treatment protection from oxidative damage in human lymphocytes induced by hydrogen peroxide in the polyphenolic-rich fractions of extracts. They found reversal of the damage occurred through elevated antioxidant enzyme activities, glutathione content, and increased levels of thiobarbituric acid-reacting substances (TBARS), used as a measure to evaluate antioxidant potential. The antiperoxidative actions of coriander seeds were studied by Chithra and Leelamma (1999) in rats fed with a high-fat diet. Lower levels of lipid peroxides, free fatty acids, and glutathione were observed. Wong and Kitts (2006) examined the antioxidant and antibacterial properties of freeze-dried and irradiated parsley and coriander leaves and stems, using methanol and water extracts. Ferrous ion-chelating activity was more pronounced in the stem methanol extracts. The total phenolic content varied between plant parts, i.e., leaf and stem, and type of extract. Aqueous extracts of the aerial parts of coriander showed pro-oxidant activity. Methanol extracts of coriander leaves exhibited higher antioxidant activity against both hydrophobic and hydrophilic radicals which the authors also linked to the higher total phenolic content.

Divya et al. (2012) evaluated different varieties of coriander for their carotenoid content and bioactivities with drying to different extents. β-carotene was found to be higher with maturation in foliage than in younger plants and in seeds. Differences existed among varieties, of which GS4 multi-cut produced the highest biomass, total carotenoids, and β-carotene content before flowering. Extracts had a high antioxidant activity, which was more effective in scavenging hydroxyl radicals and protecting DNA than standard gallic acid. Microwave drying of foliage was found to be a more effective method for preserving pigments with better retention of bioactivities than oven drying, yielding intact *trans*-β-carotene and a higher yield of pigments. They suggested coriander as a source of pro-vitamin A for combating vitamin A deficiency in populations in tropical countries. Wang et al. (2018) investigated the protective effects of natural antioxidants of unsaturated fatty acids and phospholipids in an emulsion. The authors used rosemary extract as a water-soluble antioxidant and coriander extract as an oil-soluble antioxidant to fortify both the aqueous and oil phases of the emulsion.They found dual protection of the unsaturated fatty acids and phospholipids. The effectiveness of the antioxidants towards peroxidation was indicated by lower levels of MDA and conjugated dienes in the emulsion. Hajlaoui et al. (2021) evaluated the antioxidant properties of EOs of caraway (*Carum carvi* L.) and coriander alone and in combination. Coriander EO constituted 27 components representing 99.46% with high levels of linalool (76.41%), followed, respectively, by γ-terpinene (5.35%), and α-pinene (4.44%). Antioxidant assays of the EOs were employed to detect free radical scavenging of the oil which was attributed to linalool and α-pinene in EO. The combination of the two EOs had a synergistic effect on antioxidant activity. The antioxidant activity of other reports is mentioned in Table 15.2. Such antioxidant protection from induced damage in systemic functions are mentioned under different organ-protective actions.

15.4.4 Neuroprotective Activity

Spices are known to contain antioxidants and other bioactive compounds that can positively impact our health, even before they reach the stomach. Neuroprotective effects are conceived to be those that act like a sedative, anxiolytic, analgesic, or antimigraine-like activity in different animal studies. Different animal models of those diseases were used where the coriander extracts or the EOs were found to elicit the actions. Kunzemann and Hermann (1977) found that the isolated flavonoids, quercetin and isoquercetin, from seeds have sedative effects. Quercetin also shows anti-inflammatory, antioxidant, and analgesic properties. The activation of α_2 adrenergic receptors in producing the analgesic effect of quercetin in mice was reported by Kaur et al. (2005). The study describes an antinociceptive effect in mice by quercetin, which involves the activation of receptors. Picq et al. (1991) also had the same observation on quercetin and isoquercetin glycosides.

Cha et al. (2020) conducted a study on the methanol extract of aerial parts and identified seven compounds, including two new isocoumarin glycosides and a new phenolic glycoside through phytochemical investigation. The researchers assessed all the isolated compounds for their potential neurotrophic and antineuroinflammatory activities. The results showed that these four compounds significantly stimulated nerve growth factor (NGF) release, with levels ranging from 120.94% to 128.22%. The aglycones of two of these compounds showed more potent NGF secretion activity and antineuroinflammatory effects than their glycosides. In addition, the studies of Delavar (2016) and Behrendt (2004) which found extracts to relieve migraine are listed in Table 15.2.

Patients with Alzheimer's disease (AD) often exhibit anxiety and depression as psychiatric symptoms which together contribute to cognitive decline and dementia (Craig et al., 2005). Soluble β-amyloid (Aβ) induces behavioural and biochemical changes, which relate to pathogenesis of the depressive-like profile. In a rat model of AD induced by Aβ plaques, Cioanca et al. (2014) reported that inhaled volatile oil (EO) of coriander has potent antioxidant, anxiolytic, and antidepressant effects. They deciphered that CAT inhibition and increased glutathione levels in the hippocampus contributed to mitigating anxiety, depression, and oxidative stress in these Alzheimer rats. The EO has analgesic properties that can help to reduce pain and inflammation (Ravi et al., 2007).

Emamghoreishi et al. (2005) evaluated the anxiolytic effects of a 100 mg/kg aqueous extract of seeds using anxiety models in mice where they found that the extracts reduced spontaneous activities (a sedative effect, they believed). The researchers used diazepam as a positive control and opined that the extract's anxiolytic effects were due to the action on the γ-aminobutyric acid (GABA)-A receptor complex. The central effects were attributed to quercetin and isoquercetin found in seeds. In a similar study by Emamghoreishi and Heidari-Hamedani (2006), intraperitoneal administration of aqueous, alcoholic extracts, or EO was given 30 min before a sedative injection. This led to a reduced latency to sleep and increased sleep duration. Linalool, the major monoterpenoid compound of Eos, had contributed to these effects. According to a study by Rakhshandeh et al. (2012), the hydro-alcohol extract derived from the aerial parts of a plant, as well as its ethyl acetate and butanol fractions, were found to have hypnotic effects in mice. This was evidenced by an increase in sleeping time observed in the study. Studies have revealed a decrease in the secretion of monoamines (norepinephrine, serotonin, and dopamine) in the hippocampus and cortex of mice that are treated with linalool. Coriander EO can be diffused or added to a vaporizer to create a calming and relaxing atmosphere referred to as aromatherapy. It helps to reduce anxiety and stress and promotes restful sleep. The other studies are given in Table 15.2.

One of the best-studied phytochemicals from the plant which has been shown to have anti-inflammatory, antioxidant, analgesic (pain-relieving), anxiolytic, and antidepressant properties in preclinical studies is linalool. Linalool ($C_{10}H_{18}O$) is a monoterpenoid alcohol, an octa-1,6-diene substituted by methyl groups at positions 3 and 7, with a hydroxy group at position 3. It has two enantiomeric forms in nature: (*S*)-linalool is found in the EO of coriander (70–80% of the oil) (Sugawara et al., 1998). The *S* form is perceived as sweet, floral, petitgrain-like (odour threshold 7.4 ppb), whereas the R form is a woodier and lavender-like (odour threshold 0.8 ppb) fragrance. Linalool is used in the pharmaceutical industry as an active ingredient in topical creams and gels for the treatment of skin inflammation and pain (ChemSpider ID, 60523). Linalool is the principal component of the seed EO (70%) (which has yields of only 0.18–0.39%). The effects on the central nervous system, like the anxiolytic and sedative effects, are mostly produced by this compound (Sugawara et al., 1998; Blumenthal et al., 2000; Gastón et al., 2016). Anaeigoudari et al. (2016) observed that, in the rat cortex, the attachment of glutamate is inhibited by linalool, resulting in prolonged seizure latencies. Glutamatergic and nicotinic acid receptor bindings appear to be responsible for the anxiolytic effect of linalool. The muscle relaxant effects of linalool were similar to those of benzodiazepines. It enhances the effect of GABA binding to receptors, resulting in anxiolytic, muscle relaxant, anticonvulsant, sedative, and hypnotic effects (Elisabetsky et al., 1999). Hypnotic and anticonvulsant properties of linalool have been established in *in vivo* studies in mice. Glutamatergic transmission is inhibited by blocking glutamate binding (arresting the induced convulsions by quinolinic acid) on the *N*-methyl-D-aspartate

(NMDA) receptors responsible for seizures.A simultaneous lowering of cyclic adenosine monophosphate levels was observed (Elisabetsky et al., 1995; Re et al., 2000). It is especially helpful for treating headaches, migraines, and muscle pain. Inhalation of linalool has been shown to improve social interactions and reduce vascular infractions (Souto-Maior et al., 2011). Linalool has been found to increase dopamine levels in the striatum, suggesting that it may be beneficial for treating Parkinson's disease (Cheng et al., 2015). Re et al. (2000) reported that linalool had effects that modify the kinetics of nicotinic acetylcholine receptor-ion channel in mouse neuromuscular junction, a model system for studying the mechanism of action of muscle relaxants and anaesthetics. The effect was similar to that of known muscle relaxants and anaesthetics, such as d-tubocurarine and lidocaine. Batista et al. (2008) found ionotropic glutaminergic receptors (NMDA/AMPA/kainate receptor) to be contributing to the analgesic effect of linalool in mice models. Acetic acid was administered in the peritoneal cavity to generate a nociceptive response of abdominal writhing. Dose-dependent linalool administration lowered acetic acid-induced writhing episodes, suggesting its antinociceptive effect. The activation of ionotropic glutaminergic receptors particularly is responsible for the antinociceptive effect in mice. A study conducted by Nakamura et al. (2009) found that exposing stressed rats to linalool fragrance had a significant impact on their hormone levels, white blood cell count, and gene activation levels compared with the control group. The results suggested that inhaling aromatic compounds like linalool can have both physiological and psychological effects, highlighting the synergism of the nervous and immune systems. The study also found that, although stress hormone levels were raised in both the control and treatment groups, the rats exposed to linalool maintained their white blood cell counts and had fewer activated 'stress genes' than the control group. Recently, neuropharmacological effects against AD models have been observed (Hosseini et al., 2021).

15.4.5 Hepatoprotective Activity

Coriander EO can be used to support digestive health. It has carminative properties that help to relieve gas, bloating, and indigestion. It can also stimulate appetite and improve nutrient absorption. Oral administration of ethanolic extracts of leaves and stem to animals with carbon tetrachloride-induced liver injury lowered clinical markers of hepatic injury like AST and ALT with an increase in antioxidant marker enzymes of the liver. Hepatocyte regeneration and fatty changes in the liver associated with necrosis were reversed (Sreelatha et al., 2009). Pandey et al. (2011) observed a reversal of hepatotoxicity by the leaf ethanolic extracts, with a gain in liver weight, enzymes, and bilirubin in the blood. Pandey et al. (2011) observed ethanolic extracts of leaves to protect against carbon tetrachloride-induced liver injuries which they attributed to the presence of quercetin and isoquercetin and other antioxidant phenolic compounds. Samojlik et al. (2010) pre-treated animals with EO aetheroleum and then induced liver damage where they found the EOs to be protective with pro-oxidant activities. Kükner et al. (2021) observed a reversal of liver necrosis with a lowering of markers of inflammation like tumour necrosis factor (TNF)-α, NF-κB, and caspase (apoptotic marker) after administration of extracts through oral gavage after an induced ischaemic injury. Compared with the protective effects of silymarin against intraperitoneal CCl_4-induced liver damage, treatment with EO (0,6 ml/kg) fared well. The study evidenced lower hepatic injury (through AST and ALT markers) antioxidant protection (through reduced glutathione (GSH), glutathione peroxidase (GP_X) and catalase (CAT)) and lipid peroxidation (through MDA) with further histopathological confirmations (Özbek et al., 2016; Altınok-Yipel et al., 2020). Briefly, all these studies are described in Table 15.2. Hepatoprotection is attributed to the principal components of EO, like linalool, pinene, geraniol, geranyl acetate, camphor, and terpinene.

15.4.6 Anticancer Activity

Chithra and Leelamma (2000) showed that experimentally induced (dimethyl hydrazine) colon cancer in mice was alleviated through feeding with coriander seeds. Higher phospholipid levels and lowered cholesterol levels in the cancer tissues were noticed; concomitantly, there was a healthy increase in faecal metabolite contents and faecal weight. Cortes-Esalva et al. (2004), through mutagenicity assays in bacteria induced by aromatic amines, showed that these mutagenic effects were inhibited by the chlorophyll contents of the juice. Assessment of the antioxidant potential (phenolic contents analyzed) of the variable solvent-extracted materials sequentially combined with their application to breast cancer cell lines was employed as a strategy to implicate the phytochemical protection in cancer (Tang et al., 2013). They found antiproliferative activity with caspase activation, lowered DNA damage, cell cycle progression, and apoptotic activations. Iwasaki et al. (2016) showed linalool (in EO of coriander) to lower oxidative stress in human colon cancer cell lines and induced apoptosis. Tumour sizes diminished (in *in vivo* xenografts) with lowered tumour lipid peroxidations indicating a cancer-specific lowering of antioxidant effects of linalool. Mechchate et al. (2021) observed antileukaemic effects of polyphenolic extracts of seeds *in vitro* on myeloid leukaemia and Vero cell lines and found them to be nontoxic, both *in vivo* and *in vitro*, to these cell lines. The components found in these extracts were further analyzed *in silico* to determine the possible receptor interactions. The components showed interactions with numerous proteins implicated in neoplastic transformation, like ABL kinase, BCL2, FLT3, and ABL1.

15.4.7 Anti-inflammatory Activity

In several studies of carrageenan-induced paw oedema, various parts of the coriander plant have shown potential anti-inflammatory effects. Extracts were able to reduce paw oedema in rat and mice models and generated fewer cytokines in the *in vitro* studies. Sari et al. (2021) reviewed the anti-inflammatory effects of fruit, leaf, and seed hydroalcoholic extracts. Sonika et al. (2010) found fruit ethanolic extracts (at a dose of 200 mg/kg) to mediate 40% inhibition of oedema after the third hour at the given dose. Nair et al. (2013) conducted such a study using hydroalcoholic extracts, which showed a reduction in oedema and dry granuloma weight with downregulations of inflammatory cytokines like interleukin (IL)-6 and IL-1β. Begnami et al. (2018) evaluated the ethyl acetate fraction obtained from dried leaves and stems in mice. The results were obtained in two phases. Early-phase pro-inflammatory mediators like histamine, serotonin, and prostaglandins were found to be downregulated. Seed ethanol extract at doses of 250 mg/kg or 500 mg/kg were effective according to the study of Bhat (2014). Ethanolic extracts of the aerial parts also showed a significant decrease in lipopolysaccharide (LPS)-induced NO and prostaglandin (PG)-E2 production and downregulated inflammatory regulators such as inducible nitric oxide synthase (iNOS), cyclooxygenase-2 (COX-2), and pro-IL-1β, while inhibiting the phosphorylation of IκB-α and the expression of nuclear p65 protein (Wu et al., 2010). Moreover, studies have confirmed the involvement of TLR4, MAPK pathway, and ROS generation in inflammatory responses, and the potential of restricting these pathways as a strategy to reduce inflammatory diseases. In contrast, there was an insignificant reduction in paw oedema with the administration of SO at a dose of 400 mg/kg (positive controls of indomethacin) (Ibrahim et al., 2017). A membrane stabilization effect (by induced tonicity alteration of human RBCs) was also observed with extracts (Kothari et al., 2017). A compound, lacton C, isolated from coriander fruits (extracted with ether) reduced NO levels, ROS, and inflammatory cytokines, and suppressed the expression of inflammatory mediators such as COX-2 and iNOS. MAPK and NF-κB pathways were seen to be regulated by the extracts (Yuan et al., 2020).

15.4.8 Dermaprotective Activity

Coriander EO is often used in skincare and personal care products. Coriander oil has anti-inflammatory potential and can be effective in treating UVB-induced erythema (Reuter et al., 2008). Skin lesions resembling contact dermatitis induced in animal models by agents like 2,4-dinitrochlorobenzene seem to be protected. Topical administration of the extracts prevented lesion development and reduced the levels of cytokines like TNF-α, IL-1, IL-4, IL-13 IFN-γ, and IgE in mice, suggesting its potential as an alternative therapy for the disease (Park et al., 2014). Leaf extract-coated silver nanoparticles have been found to be effective against skin diseases like acne and dandruff. *In vitro* studies have shown silver nanoparticles synthesized with leaf extracts have antimicrobial effects against *Propionibacterium acne* and *Malassezia furfur*, with minimum inhibitory concentrations (MIC) of 3.1 μg/mL and 25 μg/mL, respectively (Sathishkumar et al., 2016). Extracts have been evaluated for their efficacy in treating infant diaper dermatitis (Dastgheib et al., 2017). The calming effects of SO have also been investigated and found to control the activation of the protein TRPA1 (a membrane sensor of nociception, activated through allyl isothiocyanate) in co-cultured keratinocytes and neurons, suggesting its potential as a soothing agent in sensitized skin. The researchers also found lowered inflammatory signals (NFκB) generated by the activation of cultured keratinocytes (by TNF-α) which leads to a suppression of neuronal sensors (in the co-cultured neurons) of nociception (Kern et al., 2020). Sunlight exposure is a major cause of photoageing, which can be prevented by using appropriate sunscreen. In a study by Hwang et al. (2014), treatment with ethanol extracts of coriander leaves following UVB exposure of normal human dermal fibroblasts increased collagen precursors and decreased matrix metalloproteinase-1 (MMP-1), which are thought to stabilize epidermal health. Skin photoageing is typically associated with a decline in the levels of functional elastin with a concomitant increase of non-functional elastin fibres. UV radiation triggers the activation of the elastin promoters in DNA, which can result in the accumulation of poorly organized elastin in the skin, a condition commonly known as solar elastosis. Coriander EO, where linalool was the major component (81.29%), showed inhibitions of enzymes like elastase collagenase, hyaluronidase, and tyrosinase inhibitory activities (compared with other Apiaceous oils of fennel, anise, or cumin). The application of the oil led to a substantial elevation in collagen and functional elastin content. Significant decreases in the expression of MMP-1 and transcriptional mediators like AP-1 and JNK were noted, resulting in the improvement of constituent skin proteins for reversing photoageing. The cytokine transforming growth factor-β (TGFβ) promotes collagen production and downregulates the expression of MMPs. EO significantly augmented the expression levels of TGFβ, TGFβII, and SMAD3 proteins. Various cell signalling pathways were inhibited. Notable anti-inflammatory properties were evidenced by downregulations of COX-2 and PGE2, accompanied by lowering of MDA levels. The results suggest that the oil may have promising applications in preventing skin ageing and hyperpigmentation by preventing the synthesis of melanin (Salem et al., 2022).

Antiseptic and anti-inflammatory properties make coriander beneficial for skincare as well. Coriander EO can be applied topically to treat skin infections caused by bacteria or fungi, which include *Tenia pedis* (Beikert et al., 2013) and *Candida* sp. (Silva et al., 2011; Freires et al., 2014; Laribi et al., 2015). It has been found to be effective against common skin pathogens such as *Staphylococcus aureus* and *Candida albicans*. Aqueous extracts applied to the skin have been found to hinder the growth of *Propionibacterium acne* and *Staphylococcus epidermidis* with MICs of 1.7 mg/mL and 2.1 mg/mL, respectively (Vats & Sharma, 2012), suggesting that it may treat skin conditions such as acne, eczema, and psoriasis in humans.

15.4.9 Antimicrobial Activity

Spices are a natural source of antimicrobial compounds and may be used in place of synthetic preservatives like BHA (butylated hydroxyanisole) and butylated hydroxytoluene (BHT) (Gottardi et al., 2016).The principal bioactive compounds, such as linalool, α-pinene, and limonene, present in the extracts and in the EOs from the seeds, leaves, and stems, are the primary contributors to its antimicrobial activity. Singh et al. (2006) found the coriander EO to be effective against *Curvularia palliscens*, *Fusarium oxysporum*, *Fusarium moniliforme*, and *Aspergillus terreus* (using the inverted Petri plate method) and on the growth *of Aspergillus terreus*, *Aspergillus niger*,

Fusarium graminearum, and *Fusarium oxysporum* by their food poisoning technique. Wong and Kitts (2006) found that stem extracts (due to their higher phenol content) exhibited stronger antibacterial activity. Matasyoh et al. (2009) found EO from leaves to be effective against *S. aureus*, *Bacillus* sp., *Escherichia coli*, *Salmonella typhi*, *Klebsiella pneumoniae*, *Proteus mirabilis*, *Pseudomonas aeruginosa*, and the fungus *C. albicans*. EO of coriander shows a synergistic effect with six distinct antibacterial medications, including cefoperazone, ciprofloxacin, gentamicin, chloramphenicol, tetracycline, and piperacillin tested against *Acinetobacter baumannii*. Synergism in antibacterial effectiveness between coriander EO and certain antibiotics like chloramphenicol, ciprofloxacin, gentamicin, and tetracycline was also observed (Duarte et al., 2012). (2*E*)-dodecanal is one of several volatile compounds from coriander leaves that showed a maximum inhibitory effect on *Salmonella choleraesuis* (Kubo et al., 2004). Duarte et al. (2007) found antimicrobial activity against *Pityrosporum ovale*, whereas Wong and Kitts (2006) found it against *Bacillus subtilis* and *E. coli* from the ethanolic and aqueous extracts. Coriander EO has been shown to inhibit the growth of both Gram-positive and Gram-negative bacteria, which include *S. aureus*, *Streptococcus pyogenes*, *Bacillus* spp., *Listeria monocytogenes*, *E. coli*, *K. pneumoniae*, *P. aeruginosa*, *Proteus mirabilis*, *Campylobacter jejuni*, *Campylobacter coli*, and *S. typhi* (Delaquis et al., 2002; Rattanachaikunsopon & Phumkhachorn, 2010; Mohammadi Bazargani & Rohloff, 2016; Kačániová et al., 2020).

Dodecenal and 1-decanol found in the EO of the leaf were active against *Staphylococcus epidermidis* and *S. aureus* (Bharti et al., 2012). Long coriander (*Eryngium foetidum*, also from the Apiaceae) EO containing *E*-2-dodecenal ("eryngial") was active against *S. aureus* and *Bacillus subtilis* (Shavandi et al., 2012) and dodecanoic acid against *Listeria monocytogenes* (Ngang et al., 2014). The oil has also been shown to have antifungal activity against *C. albicans* and other *Candida* species. Coriander extracts from fresh leaves were active against various *Candida* species like *C. krusei*, *C. parapsilosis*, and *C. dubliniensis*. The chemical fractions of the EO exhibited antimicrobial activity comparable to standard antibiotics. The potent antimicrobial effects of the EO chemical fractions could be attributed to their abundance of alcohols and suggested its use to treat or prevent *Candida* and yeast infections (Lo Cantore et al., 2004; Msaada et al., 2007; Begnami et al., 2010; Soares et al., 2012). Oral candidiasis infections are increasing worldwide. Freires et al. (2014) investigated the anticandidal effectiveness of EO from leaves. They proposed a mode of action of the EO containing monoterpenes and sesquiterpenes which, by binding to membrane ergosterol, increases ionic permeability and that the membrane damage was detrimental to the survival of the yeast cells. The EO inhibited *Candida* biofilm adherence to polystyrene surfaces and lowered the proteolytic activity of *C. albicans*. Molina et al. (2020) showed the antimicrobial, antibiofilm, antiquorum sensing, and anti-swarming properties of hexane, chloroform, ethyl acetate, methanolic, and total methanolic extracts of coriander grown in Argentina against different bacteria, including *E. coli*, *Salmonella enterica*, *P. aeruginosa*, and *S. aureus*, with the observation that chloroform extracts of coriander were the most effective in inhibiting the genesis and characteristic metabolism of biofilms, especially for *S. aureus*. The methanolic extract of coriander leaves and stems is a more potent anti-bacterial agent than aqueous extracts.

15.4.10 Other Bioactivities

EO is used in oral care products to prevent the growth of bacteria that cause dental plaque and gum disease. It has been found to be effective against the bacteria that cause cavities, as well as the fungi that cause oral thrush (Sobhani et al., 2022). Extracts have been found to have lead detoxifying potential by reducing the deposition of lead in the femur and preventing renal injury caused by lead (Aga et al., 2001). Coriander has been found to have insecticidal effects against different types of insects, including mosquitoes, flies, and termites, which are listed in Table 15.2.

TABLE 15.2
Pharmacokinetic Activities from Animal Studies which Are Found in Extracts from Different Plant Parts

Pharmacokinetic Activities	Extracts from Different Plant Parts (Leaf/Stem/Seed/Fruit/Root/Flower) Responsible for the Pharmacological Effects in Animal Studies	Reference
Analgesic activity		
Anaesthetic-like effect by modifying nicotinic receptor-ion channel at neuro-muscular junctions/changes in secretion of acetylcholine	Linalool	Re et al., 2000; Hosseini et al., 2021
An analgesic effect on chronic pain states and inflammation/neuropathic pain; linalool-mediated blocking of NMDA and other ionotropic glutamate receptors; inhibition of NO formation	Polyphenols and a major compound of EO namely linalool	Kaur et al., 2005; Batista et al., 2008; Batista et al., 2010
Antiatherogenic activity		
Decrease the risk of cardiovascular diseases and atherosclerotic index	Seed extract	Aissaoui et al., 2011
Prevent oxidative damage to the myocardial tissues	Polyphenol in the leaf extract	Sreelatha & Inbavalli., 2012
Lower damage in induced myocardial necrosis	Methanolic extract of seeds	Patel et al., 2012
Lowering blood pressure	Fruit extract	Jabeen et al., 2009
High-salt and -fructose diet induced hypertension	Seed extracts	Wang et al., 2022
Antilipidaemic activity	Seed extracts	Chithra & Leelamma, 2000; Kousar et al., 2011
Inhibition of scavenger receptor (SRB1) expression, decreases the number of foam cell formations, and subsequently reduces the atherogenic plaques	Phytochemicals of plant extract	Patel et al., 2013; Sharma et al., 2014
Antibacterial activity		
Inhibit bacterial growth by reacting with DNA and disrupts DNA replication	Furocoumarins from leaf and petiole	Ceska et al., 1988
Inhibition of Gram-positive (*S. aureus*, *Streptococcus pyogenes* and *Bacillus* spp., *Listeria monocytogenes*) and Gram-negative (*E. coli*, *K. pneumoniae*, *P. aeruginosa*, *Proteus mirabilis*, *C. jejuni*, *C. coli*, and *S. typhi*) bacteria	EO of seeds and leaves	Delaquis et al., 2002; Rattanachaikunsopon et al., 2010; Mohammadi Bazargani & Rohloff, 2016; Kačániová et al., 2020
Bactericidal effect against the food-borne bacterium *Salmonella choleraesuis*	Aliphatic (2*E*)-alkenals and alkanals isolated from the fresh leaves	De Marco et al., 2004
Bactericidal against *Salmonella* (two times more effective than the usually used antibiotic medicine gentamicin)	Dodecanal component of leaf EO	Kubo et al., 2004

(*Continued*)

TABLE 15.2 (CONTINUED)
Pharmacokinetic Activities from Animal Studies which Are Found in Extracts from Different Plant Parts

Pharmacokinetic Activities	Extracts from Different Plant Parts (Leaf/Stem/Seed/Fruit/Root/Flower) Responsible for the Pharmacological Effects in Animal Studies	Reference
Synergism with six separate antibacterial drugs named ciprofloxacin, chloramphenicol, gentamicin, cefoperazone, tetracycline, and piperacillin. Synergism with EO and chloramphenicol, ciprofloxacin, gentamicin, and tetracycline against *Acinetobacter baumannii*	EO	Duarte et al., 2012
Antimicrobial against human pathogenic Gram-positive bacteria, Gram-negative and against the yeast, *Saccharomyces cerevisiae*	Extracts of leaves, seeds, and EOs	Sundar et al., 2016; Silva & Domingues, 2017
Inhibitory activity against biofilm formation of *Salmonella enterica*, *E. coli*, *Stenotropomonas maltophilia*, *P. aeruginosa*, and *Staphylococcus aureus*	Hexane and chloroform extracts of fruits	Molina et al., 2020
Anticancer activity		
Protective role in the lipid metabolism of colon cancer and apoptosis of cancer cells	Fruits	Chithra et al., 2000; Iwasaki et al., 2016
Antiproliferative activity via downregulating oxidative stress. Inhibits DNA damage, cancer cell migration, and metastasis	Leaf, stem and especially root phenolic content	Tang et al., 2013
Antileukaemic activity with highest binding affinity for their targeted receptors. Catechins and rutin proposed as alternative treatment agents for chemotherapy	Polyphenolic extract of seeds named flavonoids, catechins and rutin	Mechchate et al., 2021
Antidiabetic activity		
Reduce production of advanced glycation end-products (AGEs) and prevention of diabetic nephropathy; quercetin has strong anti-obesity property due to lipase inhibition	Seed extract rich in linalool, ascorbyl palmitate, petroselinic acid, and kaempferol, caffeic acid and quercetin	Kajal et al., 2019; Suttisansanee et al., 2021
Antihyperglycaemic activities	EO components of seed namely γ-terpinene and α-pinene	Hajlaoui et al., 2021
Increase the activities of glucose-6-phosphate dehydrogenase, hexokinase, and phosphoglucomutase-increased glycolysis; decrease in activity of glycogen phosphorylase and glucose-6-phosphatase-dehydrogenase; reduction in glycogenolysis and gluconeogenesis	Seed extract	Chithra&Leelamma, 1999
Improve insulin sensitivity, inhibit α-amylase activity, α-glucosidase activity, glucose transport, and absorption and increase glucose uptake in liver and peripheral tissues	Quercetin-rich extracts	Gray & Flatt, 1999; Eidi et al., 2009; Aissaoui et al., 2011

(*Continued*)

TABLE 15.2 (CONTINUED)
Pharmacokinetic Activities from Animal Studies which Are Found in Extracts from Different Plant Parts

Pharmacokinetic Activities	Extracts from Different Plant Parts (Leaf/Stem/Seed/Fruit/Root/Flower) Responsible for the Pharmacological Effects in Animal Studies	Reference
Hypoglycaemic effect in type 2 diabetes mellitus	Extracts of seeds, leaves, and stems	Waheed et al., 2006
Prevent high-fat diet (HFD) inducing increase in bodyweight gain, food intake, feed efficiency, fasting blood glucose, plasma insulin, fasting insulin resistance index (FIRI)	Seed extract	Patel et al., 2011
Manage high fasting blood glucose in diabetes	Polyphenol fraction of seeds	Mechchate et al., 2021
Antidiuretic activity		
Improve diuresis and urinary flow of sodium, potassium, chloride, and glomerular filtration rate in a dose-dependent way	Seed extract	Aissaoui et al., 2008
Increased sodium excretion in the urine more than potassium	Leaf extract	Thuraisingam et al., 2019
Antifungal activity		
EOs bind to membrane ergosterol of the fungus *Candida albicans*, increases ionic permeability and causing membrane damage leading to cell death and inhibition of *Candida* species biofilm adherence	EO from leaf and seed extract, mainly 2-hexen-1-ol, 3-hexen-1-ol, and cyclodecane	Lo Cantore et al., 2004; Msaada et al., 2007; Begnami et al., 2010, Furletti et al., 2011; Freires et al., 2014
Inhibitory to *Fusarium oxysporum*, *Curvularia palliscens*, *Fusarium moniliforme*, *Aspergillus terreus*	EO of seeds	Singh et al., 2006
Inhibition of *Microsporum canis* strains	EO of fruits	Soares et al., 2012
Anthelmintic activity		
Destroy adult tapeworm, *Hymenolepis nana*, in a dose-dependent manner	Seed extract	Hosseinzadeh et al., 2016
Nematicidal activity against plant-parasite nematode *Bursaphelenchus xylophilus* and animal-parasitic nematode *Haemonchus contortus*	EO	Eguale et al., 2007; Kim et al., 2008
Anti-inflammatory activity		
Inhibiting NO can reduce ROS, IL-6, TNF-α, and expression of inflammatory cytokines like iNOS and COX-2, leading to enhanced membrane stability	(7α,8α)–3α-hydroxyl-12,13α-dimethyl-5(6)-enbicyclo [5,3,0]caprolactone isolated from seeds	Kothalawala et al., 2020; Yuan et al., 2020
Decrease LPS-induced NO, PGE2 production; lowers iNOS, COX-2, and pro-IL-1β expression; reduces other pro-inflammatory mediator expression by suppressing NF-κB activation and MAPK signal transduction pathway in LPS-induced macrophages	Both stem and leaf extract	Wu et al., 2010

(*Continued*)

TABLE 15.2 (CONTINUED)
Pharmacokinetic Activities from Animal Studies which Are Found in Extracts from Different Plant Parts

Pharmacokinetic Activities	Extracts from Different Plant Parts (Leaf/Stem/Seed/Fruit/Root/Flower) Responsible for the Pharmacological Effects in Animal Studies	Reference
Anti-arthritic impacts like substantial decrease in joint diameter, paw volume, and TNF-α levels	EO of seeds	Deepa et al., 2020
Antimigraine activity		
It is an antagonist of transient receptor potential M8 (TRPM8) channel which increases in migraine and the associated pain	Linalool	Behrendt et al., 2004
Anti-inflammatory and analgesic effects, leading to migraine relief	Linalool	DelavarKasmaei et al., 2016
Antimutagenic activity		
A positive correlation between its concentration and antimutagenic effect; significantly decreases the mutagenicity of metabolized aromatic amines	Coriander juice and chlorophyllin (used as a subrogate molecule of chlorophyll)	Cortés-Eslava J. et al., 2004
Reduce the axial skeleton and affects the neural tube, cardiovascular structures, and eye	Extract of leaves	Reyes et al., 2010
Antioxidant activity		
Detoxification and ROS neutralization in different *in vivo* and *in vitro* assays	Green leaf volatile oils like alkenals	Potter, 1996
Decrease in the levels of lipid peroxides, free fatty acids, and glutathione; increase in the activity of antioxidant enzymes like superoxide dismutase and glutathione reductase	Phenolic content of seeds	Chithra&Leelamma, 1999
DPPH radical scavenging activity, lipoxygenase inhibition, iron chelating activity, hydroxyl radical scavenging activity	Extracts from both leaves and seeds, mainly flavonoids and phenolic compounds	Wangensteen et al., 2004; Hashim et al., 2005; Wong et al., 2006; Misharina et al., 2008; Sriti et al., 2011; Reddy et al., 2012
Scavenging the reactive hydroxyl radicals and protecting DNA	Carotenoids extract from foliage	Divya et al., 2012
Protect the unsaturated fatty acids and phospholipids from peroxidation	Bioactive sterols, which contain mono- and polyunsaturatedfatty acids, such as stigmasterols, sitosterols and polar lipid tocopherols and phenolics	Wang et al., 2018
Reduce ROS production	Seed extraction rich in linalool, ascorbyl palmitate and petroselinic acid	Kajal et al., 2019
Protective agent against lipid peroxidation	Carvone, γ-terpinene and *p*-cymene from EO	Hajlaoui et al., 2021
Antiulcer activity		

(*Continued*)

TABLE 15.2 (CONTINUED)
Pharmacokinetic Activities from Animal Studies which Are Found in Extracts from Different Plant Parts

Pharmacokinetic Activities	Extracts from Different Plant Parts (Leaf/Stem/Seed/Fruit/Root/Flower) Responsible for the Pharmacological Effects in Animal Studies	Reference
Protective action against ulcerogenic effects of salt, sodium hydroxide, ethanol, indomethacin, and pylorus ligation in a dose-dependent manner	Fruits	Abascal et al., 2012
Inhibit ulcers by the formation of a defensive coating of either one or more of these compounds by hydrophobic exchanges.	Antioxidant constituents named linanool, flavonoids, coumarins, catechins, terpenes and polyphenolic compounds	Al-Mofleh et al., 2006
Anxiolytic activity		
Enhance GABA effect on the receptor of linalool and exerts antidepressant and muscle-relaxant effects	Linalool and EO	Emamghoreishi et al., 2005; Linck et al., 2009
Reduce activity, locomotor activity, and neuromuscular coordination	Seed extract	Emamghoreishi et al., 2007
Protection against lipid peroxidation through reducing ROS production	Seed extraction rich in linalool, ascorbyl palmitate, and petroselinic acid	Kajal et al., 2019; Hosseini et al., 2021
Hepatoprotective activity		
Decrease AST and ALT; increase hepatic antioxidant enzymes like SOD, CAT and GPx; regenerate hepatocytes; normalize fatty changes and necrosis in liver	Aerial parts	Sreelatha et al., 2009
Increase antioxidant or detoxification agent in tissues and reduce oxidative stress, lipid peroxidation in hepatocytes	EOs and seed powder	Samojlik et al., 2010
Decrease TNF-α, NF-κB, caspase 3, and necrosis in hepatic ischaemia reperfusion injury	Seed extract	Kükner et al., 2021
Hypocholesterolaemic activity		
Lowered absorption of dietary cholesterol	Stigmasterol and β-sitosterol	Chithra et al., 1997; Sriti et al., 2010
Reduced cholesterol levels in the blood and cholesterol deposition in vessels	Linoleic acid, oleic acid, palmitic acid, stearic acid and ascorbic acid	Ertas et al., 2005
Increase the enzyme activity of HMG-CoA reductase, which suppresses cholesterol absorption and enhances hepatic bile acids synthesis, and thus increases the degradation of cholesterol	Phytosterols, especially stigmasterol	Batta et al., 2006
Increase in HMG CoA reductase and plasma LCAT activity; enhance degradation of cholesterol to faecal bile acids and neutral sterols	Seed extract and plant sterols and stanols	Dhanapakiam P et al., 2008

(Continued)

TABLE 15.2 (CONTINUED)
Pharmacokinetic Activities from Animal Studies which Are Found in Extracts from Different Plant Parts

Pharmacokinetic Activities	Extracts from Different Plant Parts (Leaf/Stem/Seed/Fruit/Root/Flower) Responsible for the Pharmacological Effects in Animal Studies	Reference
Decrease level of TC, TG, LDL, and very-low-density lipoprotein (VLDL) cholesterol; increasein HDL cholesterol	SO with stigmasterol, lanosterol, β-sitosterol, Δ^5-avenasterol, sitostanol, Δ^7-stigmastenol, Δ^7-avenasterol, tocopherols	Ramadan et al., 2008
Hypolipidemic activity		
Reducing fatty acid synthesis in the liver and fat accumulation	Flavonoids and polyphenols	Bao et al., 2016
Hypolipidaemic activities	Seed extraction rich in linalool, ascorbyl palmitate, and petroselinic acid	Chithra & Leelamma, 2000; Kousar et al., 2011; Kajal et al., 2019
Hypotensive activity		
Hypotensive activity	Pinocembrin, Apigenin, pseudobaptigenin, galangin-5-methyl ether, quercetin, baicalein, kaempferol, pinobanksin-glycosides, rutin, isorhamnetin, daidzein, luteolin, pectolinarigenin	Rajeshwari et al., 2011; Nazni et al., 2014; Hussain et al., 2018
Anti-arrhythmic activity by reducing pulse rate as well as normalizing ECG patterns and cardiac biomarker levels, including LDH, CK-MB, AST, and ALT	Seed extract	Rehman et al., 2016
ACE inhibition as a fundamental regulator of blood pressure	Flavonoids and isocoumarin glycosides-containing fractions and polyphenolic-rich extract	Luo et al., 2017; Hussain et al., 2018
Improves cardiac haemodynamic parameters, namely systolic blood pressure, diastolic blood pressure, heart rate, mean arterial pressure, and cardioprotective efficacy by endothelin dysfunction; preventive role against oxidative stress in heart failure	Seed extract	Dhyani et al., 2020
Insecticidal activity		
Inhibitory effect on storage grain insect, *Sitophilus granarius*	EO of seeds	Zoubiri&Baaliouamer, 2010
Immunotoxicity on yellow fever mosquito, *Aedes aegypti*	EO of seeds, leaf, stem	Chung et al., 2012
Neuroprotective activity		
Block the glutamatergic transmission by inhibiting L-[3H] glutamate binding, preventing quinolinic acid-induced convulsions, delaying the Induced-seizure models through NMDA receptor antagonism	Linalool	Elisabetsky et al., 1999

(*Continued*)

TABLE 15.2 (CONTINUED)
Pharmacokinetic Activities from Animal Studies which Are Found in Extracts from Different Plant Parts

Pharmacokinetic Activities	Extracts from Different Plant Parts (Leaf/Stem/Seed/Fruit/Root/Flower) Responsible for the Pharmacological Effects in Animal Studies	Reference
Improve memory function by decreasing oxidative markers like MDA, SOD, and H_2O_2 levels and increasing GPx activity in the hippocampus, effective in treating anxiety and depression-like actions in Alzheimer's disease (AD)	Linalool and volatile oils	Cioanca et al., 2014; Hosseini et al., 2021
Decrease MDA level and increase total thiol content, thus improving the antioxidant capacity of the brain; prolonging seizure latencies	Various extracts of aerial parts	Anaeigoudari et al., 2016
Stimulate neurotrophic factor (NGF), helping in nerve growth, survival, and anti-neuroinflammatory consequences	Isocoumarin glycosides, cilantroside A and B, phenolic glycosides, daphnin, benzyl-O-β-d-glucoside, and aglycones of the isocoumarins	Cha et al., 2020
Sedative-hypnotic activity		
Sedative-hypnotic activity by increasing sleeping time	Extracts and EO of seeds	Rakhshandeh et al., 2012
Sedative activity in general	Linalool	Gastón et al., 2016; Hosseini et al., 2021
Sedative-hypnotic activity	EO of seeds	Emamghoreishi&Heidari-Hamedani, 2006
Sedative effect on the central nervous system in mice	Quercetin and isoquercetin glycosides	Picq et al. (1991)

15.5 OTHER POTENTIAL USES OF CORIANDER

Coriander is one of the most widely used herbs worldwide and plays a significant role in international trade. Coriander SO is composed primarily of fatty acids having a mild, sweet, and spicy aroma, whereas coriander EO is composed of a range of chemical compounds, including terpenes, ketones, and alcohols and has a strong, sweet, and floral aroma (Kumar, 2022). The SO is often used in cooking and can be found in many culinary dishes, while the EO is used primarily for aromatherapy. According to Lawrence (1993), coriander oil is the second most valuable EO globally, behind orange oil, with an estimated production value of around 50 million dollars annually. Coriander EO is also used as a flavouring agent in food and beverages. SO is typically extracted through cold pressing the seeds while EO is typically extracted through steam distillation of the seeds and aerial parts (the latter being called the leaf volatiles). It is important to note that coriander oil and coriander EO are not interchangeable, and care should be taken to ensure that the correct product is used for the intended purpose. EO is not generally considered safe for internal use.

Efforts are underway to maximize the yields of vital components of coriander EO like linalool, phenolic antioxidants, and other terpenoids, which are relatively lipophilic in nature. Methyl jasmonate has been used to upregulate genes involved in the synthesis of linalool. A 1.5-fold increase in linalool yield, with increases in terpinene, and phenolic contents, was observed in Iranian coriander cultivars (Kianersi et al., 2022). The contents of linalool and other EO components were increased with foliar application of fertilizers by substantial amounts (2.9–9.6%). With such increases, coriander has the potential to be a biofactory of EO (Georgieva et al., 2022). A study by Lin et al. (2022) showed that light conditions change the functional bioactive attributes and phytochemical potentials of coriander-derived products, with blue light increasing the antioxidant potential, suggesting optimizations in lighting conditions in biofactories might enhance phytochemical yields of commercial value. Variations in percentage yield of EO as well as variations in yield of individual components of EOs have triggered research on grouping and clustering coriander varieties into distinct groups of germplasms. Saxena et al. (2022) observed linalool as the most abundant terpenoid but variations of its content were as much as 16.59% to 96.69% among different germplasms. Alpha-pinene, β-pinene, camphene, myrcene, cymene, and γ-terpinene contents represented up to 13.32% of EO fractions. Geranyl acetate content varied from 0.085% to 60.95% with an inverse effect seen on linalool productions in the geranyl acetate-producing varieties from India. Subcritical water extraction (SWE) of monoterpenes from coriander seeds and the effect of extraction conditions were investigated, at varying temperatures (170–200°C) and extraction times (5–20 min). Monoterpene extraction was influenced and a variation in their structures was observed. The extraction efficiency of linalool was lowered with increasing extraction temperature and time, while linalool oxide increased (which might be due to conversion to linalool oxide by hydrothermal oxidation) (Song et al., 2022).

15.5.1 Coriander as a Spice Ingredient to Flavour Various Foods and Beverages

Coriander is a commonly used herb in North American cuisine, particularly in Mexican and Tex-Mex dishes. It is often used in Indian dishes from curries to chutneys, in combination with other spices such as cumin and turmeric. Leaves are used as a garnishing herb in soups, stews, curries, and noodle dishes. They are also commonly used in salads and as a topping for rice and noodle bowls. It is an essential ingredient in many Latin American dishes such as *salsa*, *guacamole*, and *ceviche*, and used as a garnish for *tacos*, *enchiladas*, *guacamole*, *salsas*, and other Mexican dishes, adding freshness, colour, and flavour to the dish (https://www.mexicali-blue.com/exploring-the-versatility-of-cilantro-an-essential-ingredient-in-latin-american-cuisine/). Many Mediterranean dishes such as *falafel*, *tabbouleh*, and *hummus* use the herb and seeds. Leaves are used in marinades for meats and as a garnish on grilled vegetables (https://www.themediterraneandish.com/). In recent years, coriander has become more popular in European cooking, particularly in fusion cuisine and modern interpretations of traditional dishes. It is used in dishes such as *chili*, *fajitas*, and *quesadillas*, and in North American dishes, such as salads, marinades in meats, and in sauces for fish and seafood. Many Middle Eastern dishes use the seeds, particularly in spice blends such as *raselhanout and baharat*. It is also used in dishes such as *hummus*, *falafel*, *tabbouleh*, and *shakshuka*. Coriander is a staple herb in Thai cuisine, where it is known as '*pak chi*', where it is often used in curries, stir-fries, and salads in dishes such as green curry and tom yum soup. It is often used in combination with other herbs, such as lemongrass and Thai basil. Roots of the plant are used in Thai cuisine in some dishes (https://thaifoodparadise.com/what-is-cilantro-used-for/). Coriander is not as widely used in Russian cuisine as in some other cuisines, but it is still used in a variety of dishes and as a seasoning for meats and vegetables; it is sometimes used as a seasoning in borscht, a traditional Russian soup; seeds are used as an ingredient in spice blends such as Georgian *khmeli-suneli* (https://happykitchen.rocks/khmeli-suneli-georgian-spice-mix/). Coriander seeds are often used in the brine for pickled vegetables such as cucumbers and cabbage and used as a seasoning for meat dishes such as beef stroganoff and lamb *kebabs*. It is not a traditional ingredient in Chinese cuisine, but has become more popular in recent years, particularly in the north of China where it is known as '*xiangcai*', used as a filling for dumplings, and paired with pork, and sometimes used as a garnish for noodle dishes, such as *dan dan* and *zhajiangmian*; it is added to Chinese soups such as egg drop soup and hot and sour soup. In Vietnamese cuisine, it is used in dishes such as *pho*, spring rolls, and *banh mi* sandwiches (https://www.yummly.com/recipes/vietnamese-coriander). In Caribbean cuisine, where it is often used in dishes such as jerk chicken and fish, and used in sauces and marinades. Coriander is a common ingredient in Belgian beer brewing; combined with orange peel, it can create a complex and refreshing flavour profile which is used in light and flavourful beer styles (Millam et al., 1997; Al-Mofleh et al., 2006; Asfaw, 2021). Coriander oil also can be used as a flavouring agent in a variety of dishes. It is important to choose a high-quality, food-grade coriander oil that is safe for consumption.

15.5.2 Coriander as a Natural Preservative

Coriander oil is used as a natural food preservative due to its antioxidant properties. Wangensteen et al. (2004) studied the antioxidant activity of extracts of coriander leaves and seeds, using bioassays of the 2,2-diphenyl-1-picrylhydrazyl (DPPH) scavenging radical method and inhibition of 15-lipoxygenase. They found a positive correlation between the total phenolic content and antioxidant activity of ethyl acetate extracts of seeds and leaves (leaves had greater antioxidant potential). They concluded that adding coriander to dishes may increase the antioxidant content of the food to prevent food spoilage. EO is used as protection against seed spoilage caused by *Aspergillus flavus* in chickpea (50% protection) (Prakash et al., 2012). EO is used as a natural preservative in food products (Silva et al., 2017; Sobhani, 2022). Its antimicrobial properties help to inhibit the growth of harmful bacteria, making it an effective way to extend the shelf life of foods. Coriander extracts were effective against food fungi, such as *Saccharomyces cerevisiae* (brewer's yeast) (Sundar et al., 2016; Silva & Domingues, 2017). Mehri et al. (2015) reported lowered acrylamide exposure toxicity in Wistar rats as a result of decreased lipid peroxidation and MDA levels and an increase in glutathione. Rats were treated with linalool administration in food three days before acrylamide-induced neurotoxicity. Linalool had protective effects on acrylamide toxicity from fried foods and acted as a food preservative. Darughe et al. (2012) found that the EO from coriander at concentrations of 0.05% to 0.15% inhibited the formation of primary and secondary oxidation products in cake, and its effects were like those of the synthetic preservative BHA (0.02%) during a 30-day storage period. The acidity of the control and cake samples containing BHA (0.01% or 0.02%) was higher than cakes with different concentrations of coriander EO (0.05% to 0.15%). In another study, it was observed that the addition of the EO to Italian processed food, specifically salami, helped to maintain the peroxide value and TBARS values throughout the entire storage period. It had a stronger effect than BHT in retarding lipid oxidation, which increased the product's shelf life.Several studies found that coriander extracts exhibited a significant inhibitory effect on the biofilm of important pathogenic bacteria, particularly against *S. aureus* (Molina et al., 2020), *Campylobacter jejuni*, and *C. coli* strains (Duarte et al., 2016) by either decreasing their formation or interfering with the quorum sensing system. These findings suggest that the extracts could be potentially used as a natural alternative to synthetic antimicrobial agents or as antioxidants in the food industry to control foodborne pathogens and prevent spoilage (Kačániová et al., 2020).

15.5.3 Coriander as a Natural Fragrance in Perfume and Cosmetic Industries

Plants are widely used in the production of EOs for perfumes in the cosmetic industry (Zhang et al., 2006). Coriander oil has uses in consumer products like creams, lotions, soap, and perfumes indirectly or directly as an aromatherapeutic oil (Opdyke, 1973; Cooksley, 2003). EO derived from coriander is used as a fragrance and blends well with other Eos, such as bergamot, lavender, and lemon (Önder, 2018).

15.5.4 Agronomic Significance

EO containing aliphatic aldehydes are used for their potential to control plant pathogens, pests, weeds, and for post-harvest uses such as food grain protection. It also has insecticidal (against flour beetles) properties and is used as an insect repellent (against *Aedes aegypti*) (Islam et al., 2009; Zoubiri&Baaliouamer, 2010; Chung et al., 2012; Laribi et al., 2015). Raveau et al. (2022) addressed the issues related to the stability and persistence of EO effects from plants grown in polluted waste soils. The EO was found to have resisted phytopathogens like *Zymoseptoria tritici* and *Fusarium culmorum* (fungi), displaying IC_{50} ranging from 0.46 to 2.08 g/L; displayed weed anti-germination properties, and herbicidal and fumigant effects.According to a study by Al-Jaff et al. (2011), incorporating coriander seeds at a 2% level in broiler poultry diets can have a positive impact on their performance, blood profile, and immune system, particularly during heat stress. In another study, the EO extracted from coriander was tested for its ability to control the pinewood nematode, *Bursaphelenchus xylophilus*, and showed promising nematicidal activity (Kim et al., 2008). Coriander extract or powder can be used as an antibiotic alternative in broiler feeds (Hosseinzadeh et al., 2016). In another study, after heavy metal exposure in belugas (*Husohuso*), flushing of heavy metals was reported with a coriander diet of 10 mg/kg. Bahrekazemi et al. (2022) found that carcass protein, lipid content, and ash contents increased in beluga in response to the seed feeding, both before and after induced heavy metal toxicity.

15.5.5 Coriander as a Soil Remediation Crop

The exclusive use of aromatic plants for the extraction of EO has been suggested as an innovative and economically feasible solution for remediating areas contaminated with trace elements. Coriander has shown potential, particularly when combined with steel slag, for the phytomanagement of soils polluted with trace elements. This approach can improve soil quality and the structure of bacterial communities in the rhizosphere, ultimately benefiting both the environment and economy. By cultivating coriander and using arbuscular mycorrhizal fungi in combination with mineral amendments like steel slag, the negative effects of mineral amendments on rhizosphere bacterial communities and their activities can be reversed (Langrand et al., 2023).

15.5.6 Other Industrial Uses and Safety Assessments

Additionally, EO has industrial uses as medicine vehicles, as repellents, and for dyeing fabrics. It has even been used for smoking, in charms and other cultural practices, as well. It is a melliferous plant, being used in the production of honey. Coriander has been found to have refrigerant properties as well. Coriander EO can be diffused into the air to purify the environment and kill airborne pathogens. Its antimicrobial properties can help to reduce the spread of infectious diseases in enclosed spaces (Abou-Shaara, 2015; Asfaw, 2021).

Regulatory agencies, such as FDA (US Food and Drug Administration), CFR (Code of Federal Regulations), CGMP (Current Good Manufacturing Practices), CoE (Council of Europe), IOFI (International Organisation of flavour Industries), and FEMA (Flavour and Extract Manufacturers' Association), have all assessed products derived from coriander itself, such as oils from seeds or its volatile oils obtained through steam distillation of seeds and aerial parts for human uses. Its oil is approved as a safe food ingredient by the FDA, FEMA, and the CoE. The FDA has granted the Generally Recognized as Safe (GRAS) status to coriander oil for use as a flavouring agent without any limitations and adjuvant without any specified limitations but not for any direct use in treatment. FEMA has also classified it as generally safe for use in food as a flavouring ingredient. The Council of Europe has included coriander oil in its list of acceptable substances, spices, and seasonings, with possible limitations on the active principle in final products. The International Organization of Flavour Industries categorizes coriander oil as nature-identical, and it is included in the EPA Toxic Substances Control Act and FDA database of inactive ingredients for currently marketed drug products. The CoE approves the use of coriander for dyspepsia and appetite loss. The American Herbal Products Association categorizes coriander fruit as Class 1, which means it can be safely consumed when used appropriately. FEMA has categorized the permissible limits of coriander oil in parts per million for different food categories. It should be noted that the EO is only approved as a substance for flavouring and not for any direct consumption as a food oil. Studies have documented the presence of mycotoxins, pesticides, and other contaminants in the seed and its derived products (Burdock et al., 2009).

15.6 CONCLUSION

This chapter tries to provide valuable insights into the ethnomedicinal uses of coriander, a popular culinary herb that has been widely used in traditional medicine for centuries. Coriander possesses a wide range of pharmacokinetics which has largely been investigated in recent times through animal models due to interest generated by its traditional medicinal uses. But further research is called for to fully explore its potential as a therapeutic agent, as well as human studies to confirm herbal medicine roles. Phytochemicals like linalool and quercetin from coriander need to be explored through clinical trials.

LIST OF ABBREVIATIONS

ACE: Angiotensin-converting enzyme
ALT: Alanine transaminase
AST: Aspartate transaminase
BHA: Butylated hydroxyanisole
BHT: Butylated hydroxytoluene
CAT: Catalase
CK-MB: Creatine kinase-MB
COX: Cyclooxygenase
EO: Essential oil
GABA: γ-Aminobutyric acid
GPx: Glutathione peroxidase
HDL: High-density lipoprotein
HMG-CoA reductase: 3-Hydroxy-3-methyl-glutaryl-coenzyme A reductase
H_2O_2: Hydrogen peroxide
IL: Interleukin
iNOS: Inducible nitric oxide synthase

LDH: Lactate dehydrogenase
LDL: Low-density lipoprotein
LPS: Lipopolysaccharide
MDA: Malondialdehyde
MMP-1: Metalloproteinase
NGF: Nerve growth factor
NO: Nitric oxide
ROS: Reactive oxygen species
SOD: Superoxide dismutase
SO: Seed oil
TBARS: Thiobarbituric acid reactive substances
TG: Triglyceride
VLDL: Very-low-density lipoprotein

REFERENCES

Abascal, K., & Yarnell, E. 2012. Cilantro—Culinary herb or miracle medicinal plant? *Alternative and Complementary Therapies*, 18(5), 259–264.

Abou-Shaara, H. F. 2015. Potential Honey bee plants of egypt. *CercetăriAgronomiceîn Moldova*, 3(2), 99–108.

Aga, M., Iwaki, K., Ueda, Y., Ushio, S., Masaki, N., Fukuda, S., Kimoto, T., Ikeda, M., & Kurimoto, M. 2001. Preventive effect of *Coriandrum sativum* (Chinese Parsley) on localized lead deposition in ICR mice. *Journal of Ethnopharmacology*, 77(2–3), 203–268.

Aissaoui, A., Zizi, S., Israili, Z. H., & Lyoussi, B. 2011. Hypoglycemic and hypolipidemic effects of *Coriandrum sativum* L. in Merionesshawi rats. *Journal of Ethnopharmacology*, 137(1), 652–661.

Aissaoui, A., El-Hilaly, J., Israili, Z. H., & Lyoussi, B. 2008. Acute diuretic effect of continuous intravenous infusion of an aqueous extract of *Coriandrum sativum* L. in anesthetized rats. *Journal of Ethnopharmacology*, 115(1), 89–95. https://doi.org/10.1016/j.jep.2007.09.007.

Aligita, W., Susilawati, E., Septiani, H., & Atsil, R. 2018. Antidiabetic activity of coriander (*Coriandrum sativum* L) leaves' ethanolic extract. *International Journal of Pharmaceutical and Phytopharmacological Research*, 8(2), 59–63.

Al-Jaff, F. K. 2011. Effect of coriander seeds as diet ingredient on blood parameters of broiler chicks raised under high ambient temperature. *International Journal of Poultry Science*, 10(2), 82–86.

Al-Khayri, J. M., Banadka, A., Nandhini, M., Nagella, P., Al-Mssallem, M. Q., & Alessa, F. M. 2023. Essential Oil from *Coriandrum sativum*: A review on Its Phytochemistry and Biological Activity. *Molecules*, 28(2), 696. https://doi.org/10.3390/molecules28020696.

Al-Mofleh, I. A., Alhaider, A. A., Mossa, J. S., Al-Sohaibani, M. O., Rafatullah, S., & Qureshi, S. 2006. Protection of gastric mucosal damage by *Coriandrum sativum* L. pretreatment in Wistar albino rats. *Environmental Toxicology and Pharmacology*, 22(1), 64–69.

Al-Said, M. S., Al-Khamis, K. I., Islam, M. W., Parmar, N. S., Tariq, M., & Ageel, A. M. 1987. Post-coital antifertility activity of the seeds of *Coriandrum sativum* in rats. *Journal of Ethnopharmacology*, 21(2), 165–173.

Al-Snafi, A. E. 2016. A review on chemical constituents and pharmacological activities of *Coriandrum sativum*. *Journal of Pharmacy*, 6(7), 17–42. https://doi.org/10.9790/3013-067031742.

Altınok-Yipel, F., Ozan Tekeli, I., Özsoy, S. Y., Güvenç, M., Kaya, A., & Yipel, M. 2020. Hepatoprotective activity of linalool in rats against liver injury induced by carbon tetrachloride. *International Journal for Vitamin and Nutrition Research*, 90, 302–308. https://doi.org/10.1024/0300-9831/a000650.

Anaeigoudari, A., Hosseini, M., Karami, R., Vafaee, F., Mohammadpour, T., Ghorbani, A., & Sadeghnia, H. R. 2016. The effects of different fractions of *Coriandrum sativum* on pentylenetetrazole-induced seizures and brain tissues oxidative damage in rats. *Avicenna Journal of Phytomedicine*, 6(2), 223–235.

Anwar, F., Sulman, M., Hussain, A. I., Saari, N., Iqbal, S., & Rashid, U. 2011. Physicochemical composition of hydro distilled essential oil from coriander (*Coriandrum sativum* L.) seeds cultivated in Pakistan. *Journal of Medicinal Plants Research*, 5, 3537–3544.

Asfaw, M. D. 2021. A review on the chemical constituent, pharmacological and medicinal properties of *Coriandrum sativum*. *Natural Products Chemistry and Research*, 09, 419.

Bahrekazemi, M., Eslami, M., & Nikbakhsh, J. 2022. The effect of dietary coriander supplementation on growth performance, biochemical responses, carcass proximate composition, and heavy metal accumulation in beluga, Huso huso. *Journal of Applied Aquaculture*, 34(1), 23–42. https://doi.org/10.1080/10454438.2020.1782798.

Bao, L., Hu, L., Zhang, Y., & Wang, Y. I. 2016. Hypolipidemic effects of flavonoids extracted from Lomatogoniumrotatum. *Journal of Experimental Medicine*, 11(4), 1417–1424. https://doi.org/10.3892/etm.2016.3038.

Bashir, S., & Safdar, A. 2020. Coriander seeds: Ethno-medicinal, phytochemical and pharmacological profile. Sci. *Spices Culin. Herbs-Latest Lab. Pre-Clin. Clin. Stud.*, 2, 39–64.

Batista, P. A., de Paula Werner, M. F., Oliveira, E. C., Burgos, L., Pereira, P., da Silva Brum, L. F., & Dos Santos, A. R. S. 2008. Evidence for the involvement of ionotropic glutamatergic receptors on the antinociceptive effect of (-)-linalool in mice. *Neuroscience Letters*, 440(3), 299–303. https://doi.org/10.1016/j.neulet.2008.05.092.

Batista, P. A., de Paula Werner, M. F., Oliveira, E. C., Burgos, L., Pereira, P., da Silva Brum, L. F., Story, G. M., & Santos, A. R. S. 2010. The antinociceptive effect of (-)-linalool in models of chronic inflammatory and neuropathic hypersensitivity in mice. *The Journal of Pain*, 11, 1222–1229. https://doi.org/10.1016/j.jpain. 2010.02.022.

Batta, A. K., Xu, G., Honda, A., Miyazaki, T., & Salen, G. 2006. Stigmasterol reduces plasma cholesterol levels and inhibits hepatic synthesis and intestinal absorption in the rat. *Metabolism*, 55(3), 292–299. https://doi.org/10.1016/j.metabol.2005.08.024.

Begnami, A. F., Duarte, M. C. T., Furletti, V., & Rehder, V. L. G. 2010. Antimicrobial potential of *Coriandrum sativum* L. against different Candida species in vitro. *Food Chemistry*, 118(1), 74–77. https://doi.org/10.1016/j.foodchem.2009.04.089.

Begnami, A. F., Spindola, H. M., Ruiz, A. L. T. G., de Carvalho, J. E., Groppo, F. C., & Rehder, V. L. G. 2018. Antinociceptive and anti-edema properties of the ethyl acetate fraction obtained from extracts of *Coriandrum sativum* Linn. leaves. *Biomedicine and Pharmacotherapy*, 103(April), 1617–1622. https://doi.org/10.1016/j.biopha.2018.04.196.

Behrendt, H. J., Germann, T., Gillen, C., Hatt, H., & Jostock, R. 2004. Characterization of the mouse cold-menthol receptor TRPM8 and vanilloid receptor type-1 VR1 using a fluorometric imaging plate reader (FLIPR) assay. *British Journal of Pharmacology*, 141(4), 737–745. https://doi.org/10.1038/sj.bjp.0705652.

Beikert, F. C., Anastasiadou, Z., Fritzen, B., Frank, U., & Augustin, M. 2013. Topical treatment of tinea pedis using 6% coriander oil in unguentumleniens: A randomized, controlled, comparative pilot study. *Dermatology*, 226(1), 47–51.

Bharti, R., Ahuja, G., Ganapathy, S., & Shruthi, S. D. 2012. A review on medicinal plants having Antioxidant potential. *The Journal of Pharmacy*, 5, 4278–4287.

Bhat, S. P., Rizvi, W., & Kumar, A. 2014. *Coriandrum sativum* on pain and inflammation. *International Journal of Research in Pharmacy and Chemistry*, 4(4), 939–945. Retrieved from http://www.ijrpc.com/files/25-4196.pdf.

Bhuiyan, N. I., Begum, J., & Sultana, M. 2009. Chemical Composition of Leaf and Seed Essential Oil of *Coriandrum sativum* L. from Bangladesh. *Bangladesh Journal of Pharmacology*, 4, 150–153. https://doi.org/10.3329/bjp.v4i2.2800.

Blumenthal, M., Goldberg, A., & Brinkmann, J. 2000. *Coriander Seed. Herbal Medicine-Expanded Commission E Monographs*. Integrative Medicine Communications, Newton, MA, 75–77.

Bruneton, J. 1995. *Pharmacognosy, Phytochemistry, Medicinal Plants*. Lavoisier Publishing.

Burdock, G. A., & Carabin, I. G. 2009. Safety assessment of coriander (*Coriandrum sativum* L.) essential oil as a food ingredient. *Food and Chemical Toxicology*, 47(1), 22–34. https://doi.org/10.1016/j.fct.2008.11.006.

Ceska, O., Chaudhary, S. K., Warrington, P., Ashwood-Smith, M. J., Bushnell, G. W., & Poultont, G. A. 1988. Coriandrin, a novel highly photoactive compound isolated from *Coriandrum sativum. Phytochemistry* January 1, 27(7), 2083–2087.

Cha, J. M., Yoon, D., Kim, S. Y., Kim, C. S., & Lee, K. R. 2020. Neurotrophic and anti-neuroinflammatory constituents from the aerial parts of *Coriandrum sativum. Bioorganic Chemistry*, 105, 104443.

Chahal, K. K., Singh, R., Kumar, A., & Bhardwaj, U. 2017. Chemical composition and biological activity of *Coriandrum sativum* L.: A review. *Indian Journal of Natural Products and Resources*, 8(3), 193–203. http://nopr.niscair.res.in/handle/123456789/43357.

Cheng, B. H., Sheen, L. Y., & Chang, S. T. 2015. Evaluation of anxiolytic potency of essential oil and S-(+)-linalool from Cinnamomum osmophloeum ct. linalool leaves in mice. *Journal of Traditional and Complementary Medicine*, 5(1), 27–34. https://doi.org/10. 1016/j.jtcme.2014.10.007.

Chithra, V., & Leelamma, S. 1999. *Coriandrum sativum* changes the levels of lipid peroxides and activity of antioxidant enzymes in experimental animals. *Indian Journal of Biochemistry and Biophysics*, 36(1), 59–61.

Chithra, V., & Leelamma, S. 2000. *Coriandrum sativum* - Effect on lipid metabolism in 1, 2-dimethyl hydrazine induced colon cancer. *Journal of Ethnopharmacology*, 71(3), 457–463.

Chung, I. M., Ahmad, A., Kim, E. H., Kim, S. H., Jung, W. S., Kim, J. H., Nayeem, A., & Nagella, P. 2012. Immunotoxicity activity from the essential oils of coriander (*Coriandrum sativum*) seeds. *Immunopharmacology and Immunotoxicology*, 34(3), 499–503.

Cioanca, O., Hritcu, L., Mihasan, M., Trifan, A., & Hancianu, M. 2014. Inhalation of coriander volatile oil increased anxiolytic-antidepressant-like behaviors and decreased oxidative status in beta-amyloid (1–42) rat model of Alzheimer's disease. *Physiology and Behavior*, 131, 68–74. https://doi.org/10.1016/j.physbeh.2014.04.021.

Cooksley, V. 2003. An integrative aromatherapy intervention for palliative care. *The International Journal of Aromatherapy*, 13(2–3), 128–137.

Cortés-Eslava, J., Gómez-Arroyo, S., Villalobos-Pietrini, R., & Espinosa-Aguirre, J. J. 2004. Antimutagenicity of coriander (*Coriandrum sativum*) juice on the mutagenesis produced by plant metabolites of aromatic amines. *Toxicology Letters*, 153(2), 283–292.

Craig, D., Mirakhur, A., Hart, D., McIlroy, S., & Passmore, A. 2005. A cross-sectional study of neuropsychiatric symptoms in 435 patients with Alzheimer's disease. *American Journal of Geriatric Psychiatry*, 13(6), 460–468.

Darughe, F., Barzegar, M., & Sahari, M. A. 2012. Antioxidant and antifungal activity of Coriander (*Coriandrum sativum* L.) Essential oil in cake. *International Food Research Journal*, 19, 1253–1260.

Dastgheib, L., Pishva, N., Saki, N., Khabnadideh, S., Kardeh, B., Torabi, F., Arabnia, S., & Heiran, A. 2017. Efficacy of topical *Coriandrum sativum* extract on treatment of infants with diaper dermatitis: A single blinded non-randomised controlled trial. *The Malaysian Journal of Medical Sciences*, 24(4), 97–101. https://doi.org/ 10.21315/mjms2017.24.4.11.

De Marco, A., Senatore, F., Capasso, F., Iacobellis, N. S., & Lo Cantore, P. 2004. Antibacterial activity of *Coriandrum sativum* L. and Foeniculum vulgare Miller Var. vulgare (Miller) essential oils. *Journal of Agriculture and Food Chemistry*, 52(26), 7862–7866.

Deepa, B., Acharya, S., & Holla, R. 2020. Evaluation of antiarthritic activity of coriander seed essential oil in Wistar albino rats. *Research Journal of Pharmacy and Technology*, 13(2), 761–766. https: //doi.org/10.5958/0974-360X.2020.00144.4.

Delaquis, P. J., Stanich, K., Girard, B., & Mazza, G. 2002. Antimicrobial activity of individual and mixed fractions of Dill, Cilantro, Coriander and eucalyptus essential oils. *International Journal of Food Microbiology*, 74(1–2), 101–109. https://doi.org/10.1016/S0168-1605(01)00734-6.

DelavarKasmaei, H., Ghorbanifar, Z., Zayeri, F., Minaei, B., Kamali, S. H., Rezaeizadeh, H., Amin, G., Ghobadi, A., & Mirzaei, Z. 2016. Effects of *Coriandrum sativum* syrup on migraine: A randomized, triple-blind, placebo-controlled trial. *Iranian Red Crescent Medical Journal*, 18(1), e20759. https://doi.org/10.5812/ircmj.20759.

Dhanapakiam, P., Joseph, J. M., Ramaswamy, V. K., Moorthi, M., & Kumar, A. S. 2008. The cholesterol lowering property of coriander seeds (*Coriandrum sativum*): Mechanism of action. *Journal of Environmental Biology*, 29(1), 53–56.

Dharmalingam, R., & Nazni, P. 2013. Phytochemical evaluation of Coriandrum L flowers. *International Journal of Food and Nutritional Science*, 2, 34–39.

Dhyani, N., Parveen, A., Siddiqi, A., Hussain, M. E., & Fahim, M. 2020. Cardioprotective efficacy of *Coriandrum sativum* (L.) seed extract in heart failure rats through modulation of endothelin receptors and antioxidant potential. *Journal of Dietary Supplements*, 17(1), 13–26.

Divya, P., Puthusseri, B., & Neelwarne, B. 2012. Carotenoid content, its stability during drying and the antioxidant activity of commercial coriander (*Coriandrum sativum* L.) varieties. *Food Research International*, 45(1), 342–350.

Duarte, A., Ferreira, S., Silva, F., & Domingues, F. C. 2012. Synergistic activity of coriander oil and conventional antibiotics against Acinetobacter baumannii. *Phytomedicine*, 19(3–4), 236–238.

Duarte, A., Luís, Â., Oleastro, M., & Domingues, F. C. 2016. Antioxidant properties of coriander essential oil and linalool and their potential to control Campylobacter spp. *Food Control*, 61, 115–122. https://doi.org/10.1016/j.foodcont.2015.09.033.

Duarte, M. C. T., Figueira, G. M., Delarmelina, C., & Sartoratto, A. 2007. Investigação da atividade do oleo essencial de duas variedades de *Coriandrum sativum* contra microrganismosenvolvidos com patologiasdérmicas. *HorticulturaBrasileira*, 25(1), s36.

Eguale, T., Tilahun, G., Debella, A., Feleke, A., & Makonnen, E. 2007. In vitro and in vivo anthelmintic activity of crude extracts of *Coriandrum sativum* against Haemonchuscontortus. *Journal of Ethnopharmacology*, 110(3), 428–433.

Eidi, M., Eidi, A., Saeidi, A., Molanaei, S., Sadeghipour, A., Bahar, M., & Bahar, K. 2009. Effect of coriander seed (*Coriandrum sativum* L.) ethanol extract on insulin release from pancreatic beta cells in streptozotocin-induced diabetic rats. *Phytotherapy Research*, 23(3), 404–406. https://doi.org/10.1002/ptr.2642.

Elisabetsky, E., Marschner, J., & Souza, D. O. 1995. Effects of linalool on glutamatergic system in the rat cerebral cortex. *Neurochemical Research*, 20(4), 461–465. https://doi.org/10.1007/BF0097.

Elisabetsky, E., Silva Brum, L. F., & Souza, D. O. 1999. Anticonvulsant properties of linalool in glutamate-related seizure models. *Phytomedicine*, 6, 107–113. https://doi.org/10.1016/S0944-7113(99)80044-0.

Emamghoreishi, M., & Heidari-Hamedani, G. 2006. Sedative–Hypnotic activity of extracts and essential oil of coriander seeds. *IJMS*, 31, 22–27.

Emamghoreishi, M., Khasaki, M., & Aazam, M. F. 2007. *Coriandrum sativum* has anxiolytic and potentially sedative and muscle relaxant effects. *Molecular Cancer Therapeutics*, 6(3), 1013–1021.

Emamghoreishi, M., Khasaki, M., & Aazam, M. F. 2005. *Coriandrum sativum*: Evaluation of its anxiolytic effect in the elevated plus-maze. *Journal of Ethnopharmacology*, 96(3), 365–370.

Eriksson, N., Wu, S., Do, C., Kiefer, A., Tung, J., Mountain, J., & Francke, U. 2012. A genetic variant near olfactory receptor genes influences cilantro preference. *Flavour*, 1(1), 22. https://doi.org/10.1186/2044-7248-1-22.

Ertas, O. N., Guler, T., Ciftci, M., Dalkilic, B., & Ylmaz, O. 2005. The effect of a dietary supplement coriander seeds on the fatty acid composition of breast muscle in Japanese quail. *Revue de MédecineVétérinaire*, 156(10), 514–518.

Eyres, G., Dufour, J. P., Hallifax, G., Sotheeswaran, S., & Marriott, P. J. 2005. Identification of character-impact odorants in coriander and wild coriander leaves using Gas Chromatography-Olfactometry (GCO) and comprehensive two-dimensional Gas Chromatography-Time-Off Light Mass Spectrometry (GCxGC-TOFMS). *Journal of Separation Science* 28, 1061–1074. https://doi.org/10.1002/jssc.200500012.

Freires, I. D. A., Murata, R. M., Furletti, V. F., Sartoratto, A., de Alencar, S. M., Figueira, G. M., Rodrigues, J. A. D. O., Duarte, M. C. T., & Rosalen, P. L. 2014. *Coriandrum sativum* L. (Coriander) essential oil: Antifungal activity and mode of action on Candida spp., and molecular targets affected in human whole-genome expression. *PLOS One*, 9(6), e99086.

Furletti, V. F., Teixeira, P., Obando-Pereda, G., Mardegan, R. C., Sartoratto, A., Figueira, G. M., Duarte, R. M. T., Rehder, V. L. G., Duarte, M. C. T., & Hofling, J. F. 2011. Action of *Coriandrum sativum* L essential oil upon oral Candida albicans biofilm formation. *Evidence-Based Complementary and Alternative Medicine: eCAM*, 20(11), 1–9.

Gastón, M. S., Cid, M. P., Vázquez, A. M., Decarlini, M. F., Demmel, G. I., Rossi, L. I., Aimar, M. L., & Salvatierra, N. A. 2016. Sedative effect of central administration of *Coriandrum sativum* essential oil and its major component linalool in neonatal chicks. *Pharmaceutical Biology*, 54(10), 1954–1961. https://doi.org/10.3109/ 13880209.2015.1137602.

Georgieva, R., Delibaltova, V., & Chavdarov, P. 2022. Change in agronomic characteristics and essential oil composition of coriander after application of foliar fertilizers and biostimulators. *Industrial Crops and Products*, 181, 114819. https://doi.org/10.1016/j.indcrop.2022.114819.

Gottardi, D., Bukvicki, D., Prasad, S., & Tyagi, A. K. 2016. Beneficial effects of spices in food preservation and safety. *Frontiers in Microbiology*, 7, 1394.

Gray, A. M., & Flatt, P. R. 1999. Insulin-releasing and insulin-like activity of the traditional anti-diabetic plant *Coriandrum sativum* (coriander). *The British Journal of Nutrition*, 81(3), 203–209. https://doi.org/10.1017/s0007114599000392.

Grieve, M. 2013. *A Modern Herbal*, vol. 2. Courier Corporation, North Chelmsford, MA.

Happy Kitchen Rocks. 2018, August 22. *KhmeliSuneli* (Georgian spice mix) [Blog post]. Retrieved April 2, 2023, from https://happykitchen.rocks/khmeli-suneli-georgian-spice-mix/.

Hashim, M. S., Lincy, S., Remya, V., Teena, M., & Anila, L. 2005. Effect of polyphenolic compounds from *Coriandrum sativum* on H_2O_2-induced oxidative stress in human lymphocytes. *Food Chemistry*, 92(4), 653–660.

Hassar, M. 1999. La phytothérapie au Maroc. *EspéranceMédicale*, 6, 83–85.

Heider, M. 1992. *Coriandrum sativum. Applications of Plants Prevention and Treatment of Illnesses. Persian*, 1(257), 252.

Helal, M. A., Abdel-Gawad, A. M., Kandil, O. M., Khalifa, M. M. E., Cave, G. W. V., Morrison, A. A., Bartley, D. J., & Elsheikha, H. M. 2020. Nematocidal effects of a coriander essential oil and five pure principles on the infective larvae of major ovine gastrointestinal nematodes in vitro. *Pathogens (Basel, Switzerland)*, 9(9), 740. https://doi.org/10.3390/pathogens9090740.

Hosseini, M., Boskabady, M. H., & Khazdair, M. R. 2021. Neuroprotective effects of *Coriandrum sativum* and its constituent, linalool: A review. *Avicenna Journal of Phytomedicine*, 11(5), 436–450. https://doi.org/10.22038/AJP.2021.55681.2786.

Hosseinzadeh, S., Ghalesefidi, M. J., Azami, M., Mohaghegh, M. A., Hejazi, S. H., & Ghomashlooyan, M. 2016. In vitro and in vivo anthelmintic activity of seed extract of *Coriandrum sativum* compared to Niclosamid against Hymenolepis nana infection. *Journal of Parasitic Diseases: Official Organ of the Indian Society for Parasitology*, 40(4), 1307–1310. https://doi.org/10.1007/s12639-015-0676-y.

Hussain, F., Jahan, N., Rahman, K. U., Sultana, B., & Jamil, S. 2018. Identification of hypotensive biofunctional compounds of *Coriandrum sativum* and evaluation of their angiotensin-converting enzyme (ACE) inhibition potential. *Oxidative Medicine and Cellular Longevity*, 2018, 4643736. https://doi.org/10.1155/2018/4643736.

Hwang, E., Lee, D. G., Park, S. H., Oh, M. S., & Kim, S. Y. 2014. Coriander leaf extract exerts antioxidant activity and protects against UVB-induced photoaging of skin by regulation of procollagen type I and MMP-1 expression. *Journal of Medicinal Food*, 17(9), 985–995. https://doi.org/10.1089/jmf.2013.2999.

Ibrahim, F. M., Attia, H. N., Maklad, Y. A. A., Ahmed, K. A., & Ramadan, M. F. 2017. Biochemical characterization, anti-inflammatory properties and ulcerogenic traits of some cold-pressed oils in experimental animals. *Pharmaceutical Biology*, 55(1), 740–748. https://doi.org/10.1080/13880209.2016.1275705.

Islam, M. S., Hasan, M. M., Xiong, W., Zhang, S. C., & Lei, C. L. 2009. Fumigant and repellent activities of essential oil from *Coriandrum sativum* (L.) (Apiaceae) against red flour beetle Triboliumcastaneum (Herbst) (Coleoptera: Tenebrionidae). *Journal of Pest Science*, 82(2), 171–177.

Iwasaki, K., Zheng, Y. W., Murata, S., Ito, H., Nakayama, K., Kurokawa, T., Sano, N., Nowatari, T., Villareal, M. O., Nagano, Y. N., Isoda, H., Matsui, H., & Ohkohchi, N. 2016. Anticancer effect of linalool via cancer-specific hydroxyl radical generation in human colon cancer. *World Journal of Gastroenterology*, 22(44), 9765–9774. https://doi.org/10.3748/wjg.v22.i44.9765.

Jabeen, Q., Bashir, S., Lyoussi, B., & Gilani, A. H. 2009. Coriander fruit exhibits gut modulatory, blood pressure lowering and diuretic activities. *Journal of Ethnopharmacology*, 122(1), 123–130. https://doi.org/10.1016/j.jep.2008.12.016.

Jesuthasan, A. S., & Uluwaduge, D. I. 2017. Ethnobotanics used in folk medicine of Tamil culture in Sri Lanka: A scientific review. *Journal of Integrative Medicine* January 1, 15(1), 19–26.

Kačániová, M., Galovičová, L., Ivanišová, E., Vukovic, N. L., Štefániková, J., Valková, V., Borotová, P., Žiarovská, J., Terentjeva, M., Felšöciová, S., & Tvrdá, E. 2020. Antioxidant, antimicrobial and antibiofilm activity of coriander (*Coriandrum sativum* L.) essential oil for its application in foods. *Foods*, 9(3), 282. https: //doi.org/10.3390/foods9030282.

Kajal, A., & Singh, R. 2019. *Coriandrum sativum* seeds extract mitigate progression of diabetic nephropathy in experimental rats via AGEs inhibition. *PLOS ONE*, 14(3), e0213147. https://doi.org/10.1371/journal.pone.0213147.

Kaur, R., Singh, D., & Chopra, K. 2005. Participation of α2 receptors in the antinociceptive activity of quercetin. *Journal of Medicinal Food*, 8(4), 529–532. https://doi.org/10.1089/jmf.2005.8.529.

Kern, C., Gombert, C., Roso, A., & Garcia, C. 2020. Soothing effect of virgin coriander seed oil on sensitive skin. OCL. *Oilseeds and Fats, Crops and Lipids*, 27, 49. https://doi.org/10.1051/ocl/2020043.

Khare, C. P. 2008. *Indian Medicinal Plants: An Illustrated Dictionary*. Springer Science & Business Media, Berlin/Heidelberg.

Kianersi, F., Amin Azarm, D., Fatemi, F., Pour-Aboughadareh, A., & Poczai, P. 2022. Methyl jasmonate induces genes involved in linalool accumulation and increases the content of phenolics in two Iranian coriander (*Coriandrum sativum* L.) ecotypes. *Genes*, 13(10), 1717. https://doi.org/10.3390/genes13101717.

Kim, E. J., Martinez Amezcua, C., Utterback, P. L., & Parsons, C. M. 2008. Phosphorus bioavailability, true metabolizable energy, and amino acid digestibilities of high protein corn distillers dried grains and dehydrated corn germ. *Poultry Science*, 87(4), 700–705.

Kothalawala, S. D., Edward, D., Harasgama, J. C., Ranaweera, L., Weerasena, O. V. D. S. J., Niloofa, R., Ratnasooriya, W. D., Premakumara, G. A. S., & Handunnetti, S. M. 2020. Immunomodulatory activity of a traditional Sri Lankan concoction of *Coriandrum sativum* L. and Cosciniumfenestratum G. *Evidence-Based Complementary and Alternative Medicine*, 2020, 9715060. https://doi.org/10.1155/2020/9715060.

Kothari, S., Priya, V. V., & Gayathri, R. 2017. Anti-inflammatory activity of *Coriandrum sativum* using HRBC membrane stabilizing method. *International Journal of Pharmaceutical Sciences Review and Research*, 43(2), 68–70.

Kousar, S., Jahan, N., Khalil Ur, R., & Nosheen, S. 2011. Antilipidemic activity of *Coriandrum sativum. Bioscience Research*, 8, 8–14.

Kubo, I., Fujita, K.-I., Kubo, A., Nihei, K.-I., & Ogura, T. 2004. Antibacterial activity of coriander volatile compounds against Salmonella choleraesuis. *Journal of Agriculture and Food Chemistry*, 52(11), 3329–3332.

Kükner, A., Soyler, G., Toros, P., Dede, G., Meriçli, F., Işık, S., Edebal, O., & Özoğul, C. 2021. Protective effect of *Coriandrum sativum* extract against inflammation and apoptosis in liver ischemia reperfusion injury. *Folia Morphologica*, 80(2), 363–371. https://doi.org/10.5603/FM.a2020.0060.

Kumar, V. 2020. Seven spices of India—from kitchen to clinic. *Journal of Ethnic Foods* December, 7(1), 1–6.

Kumar, S., Ahmad, R., Saeed, S., Azeem, M., Mozuraitis, R., Borg-Karlson, A.-K., & Zhu, G. 2022. Chemical composition of fresh leaves headspace aroma and essential oils of four coriander cultivars. *Frontiers in Plant Science*, 13, 820644. https://doi.org/10.3389/fpls.2022.820644.

Kunzemann, J., & Hermann, K. 1977. Isolation and identification of flavon(ol)-O-glycosides in caraway (Carum carvi L.), fennel (Foeniculum vulgare Mill.), anise (Pimpinella anisum L.), and coriander (*Coriandrum sativum* L.), and of flavon-C-glycosides in anise I. Phenolic of spices. *Z. LebensumUntersForsch*, 164(3), 194–200.

Langrand, J., Lounès-Hadj Sahraoui, A., Duclercq, J., Raveau, R., Laruelle, F., Bert, V., Facon, N., Tisserant, B., & Fontaine, J. 2023. Coriander (*Coriandrum sativum*) cultivation combined with arbuscular mycorrhizal fungi inoculation and steel slag application influences trace elements-polluted soil bacterial functioning. *Plants*, 12(3), 618. https://doi.org/10.3390/plants12030618.

Laribi, B., Kouki, K., M'Hamdi, M., & Bettaieb, T. 2015. Coriander (*Coriandrum sativum* L.) and its bioactive constituents. *Fitoterapia*, 103, 9–26. https://doi.org/10.1016/j.fitote.2015.03.012.

Lawrence, B. M. 1993. A planning scheme to evaluate new aromatic plants from the flavor and fragrance industries. In: J. Janick & J. E. Simon (Eds.), *New Crops Exploration, Research and Commercialization*. John Wiley and Sons, Inc., New York, 620–627.

Lin, H. H., Lin, K. H., Yang, M. J., Nguyen, H. C., Wang, H. J., Huang, H. X., & Huang, M. Y. 2022. Physiological responses and antioxidant properties of coriander plants (*Coriandrum sativum* L.) under different light intensities of red and blue lights. *Scientific Reports*, 12(1), 21139. https://doi.org/10.1038/s41598-022-25749-3.

Linck, V. D. M., da Silva, A. L., Figueiró, M., Piato, A. L., Herrmann, A. P., Dupont Birck, F., Caramão, E. B., Sávio Nunes, D., Moreno, P. R. H., & Elisabetsky, E. 2009. Inhaled linalool-induced sedation in mice. *Phytomedicine*, 16(4), 303–307. https://doi.org/10.1016/j. phymed.2008.08.001.

Lo Cantore, P., Iacobelli, N. S., Marco, A., Capasso, F., & Senatore, F. 2004. Antibacterial activity of *Coriandrum sativum* L., and Foeniculum vulgare Miller Var vulgare (Miller) essential oils. *Journal of Agriculture and Food Chemistry*, 52(26), 7862–7866.

Luo, J., Zhang, C., Liu, Q., Ou, S., Zhang, L., & Peng, X. 2017. Combinative effect of sardine peptides and quercetin alleviates hypertension through inhibition of angiotensin I converting enzyme activity and inflammation. *Food Research International*, 100(1), 579–585. https://doi.org/10.1016/j.foodres.2017.07.019 PubMed.

Mahendra, P., & Bisht, S. 2011. *Coriandrum sativum*: A daily use spice with great medicinal effect. *Pharmacognosy Journal*, 3(21), 84–88. https://doi.org/10.5530/pj.2011.21.16.

Mahleyuddin, N. N., Moshawih, S., Ming, L. C., Zulkifly, H. H., Kifli, N., Loy, M. J., Sarker, M. M. R., Al-Worafi, Y. M., Goh, B. H., Thuraisingam, S., et al. 2022. *Coriandrum sativum* L.: A review on Ethnopharmacology, Phytochemistry, and cardiovascular Benefits. *Molecules*, 27(1), 209. https://doi.org/10.3390/molecules27010209.

Mahmoud, M. F., Ali, N., Mostafa, I., Hasan, R. A., & Sobeh, M. 2022. Coriander oil reverses dexamethasone-induced insulin resistance in rats. *Antioxidants*, 11(3), 441. https://doi.org/10.3390/antiox11030441.

Mandal, S., & Mandal, M. 2015. Coriander (*Coriandrum sativum* L.) essential oil: Chemistry and biological activity. *Asian Pacific Journal of Tropical Biomedicine*, 5(6), 421–428. https://doi.org/10.1016/j.apjtb.2015.02.006.

Marichali, A., Dallali, S., Ouerghemmi, S., Sebei, H., & Hosni, K. 2014. Germination, morpho-physiological and biochemical responses of coriander (*Coriandrum sativum* L.) to zinc excess. *Industrial Crops and Products*, 55, 248–257. https://doi.org/10.1016/ j.indcrop.2014.02.033.

Matasyoh, J. C., Maiyo, Z. C., Ngure, R. M., & Chepkorir, R. 2009. Chemical composition and antimicrobial activity of the essential oil of *Coriandrum sativum*. *Food Chemistry*, 113(2), 526–529. https://doi.org/10.1016/j.foodchem.2008.07.097.

Mechchate, H., Costa de Oliveira, R., Es-safi, I., Vasconcelos Mourão, E. M., Bouhrim, M., Kyrylchuk, A., Soares Pontes, G., Bousta, D., & Grafov, A. 2021. Antileukemic activity and molecular docking study of a polyphenolic extract from coriander seeds. *Pharmaceuticals*, 14(8), 770. https://doi.org/ 10.3390/ph14080770.

Mehri, S., Meshki, M. A., & Hosseinzadeh, H. 2015. Linalool as a neuroprotective agent against acrylamide-induced neurotoxicity in Wistar rats. *Drug and Chemical Toxicology*, 38(2), 162–166.

Mexicali Blue. n.d. Exploring the versatility of cilantro, an essential ingredient in Latin American cuisine. Retrieved April 2, 2023, from https://www.mexicali-blue.com/exploring-the-versatility-of-cilantro-an-essential-ingredient-in-latin-american-cuisine/.

Millam, S., Mitchell, S., Craig, A., Paoli, M., Moscheni, E., & Angelini, L. 1997. In vitro manipulation as a mean for accelerated improvement of some new potential oil crop species. *Industrial Crops and Products*, 6(3–4), 213–219.

Miller, M. 2009. Dyslipidemia and cardiovascular risk: The importance of early prevention. *QJM*, 102(9), 657–667. https://doi.org/10. 1093/qjmed/hcp065.

Misharina, T. A., & Samusenko, A. L. 2008. Antioxidant properties of essential oils from lemon, grapefruit, coriander, clove, and their mixtures. *Applied Biochemistry and Microbiology*, 45(4), 438–442.

Mohammadi Bazargani, M., & Rohloff, J. 2016. Antibiofilm activity of essential oils and plant extracts against Staphylococcus aureus and Escherichia coli biofilms. *Food Control*, 61, 156–164. https://doi. https://doi.org/10.1016/j.foodcont.2015.09.036.

Molina, R. D. I., Campos-Silva, R., Macedo, A. J., Blázquez, M. A., Alberto, M. R., & Arena, M. E. 2020. Antibiofilm activity of coriander (Coriander sativum L.) grown in Argentina against food contaminants and human pathogenic bacteria. *Industrial Crops and Products*, 151(March), 112380. https://doi.org/10.1016/j.indcrop.2020.112380.

Momin, A. H., Acharya, S. S., & Gajjar, A. V. 2012. *Coriandrum sativum*-review of advances in phytopharmacology. *International Journal of Pharmaceutical Sciences and Research*, 3, 1233.

Msaada, K., Taarit, M. B., Hosni, K., Hammami, M., & Marzouk, B. 2009. Regional and maturational effects on essential oils yields and composition of coriander (*Coriandrum sativum* L.) fruits. *Scientia Horticulturae*, 122, 116–124. https://doi.org/10.1016/j.scienta.2009.04.008.

Msaada, K., Hosni, K., Taarit, M. B., Chahed, T., Kchouk, M. E., & Marzouk, B. 2007. Changes on essential oil composition of coriander (*Coriandrum sativum* l.) fruits during three stages of maturity. *Food Chemistry*, 102(4), 1131–1134. https://doi.org/10.1016/j.foodchem.2006.06.046.

Nair, V., Singh, S., & Gupta, Y. K. 2013. Anti-granuloma activity of *Coriandrum sativum* in experimental models. *Journal of Ayurveda and Integrative Medicine*, 4(1), 13–18. https://doi.org/10.4103/0975-9476.109544.

Nakamura, A., Fujiwara, S., Matsumoto, I., & Abe, K. 2009. Stress repression in restrained rats by linalool inhalation and gene expression profiling of their whole blood cells. *Journal of Agriculture and Food Chemistry*, 57(12), 5480–5485.

Nambiar, V. S., Daniel, M., & Guin, P. 2010. Characterization of polyphenols from coriander leaves (*Coriandrum sativum*), red Amaranthus (A. paniculatus) and green Amaranthus (A.frumentaceus) using paper chromatography: And their health implications. *Journal of Herbal Medicine and Toxicology*, 4, 173–177.

Nazni, P., & Dharmaligam, R. 2014. Isolation and separation of phenolic compound from coriander flowers. *International Journal of Agriculture and Food Science*, 4, 13–21.

Neffati, M., Sriti, J., Hamdaoui, G., Kchouk, M. E., & Marzouk, B. 2011. Salinity impact on fruit yield, essential oil composition and antioxidant activities of *Coriandrum sativum* fruit extracts. *Food Chemistry*, 124(1), 221–225. https://doi.org/10.1016/j.foodchem.2010. 06.022.

Ngang, J. J., Nyegue, M. A., Ndoye, F. C., TchuenchieuKamgain, A. D., Sado Kamdem, S. L., Lanciotti, R., Gardini, F., & Etoa, F. X. 2014. Characterization of Mexican coriander (Eryngium foetidum) essential oil and its inactivation of Listeria monocytogenes in vitro and during mild thermal pasteurization of pineapple juice. *Journal of Food Protection*, 77(3), 435–443. https://doi.org/10.4315/0362-028X.JFP-13-323.

Oganesyan, E., Nersesyan, Z., & Parkhomenko, A. Y. 2007. Chemical composition of the above-ground part of *Coriandrum sativum*. *Pharmaceutical Chemistry Journal*, 41(3), 149–153. https://doi.org/10.1007/s11094-007-0033-2.

Önder, A. 2018. Coriander and its phytoconstituents for the beneficial effects. *Potential of Essential Oils*, 165–185. http://doi.org/10.5772/intechopen.78656.

Opdyke, D. L. J. 1973. Monographs on fragrance raw materials: Coriander oil. *Food and Cosmetics Toxicology*, 11(6), 1077.

Orav, A., Arak, E., & Raal, A. 2011. Essential oil composition of *Coriandrum sativum* L. fruits from different countries. *Journal of Essential Oil Bearing Plants*, 14(1), 118–123. https://doi.org/10.1080/0972060X.2011.10643910.

Özbek, H., Kırmızı, N.˙I., Cengiz, N., & Erdo ˘gan, E. 2016. Hepatoprotective effects of *Coriandrum sativum* essential oil against acute hepatotoxicity induced by carbon tetrachloride on rats. *Acta Pharmaceutica Sciencia*, 54, 35. https://doi.org/10.23893/1307-2080.APS.05404.

Pandey, A., Bigoniya, P., Raj, V., & Patel, K. K. 2011. Pharmacological screening of *Coriandrum sativum* Linn. for hepatoprotective activity. *Journal of Pharmacy and Bioallied Sciences*, 3(3), 435–441.

Paniagua-Zambrana, N. Y., Bussmann, R. W., & Romero, C. 2020. *Coriandrum sativum* L. A piaceae. *Ethnobot Andes*, 1–7 [CrossRef].

Park, G., Kim, H. G., Lim, S., Lee, W., Sim, Y., & Oh, M. S. 2014. Coriander alleviates 2,4-dinitrochlorobenzene-induced contact dermatitis-like skin lesions in mice. *Journal of Medicinal Food*, 17(8), 862–868. https://doi.org/10.1089/jmf.2013.2910.

Patel, D. K., Desal, S. N., Devkar, R. V., & Ramachandran, A. V. 2011. *Coriandrum sativum* L. aqueous extract mitigates high fat diet induced insulin resistance by controlling visceral adiposity. *Boletínlatinoamericano y del Caribe de Plantasmedicinales y Aromáticas*, 10(2), 127–135.

Patel, D. K., Desai, S. N., Gandhi, H. P., Devkar, R. V., & Ramachandran, A. V. 2012. Cardio protective effect of *Coriandrum sativum* L. on isoproterenol induced myocardial necrosis in rats. *Food and Chemical Toxicology*, 50(9), 3120–3125. https://doi.org/10.1016/j.fct.2012.05.031.

Patel, D., Desai, S., Gajaria, T., Devkar, R., & Ramachandran, A. V. 2013. *Coriandrum sativum* L. seed extract mitigates lipotoxicity in raw 264.7 cells and prevents atherogenic changes in rats. *EXCLI Journal*, 12, 313–334.

Pawłowska, K. A., Baracz, T., Skowrońska, W., Piwowarski, J. P., Majdan, M., Malarz, J., Stojakowska, A., Zidorn, C., & Granica, S. 2022. The contribution of phenolics to the anti-inflammatory potential of the extract from Bolivian coriander (Porophyllumruderale subsp. ruderale). *Food Chemistry*, 371, 131116. https://doi.org/10.1016/j.foodchem.2021.131116.

Picq, M., Cheav, S. L., & Prigent, A. F. 1991. Effect of two flavonoid compounds on central nervous system. Analgesic activity. *Life Sciences*, 49(26), 1979–1988.

Pieroni, A., & Gray, C. 2008. Herbal and food folk medicines of the Russlanddeutschen living in Künzelsau/Taläcker, South- Western Germany. *Phytotherapy Research* July, 22(7), 889–901.

Plantura. n.d. Coriander – An overview of the popular herb. Retrieved April 2, 2023, from https://plantura.garden/uk/herbs/coriander/coriander-overview.

Platel, K., & Srinivasan, K. 2004. Digestive stimulant action of spices: A myth or reality? *Indian Journal of Medical Research*, 119(5), 167–179.

Potter, T. L. 1996. Essential oil composition of cilantro. *Journal of Agriculture and Food Chemistry*, 44(7), 1824–1826. https://doi.org/10.1021/jf950814c.

Raal, A., Arak, E., & Orav, A. 2004. Chemical composition of coriander seed essential oil and their conformity with EP standards. *Agraarteadus*, 15, 234–239.

Rajeshwari, C., & Andallu, B. 2011. Isolation and simultaneous detection of flavonoids in the methanolic and ethanolic extracts of *Coriandrum sativum* L. seeds by RP-HPLC. *Pakistan Journal of Food Sciences*, 21, 13–21.

Rakhshandeh, H., Sadeghnia, H. R., & Ghorbani, A. 2012. Sleepprolonging effect of *Coriandrum sativum* hydro-alcoholic extract in mice. *Natural Product Research*, 26(22), 2095–2098. https://doi.org/10.1080/14786419.2011.613388.

Ramadan, M. F., Amer, M. M. A., & Awad, A. E. S. 2008. Coriander (*Coriandrum sativum* L.) seed oil improves plasma lipid profile in rats fed a diet containing cholesterol. *European Food Research and Technology*, 227(4), 1173–1182. https://doi.org/10.1007/s00217-008-0833-y.

Rattanachaikunsopon, P., & Phumkhachorn, P. 2010. Potential of coriander (*Coriandrum sativum*) oil as a natural antimicrobial compound in controlling Campylobacter jejuni in raw meat. *Bioscience, Biotechnology, and Biochemistry*, 74(1), 31–35. https://doi.org/10.1271/bbb.90409.

Raveau, R., Fontaine, J., Soltani, A., Ben Jemâa, J. M., Laruelle, F., & Lounès-Hadj Sahraoui, A. 2022. In vitro potential of clary sage and coriander essential oils as crop protection and post-harvest decay control products. *Foods*, 11(3), 312. https://doi.org/10.3390/foods11030312.

Ravi, R., Prakash, M., & Bhat, K. K. 2007. Aroma characterization of coriander (*Coriandrum sativum* L.) oil samples. *European Food Research and Technology*, 225(3), 367–374.

Re, L., Barocci, S., Sonnino, S., Mencarelli, A., Vivani, C., Paolucci, G., Scarpantonio, A., Rinaldi, L., & Mosca, E. 2000. Linalool modifies the nicotinic receptor-ion channel kinetics at the mouse neuromuscular junction. *Pharmacological Research*, 42(2), 177–181. https://doi.org/10.1006/phrs.2000.0671.

Reddy, L. H., Jalli, R. D., Jose, B., & Gopu, S. 2012. Evaluation of antibacterial and DPPH radical scavenging activities of the leaf extracts and essential oil of *Coriandrum sativum* Linn. *World Journal of Pharmaceutical Research*, 1(3), 705–716.

Rehman, N., Jahan, N., ul-Rahman, K., Khan, K. M., & Zafar, F. 2016. Anti-arrhythmic potential of *Coriandrum sativum* seeds in salt induced arrhythmic rats. *Pakistan Veterinary Journal*, 36(4), 465–471.

Reuter, J., Huyke, C., Casetti, F., Theek, C., Frank, U., Augustin, M., & Schempp, C. 2008. Anti-inflammatory potential of a lipolotion containing coriander oil in the ultraviolet erythema test. *JDDG: Journal der Deutschen Dermatologischen Gesellschaft*, 6(10), 847–851. https://doi.org/10.1111/j.1610-0387.2008.06704.x.

Reyes, M. R., Reyes-Esparza, J., Angeles, O. T., & Rodríguez-Fragoso, L. 2010. Mutagenicity and safety evaluation of water extract of *Coriandrum sativum* leaves. *Journal of Food Science*, 75(1), T6–T12.

Sachan, A. K., Kumar, S., Kumari, K., & Singh, D. 2018. Medicinal uses of spices used in our traditional culture: Worldwide. *Journal of Medicinal Plants Studies*, 6(3), 116–122.

Sahib, N. G., Anwar, F., Gilani, A. H., Hamid, A. A., Saari, N., & Alkharfy, K. M. 2013. Coriander (*Coriandrum sativum* L.): A potential source of high-value components for functional foods and nutraceuticals--A review. *Phytotherapy Research: PTR*, 27(10), 1439–1456. https://doi.org/10.1002/ptr.489.

Salem, M. A., Manaa, E. G., Osama, N., Aborehab, N. M., Ragab, M. F., Haggag, Y. A., Ibrahim, M. T., & Hamdan, D. I. 2022. Coriander (*Coriandrum sativum* L.) essential oil and oil-loaded nano-formulations as an anti-aging potentiality via TGFβ/SMAD pathway. *Scientific Reports*, 12(1), 6578. https://doi.org/10.1038/s41598-022-10494-4.

Samojlik, I., Lakic, N., Mimica-Dukic, N., Dakovic-Svajcer, K., & Bozin, B. 2010. Antioxidant and hepatoprotective potential of essential oils of coriander (*Coriandrum sativum* L.) and caraway (Carum carvi L.) (Apiaceae). *Journal of Agricultural and Food Chemistry*, 58(15), 8848–8853.

Sari, D. P. R., Bellatasie, R., & Ifora, I. 2021. Anti-inflammatory properties of *Coriandrum sativum* L.: A review. *International Research Journal of Pharmacy and Medical Sciences (IRJPMS)*, 4(2), 34–38.

Sasongko, P., Laohankunjit, N., & Kerdchoechuen, O. 2011. Persicaria odorata antibacterial activity of the essential oil from Persicaria odorata leaves.

Sathishkumar, P., Preethi, J., Vijayan, R., Yusoff, A. R. M., Ameen, F., Suresh, S., Balagurunathan, R., & Palvannan, T. 2016. Antiacne, anti-dandruff and anti-breast cancer efficacy of green synthesised silver nanoparticles using *Coriandrum sativum* leaf extract. *Journal of Photochemistry and Photobiology B: Biology*, 163, 69–76. https://doi.org/10.1016/j.jphotobiol.2016.08.005.

Saxena, S. N., Swarup Meena, R., Vishal, M. K., John, S., Kumar Sharma, L., Mishra, B. K., & Agarwal, D. 2022. Variation in essential oil constituents of coriander (*Coriandrum sativum* L.) germplasm across coriander growing regions in India. *Journal of Essential Oil Research*, 34(2), 173–180. https://doi.org/10.1080/10412905.2022.2036644.

Sayed-Ahmed, B., Talou, T., Saad, Z., Hijazi, A., & Merah, O. 2017. The Apiaceae: Ethnomedicinal Family as source for industrial uses. *Industrial Crops and Products*, 109, 661–671. https://doi.org/10.1016/j.indcrop.2017.09.027.

Sharma, N., Sharma, P., Jasuja, N. D., & Joshi, S. C. 2014. Ameliorative efficiency of *Coriandrum sativum* seed extract on atherosclerosis and oxidative stress in male albino hyperlipidemic rabbits. *Research Journal of Pharmaceutical, Biological and Chemical Sciences*, 5, 26–39.

Sharma, P. V. 1995. *DravyagunaVijnana*, vol. 2. Chaukhamba Bharati Academy, Varanasi, 69, 72, 138, 152, 162, 246, 250, 314, 331, 332, 350, 362, 365, 394, 401, 458, 494, 596, 642, 719, 729, 823.

Shavandi, M. A., Haddadian, Z., & Ismail, M. H. S. 2012. Eryngium foetidum L. *Coriandrum sativum* and Persicaria odorata L.: A review. *Journal of Asian Scientific Research*, 2(8), 410–426.

Silva, F., & Domingues, F. C. 2017. Antimicrobial activity of coriander oil and its effectiveness as food preservative. *Critical Reviews in Food Science and Nutrition*, 57(1), 35–47.

Silva, F., Ferreira, S., Duarte, A., Mendonça, D. I., & Domingues, F. C. 2011. Antifungal activity of *Coriandrum sativum* essential oil, its mode of action against Candida species and potential synergism with amphotericin B. *Phytomedicine*, 19(1), 42–47.

Singh, Y. N. 1986. Traditional medicine in Fiji: Some herbal folk cures used by Fiji Indians. *Journal of Ethnopharmacology* January 1, 15(1), 57–88.

Singh, G., Maurya, S., De Lampasona, M. P., & Catalan, C. A. N. 2006. Studies on essential oils, Part 41. Chemical composition, antifungal, antioxidant and sprout suppressant activities of coriander (*Coriandrum sativum*) essential oil and its oleoresin. *Flavour and Fragrance Journal*, 21(3), 472–479. https://doi.org/10.1002/ffj.1608.

Singletary, K. 2016. Coriander: Overview of potential health benefits. *Nutrition Today*, 51(3), 151–161. https://doi.org/10.1097/NT.0000000000000159.

Soares, B. V., Morais, S. M., Fontenelle, D. S. R. O., Queiroz, V. A., Vila-Nova, N. S., Pereira, C. M. C., Brito, E. S., Neto, M. A. S., Brito, E. H. S., Cavalcante, C. S. P., Castelo-Branco, D. S. C. M., & Rocha, M. F. G. 2012. Antifungal activity, toxicity and chemical composition of the essential oil of *Coriandrum sativum* L. fruits. *Molecules* (Basel, Switzerland), 17(7), 8439–8448. https://doi.org/10.3390/molecules17078439.

Sobhani, Z., Mohtashami, L., Amiri, M. S., Ramezani, M., Emami, S. A., & Simal-Gandara, J. 2022. Ethnobotanical and phytochemical aspects of the edible herb *Coriandrum sativum* L. *Journal of Food Science*, 87(4), 1386–1422. https://doi.org/10.1111/1750-3841.16085.

Song, E. J., & Ko, M. J. 2022. Extraction of monoterpenes from coriander (*Coriandrum sativum* L.) seeds using subcritical water extraction (SWE) technique. *The Journal of Supercritical Fluids*, 188, 105668. https://doi.org/10.1016/j.supflu.2022.105668.

Sonika, G., Manubala, R., & Deepak, J. 2010. Comparative studies on anti-inflammatory activity of *Coriandrum sativum*, Datura stramonium and Azadirachta indica. *Asian Journal of Experimental Biological Sciences*, 1(1), 151–154 [Online]. Available: http://www.ajebs.com/vol1-2010/Society of Applied Sciences151-154.pdf.

Souto-Maior, F. N., de Carvalho, F. L., de Morais, L. C. S. L., Netto, S. M., de Sousa, D. P., & de Almeida, R. N. 2011. Anxiolytic-like effects of inhaled linalool oxide in experimental mouse anxiety models. *Pharmacology, Biochemistry, and Behavior*, 100(2), 259–263.

Sreelatha, S., & Inbavalli, R. 2012. Antioxidant, antihyperglycemic, and antihyperlipidemic effects of *Coriandrum sativum* leaf and stem in alloxan-induced diabetic rats. *Journal of Food Science*, 77(7), T119–T123.

Sreelatha, S., Padma, P. R., & Uma Devi, M. 2009. Protective effects of *Coriandrum sativum* extracts on carbon tetrachloride induced hepatotoxicity in rats. *Food and Chemical Toxicology*, 47(4), 702–708.

Sriti, J., Talou, T., Wannes, W. A., Cerny, M., & Marzouk, B. 2009. Essential oil, fatty acid and sterol composition of Tunisian coriander fruit different parts. *Journal of the Science of Food and Agriculture*, 89(10), 1659–1664.

Sriti, J., Wannes, W. A., Talou, T., Mhamdi, B., Handaoui, G., & Marzouk, B. 2010. Lipid fatty acidandtocol distribution of coriander fruits' different parts. *Industrial Crops and Products*, 31(2), 294–300.

Sugawara, Y., Hara, C., Tamura, K., Fujii, T., Nakamura, K.-I., Masujima, T., & Aoki, T. 1998. Sedative effect on humans of inhalation of essential oil of linalool. Sensory evaluation and physiological measurement using optically active linalool. *Analytica Chimica Acta*, 365(1–3), 293–299.

Sundar, S., Padmalatha, K., Helasri, G., Vasanthi, M., Sai Narmada, B., Lekhya, B., Naga Jyothi, R., & Sravani, T. 2016. Anti-microbial activity of aqueous extract of natural preservatives—Cumin, cinnamon, coriander and mint. *Research Journal of Pharmacy and Technology*, 9, 843–847. https://doi.org/10.5958/0974-360X.2016. 00159.1.

Taherian, A. A., Vafaei, A. A., & Ameri, J. 2012. Opiate system mediate the antinociceptive effects of *Coriandrum sativum* in mice. *Iranian Journal of Pharmaceutical Research*, 11(2), 679.

Tang, E. L., Rajarajeswaran, J., Fung, S. Y., & Kanthimathi, M. S. 2013. Antioxidant activity of *Coriandrum sativum* and protection against DNA damage and cancer cell migration. *BMC Complementary and Alternative Medicine*, 13(1), 347.

Taniguchi, M., Yanai, M., Xiao, Y. Q., Kido, T. I., & Baba, K. 1996. Three isocoumarins from *Coriandrum sativum*. *Phytochemistry*, 42(3), 843–846. https://doi.org/10.1016/0031-9422(95)00930-2.

Thai Food Paradise. n.d. What is cilantro used for? Retrieved April 2, 2023, from https://thaifoodparadise.com/what-is-cilantro-used-for/.

Thakur, S., Kumari, I., Chaudhary, S., Madhusudan, S., Kaurav, H., & Chaudhary, G. 2021. Coriander (*Coriandrum sativum*): A common Indian traditional spice and ayurvedic remedy. *Journal of Drug Delivery and Therapeutics*, 11(2), 146–151.

The Mediterranean Dish. n.d. *Recipes*. Retrieved April 2, 2023, from https://www.themediterraneandish.com/recipes/.

Ugulu, I., Baslar, S., Yorek, N., & Dogan, Y. 2009. The investigation and quantitative ethnobotanical evaluation of medicinal plants used around Izmir Province, Turkey. *Journal of Medicinal Plants Research* May 31, 3(5), 345–367.

United Nations Comtrade. 2019. UN Comtrade database. Retrieved April 2, 2023, from https://wits.worldbank.org/trade/comtrade/en/country/All/year/2019/tradeflow/Exports/partner/WLD/product/090920.

USDA Food Data Central. n.d. Nutrient data for 169997, Coriander (cilantro) leaves, raw. Retrieved April 2, 2023, from https://fdc.nal.usda.gov/fdc-app.html#/food-details/169997/nutrients.

Varier, P. S. 1994. *Coriandrum sativum in Indian Medicinal Plants: A Compendium of 500 Species (2).* Orient Longman LTD, Chennai, 416–417.

Vats, A., & Sharma, P. 2012. Formulation and evaluation of topical anti acne formulation of coriander extract. *International Journal of Pharmaceutical Sciences Review and Research*, 16, 97–103.

Waheed, A., Miana, G. A., Ahmad, S. I., & Khan, M. A. 2006. Clinical investigation of hypoglycemic effect of *Coriandrum sativum* in type-2 (NIDDM) diabetic patients. *Pakistan Journal of Pharmacology*, 23(1), 7–11.

Wang, X., Liu, Y., Wang, Y., Dong, X., Wang, Y., Yang, X., Tian, H., & Li, T. 2022. Protective effect of coriander (*Coriandrum sativum* L.) on high-fructose and high-salt diet-induced hypertension: Relevant to improvement of renal and intestinal function. *Journal of Agricultural and Food Chemistry*, 70(12), 3730–3744. https://doi.org/10.1021/acs.jafc.2c00267.

Wang, X., Zhu, C., Peng, T., Zhang, W., Zhang, J., Liu, H., Wu, C., Pan, X., & Wu, C. 2018. Enhanced stability of an emulsion enriched in unsaturated fatty acids by dual natural antioxidants fortified in both the aqueous and oil phases. *Food Hydrocolloids*, 82, 322–328. https://doi.org/10.1016/j.foodhyd.2018.02.012.

Wangensteen, H., Samuelsen, A. B., & Malterud, K. E. 2004. Antioxidant activity in extracts from coriander. *Food Chemistry*, 88(2), 293–297. https://doi.org/10.1016/j.foodchem.2004.01.047.

Wichtl, M. 2004. *Herbal Drugs and Phytopharmaceuticals: A Handbook for Practice on a Scientific Basis.* M. Wichtl (Ed.). Medpharm GmbH Scientific Publishers, Stuttgart.

Wong, P. Y. Y., & Kitts, D. D. 2006. Studies on the dual antioxidant and antibacterial properties of parsley (Petroselinum crispum) and cilantro (*Coriandrum sativum*) extracts. *Food Chemistry*, 97(3), 505–515.

Wu, T., He, S., Li, Y., Xu, G., & Li, C. 2010. Suppressive effects of extracts from the aerial part of *Coriandrum sativum* L. on LPS-induced inflammatory responses in murine RAW 264.7 macrophages. *Journal of the Science of Food and Agriculture*, 90(11), 1846–1854. https://doi.org/10.1002/jsfa.4023.

Yuan, R., Coskuner, Y., & Karababa, E. 2020. Novel compounds in fruits of Coriander (Coşkuner & Karababa) with anti-inflammatory activity. *Journal of Functional Foods*, 73(July), 1–9. https://doi.org/10.1016/j.jff.2020.104145.

Yummly. n.d. Vietnamese coriander recipes. Retrieved April 2, 2023, from https://www.yummly.com/recipes/vietnamese-coriander.

Zoubiri, S., & Baaliouamer, A. 2010. Essential oil composition of *Coriandrum sativum* seed cultivated in Algeria as food grains protectant. *Food Chemistry*, 122(4), 1226–1228.

16 Cumin Seeds (*Cuminum cyminum*)

Leila Hakimi, Mohammad Habiib Abasss Alfaham, and Naser Nasrollahzadeh-asl

16.1 INTRODUCTION

Spices are dried, fragrant plant pieces that have been used for centuries to season food and preserve it. They are commonly used to improve the shelf life of foods, improve their taste, and give them colour. In particular, they are a rich source of protein, tannins, saponins, terpenes, flavonoids, alkaloids, tannic acids, anthocyanins, coumarins, glucosides, and other medicinal bioactive components. They have positive effects on the body's defense mechanisms against microbial infection, oxidative stress, inflammation, diabetes, cardiovascular diseases, neurodegenerative disorders, wound healing, reproductive abnormalities, cancer, and other conditions (Singh et al., 2021; Husen, 2021, 2022a, 2022b, 2023). These bioactive substances are also recommended for the treatment of rheumatic pain, cognitive difficulties, convulsion and inflammation, leprosy, colds, coughs, appetite loss, and many other conditions. More than 100 plants are utilized as spices and condiments on a global scale. Cumin, which is also known as zeera or jeera in South Asia, is one of the most widely used medicinal seed spices, and belongs to the Apiaceae family. Because it is a plentiful source of a potent and distinct scent, it is frequently employed as a culinary spice and condiment in the creation of several cuisines from various civilizations throughout the world (Koohsari et al., 2020). For many types of cookery, cumin seeds are primarily favoured as a garnish or flavoring ingredient. The seed oil is widely used in the manufacture of chutneys, pickles, curries, meats, sausages, and cheese. The seeds are used extensively in the fragrance business in tiny quantities because of their strong scent. Additionally, it contains a large number of medicinally valuable bioactive compounds that are widely used for a variety of medical applications (Goodarzi et al., 2020; Ghasemi et al., 2021; Singh et al., 2021). The plant is mainly well known for its therapeutic potential, possessing properties such as antimicrobial, antioxidant, antidiabetic, immunomodulatory, anti-osteoporotic, anti-inflammatory, anti-asthmatic, anti-tussive, anti-infertility, anticancer, anti-stress, antipyretic, hypotensive, antiplatelet coagulation, hepatoprotective, analgesic, etc. (Singh et al., 2021). Cumin is widely recognized for its many phytoconstituents with a variety of health benefits. The primary component of cumin essential oil (CEO), cuminaldehyde, is what gives the oil its distinct and recognizable scent (Koohsari et al., 2020; Nirmala et al., 2020). Cumin seeds have several reported therapeutic potentials involving different biochemical compounds. One of the main factors contributing to the spice's extraordinary pharmacological and therapeutic qualities may be the synergistic interactions between these substances. Accordingly, the current study takes into account all pertinent information that is currently known on the therapeutic capabilities, phytochemistry, and ethnomedicinal properties of *Cuminum cyminum* L. Nigella (*Nigella sativa* L.), sometimes known as black cumin, has been reported by the Food Safety and Standards Authority of India (SSAI) under the cumin category. However, in this chapter, we focus on only cumin (*C. cyminum*).

DOI: 10.1201/9781003387046-16

16.2 BOTANICAL CHARACTERIZATION

Cumin (*Cuminum cyminum* L.) is an annual, dicotyledonous, and fragrant flowering plant of the Apiaceae /Umbelliferae family. The fragile, glabrous plant grows to a height of 10 to 50 cm. The base of the forked, branching stem is covered in glabrous, delicately pinnatifid leaves. The alternating, dissected, sparsely haired, 5-cm-long leaves are bluish-green and have thread-like leaflets. Each umbel flower has a cluster of three to five blossoms (Kemal et al., 2023). The petals have a strong border and a long, indented tip. They are rectangular in form and either white or pink in colour. Long and straight involucral bracts contrast with a short, abruptly curved outer style. There are three locules in the inferior ovary. The fruits, which are schizocarp and brown, yellowish-green, or both, are 6 mm in length and 1.5 mm in width. They have awl-shaped calyx points on their crowns. The nine protuberances on the seeds are elongated or oblong in form, and golden grey in colour (Singletary et al., 2021).

16.3 CUMIN CULTIVATION

Cumin is widely grown as a cold-season crop in lowland regions and as a summer crop in the hills. Like several species in the Apiaceae family, the tiny white or pink blooms are borne on little compound umbels. It is the primary crop in a dry agricultural system in the southern portion of Iran's Khorasan region, with an annual yield of roughly 12,000 tons seed on average. Eastern Europe and northern India (Himalayas, Punjab, Baluchistan, and Kashmir) are other places where it is grown. The three main varieties of cumin seed available on the market – Iranian, Indian, and Middle Eastern – differ in terms of the colour of the seeds, the amount of essential oil, and the flavour. The seeds are sown in early April for the summer crop. To get the plants started, they should be seeded in little pots with light soil and put first in a reasonably warm bed. They can then be progressively hardened off in an open frame before being planted in a warm border of healthy soil. The plants were harvested after 85% flowering in June and July. After being spread out on fabric to dry, the fresh seeds are placed in cotton bags for storage. Cumin seeds and its powder should be stored in airtight glass containers in cool and dark conditions. The entire seeds of cumin will remain fresh for around a year, while the ground cumin can stay fresh for roughly six months (Kumar et al., 2021).

16.4 SEED OIL PRODUCTION

The ripe seeds, either whole or roughly crushed, are used to extract the oil. The entire seeds must be utilized if an oil that is easily soluble in alcohol is needed. Essential oils are extracted by hydrodistillation, which yields an oily liquid that is either colorless or light yellow and has a strong scent. Depending on whether the full seed or a coarsely milled version of the seed is distilled, the yield for oil production ranges from 2.5–4.5%. Thymol may be produced artificially from cumin oil with ease. The volatile oil should be stored in aluminium cans or dark bottles with tight lids.

16.5 SEED SPECIFICATION

The powdered seeds have a fragrant, somewhat camphoraceous flavor and are a yellowish-brown colour.

- The epicarp is a layer of colourless cells that are polygonal in surface view with thin sinuous walls and a weakly and irregularly striated cuticle. Stomata are quite common, and cicatrices may very rarely be present.
- The covering trichomes are multicellular, multiseriate, and rounded at the apex; they vary in length and are made of fairly thick-walled cells that occasionally display faint, scattered

FIGURE 16.1 Images of mature cumin plant and seeds.

cuticular striations; underneath the epicarp, the thin-walled cells of the palisade are occasionally visible.

- The very abundant pale yellowish-brown pieces of the vittae (oil tubes) are formed of polygonal, rather big cells with thin walls. The shards that span the whole width of the vittae show that they are broader than in most of the other fruits of the Apiaceae.
- The mesocarp has two primary kinds of sclereids. The first type is composed of a single layer of longitudinally elongated cells with somewhat thickened walls and many evenly spaced, clearly defined pits. The second type is found in tiny groups and is made up of significantly elongated cells that are positioned nearly end-to-end in a longitudinal manner. They are typically connected to vascular tissue or vitta fragments. Although there are occasionally clusters of these sclereids with walls that are more extensively thickened and striated, there aren't many pits in the majority of these sclereids (Amin et al., 2012) (Figure 16.1).

16.6 NUTRITIONAL COMPOSITION

Cumin seeds are commonly utilized as a natural health enhancer and have a strong potential as nutraceuticals. They are used for a variety of things, such as vegetables, cosmetics, food, drinks, medicines, and other industrial uses. The seeds contain a healthy number and concentration of vitamins, minerals, amino acids, and other nutrients that are beneficial for our health (Mughal et al., 2022). This demonstrates that cumin includes a nutritionally significant (both qualitatively and quantitatively) composition that not only gives the plant its distinctive flavours and colours but also aids in the creation of phytomedicines for human health (Singh et al., 2021).

16.7 TRADITIONAL USES

Natural products continue to be the source of many bioactives, which have been used for important medicinal purposes throughout history. Cumin seeds have been utilized for their distinctive and strong scent as well as a number of therapeutic characteristics in different traditional formulations and culinary preparation since the dawn of time. According to the Ayurvedic system of medicine, the seeds have carminative, emmanogogic, antispasmodic, astringent, epileptic, abortifacient, and diuretic properties. According to traditional medicines, the plant can be used to treat a variety of ailments, including diarrhoea, jaundice, weight loss, hoarseness, hypertension, dyspepsia, indigestion, and flatulence (Kang et al., 201).

TABLE 16.1
Traditional Use of Cumin Seeds

Properly	Description	Key References
Regulation of blood cholesterol	Consuming cumin powder per day reduces low-density lipoprotein in the blood.	Zare et al. (2014)
Control of diabetes	The powdered seed can control the diabetes	Mnif and Aifa (2015); Srinivasan et al. (2018)
Improvement of memory performance and anti-Parkinson's	The use of seed powder as a spice	Omari et al. (2021)
Anti-inflammatory	The use of seed powder as a spice	Mnif and Aifa (2015); Kang et al. (2019)
Anticancer	The use of seed powder as a spice	Prakash and Gupta (2014)
Treatment of respiratory disorders	Cumin tea in the morning can help control respiratory disorders	Srinivasan et al. (2018)
Fortifying energy	Flavouring of foods, especially soft dates	Petretto et al. (2018)

The digestive process is accelerated by cumin's stimulation of bile production (Srinivasan, 2018). Cumin fruit has been used in Unani medicine to treat corneal opacities, ulcers, boils, styes, coughs, and inflammation. Cumin paste is used in Indonesia to treat rheumatism and prevent bloody diarrhoea, in addition to relieving headaches (Tahir et al., 2016). According to Tabarsa et al. (2020), the herb is historically used in Iran to promote the production of milk, relieve discomfort and flatulence, and act as a potent antispasmodic. Additionally, some clinical research has been done to support the ethnopharmacological benefits of cumin. Omari et al. (2021) represented that consuming cumin powder each day reduces the concentration of low-density lipoproteins in the blood. Prakash and Gupta (2014) showed that the use of the seed powder as a spice is also important in treating a wide range of cancers due to its high antioxidant activity. Although parts of this herbal formulation's traditional use as a medicine have received scientific validation, others still lack the necessary backing in terms of safe dosage and time frames. Establishing the relationship between various age groups and the advised safe dose in connection to the specific ethnomedicinal usage is also necessary. Table 16.1 shows some traditional uses of cumin seeds.

16.8 PHYTOCHEMICAL COMPOUNDS

Alkaloids, coumarins, anthraquinones, flavonoids, glycosides, proteins, resins, saponins, tannins, steroids, dietary fibres, minerals, fats (particularly monounsaturated fats), B vitamins (thiamin, vitamin B_6, niacin, and riboflavin), vitamin A, vitamin C, and vitamin E, among other nutrients, are abundant in cumin seeds. Cumin seeds also include organic acids, such as aspartic, citric, malic, tartaric, propionic, ascorbic, oxalic, maleic, and fumaric acids (Belal et al., 2017). Luteolin, apigenin, and oestrogenic isoflavonoids are present in the fruits. The seeds contain a significant number and concentration of flavonoids (4.1–5.8%), with luteolin, apigenin, and quercetin being the three most abundant types (Kang et al., 2019). Dubey et al. (2017) reported that the essential oil of seed cumin consisted of α-terpinene, α-pinene, santolina triene, γ-terpinene, β-pinene, and α-thujene. The main alcohols were cumic alcohol (10.33–14.40%), and anethole +estragole (4.03–9.15%), with traces of cymol (2.9–4.26%) and geraniol (0.02–2.39%), while the flavonoid concentration in the AESRs ranged between 27.14 and 36.03 mg QE g^{-1} seeds. In addition, cumin seed oils contain *p*-cymene, cuminic alcohol (cuminol), γ-terpinene, safranal, limonene, eugenol, β-myrcene, α-phellandrene, β-phellandrene, and α- and β-pinene (Nadeem & Riaz, 2012; Singh et al., 2021).

The main compounds of cumin seed essential oils are listed in Table 16.2. Different phytoconstituents are associated with various pharmacological effects, although occasionally a single phytocompound might provide many therapeutic effects. Some of the main phytochemical are listed in Table 16.3. The relationship between different phytocompounds and various pharmacologies is well established. Cuminaldehyde has been more thoroughly investigated for various pharmacological properties than have other bioactives. The pharmacological effects of the various cumin bioactives should thus also be evaluated. According to Srinivasan (2018), substituted pyrazines, such as 2-ethoxy-3-isopropylpyrazine, 2-methoxy-3-sec-butylpyrazine, and 2-methoxy-3-methylpyrazine, are what give roasted cumin seeds their strong scent. In addition, it has been shown that the majority of cumin's biological activities are mainly dependent on cuminaldehyde. Although this chemical makes a significant contribution to numerous ethnomedicinal, pharmacological, and clinical studies, parallel studies on the medicinal properties of other bioactives should also be conducted (Singh et al., 2021).

TABLE 16.2
Essential Oil Constituents of Cumin Seeds

	Compound	Class of Compound	Key References
1	γ-Terpinine	Monoterpene	Moghaddam et al. (2015); Adel et al. (2016)
2	α-Sabinin	Monoterpene	Adel et al. (2016)
3	*p*-Cymene	Monoterpene	Moghaddam et al. (2015)
4	α-Pinene	Monoterpene	Abbdellaoui et al. (2019)
5	(−)-β-Pinene	Monoterpene	Abbdellaoui et al. (2019)
6	α-Phellandrene	Monoterpene	Abbdellaoui et al. (2019)
7	α-Terpinene	Monoterpene	Abbdellaoui et al. (2019)
8	α-Terpineol	Monoterpene	Sadeghi et al. (2019)
9	Safranal	Monoterpene	Sadeghi et al. (2019)
10	Limonene	Monoterpene	Sadeghi et al. (2019)
11	1,8-Cineole	Monoterpene	Moghaddam et al. (2015); Abbdellaoui et al. (2019)
12	Linalool	Monoterpene	Abbdellaoui et al. (2019)
13	Linalyl acetate	Monoterpene	Abbdellaoui et al. (2019); Allaq et al. (2020)
14	α-Terpineol acetate	Monoterpene	Abbdellaoui et al. (2019); Allaq et al. (2020)
15	Geraniol	Monoterpene	Moghaddam et al. (2015; Mayank et al. (2022)
16	Methyl eugenol	Monoterpene	Rebey et al. (2012); Allaq et al. (2020)
17	Sabinene	Monoterpene	Allaq et al. (2020)
18	Terpinolene	Monoterpene	Allaq et al. (2020)
19	α-Thujene	Monoterpene	Rebey et al. (2012; Allaq et al. (2020)
20	Myrcene	Monoterpene	Allaq et al. (2020)
21	γ-Terpineol	Monoterpene	Sadeghi et al. (2019)
22	d3-Carene	Sesquiterpene	Sadeghi et al. (2019)
23	Pinocarvone	Sesquiterpene	Sadeghi et al. (2019)
24	Cariophyllene	Sesquiterpene	Sadeghi et al. (2019)
25	Farnesene-(Z)-β	Sesquiterpene	Allaq et al. (2020)
26	Germacrene D	Sesquiterpene	Allaq et al. (2020)
27	α-Acoradiene	Sesquiterpene	Allaq et al. (2020)
28	Carotol	Sesquiterpene	Allaq et al. (2020)

TABLE 16.3
Main Chemical Compounds of Cumin Seeds

	Compound	Class of Compound	Key References
1	Linalyl acetate	Esters	Metin et al. (2022)
2	Methyl geranate	Esters	Adel et al. (2016)
3	Citronellyl acetate	Esters	Metin et al. (2022)
4	Pyrazine	Pyrazine	Srinivasan (2018)
5	2-Ethoxy-3-isopropylpyrazine	Pyrazine	Srinivasan (2018)
6	2-Methoxy-3-sec-butylpyrazine	Pyrazine	Srinivasan (2018)
7	2-Methoxy-3-methylpyrazine	Pyrazine	Srinivasan (2018)
8	Apigenin	Flavonoids	Goodarzi et al. (2020)
9	Apigenin-7-*O*-glucoside	Flavonoids	Ravi et al., (2013)
10	Luteolin-7-*O*-glucoside	Flavonoids	Ravi et al., (2013)
11	Cumene	Isopropylbenzene	Volianiuk et al. (2022)
12	Cumic acid	Benzoic acid	Nirmala et al. (2020
13	Octanal	Aldehyde	Abu Ahmed et al. (2022)

16.8.1 Pharmacological Activities

Cumin seeds have a number of health benefits that have been well documented, including antioxidant, antibacterial, insecticidal, anti-inflammatory, pain-relieving, antitumour, platelet disaggregation, antiallergenic, antihypertension, hyperglycaemic, hypoglycaemic, antihyperlipidaemic, bronchodilatory, immunomodulatory, contraceptive, anti-amyloidogenic, anti-osteoporotic, and antifungal. Properties. Phenols, terpenoids, and flavonoids are primarily responsible for cumin's remarkable pharmacological properties (Tabarsa et al., 2020). According to Singh et al. (2021), cumin's immunomodulatory effects are caused by its flavonoid glycosides, high iron concentration, vitamin C content, and vitamin A content. The fruits are useful in the treatment of cardiovascular diseases, osteoporosis, and menopausal issues. Additionally, cumin fruits change how carcinogens are metabolized and are used to treat a number of cancers, including those of the stomach, colon, uterus, and cervix, hepatocellular cancer, and so on. Despite its enormous therapeutic potential, there have not been enough scientific investigations on cumin, mostly because there aren't enough relevant experimental data and validating standards (Izzo et al., 2020).

16.8.2 Antimicrobial Activity

The symptoms of microbial infections include pain, fever, chills and sweating, headache, swelling, vomiting, nausea, diarrhoea, abdominal pain, swollen lymph nodes, skin irritations, rashes, and many more symptoms. With the aid of the appropriate antimicrobial medications, these signs of microbial infection can be avoided. Antimicrobial medications destroy (biocidal activity) or halt the development (biostatic activity) of microbial pathogens to lessen their harmful effects. Significant antibacterial capabilities are displayed by cumin oil. Abbaszadegan et al. (2016) showed that all cumin preparations totally inhibited *Enterococcus faecalis* growth after 24 h. Belal et al. (2017) evaluated 10 µg of cumin oil showed antibacterial activity against Gram-positive and Gram-negative bacteria. Hajib et al. (2020) indicated that terpinen-7-al (37.2%), cuminaldehyde (29.1%), terpinene (10.0%), pinene (7.1%), and *p*-cymene (6.1%) made up the main part of the cumin seed essential oil. Twenty-one compounds were found in the seed oil of the North African medicinal plant of the Apiaceae, *Ammodaucus leucotrichus*, accounting for 98.4% of the total seed oil, of which perilla

TABLE 16.4
Antibacterial and Antifungal Activities of Cumin Seed

Properties	Bacterial/Fungal Species	Results	Key References
Antibacterial activity	*Enterococcus faecalis*	The results of time-kill assay revealed that all preparations totally inhibited the bacterial growth after 24 h	Abbaszadegan et al. (2016)
	Klebsiella pneumoniae	The cumin alcoholic extract had no effect on biofilm formation ability of *K. pneumoniae*.	Derakhshan et al. (2017)
	Proteus vulgaris, Klebsiella pneumoniae, Enterococcus faecalis and *Staphylococcus aureus*	10 µg of cumin oil showed antibacterial activity against gram positive and Gram-negative bacteria.	Belal et al. (2017)
	Enterococcus faecalis, Salmonella sp,, *Staphylococcus aureus, Klebsiella pneumoniae, Acinetobacter baumannii, Staphylococcus epidermis, Citrobacter freundii* and *Escherichia coli*	High antibacterial potential of cumin essential oil	Hajib et al. (2020)
Antifungal activity	*Botrytis cinerea, Aspergillus niger* and *Penicillium expansum*	Essential oil at concentrations of $\geq$ 750 µL/L inhibited the mycelial growth of the tested fungi	Ghasemi et al. (2019)
	Aspergillus aculeatus	The EC_{50} values against the conidial germination were estimated. Significant reduction of conidia germination was also detected in *C. cyminum* essential oil and cumin aldehyde at a concentration of 1,000 and 100 µg/mL, respectively.	Tanapichatsakul et al. (2020)

aldehyde (78.2%) and limonene (9.5%) were the main components. The capacity of isolated essential oils to scavenge radicals using the stable DPPH and ABTS assays, to have reducing power using a test based on the reduction of ferric cations, and to have the ability to prevent lipid peroxidation using the β-carotene bleaching assay were all investigated. Using the agar diffusion method, the antibacterial activity of each essential oil was assessed against samples of Gram-positive and Gram-negative bacteria. Ghasemi et al. (2020) reported that essential oil of cumin seeds at concentrations of $\geq$ 750 µL/L inhibited the mycelial growth of the fungi *Botrytis cinerea, Aspergillus niger*, and *Penicillium expansum.* In addition, Tanapichatsakul et al. (2020) investigated the EC_{50} values against conidial germination. Cumin aldehyde and essential oil at concentrations of 1,000 and 100 g/mL, respectively, reduced the germination of the conidia. Table 16.4 shows the antibacterial and antifungal activities of cumin seeds

16.8.3 Anti-inflammatory Activity

Nitric oxide (NO) production is high and there is an increase in pro-inflammatory cytokines under conditions of inflammation. Anti-inflammatory drugs may change the signalling pathways involved in NO generation and decrease the production of pro-inflammatory cytokines (Bagirova et al., 2020). Cumin seeds are reported to have strong anti-inflammatory properties. The anti-inflammatory potential of several novel compounds that Wei et al. (2015) identified from cumin seeds was compared with those of compounds already known to exist in cumin. Molecule 2, also known as cuminin or 8-(amino[4-isopropylphenyl]methyl), was the second novel molecule discovered.-5-hydrox y-2-(4-hydroxyphenyl). Using a lipopolysaccharide (LPS)-stimulated RAW264.7 cell test, it was discovered that 7-methoxy-4-oxo-4*H*-chromene-6-carboxylic acid (an unusual nitrogen-containing flavonoid molecule) had the strongest anti-inflammatory effect, with an IC_{50} value of 5.25 M. They found that lipopolysaccharide activated NK-92 cells by causing the generation of tumour necrosis factor (TNF), interferon (IFN), and perforin, whereas RAW264.7 cells were activated through NF-B and MAPK signalling. The mitochondrial-respiration-dependent 3-(4,5-dimethylthiazol-2-yl)-2,5-diphenyltetrazolium (MTT) reduction assay showed that cumin essential oil did not exhibit any cytotoxic effect at the concentrations (0.0005–0.01%) employed. Real-time quantitative PCR tests showed that the essential oil significantly inhibited the mRNA abundance from the genes encoding inducible nitric oxide synthase (iNOS), cyclooxygenase (COX-2), interleukin- (IL-) 1, and IL-6. Moreover, western blotting analysis revealed that the cumin essential oil blocked LPS-induced transcriptional activation of nuclear factor-kappa B (NF-κ B) and inhibited the phosphorylation of extracellular signal-regulated kinase (ERK) and c-Jun N-terminal kinase (JNK). These findings suggested that cumin essential oil inhibited NF-κ B and mitogen-activated protein kinases ERK and JNK signaling in LPS-stimulated RAW 264.7 cells, exerting anti-inflammatory effects. The chemical may be used as a source of anti-inflammatory agents as well as a dietary supplement for health promotion. Kang et al. (2019) determined methyl 3-((7*H*-purin-2-yl) amino)-3-(4-isopropylphenyl) propanoate (**1**), 8-(amino(4-isopropylphenyl)methyl)-5-hydroxy-2-(4-hydroxyphe nyl)-7-methoxy-4-oxo-4*H*-chromene-6-carboxylic acid (**2**), (3,4,5-trihydroxy-6-((4-isopropylbe nzyl)oxy)tetrahydro-2*H*-pyran-2-yl)methyl (*E*)-3-(4-propoxyphenyl)acrylate (**3**), and (3,4,5-trihydro xy-6-((5-hydroxy-2-(4-hydroxyphenyl)-4-oxo-4*H*-chromen-7-yl)oxy)tetrahydro-2*H*-pyran-2-yl) methyl 3-(4 isopropylphenyl)-2-methoxypropanoate (**4**). Compound **2**, an atypical nitrogen-containing flavonoid, exhibited the most active inhibitory effect on nitric oxide, with an IC_{50} of 5.25 μM in the lipopolysaccharide-stimulated RAW264.7 cell assay. Compound **2** was found to downregulate the expression of the genes encoding inducible nitric oxide synthase and cyclooxygenase-2. Furthermore, it was revealed that both nuclear factor κB and the mitogen-activated protein kinases were involved in the anti-inflammatory process of compound **2**. Some anti-inflammatory, antioxidant, and anticancer properties of cumin are reported in Table 16.5.

16.8.4 Anticancer Activity

Uncontrolled cell development, leading to tumour formation, cell proliferation, and metastasis, is how cancer is defined. A substance with high potential for cytotoxicity can be used as a powerful anti-cancer agent (Dinparvar et al., 2020). It was claimed that cumin was a potent anti-cancer substance. The Brine Shrimp Test (BST), used by Goodarzi et al. (2020), was used to assess the cytotoxicity impact of the ethyl acetate and hexane fractions and sub-fractions of cumin seeds. They also used the MTT assay to test the separated chemicals for their ability to prevent breast cancer. They reported good IC_{50} values, which support the cytotoxic effectiveness of cumin fruits. Dinparvar et al. (2020) showed that the seed extract did not show any inhibitory effect on J774 macrophage cells, but killing 50% of the MCF-7 cells at the concentrations used. In addition, Siddiqui et al. (2020) reported that the cytotoxic activity of seed extracts against the Hep–2 cell line occurred as a result of the production of reactive oxygen species.

16.8.5 Antidiabetic Activity

Diabetes is a chronic condition brought on by either insufficient insulin production by the pancreas or inefficient insulin use by the body. This hormone called insulin controls blood sugar levels. Uncontrolled diabetes frequently causes hyperglycaemia, also known as high blood glucose or

TABLE 16.5
Some Pharmacological Attributes of Cumin Seeds

Properties	Results	Key References
Antioxidant	Essential oil of cumin seedshad high antioxidant activity [150 μL exhibiting strong reducing power; 2200 (FRAP) μmol/L Fe^{2+} for 15 min and ~ 89 DPPH % at 60 min].	Ghasemi et al. (2019)
	High potential of essential oil of cumin in scavenging free radicals	Mohamed et al. (2018)
	The essential oil had β-carotene bleaching test with IC_{50} of 12.47± 0.24 μg/mL, lipid peroxidation inhibition with 95.71± 0.87 μg/mL, and reducing power with EC_{50} of 12.45 ± 0.31 μg/mL	Ladan Moghadam (2016)
Anticancer	Seed extract did not show any inhibitory effect on J774 macrophage cells, but killed 50% of the MCF-7 cells at the investigated concentrations	Dinparvar et al. (2020)
	Cytotoxic activity of seed extracts against Hep–2 cell line by the formation of reactive oxygen was confirmed	Siddiqui et al. (2020)
	Dietary interventions using nutraceutical food formulation made out of spent cumin play a significant role in the prevention and management of degenerative diseases.	Arun et al. (2016)
Anti-inflammatory	High inflammatory potential due to 8-(amino(4-isopropylphenyl)methyl)-5-hydroxy-2-(4-hydroxyphenyl)-7-methoxy-4-oxo-4*H*-chromene-6-carboxylic acid	Kang et al. (2019)
	Essential oil anti-inflammatory effects in LPS-stimulated RAW 264.7 cells via inhibition of NF-κB and mitogen-activated protein kinases ERK and JNK signalling; the chemical could be used as a source of anti-inflammatory agents as well as dietary complement for health promotion	Wei et al. (2015)

raised blood sugar, which, over time, can seriously harm many different bodily systems, including the neurons and blood vessels. Mohamed et al. (2018) showed that total phenolic and total flavonoids were present in the crude ethanol extract of cumin seeds (CEEC) at concentrations of 23.02±0.045 mg gallic acid-equivalents (GAE) g^{-1} extract and 19±0.132 mg QE g^{-1} extract, respectively. CEEC showed free radical scavenging activity. Acute toxicity assays revealed that CEEC was completely safe up to 12 g kg^{-1} mice. Oral administration of CEEC reduced glycaemic levels. In comparison with diabetic control groups, CEEC dramatically increased insulin levels and decreased plasma glucose levels in diabetic rats. In diabetic rats, CEEC decreased inflammation, oxidative stress, and the plasma lipid profile. Additionally, through α-amylase inhibition studies, Tahir et al. (2016) assessed the anti-diabetic potential of various emulsions (B1–B5) of cumin seed and discovered that the B5 emulsion exhibited the greatest decrease (83.09%) in α-amylase activity. To demonstrate the antidiabetic effects of various dosages of cuminaldehyde (2.5, 5, or 10 mg kg^{-1}), Kannan et al. (2018) studied high-fat-diet-induced type 2 diabetes in mice. The concentrations of the pathogenic biomarker glycoproteins sialic acid, fucose, and hexose/hexosamine dramatically decreased after 35 days of therapy with cuminaldehyde. As a result, the antidiabetic potential of cumin can be used to create novel cumin-enriched dietary supplements to manage diabetes. But before they are marketed, these artificial dietary supplements need to be clinically verified for this specific pharmacological effect.

16.8.6 Anti-obesity Activity

Obesity results from the accumulation of excessive dietary energy, which is eventually stored as visceral fat. According to Karpagam and Priya (2019), it is identifiable when the body mass index (BMI) is more than 25. In this regard, Haque and Ansari (2018) demonstrated an anti-obesity effect of cumin seed essential oil on Wistar rats in a clinical study.

16.9 CONCLUSIONS

The enormous potential of cumin seeds is illustrated in the current chapter. The primary active chemicals were discovered in cumin seed essential oils. The high antibacterial, anti-inflammatory, antioxidant, anticancer, and antidiabetic properties of cumin seed were discussed. Therefore, cumin seeds are a natural source for treating a variety of ailments and managing the cultivation of this plant as a crop might be crucial from an economic standpoint. These seeds can replace certain toxic chemical medications that are harmful to humans. For additional research, a comparison of the chemical composition and pharmacological properties of green and black cumin seeds would be beneficial.

REFERENCES

Abbaszadegan, A., Gholami, A., Ghahramani, Y., Ghareghan, R., Ghareghan, M., Kazemi, A. ... Ghasemi, Y. (2016). Antimicrobial and cytotoxic activity of *Cuminum cyminum* as an intracanal medicament compared to chlorhexidine gel. *Iranian Endodontic Journal, 11*(1), 44.

Abbdellaoui, M., Bouhlali, E. D. T., & Rhaffari, L. E. (2019). Chemical composition and antioxidant activities of the essential oils of cumin (*Cuminum cyminum*) conducted under organic production conditions. *Journal of Essential Oil Bearing Plants, 22*(6), 1500–1508.

Abu Ahmed, S. E. S., Alshiteewi, S. A., Moawed, E. A., & Ayyad, S. N. (2022). Phytochemical composition of *Cuminum cyminum* and its biological activity. *Scientific Journal for Damietta Faculty of Science, 12*(1), 29–38.

Adel, M., Safari, R., Ghitanchi, A. H., & Zorriehzahra, M. J. (2016). Chemical composition and in vitro antimicrobial activity of some Iranian medical herbs against Yersinia ruckeri. *Iranian Journal of Fisheries Sciences, 15*(3), 1108–1123.

Allaq, A. A., Sidik, N. J., Abdul-Aziz, A., & Ahmed, I. A. (2020). Cumin (*Cuminum cyminum* L.): A review of its ethnopharmacology, phytochemistry. *Biomedical Research and Therapy, 7*(9), 4016–4021.

Amin, G. (2012). Cumin. In *Handbook of Herbs and Spices*, pp. 250–259. Woodhead Publishing. Woodhead Publishing Series in Food Science, Technology and Nutrition. Sawston, Cambridge.

Arun, K. B., Aswathi, U., Venugopal, V. V., Madhavankutty, T. S., & Nisha, P. (2016). Nutraceutical properties of cumin residue generated from Ayurvedic industries using cell line models. *Journal of Food Science and Technology*, *53*(10), 3814–3824.

Bagirova, M., Dinparvar, S., Allahverdiyev, A. M., Unal, K., Abamor, E. S., & Novruzova, M. (2020). Investigation of antileshmanial activities of *Cuminum cyminum* based green silver nanoparticles on L. tropica promastigotes and amastigotes in vitro. *Acta Tropica*, *208*, 105498.

Belal, A. A., Ahmed, F. B., & Ali, L. I. (2017). Antibacterial activity of *Cuminum cyminum* L. oil on six types of bacteria. *American Journal of Bioscience*, *5*(4), 70–73.

Derakhshan, S., Navidinia, M., & Ahmadi, A. (2017). Antibacterial activity of Dill (*Anethum graveolens*) essential oil and antibiofilm activity of Cumin (*Cuminum cyminum*) alcoholic extract. *Infection Epidemiology and Microbiology*, *3*(4), 122–126.

Dinparvar, S., Bagirova, M., Allahverdiyev, A. M., Abamor, E. S., Safarov, T., Aydogdu, M., & Aktas, D. (2020). A nanotechnology-based new approach in the treatment of breast cancer: Biosynthesized silver nanoparticles using *Cuminum cyminum* L. seed extract. *Journal of Photochemistry and Photobiology, B: Biology*, *208*, 111902.

Dubey, P. N., Saxena, S. N., Mishra, B. K., Solanki, R. K., Vishal, M. K., Singh, B. … Yogi, A. (2017). Preponderance of cumin (*Cuminum cyminum* L.) essential oil constituents across cumin growing Agro-Ecological Sub Regions, India. *Industrial Crops and Products*, *95*, 50–59.

Ghasemi, G., Fattahi, M., Alirezalu, A., & Ghosta, Y. (2019). Antioxidant and antifungal activities of a new chemovar of cumin (*Cuminum cyminum* L.). *Food Science and Biotechnology*, *28*(3), 669–677.

Goodarzi, S., Tabatabaei, M. J., Mohammad Jafari, R., Shemirani, F., Tavakoli, S., Mofasseri, M., & Tofighi, Z. (2020). Cuminum cyminum fruits as source of luteolin-7-O-glucoside, potent cytotoxic flavonoid against breast cancer cell lines. *Natural Product Research*, *34*(11), 1602–1606.

Hajib, A., Nounah, I., Oubihi, A., Harhar, H., Gharby, S., Kartah, B. … Charrouf, Z. (2020). Chemical composition and biological activities of essential oils from the fruits of *Cuminum cyminum* L. and Ammodaucus leucotrichus L.(Apiaceae). *Journal of Essential Oil Bearing Plants*, *23*(3), 474–483.

Haque, M. R., & Ansari, H. S. (2018). Anti-obesity effect of arq zeera and its main components thymol and cuminaldehyde in high fat diet induced obese rats. *Drug Research*, *68*(11), 637–647.

Husen, A. (2021). *Traditional Herbal Therapy for the Human Immune System*. CRC Press, Boca Raton, FL. https://doi.org/10.1201/9781003137955.

Husen, A. (2022a). *Herbs, Shrubs and Trees of Potential Medicinal Benefits*. CRC Press, Boca Raton, FL. https://doi.org/10.1201/9781003205067.

Husen, A. (2022b). *Exploring Poisonous Plants: Medicinal Values, Toxicity Response and Other Potential Uses*. CRC Press, Boca Raton, FL. https://doi.org/10.1201/b23017.

Husen, A. (2023). *Antidiabetic Medicinal Plants and Herbal Treatments*. CRC Press, Boca Raton, FL. https://doi.org/10.1201/b23347.

Izzo, L., Rodríguez-Carrasco, Y., Tolosa, J., Graziani, G., Gaspari, A., & Ritieni, A. (2020). Target analysis and retrospective screening of mycotoxins and pharmacologically active substances in milk using an ultra-high-performance liquid chromatography/high-resolution mass spectrometry approach. *Journal of Dairy Science*, *103*(2), 1250–1260.

Kang, N., Yuan, R., Huang, L., Liu, Z., Huang, D., Huang, L. … Yang, S. (2019). Atypical nitrogen-containing flavonoid in the fruits of cumin (*Cuminum cyminum* L.) with anti-inflammatory activity. *Journal of Agricultural and Food Chemistry*, *67*(30), 8339–8347.

Kang, N., Yuan, R., Huang, L., Liu, Z., Huang, D., Huang, L. … Yang, S. (2019). Atypical nitrogen-containing flavonoid in the fruits of cumin (*Cuminum cyminum* L.) with anti-inflammatory activity. *Journal of Agricultural and Food Chemistry*, *67*(30), 8339–8347.

Kannan, D., Palani, L., Gunaseelan, T., Shyamaladevi, B., & Madhan, M. (2018). Changes in glycoprotein components in high-fat diet induced type 2 diabetes: Influence of cumin aldehyde. *Drug Invention Today*, *10*, 1–6.

Karpagam, K., & Priya, U. (2019). Effectiveness of cumin cyminum plus lime administration on obesity among womens. *International Journal of Research in Pharmaceutical Sciences*, *10*(4), 2965–2969.

Kemal, M., Esertaş, Ü. Z. Ü., Kanbur, E. D., Kara, Y., Özçelik, A. E., Can, Z., & Kolaylı, S. (2023). Characterization of the black cumin (*Nigella sativa* L.) honey from Türkiye. *Food Bioscience*, *53*, 102760.

Koohsari, S., Sheikholeslami, M. A., Parvardeh, S., Ghafghazi, S., Samadi, S., Poul, Y. K., … Amiri, S. (2020). Antinociceptive and antineuropathic effects of cuminaldehyde, the major constituent of Cuminum cyminum seeds: Possible mechanisms of action. *Journal of Ethnopharmacology, 255*, 112786.

Kumar, M., Sahoo, P. K., Kushwaha, D. K., Saxena, S. N., Mani, I., Singh, G., & Pradhan, N. C. (2021). Cumin cultivation: Present status and future prospects. *The Pharma Innovation Journal, 10*, 1121–1123.

Ladan Moghadam, A. L. (2016). Efficacy of chemically characterized *Cuminum cyminum* essential oil as an antioxidant and lipid peroxidation inhibitor. *Journal of Essential Oil Bearing Plants, 19*(1), 134–139.

Mayank, G., Mehta, N., Panwar, H., Malav, O. P., & Bedi, J. S. (2022). *In-vitro* Assessment of antibacterial and antioxidant capacity of essential oils from cumin (*Cuminum cyminum*) and Lemon (*Citrus limon*) for Future Applications in Meat Industry. *Journal of Animal Research, 12*(5), 795–802.

Metin, S., Yigit, N. O., Balkaya, M., Didinen, B. I., Didinen, H., & Ozmen, O. (2022). Efficacy of cumin (*Cuminum cyminum*) and lavender (*Lavandula angustifolia*) essential oils as anaesthetics in common carp (*Cyprinus carpio* L. 1758). *Aquaculture Research, 53*(10), 3598–3607.

Mnif, S., & Aifa, S. (2015). Cumin (*Cuminum cyminum* L.) from traditional uses to potential biomedical applications. *Chemistry and Biodiversity, 12*(5), 733–742.

Moghaddam, M., Miran, S. N. K., Pirbalouti, A. G., Mehdizadeh, L., & Ghaderi, Y. (2015). Variation in essential oil composition and antioxidant activity of cumin (*Cuminum cyminum* L.) fruits during stages of maturity. *Industrial Crops and Products, 70*, 163–169.

Mohamed, D. A., Hamed, I. M., & Fouda, K. A. (2018). Research article antioxidant and anti-diabetic effects of cumin seed crude ethanol extract. *Journal of Biological Sciences, 18*(5), 251–259.

Mohamed, D. A., Hamed, I. M., & Fouda, K. A. (2018). Research article antioxidant and anti-diabetic effects of cumin seeds crude ethanol extract. *Journal of Biological Sciences, 18*(5), 251–259.

Mughal, S. S. (2022). A review on potential antioxidant effects of Cumin (*Cuminum cyminum*), phytochemical Profile and its uses. *Authorea Preprints*. 8, 1–19.

Nadeem, M., & Riaz, A. (2012). Cumin (*Cuminum cyminum*) as a potential source of antioxidants. *Pakistan Journal of Food Sciences, 22*(2), 101–107.

Nirmala, M. J., Durai, L., Rao, K. A., & Nagarajan, R. (2020). Ultrasonic nanoemulsification of Cuminum cyminum essential oil and its applications in medicine. *International Journal of Nanomedicine*, 795–807.

Omari, Z., Kazunori, S., Sabti, M., Bejaoui, M., Hafidi, A., Gadhi, C., & Isoda, H. (2021). Dietary administration of cumin-derived cuminaldehyde induce neuroprotective and learning and memory enhancement effects to aging mice. *Aging (Albany NY), 13*(2), 1671.

Petretto, G. L., Fancello, F., Bakhy, K., Faiz, C. A., Sibawayh, Z., Chessa, M. … Pintore, G. (2018). Chemical composition and antimicrobial activity of essential oils from *Cuminum cyminum* L. collected in different areas of Morocco. *Food Bioscience, 22*, 50–58.

Prakash, E., & Gupta, D. K. (2014). Cytotoxic activity of ethanolic extract of *Cuminum cyminum* Linn against seven human cancer cell line. *Universal Journal of Agricultural Research, 2*(1), 27–30.

Ravi, R., Prakash, M., & Bhat, K. K. (2013). Characterization of aroma active compounds of cumin (*Cuminum cyminum* L.) by GC-MS, E-Nose, and sensory techniques. *International Journal of Food Properties, 16*(5), 1048–1058.

Rebey, I. B., Jabri-Karoui, I., Hamrouni-Sellami, I., Bourgou, S., Limam, F., & Marzouk, B. (2012). Effect of drought on the biochemical composition and antioxidant activities of cumin (*Cuminum cyminum* L.) seeds. *Industrial Crops and Products, 36*(1), 238–245.

Sadeghi, R., Eshrati, M. R., Mortazavian, S. M. M., & Jamshidnia, A. (2019). The effects of the essential oils isolated from four ecotypes of cumin (*Cuminum cyminum* L.) on the blood cells of the pink stem borer, *Sesamia cretica* Ledere (Lepidoptera: Noctuidae). *Journal of the Kansas Entomological Society, 92*(1), 390–399.

Siddiqui, N., Pal, K., Karmakar, P., Islam, M. M., & Mukhopadhyay, S. (2020). A dual role of cumin-seed extract towards the silver nanoparticle synthesis and stabilisation and its potential for antibacterial and anticancer activities through oxidative damage. *Advances in Natural Sciences: Nanoscience and Nanotechnology, 11*(2), 025019.

Singh, N., Yadav, S. S., Kumar, S., & Narashiman, B. (2021). A review on traditional uses, phytochemistry, pharmacology, and clinical research of dietary spice *Cuminum cyminum* L. *Phytotherapy Research, 35*(9), 5007–5030.

Singletary, K. W. (2021). Cumin: Potential health benefits. *Nutrition Today, 56*(3), 144–151.

Srinivasan, K. (2018). Cumin (*Cuminum cyminum*) and black cumin (*Nigella sativa*) seeds: Traditional uses, chemical constituents, and nutraceutical effects. *Food Quality and Safety, 2*(1), 1–16.

Tabarsa, M., You, S., Yelithao, K., Palanisamy, S., Prabhu, N. M., & Nan, M. (2020). Isolation, structural elucidation and immuno-stimulatory properties of polysaccharides from *Cuminum cyminum*. *Carbohydrate Polymers*, *230*, 115636.

Tahir, H. U., Sarfraz, R. A., Ashraf, A., & Adil, S. (2016). Chemical composition and antidiabetic activity of essential oils obtained from two spices (*Syzygium aromaticum* and *Cuminum cyminum*). *International Journal of Food Properties*, *19*(10), 2156–2164.

Tanapichatsakul, C., Khruengsai, S., & Pripdeevech, P. (2020). *In vitro* and *in vivo* antifungal activity of *Cuminum cyminum* essential oil against *Aspergillus aculeatus* causing bunch rot of postharvest grapes. *PLOS One*, *15*(11), e0242862.

Volianiuk, K., Mitina, N., Kinash, N., Harhay, K., Dolynska, L., Nadashkevich, Z. … Zaichenko, A. (2022). Telechelic oligo (*N*-Vinylpyrolydone) with cumene based terminal groups for block-copolymer and nanoparticle obtaining. *Chemistry and Chemical Technology*, *16*(1), 34–41.

Wei, J., Zhang, X., Bi, Y., Miao, R., Zhang, Z., & Su, H. (2015). Anti-inflammatory effects of cumin essential oil by blocking JNK, ERK, and NF-κB signaling pathways in LPS-stimulated RAW 264.7 cells. *Evidence-Based Complementary and Alternative Medicine*. 2015, 1–8. 43.

Zare, R., Heshmati, F., Fallahzadeh, H., & Nadjarzadeh, A. (2014). Effect of cumin powder on body composition and lipid profile in overweight and obese women. *Complementary Therapies in Clinical Practice*, *20*(4), 297–301.

17 Fenugreek Seed (*Trigonella foenum-graecum*)

Surjit Sen, Shuvadip Mondal, Arghya Naskar, Manjula Rai, Tuyelee Das, and Rupa Sanyal

17.1 INTRODUCTION

Advances in medical sciences have improved people's understanding of the medica; ailments in the past century. Plant-derived pharmacological compounds are widely available and prescribed by doctors to overcome various types of health disorders with minimum side effects and innumerable positive effects on human health (Yadav and Baquer 2014; Husen 2021, 2022a, 2022b; 2023; Rahman and Husen 2021, 2022, 2023a, 2023b). *Trigonella foenum-graecum* L. (fenugreek), a diploid annual plant belonging to the legume family Fabaceae (Yadav & Baquer, 2014; Munshi et al., 2020). This plant is known to have originated from Eastern Europe, mainly from Greece as the specific epithet '*foenum-graecum*' means 'Greek hay' (Yadav & Baquer, 2014; Syed et al., 2020). However, some investigators have claimed its origin to be in India and North Africa. It has been reported that around 70–97 different fenugreek species are cultivated around the world (Syed et al., 2020; Visuvinathan et al., 2022). The plant has different vernacular names like Hulbah (Arabic), Foingrec and Trigonelle (French), Bockshorklee (German), Bockshornklee (Swedish), Sarviapila (Finnish), Sambala and Pazhitnik (Russian), Hu-lu-ba (Chinese), Methika (Sanskrit), and Methi (Indian) (Snehlata & Payal, 2012; Sharma et al., 2020a).

This plant is cultivated extensively on the Indian subcontinent, in Mediterranean Europe and North Africa for its leaves and seeds, which are popularly utilized as vegetables and condiments (Yadav & Baquer, 2014; Oufquir et al., 2020). Records of fenugreek and its uses have been encrypted in Egyptian papyrus dating back to 1500 BC as well as being documented for use in different medicinal purposes in various traditional systems like Ayurveda, Unani, and Chinese medicinal systems (Yadav et al., 2019; Kumar et al., 2021). It is reported that fenugreek is one of the oldest medicinal plants being used in traditional practices, as well as the leaves and seeds being used as an esoteric spice agent to increase the flavour, colour, and texture of food (Wani et al., 2016; Pal & Mukherjee, 2020; Oufquir et al., 2020). Contemporary uses of fenugreek seeds and leaves as a spice and food additive, food stabilizer, and emulsifying agent due to its high content of fibre, pectin, and gum in different culinary preparations, has been observed in many countries (Visuvinathan et al., 2022). In addition to its culinary attributes, this plant has also been reported to function as a natural remedy to many kinds of daily-life disorders like cardiovascular diseases, hypercholesterolaemia, hypolipidaemia, hyperglycaemia, cancer, liver diseases, sexual disorders like testosterone deficiency syndrome, and infertility (Syed et al., 2020; Oufquir et al., 2020). The therapeutic approaches of this plant have been used in both animal and human trials by various investigators to demonstrate the antagonistic effects towards hypoglycaemia, hypolipidaemia, and hypocholesterolaemia (Visuvinathan et al., 2022).

Fenugreek seeds contain many nutritive and bioactive compounds, including carbohydrates, of which 58% is mucilaginous fibre, in which galactomannan is dominant, proteins and amino acids (20–30%) (chiefly tryptophan and lysine), 5–10% lipids and 25% fibres (Syed et al., 2020; Visuvinathan et al., 2022). Apart from seed, fenugreek leaf also contains 6%, 4.4%, and 1.15%

DOI: 10.1201/9781003387046-17

carbohydrates, proteins, and fibres, respectively (Syed et al., 2020). This plant is a good source of a variety of minerals like K (603 mg/100 g), Mg (42 mg/100 g), Ca (75 mg/100 g), Zn (2.4 mg/100 g), Mn and Cu (0.9 mg/100 g), Fe (25.8 mg/100 g), and vitamins A, B, and C (Syed et al., 2020). Active phytochemicals like pyridine alkaloids (mainly choline, trigonelline, gentianine, and carpaine), flavonoids (mainly apigenin, orientine, quercetin, vitexin, and isovitexin), and free amino acids (4-hydroxyisoleucine, arginine, lysine, and histidine) are commonly found in this plant (Visuvinathan et al., 2022). It has been reported that fenugreek leaves contain higher amount of phenolic compounds and flavonoids as compared to seeds (Kumar et al., 2021; Salam et al., 2023). Thus, the present chapter is aimed at dealing with the beneficial nutritional and pharmacological aspects of fenugreek and its wide applications for the maintenance and promotion of human health benefits.

17.2 DISTRIBUTION AND PRODUCTION

Review of the literature on fenugreek states its origin to be in India and North Africa; but it is now broadly distributed and cultivated in the regions of Mediterranean Europe, northern Africa, South Asia, western Asia, Argentina, Australia, United States, Nepal, France, Spain, and Turkey (Helambe & Dande, 2012; Rasheed et al., 2015; 52Sharma et al., 2020a; Oufquir et al., 2020). India is the major country to produce this plant commercially (Visuvinathan et al., 2022), contributing more than 80% of the global fenugreek seed production, and nearly all major consumers globally import it from India (Rasheed et al., 2015).

17.3 BOTANICAL DESCRIPTION

In Latin, the genus name of fenugreek (*Trigonella foenum-graecum*) is '*Trigonella*', that means 'little triangle' as the plant possesses yellowish-white triangular flower, whereas the species name '*foenum-graecum*' refers to 'Greek hay' (Goyal et al., 2016; Sun et al., 2021). Taxonomically, the plant belongs to the kingdom Plantae, the division Magnoliophyta, the class Magnoliopsida, the order Fabales, the family Fabaceae, and the sub-family Papilionaceae (Yadav et al., 2019; Snehlata & Payal 2012)

Fenugreek is an annual aromatic plant where seedlings can grow in an erect or semi-erect manner, attaining a height of 30–60 cm (Yadav et al., 2019). Seeds of this plant generally germinate in more-or-less three days if sown in suitable soil and under compatible weather conditions. Leaves are generally trifoliate (or three-part) in structure, petiolated, greyish-green, obovate to oblanceolate in shape with toothed margins and attaining a size of approx 20–50 × 8–17 mm (Yadav et al., 2019, Sharma et al., 2020a). Flowers are born axillary, sessile, whitish to lemon-yellow in colour and generally flourish in June-July. Flowers are bisexual in nature and pollination happens by insects. However, some investigators found fenugreek to be self-pollinated (Visuvinathan et al., 2022). Fruits are long or oblong in shape, 80–150 × 1–4 mm in size, ending in an elongated beak, with each seed containing 10–20 seeds (Sharma et al., 2020a). Seeds are brownish-yellow in colour, tiny, rhomboid shaped, hard in nature, with size 3–6 mm long, 2–5 mm wide, and 2 mm thick. Seeds are glutinous, fibre-containing, sticky, and endospermic in nature (Visuvinathan et al., 2022).

17.4 CULTIVATION

Fenugreek has an aggressive nature and can grow in various types of land, such as dry grassland, either cultivated or uncultivated land, on hill slopes, plain areas, and at edges of fields (Sharma et al., 1990; Pal & Mukherjee, 2020). It is a sun-loving plant and drought tolerant (Pal & Mukherjee, 2020). But, in tropical climates, having soft cool, humid summer weather enables this plant to grow better and the development of leaf and flower is completely temperature dependent (Smith 2003; Pal & Mukherjee, 2020). The seeds are generally sown by line-sowing (Sharma et al., 2020a). Before sowing, the ploughed land is irrigated alternately along the rows of sowing beds as needed (Sharma et al., 2020a). *Rhizobium* culture inoculation of the soaked seeds before sowing is also useful to achieve better growth of the plant. Seeds germinate within 6–8 days of sowing and harvesting of the vegetative part is carried out after 30–40 days and seed harvesting is done at 105–140 days (Sharma et al., 2020a). Seeds need to be stored in conditions below 12% humidity and cool, dry conditions (Nayak & Pal, 2016).

17.5 PHYTOCHEMISTRY

Phytochemicals found in fenugreek seeds include saponins, alkaloids, flavonoids, and phenolic acids. Many of fenugreek's health benefits and therapeutic properties are attributed to these secondary metabolite compounds. Saponins, specifically the steroidal saponin known as diosgenin, is one of the best-known and most extensively studied phytochemicals in fenugreek (Akbari et al., 2019; Arya et al., 2021). Fenugreek seeds also contain alkaloids like trigonelline, which has been shown to have anti-diabetic properties by improving insulin secretion and sensitivity (Mohamadi et al., 2021). Fenugreek also contains the alkaloids gentianine, carpaine, and choline (Tewari et al., 2020). Flavonoids found in fenugreek seeds include quercetin, kaempferol, and isorhamnetin (Mandegary et al., 2012). These compounds have antioxidant and anti-inflammatory properties, as well as anticancer activity. Fenugreek contains phenolic acids such as gallic acid, caffeic acid, and ferulic acid (Pajk et al., 2019). These compounds have antioxidant and anti-inflammatory properties, as well as anticancer activity (Mbarki et al., 2017). Overall, the phytochemicals found in fenugreek seeds have a wide range of potential health benefits, including improved insulin sensitivity, decreased inflammation, and possibly reduced cancer risk. More research is needed, however, to fully understand the mechanisms underlying these effects and to determine optimal therapeutic doses.

17.6 PHARMACOLOGICAL UTILITY OF FENUGREEK

In many ethnic cuisines, fenugreek is typically used as a spice or culinary addition. In addition to its nutritional importance, its varied medicinal effects are well-documented. Being a rich source of diverse phytoconstituents, it demonstrates numerous pharmacological activities.

17.6.1 Hypoglycaemic Activity

Diabetes is a metabolic ailment marked by elevated glucose levels, insulin resistance, and ineffective insulin production and activity (52Sharma et al., 2020b). The number of diabetic people is on the rise and is projected to reach 366 million by 2030. Many antidiabetic substances are reportedly found in fenugreek seeds (Talapatra & Talapatra, 2015).

Fenugreek has been shown to exert hypoglycaemic effects in both type 1 and type 2 diabetic people and animals. According to Raghuram et al. (1994), fenugreek improves peripheral glucose utilization and tolerance in non-insulin-dependent diabetics. Puri et al. (2011) investigated the antidiabetic effects of the water-soluble GII compound from *Trigonella* seeds on alloxan-induced diabetic rabbits. For 15 mornings, these rabbits received 50 mg/kg GII compound. The compound administration lowered total cholesterol (TC), high-density lipoprotein (HDL), low-density lipoprotein (LDL), glucose-6-phosphate, sorbitol dehydrogenase, and aldose reductase and increased liver and muscle glycogen, hexokinase, glucokinase, pyruvate kinase, malic enzyme, superoxide dismutase, and glutathione peroxidase, and normalized glucose levels. In individuals with moderate type 2 diabetes, long-term use of fenugreek seeds improves glycaemic control, according to clinical research. It is thought that the galactomannan-rich soluble fraction of fenugreek is responsible for its hypoglycaemic properties since the fibre slows stomach emptying and delays glucose absorption in the small intestine (Snehlata & Payal, 2012).

An amino acid isolated from fenugreek, 4-hydroxy isoleucine, is believed to have hypoglycaemic activity because treatment of diabetic rats at 50 mg/kg/day for 28 days restored plasma glucose and high lipid levels, including TC, HDL, LDL, and triglycerides (TG), to optimal levels (Haeri et al., 2012). Carpaine, an alkaloid found in fenugreek seeds, has a strong interaction with α-amylase (enzyme which converts complex starch to oligosaccharides) with a binding energy of –7.44 kcal/mol and possesses considerable α-amylase inhibitory activity, according to *in-silico* experiments. In addition, an *in-vitro* investigation was conducted to corroborate the *in-silico* findings, and it was discovered that the extract, at a concentration of 1 mg/mL had the maximum inhibitory action of 37.7% (Bansode et al., 2016).

Sharma et al., (2020b) showed antidiabetic activity of a *Trigonella* seed and cinnamon bark aqueous solution on alloxan-induced diabetic Wistar rats. Two weeks of supplementation with 1740 mg/kg fenugreek, 200 mg/kg cinnamon, or 0.6 mg/kg glibenclamide (a synthetic diabetes medication) reduced superoxide anion levels by 48%, 38%, and 53%, respectively. In addition, the fenugreek ethanolic extract enhanced antioxidant status and decreased DNA damage. Also, it was claimed that the fenugreek has the ability to block α-glucosidase (converts polysaccharides to glucose) enzymes, which are crucial in the development of diabetes. The pharmacokinetics of metformin, a first-line antidiabetic medicine, were recently compared with or without the simultaneous administration of fenugreek extract using a rat animal model. According to the data, co-administration of fenugreek with metformin increased metformin bioavailability and decreased drug distribution volume by 70% (Abdelwahab et al., 2021).

17.6.2 Antineoplastic Activity

Cancer is one of those illnesses with a significant death rate, and researchers are continually attempting to comprehend the disease's manifestation and to create novel medicine and therapeutic treatments. Using cancer models employing *in vitro* cancer cell lines and animal models, it has been revealed that fenugreek exhibits anticarcinogenic properties. (Shabbeer et al., 2009) reported that an ethanolic extract of fenugreek is preferentially cytotoxic to prostate cancer cell lines, breast cancer cell lines, and pancreatic cancer cell lines, but not to normal cell lines. The startling finding is that death of cancer cells occurs despite growth stimulatory pathways being concurrently upregulated (phosphorylated) by the extract. Similar research found that the fenugreek hydro-extract had a specific cytotoxic effect *in vitro* against an array of cancer cell lines, including T cell lymphoma, B cell lymphomas, thyroid papillary carcinoma (FRO), and human breast cancer (MCF7) (Alsemari et al., 2014). The effects of methanolic fenugreek extract were examined *in vitro* using breast cancer cell lines MCF-7 and SK-BR3, followed by an acute oral toxicity investigation in a Swiss albino mouse model. Both cancer cell lines responded to the fenugreek extract in a dose-dependent way with an anti-metastatic impact, an inhibition of cell migration, and induced apoptosis (Alrumaihi et al., 2021). The findings demonstrated an upregulation of p53, suggesting that the fenugreek effect is related to the signalling pathway that inhibits further DNA mutation and triggers cell death. According to the toxicological analysis, oral administration of fenugreek extract to mice had no harmful impact on the mice. Another *in-vivo* investigation employing albino mice found that treatment with an aqueous fenugreek extract boosted the mouse survival rate. Histopathological analysis revealed that the treated group had pancreatic tissue with fewer berry-sized lesions than the untreated group (Almalki & Naguib, 2022).

It is known that saponins specifically inhibit cell proliferation and promote programmed cell death (PCD) in malignant cells. Chemical analysis proved the seeds of fenugreek to be rich in sapogenin compounds (Arivalagan et al., 2013; Petit et al., 1995; Zhang et al., 2020). Jagadeesan et al. (2012) assessed the antitumour efficacy of diosgenin, a steroidal saponin derived from *Trigonella*, against *N*-methyl-*N*-nitrosourea (NMU)-induced breast cancer in female Sprague Dawley rats. They were given 20 mg diosgenin/kg of body weight each day for 45 days. Eventually, mice treated with diosgenin exhibited decreased lipid peroxidation and an enhanced antioxidant defence system.

The aforementioned research, on the other hand, shows the function of fenugreek as an anticancer agent at some level. But, in order to establish it as a clinical therapeutic antitumour agent, it is necessary to conduct more anticancer studies with identification of their respective mechanisms of action.

17.6.3 Anti-inflammatory Properties

Anti-inflammatory chemicals have the ability to suppress pro-inflammatory cytokines, which play a vital role in the development of inflammation and related illnesses. *Trigonella* secondary metabolites have powerful anti-inflammatory properties.

Sindhu et al. (2012) examined the anti-inflammatory efficacy of fenugreek seed mucilage on the actions of inflammatory enzymes such as cyclooxygenase (COX), lipoxygenase (LOX), myeloperoxidase, and levels of nitrite and C-reactive protein in rats suffering from arthritis. Fenugreek mucilage considerably suppressed the activities of COX-2 and myeloperoxidase enzymes and lowered the levels of thiobarbituric acid reactive substances (TBARS). Another group of researchers tested the anti-inflammatory effects of the ethanolic extract of fenugreek seed on rats with CFA-induced arthritis for 22 days. Rats given 200 or 400 mg/kg of extract per day exhibited substantially decreased paw oedema, interleukin (IL)-1, IL-1, IL-2, IL-6, tumour necrosis factors (TNFs), and lipid peroxidation levels, as well as elevated red blood cells (4.78 ± 0.46), hemoglobin (15.46 ± 0.158), superoxide dismutase, and reduced glutathione levels (Suresh et al., 2012). Sharififar et al. (2012) investigated the antinociceptive and anti-inflammatory effectiveness of methanolic extracts of many plants, such as *Trigonella*, *Zhumeria majdae*, *Achillea wilhelsii*, and *Viola tricolor*, using hot-plate and carrageenan-induced oedema tests. *Trigonella* extract (100 or 200 mg/kg) was the most beneficial for its antinociceptive and anti-inflammatory effectiveness, considerably reducing paw oedema and increasing pain tolerance compared with the effects of other plants examined. These studies established that several formulations derived from fenugreek were potent anti-inflammatory substances that can be used in cases of minor pain and related inflammation.

17.6.4 Antimicrobial Activity

To combat the growing problem of microbial drug resistance, there is an urgent need to discover or develop new therapeutic compounds with a wide range of antimicrobial properties (Parish, 2018). In addition to fungi, there are several plants having antimicrobial properties, including fenugreek. Dharajiya et al. (2016) investigated the antibacterial activity of fenugreek leaf extract with different solvents by determining its inhibition zone against the pathogenic microorganisms *Escherichia coli*, *Pseudomonas aeruginosa*, and *Bacillus cereus*. Among the several organic extracts, methanolic extract was found to be the most efficient against the microbes showing MIC values of 9.33, 9.50, and 11.50 mg/mL for *E. coli*, *P. aeruginosa*, and *B. cereus*, respectively. Using the well diffusion technique, Chalghoumi et al. (2016) and Sharma et al. (2017) evaluated the antibacterial activity of the fenugreek stem, seed, and leaf extract with various solvents. In a study, both methanolic and aqueous preparations of fenugreek seed were assessed against Gram-positive and Gram-negative bacteria. According to the findings, the methanolic extract showed significant antibacterial properties in a dose-dependent manner but the aqueous extract did not show any potential against the tested microbes (Al-Timimi, 2019). In relation to all of these investigations, the antimicrobial effectiveness and severity of effects vary based on the plant components and microbe species involved, as well as on the extraction medium.

17.6.5 Neuroprotective Activity

Neurological diseases, including neuropathic pain, Alzheimer's, and Parkinson's diseases, have become very prevalent these days, and empirical evidence suggests that inflammatory cytokines and microglial cells contribute to the genesis of neuropathic pain (Soleimani et al., 2019). Using animal models, researchers have identified the potential advantages of medicinal herbs for treating neurological disorders. In this context, fenugreek has also been examined as a possible medicinal plant for the intervention of neurological disorders. In a research study in which rats were given fenugreek saponins (0.05–2.0%) for 45 days, it was observed that the therapy suppressed apoptosis and acetylcholinesterase (AChE) activity, leading to neuroprotective effects (Khalil et al., 2016). Using a mouse model, Bin-Hafeez et al. (2003) investigated the neuroprotection activity of 5% fenugreek seed powder on aluminium chloride-induced neurotoxicity and concluded that fenugreek seed powder had a notable neuroprotective effect. *Trigonella* has also been shown to reduce the risk of Parkinson's disease by suppressing the rotating action of SNc (substantia nigra pars compacta) and restoring MDA (malondialdehyde) levels from the elevated levels associated with response to oxidative stress (Foltynie & Kahan, 2013).

17.6.6 Lipid Metabolism

A disorder of carbohydrate and lipid metabolism, obesity is characterized by an abnormally high amount of fat buildup in adipose tissue as well as in other internal organs. There is a correlation between being overweight or obese and insulin-resistant type 2 diabetes, cancer, coronary heart disease, and respiratory problems (McCarthy, 2010). Throughout the years, several medications have been tried to treat obesity, but the bulk of them have been discontinued owing to dangerous side effects. Several studies have revealed that fenugreek contains anti-obesity properties, making it a viable plant-based alternative for treating obesity. Fenugreek has a high amount of soluble fibre, which accelerates weight loss by boosting digestion and metabolism. Galactomannan, a water-soluble fibre found in fenugreek seeds, lowers appetite by enhancing feelings of fullness, hence promoting weight reduction. Overall, it improves glucose and lipid metabolism, insulin sensitivity, antioxidant defence, and downregulation of lipogenic enzymes (Kumar et al., 2014). Furthermore, diosgenin from fenugreek is known to inhibit cholesterol synthesis. Saponins were first thought to ameliorate dyslipidaemia by speeding up cholesterol metabolism and reversing cholesterol transport, in addition to inhibiting 3-hydroxy-3-methylglutaryl coenzyme A reductase in serum and liver (Achari & Jain, 2017).

17.6.7 Hormonal Impact

Many diseases, including infertility, diabetes, dysmenorrhoea, menopausal osteoporosis, and lactation impairment, are associated with hormonal imbalance. Hormone replacement therapy has become more popular as a treatment for hormone-related disorders. Yet, decades of study have shown the negative long-term implications of this method. Natural chemicals, particularly phytosteroids, are emerging as an effective, side effect-free alternative to manufactured synthetic medication.

The effectiveness of fenugreek in treating primary dysmenorrhoea has been the subject of several investigations. Primary dysmenorrhoea is characterized by spasmodic stomach pain during menstruation in the absence of any underlying pelvic ailment (Proctor & Farquhar, 2006). In a double-blind, randomized, placebo-controlled mice model investigation Younesy et al. (2014) examined the effects of fenugreek seed extracts on the severity of primary dysmenorrhoea among unmarried students. The fenugreek group saw a much greater reduction in pain, while the duration of discomfort decreased between the two consecutive cycles. Another related clinical research study was conducted to assess the effectiveness and safety of fenugreek seed in reducing pain intensity in individuals with primary dysmenorrhoea. The test group that took oral fenugreek supplements saw a 66.89% decrease in lower abdomen discomfort (Inanmdar et al., 2016).

Menopausal osteoporosis is an additional hormone-related condition in which oestrogen insufficiency is directly associated with bone resorption (Marcus, 2002). Hormone replacement therapy is the only successful treatment for this illness at present (Gosset et al., 2021). A study evaluated the effect of oral fenugreek seed hydro-ethanolic extract on the skeletal structure and mechanical qualities of ovari-ectomized Wistar rats. According to the findings of this investigation, supplementation with fenugreek seed extract increased the maximum flexor force necessary to fracture the femur. In addition, the treated group of ovari-ectomized rats revealed improvements in tibial dry weight, bone microstructure, and prevention of trabecular bone loss. Overall, it is believed that fenugreek's steroidal phytoestrogens may be responsible for its positive effects on bone mechanics and strength (Anjaneyulu et al., 2019).

17.7 TRADITIONAL AND OTHER USES

One of the ancient traditional systems of medicine, the Ayurveda, claimed fenugreek to be an important herb to treat different types of digestive and mucosal problems (Yadav & Baquer, 2014). In Ayurvedic scripts, the potentiality of this plant as an aphrodisiac has been recorded (Wani et al., 2016). In addition to Indian Ayurveda, the medicinal properties of fenugreek seed had also been mentioned in Greek and Latin pharmacopoeia (Wani et al., 2016). Traditional Persian medicine, originating from Iran, included this plant as the oldest medicinal herb, whereas, in the Chinese medicinal system, it was introduced during the Sung Dynasty in 1057 AD (Sun et al., 2021). The well-known Greek physician, Dioscorides, popular as the 'Father of Pharmacology', mentioned fenugreek as an active component of ointments and a spice for different culinary practices (Sun et al., 2021). In most of the traditional practices, it was long being utilized to treat common cold, coughs, bronchitis, sore throats, arthritis, and menstrual pain, and as an intestinal digestive stimulant (Kumar et al., 2021). Because of having properties like carminative, demulcent, expectorant, and laxative activities, Fenugreek seeds have been useful since ancient times (Yadav and Baquer 2014).

Local application of poultices made from the fenugreek plant can relieve swelling and a paste of leaves can be applied over burns or on the scalp to prevent premature hair-greying (Williamson, 2002; Arunabha & Bhattacharjee, 2019). Different ethno-medicinal studies have shown the utilization of fenugreek in different countries in various ways (Arunabha & Bhattacharjee, 2019). In India, the leaves are eaten as a vegetable and the seeds are added as condiments in food items like curries and chutneys (Arunabha & Bhattacharjee, 2019; Sun et al., 2021). Also, the roasted seeds are used as being complementary to coffee in India (Sun et al., 2021). The use of fenugreek seed flour, in place of wheat and maize flour, is still in practice for bread-making in Egypt (Sun et al., 2021). In ancient Rome, this plant was used to aid child-delivery and in Iran, as well as in India, it used to be applied to new mothers as a lactation stimulant (Arunabha & Bhattacharjee, 2019; Sun et al., 2021). The traditional Chinese medicinal system employs fenugreek seed as a syrup, and a treatment for instability, weakness and oedema of legs (Arunabha & Bhattacharjee, 2019). It has also been reported that, in ancient Egypt, it was used to preserve dead bodies for mummification (Sun et al., 2021).

17.8 CONCLUSION AND FUTURE REMARKS

Fenugreek is a plant that has been used for centuries in traditional medicine and as a culinary spice. Its seeds and leaves contain phytochemicals such as saponins, alkaloids, flavonoids, and phenolic acids, which have been shown to have potential health benefits such as improving insulin sensitivity, reducing inflammation, and possibly lowering the risk of cancer.

Ultimately, fenugreek is a promising plant with many potential health benefits. However, more research is required to fully understand the mechanisms underlying these effects and to determine optimal therapeutic doses. Furthermore, research is needed to investigate the safety and efficacy of fenugreek in different populations, as well as to assess its potential use in conjunction with other natural or conventional therapies. Although some progress has been made, more research is needed to fully understand the molecular mechanisms of action of fenugreek extracts and isolated individual compounds. Researchers were able to identify and characterize numerous novel phytochemicals derived from the fenugreek plant, using advanced analytical techniques. Future research directions for fenugreek may include looking into its effects on gut microbiota as well as its potential use to improve gut health via regulating intestinal microbiota composition and metabolites, inhibiting inflammatory response, and improving body immunity.

While some progress has been made, more research is needed to fully understand the molecular modes of action of fenugreek products and isolated compounds. Researchers were able to identify and characterize numerous novel phytochemicals derived from the fenugreek plant, using advanced analytical techniques. Chronic diseases such as type 2 diabetes, cardiovascular disease, and cancer are diagnosed and treated. Overall, fenugreek has a lot of potential as a natural remedy and nutritional supplement, and its use in traditional medicine and culinary practices may pave the way for more research into its health benefits.

REFERENCES

Abdelwahab, N. S., Morsi, A., Ahmed, Y. M., Hassan, H. M., & Aboulmagd, A. M. (2021). Ecological HPLC method for analyzing an antidiabetic drug in real rat plasma samples and studying the effects of concurrently administered fenugreek extract on its pharmacokinetics. *RSC Advances*, *11*(8), 4740–4750. https://doi.org/10.1039/D0RA08836F

Achari, A. E., & Jain, S. K. (2017). Adiponectin, a therapeutic target for obesity, diabetes, and endothelial dysfunction. *International Journal of Molecular Sciences*, *18*(6), 1321. https://doi.org/10.3390/IJMS18061321

Akbari, S., Abdurahman, N. H., & Yunus, R. M. (2019 Nov 1). Optimization of saponins, phenolics, and antioxidants extracted from fenugreek seeds using microwave-assisted extraction and response surface methodology as an optimizing tool. *Comptes Rendus Chimie*, *22*(11–12), 714–727.

Almalki, D. A., & Naguib, D. M. (2022). Anticancer activity of aqueous fenugreek seed extract against pancreatic cancer, histological evidence. *Journal of Gastrointestinal Cancer*, *53*(3), 683–686. https://doi.org/10.1007/S12029-021-00687-X/FIGURES/2

Alrumaihi, F. A., Khan, M. A., Allemailem, K. S., Alsahli, M. A., Almatroudi, A., Younus, H., Alsuhaibani, S. A., Algahtani, M., & Khan, A. (2021). Methanolic fenugreek seed extract induces p53-dependent mitotic catastrophe in breast cancer cells, leading to apoptosis. *Journal of Inflammation Research*, *14*, 1511. https://doi.org/10.2147/JIR.S300025

Alsemari, A., Alkhodairy, F., Aldakan, A., Al-Mohanna, M., Bahoush, E., Shinwari, Z., & Alaiya, A. (2014). The selective cytotoxic anti-cancer properties and proteomic analysis of *Trigonella foenum-graecum*. *BMC Complementary & Alternative Medicine*, *14*(1), 1–9. https://doi.org/10.1186/1472-6882-14-114/FIGURES/6

Al-Timimi, L. A. N. (2019). Antibacterial and Anticancer Activities of Fenugreek Seed Extract. *Asian Pacific journal of cancer prevention: APJCP*, 20(12), 3771–3776. https://doi.org/10.31557/APJCP.2019.20.12.3771

Anjaneyulu, K., Bhat, K. M., Srinivasa, S. R., Devkar, R. A., & Henry, T. (2019). Beneficial role of hydro-alcoholic seed extract of *Trigonella foenum-graecum* on bone structure and strength in menopause Induced osteopenia. *Ethiopian Journal of Health Sciences*, *28*(6), 787–794. https://doi.org/10.4314/ejhs.v28i6.14

Arivalagan, M., Gangopadhyay, K. K., & Kumar, G. (2013). Determination of steroidal saponins and fixed oil content in fenugreek (*Trigonella foenum-graecum*) genotypes. *Indian Journal of Pharmaceutical Sciences*, *75*(1), 110. https://doi.org/10.4103/0250-474X.113542

Arya, P., & Kumar, P. (2021 Jun 1). Comparison of ultrasound and microwave-assisted diosgenin extraction from *Trigonella foenum-graceum* seed. *Ultrasonics Sonochemistry*, *74*, 105572.

Arunabha, M., & Bhattacharjee, C. (2019). *Trigonella foenum*-graecum: A review on its traditional uses, phytochemistry and pharmacology. *International Journal of Advances in Scientific Research*, *5*(5), e5217. https://doi.org/10.7439/ijasr.v5i5.5217

Bansode, T. S., Gupta, A., Chaphalkar, S., Salalkar, B. K., Fawzy, M., & Hassanien, R. (2016). Integrating *in-silico* and *in-vitro* approaches to screen the antidiabetic drug from *Trigonella foenum- graecum* Linn. *International Journal of Biochemistry Research & Review*, *14*(3), 1–10. https://doi.org/10.9734/IJBCRR/2016/29875

Bin-Hafeez, B., Haque, R., Parvez, S., Pandey, S., Sayeed, I., & Raisuddin, S. (2003). Immunomodulatory effects of fenugreek (*Trigonella foenum-graecum* L.) extract in mice. *International Immunopharmacology*, *3*(2), 257–265. https://doi.org/10.1016/S1567-5769(02)00292-8

Chalghoumi, R., Mabrouki, S., Abdouli, H., & Line, J. E. (2016). Antibacterial activity of fenugreek seeds (*Trigonella foenum-graecum*) crude extracts against a rabbit Escherichia coli isolate. *Academia Journal of Microbiology Research*, *4*(11), 139–144.

Dharajiya, D., Jasani, H., Khatrani, T., Kapuria, M., Pachchigar, K., & Patel, P. (2016). Evaluation of antibacterial and antifungal activity of fenugreek (*Trigonella* foenum-graecum) extracts. *International Journal of Pharmacy & Pharmaceutical Science*, *8*(4), 212–217.

Foltynie, T., & Kahan, J. (2013). Parkinson's disease: An update on pathogenesis and treatment. *Journal of Neurology*, *260*(5), 1433–1440. https://doi.org/10.1007/S00415-013-6915-1/FIGURES/1

Gosset, A., Pouillès, J. M., & Trémollieres, F. (2021). Menopausal hormone therapy for the management of osteoporosis. *Best Practice & Research in Clinical Endocrinology & Metabolism*, *35*(6), 101551. https://doi.org/10.1016/J.BEEM.2021.101551

Goyal, S., Gupta, N., & Chatterjee, S. (2016). Investigating therapeutic potential of *Trigonella foenum-graecum* L. as our defense mechanism against several human diseases. *Journal of Toxicology*. http://doi.org/10.1155/2016/1250387

Haeri, M. R., Limaki, H. K., White, C. J. B., & White, K. N. (2012). Non-insulin dependent anti-diabetic activity of (2S, 3R, 4S) 4-hydroxyisoleucine of fenugreek (*Trigonella foenum-graecum*) in streptozotocin-induced type I diabetic rats. *Phytomedicine*, *19*(7), 571–574. https://doi.org/10.1016/J.PHYMED.2012.01.004

Helambe, S. S., & Dande, R. P. (2012). Fenugreek (*Trigonella foenum-graceum* L.): An overview. *International Journal of Current Pharmaceutical Review & Research*, *2*(4), 169–187.

Husen, A. (2021). *Traditional Herbal Therapy for the Human Immune System*. CRC Press, Boca Raton, FL. https://doi.org/10.1201/9781003137955

Husen, A. (2022a). *Herbs, Shrubs and Trees of Potential Medicinal Benefits*. CRC Press, Boca Raton, FL. https://doi.org/10.1201/9781003205067

Husen, A. (2022b). *Exploring Poisonous Plants: Medicinal Values, Toxicity Response and Other Potential Uses*. CRC Press, Boca Raton, FL. https://doi.org/10.1201/b23017

Husen, A. (2023). *Antidiabetic Medicinal Plants and Herbal Treatments*. CRC Press, Boca Raton, FL. https://doi.org/10.1201/b23347

Inanmdar, W., Sultana, A., Mubeen, U., & Rahman, K. (2016). Clinical efficacy of *Trigonella foenum-graecum* (fenugreek) and dry cupping therapy on intensity of pain in patients with primary dysmenorrhea. *Chinese Journal of Integrative Medicine*, 1–8. https://doi.org/10.1007/S11655-016-2259-X/METRICS

Jagadeesan, J., Nandakumar, N., Rengarajan, T., & Balasubramanian, M. P. (2012). Diosgenin, a steroidal saponin, exhibits anticancer activity by attenuating lipid peroxidation via enhancing antioxidant defense system during NMU-induced breast carcinoma. *Journal of Environmental Pathology Toxicology & Oncology*, *31*(2), 121–129. https://doi.org/10.1615/JENVIRONPATHOLTOXICOLONCOL.V31.I2.40

Khalil, W. K. B., Roshdy, H. M., & Kassem, S. M. (2016). The potential therapeutic role of fenugreek saponin against Alzheimer's disease: Evaluation of apoptotic and acetylcholinesterase inhibitory activities. *Journal of Applied Pharmaceutical Science*, *6*(09), 166–173. https://doi.org/10.7324/JAPS.2016.60925

Kumar, N., Kumar, M., Verma, M. K., Ramanarayanan, S., Ranjan, A., & Ranjan, R. (2021). Bioactive effects and safety profiles of fenugreek (*Trigonella foenum-graecum* L.) for pharmaceutical and medicinal applications. *The Pharma Innovation Journal*, *10*(12), 912–919.

Kumar, P., Bhandari, U., & Jamadagni, S. (2014). Fenugreek seed extract inhibit fat accumulation and ameliorates dyslipidemia in high fat diet-induced obese rats. *BioMed Research International*, *2014*. https://doi.org/10.1155/2014/606021

Mandegary, A., Pournamdari, M., Sharififar, F., Pournourmohammadi, S., Fardiar, R., & Shooli, S. (2012 Jul 1). Alkaloid and flavonoid rich fractions of fenugreek seeds (*Trigonella foenum-graecum* L.) with antinociceptive and anti-inflammatory effects. *Food & Chemical Toxicology*, *50*(7), 2503–2507.

Marcus, R. (2002). Post-menopausal osteoporosis. *Best Practice & Research: Clinical Obstetrics & Gynaecology*, *16*(3), 309–327. https://doi.org/10.1053/BEOG.2002.0284

Mbarki, S., Alimi, H., Bouzenna, H., Elfeki, A., & Hfaiedh, N. (2017 Apr 1). Phytochemical study and protective effect of *Trigonella foenum-graecum* (fenugreek seeds) against carbon tetrachloride-induced toxicity in liver and kidney of male rat. *Biomedicine & Pharmacotherapy*, *88*, 19–26.

McCarthy, M. I. (2010). Genomics, Type 2 diabetes, and obesity. *New England Journal of Medicine*, *363*(24), 2339–2350. https://doi.org/10.1056/NEJMRA0906948

Mohamadi, N., Sharififar, F., Ansari, M., Pournamdari, M., Rezaei, M., & Hassanabadi, N. (2021 May 4). Pharmacokinetic profile of diosgenin and trigonelline following intravenous and oral administration of fenugreek seed extract and pure compound in rabbit. *Journal of Asian Natural Products Research, 23*(5), 466–477.

Munshi, M., Arya, P., & Kumar, P. (2020). Physico-chemical analysis and fatty acid profiling of fenugreek (*Trigonella foenum-graecum*) seed oil using different solvents. *Journal of Oleo Science, 69*(11), 1349–1358. https://doi.org/10.5650/jos.ess20137

Nayak, A. K., & Pal, D. (2016). Sterculia gum-based hydrogels for drug delivery applications. In: *Polymeric Hydrogels as Smart Biomaterials* (Ed. Kalia, S.). Springer, Cham, 105–151.

Oufquir, S., Laaradia, M. A., Gabbas, Z. E., Bezza, K., Laadraoui, J., Aboufatima, R., Sokar, Z., & Chait, A. (2020). *Trigonella foenum-graecum* L. sprouted seed extract: Its chemical HPLC analysis, abortive effect, and neurodevelopmental toxicity on mice. *Hindawi Evidence-Based Complementary and Alternative Medicine.* https://doi.org/10.1155/2020/161579

Pal, D., & Mukherjee, S. (2020). Fenugreek (*Trigonella foenum-graecum*) seeds in health and nutrition. In: *Nuts and Seeds in Health and Disease Prevention*, Second Edition. https://doi.org/10.1016/B978-0-12-818553-7.00013-9

Pająk, P., Socha, R., Broniek, J., Królikowska, K., & Fortuna, T. (2019 Mar 1). Antioxidant properties, phenolic and mineral composition of germinated chia, golden flax, evening primrose, phacelia and fenugreek. *Food Chemistry, 275*, 69–76.

Parish, T. (2018). Steps to address anti-microbial drug resistance in today's drug discovery. *Expert Opinion on Drug Discovery, 14*(2), 91–94. https://doi.org/10.1080/17460441.2019.1550481

Petit, P. R., Sauvaire, Y. D., Hillaire-Buys, D. M., Leconte, O. M., Baissac, Y. G., Ponsin, G. R., & Ribes, G. R. (1995). Steroid saponins from fenugreek seeds: Extraction, purification, and pharmacological investigation on feeding behavior and plasma cholesterol. *Steroids, 60*(10), 674–680. https://doi.org/10.1016/0039-128X(95)00090-D

Proctor, M., & Farquhar, C. (2006). Diagnosis and management of dysmenorrhoea. *BMJ, 332*(7550), 1134–1138. https://doi.org/10.1136/BMJ.332.7550.1134

Puri, D., Prabhu, K. M., Dev, G., Agarwal, S., & Murthy, P. S. (2011). Mechanism of antidiabetic action of compound GII purified from fenugreek (*Trigonella foenum-graecum*) seeds. *Indian Journal of Clinical Biochemistry, 26*(4), 335–346. https://doi.org/10.1007/S12291-011-0150-2/FIGURES/3

Rahman, S., & Husen, A. (2022). *Asparagus racemosus* (Shatavari) and *Dioscorea villosa* (Wild yams). In: *Herbs, Shrubs, and Trees of Potential Medicinal Benefits* (Ed. Husen, A.). CRC Press, Boca Raton, FL.

Rahman, S., & Husen, A. (2021). Potential role of medicinal plants in the cure of liver and kidney diseases. In: *Non-timber Forest Products* (Eds. Husen, A., Bachheti, R.K., & Bachheti, A.), 229–254. Springer International Publishing AG, Cham. https://doi.org/10.1007/978-3-030-73077-2_10

Rahman, S., & Husen, A. (2023a). *Trigonella foenum-graecum* L.: A Useful Medicinal Plant for Diabetes and Other Diseases. In: *Antidiabetic Medicinal Plants and Herbal Treatments* (Ed. Husen, A.). CRC Press, Boca Raton, FL.

Rahman, S., & Husen, A. (2023b). Role of green chiretta [*Andaphis paniculata* (Burm. F.)] Nees in diabetes and other ailments. In: *Antidiabetic Medicinal Plants and Herbal Treatments* (Ed. Husen, A.). CRC Press, Boca Raton, FL.

Raghuram, T. C., Sharma, R. D., Sivakumar, B., & Sahay, B. K. (1994). Effect of fenugreek seeds on intravenous glucose disposition in non-insulin dependent diabetic patients. *Phytotherapy Research, 8*(2), 83–86. https://doi.org/10.1002/PTR.2650080206

Rasheed, S. A. A., Wankhade, V., Saifuddin, S. S. K., & Sudarshan, A. R. (2015). Physico chemical properties of fenugreek (*Trigonella foenum-graceum* L.) seeds. *International Journal of Engineering Research & Technology (IJERT), 4*(10), 68–70.

Salam, S.G.A., Rashed, M.M., Ibrahim, N.A., Rahim, E.A.A., Aly, T.A.A., & Al-Farga, A. (2023). Phytochemical screening and in-vitro biological properties of unprocessed and household processed fenugreek (Trigonella foenum-graecum Linn.) seeds and leaves. *Sci Rep.* 13(1), 7032. doi: 10.1038/s41598-023-31888-y. PMID: 37120447; PMCID: PMC10148852.

Shabbeer, S., Sobolewski, M., Anchoori, R. K., Kachhap, S., Davidson, N., Carducci, M. A., & Khan, S. R. (2009). Fenugreek: A naturally occurring edible spice as an anticancer agent. *Cancer Biology & Therapy, 8*(3), 272–278. https://doi.org/10.4161/CBT.8.3.7443

Sharififar, F., Khazaeli, P., Alli, N., Talebian, E., Zarehshahi, R., & Amiri, S. (2012). Study of antinociceptive and anti-inflammatory activities of certain Iranian medicinal plants. *Journal of Complementary Medicine Research, 1*(1), 19–24. https://doi.org/10.5455/JICE.20120227104636

Sharma, M. K., Sharma, P. K., & Sharma, J. (2020a). Phytochemical and pharmacological activity of *Trigonella foenum-graecum*: A comprehensive review. *European Journal of Molecular & Clinical Medicine*, *7*(3), 907–916.

Sharma, R. D., Raghuram, T. C., & Rao, N. S. (1990). Effect of fenugreek seeds on blood glucose and serum lipids in type I diabetes. *European Journal of Clinical Nutrition*, *44*(4), 301–306.

Sharma, S., Mishra, V., & Srivastava, N. (2020b). Protective effect of *Trigonella foenum-graecum* and *Cinnamomum zeylanicum* against diabetes induced oxidative DNA damage in rats. *Indian Journal of Biochemistry & Biophysics (IJBB)*, *57*(1), 15–26.

Sharma, V., Singh, P., & Rani, A. (2017). Antimicrobial activity of *Trigonella foenum-graecum* L. (Fenugreek). *European Journal of Experimental Biology*, *7*(1:4), 4. https://doi.org/10.21767/2248-9215.100004

Sindhu, G., Ratheesh, M., Shyni, G. L., Nambisan, B., & Helen, A. (2012). Anti-inflammatory and antioxidative effects of mucilage of *Trigonella foenum-graecum* (fenugreek) on adjuvant induced arthritic rats. *International Immunopharmacology*, *12*(1), 205–211. https://doi.org/10.1016/J.INTIMP.2011.11.012

Smith, M. (2003). Therapeutic applications of fenugreek. *Alternative Medicine Review*, *8*(1), 20–27.

Snehlata, H. S., & Payal, D. R. (2012). Fenugreek (*Trigonella foenum-graecum* L.): An Overview. *International Journal of Current Pharmaceutical Review & Research*, *2*(4), 169–187.

Soleimani, M., Sheikholeslami, M. A., Ghafghazi, S., Pouriran, R., & Parvardeh, S. (2019). Analgesic effect of α-terpineol on neuropathic pain induced by chronic constriction injury in rat sciatic nerve: Involvement of spinal microglial cells and inflammatory cytokines. *Iranian Journal of Basic Medical Sciences*, *22*(12), 1445. https://doi.org/10.22038/IJBMS.2019.14028

Sun, W., Shahrajabian, M. H., & Cheng, Q. (2021). Fenugreek cultivation with emphasis on historical aspects and its uses in traditional medicine and modern pharmaceutical science. *Mini-Reviews in Medicinal Chemistry*, 724–730. https://doi.org/10.2174/1389557520666201127104907

Suresh, P., Kavitha, C. N., Babu, S. M., Reddy, V. P., & Latha, A. K. (2012). Effect of ethanol extract of *Trigonella foenum-graecum* (fenugreek) seeds on Freund's adjuvant-induced arthritis in albino rats. *Inflammation*, *35*(4), 1314–1321. https://doi.org/10.1007/S10753-012-9444-7/FIGURES/2

Syed, Q. A., Rashid, Z., Ahmad, M. H., Shukat, R., Ishaq, A., Muhammad, N., & Rahman, H. U. U. (2020). Nutritional and therapeutic properties of fenugreek (*Trigonella foenum-graecum*): A review. *International Journal of Food Properties*. https://doi.org/10.1080/10942912.2020.1825482

Talapatra, S. K., & Talapatra, B. (2015). Chemistry of plant natural products: Stereochemistry, conformation, synthesis, biology, and medicine. In: *Chemistry of Plant Natural Products: Stereochemistry, Conformation, Synthesis, Biology, and Medicine*. Springer, Berlin Heidelberg. https://doi.org/10.1007/978-3-642-45410-3/COVER

Tewari, D., Jóźwik, A., Łysek-Gładysińska, M., Grzybek, W., Adamus-Białek, W., Bicki, J., Strzałkowska, N., Kamińska, A., Horbańczuk, O. K., & Atanasov, A. G. (2020 Aug 24). *Fenugreek (Trigonella foenum-graecum* L.) seeds dietary supplementation regulates liver antioxidant defense systems in aging mice. *Nutrients*, *12*(9), 2552.

Visuvanathan, T., Than, L. T. L., Stanslas, J., Chew, S. Y., & Vellasamy, S. (2022). Revisiting *Trigonella foenum-graecum* L.: Pharmacology and Therapeutic Potentialities. *Plants*, *11*(11). https://doi.org/10.3390/plants11111450

Wani, S. A., & Kumar, P. (2016). Fenugreek: A review on its nutraceutical properties and utilization in various food products. *Journal of the Saudi Society of Agricultural Sciences*, *22*(3), 97–106. http://doi.org/10.1016/j.jssas.2016.01.007

Williamson, E. M. (2002). *Major Herbs of Ayurveda*. Elsevier Science, Churchill Livingstone, UK, 315–320.

Yadav, S. R., Biyani, D. M., & Umekar, M. J. (2019). *Trigonella foenum-graecum*: A herbal plant review. *World Journal of Pharmaceutical Research*, *8*(12), 402–419. https://doi.org/10.20959/wjpr201912-16063

Yadav, U. C. S., & Baquer, N. Z. (2014). Pharmacological effects of *Trigonella foenum-graecum* L. in health and disease. *Pharmaceutical Biology*, *52*(2), 243–254. https://doi.org/10.3109/13880209.2013.826247

Younesy, S., Amiraliakbari, S., Esmaeili, S., Alavimajd, H., & Nouraei, S. (2014). Effects of fenugreek seed on the severity and systemic symptoms of dysmenorrhea. *Journal of Reproduction & Infertility*, *15*(1), 41. /pmc/articles/PMC3955423

Zhang, H., Xu, J., Wang, M., Xia, X., Dai, R., & Zhao, Y. (2020). Steroidal saponins and sapogenins from fenugreek and their inhibitory activity against α-glucosidase. *Steroids*, *161*, 108690. https://doi.org/10.1016/J.STEROIDS.2020.108690

18 Garlic Bulb (*Allium sativum*)

Kawtar El Kadri, Nadia Dakka, Abdelaali Balahbib, and Abdelhakim Bouyahya

18.1 INTRODUCTION

Garlic (*Allium sativum*) stands out as a highly bioactive natural source of phytochemicals. An array of pharmacological attributes has been associated with its consumption, including antidiabetic, anti-inflammatory, antioxidant, and anticarcinogenic activities (Ramirez et al. 2021). The relationship between the signalling pathways and garlic derivatives may allow a better understanding of how the pharmacological actions of garlic occur, contributing to its hypoglycaemic effect, the modulation of inflammation, its antioxidant activity, and the inhibition of different stages of cancer, including initiation, promotion, and progression. The main focus of the current chapter will be on the major bioactive compounds of garlic, their potential pharmaceutical properties, and their mechanisms of action.

The phytochemical analysis of the garlic bulb aqueous extract showed the presence of high levels of steroids and protein, moderate levels of tannins, terpenoids, and saponins, and low levels of reducing sugar and phenols (Bar et al. 2022). Garlic contains at least 33 sulphur-containing compounds, which are responsible for garlic's pungent aroma and its many health benefits. In addition, garlic contains a variety of minerals, including germanium, potassium, copper, selenium, iron, calcium, magnesium, and zinc, as well as vitamins A, B_1, and C, fibre, and water. Garlic also contains 17 amino acids, namely lysine, aspartic acid, histidine, arginine, threonine, cystine, glycine, glutamine, alanine, proline, valine, leucine, methionine, phenylalanine, isoleucine, tryptophan, and serine; these amino acids are essential for the proper functioning of the body and play an important role in various physiological processes. Garlic also contains enzymes such as peroxidase, alliinase, and tyrosinase (Choudhary et al. 2022).

Garlic's wide range of therapeutic properties are attributed to its active chemical constituents, which are represented by three main phytochemical families: organosulphur compounds, fructans, and phenolic compounds (Ramirez et al. 2021). Sulphur-containing compounds are the primary bioactives of garlic and its preparations, and include allicin, ajoene, 2-vinyldithiins, diallyl sulphide (DAS), diallyl trisulphide (DATS), and diallyl disulphide (DADS) (Subroto et al. 2021). Alliin is the predominant organosulphur compound, which is converted to allicin by the enzyme alliinase. Ajoene is a secondary compound that is formed after alliin transformation (Ţigu et al. 2021).

These bioactive sulphur compounds have been the focus of numerous studies due to their antitumour effects; DATS, for instance, can promote apoptosis by stimulating cytotoxicity via the ROS-dependent caspase pathway, which targets reactive oxygen species (ROS). Additionally, sulphur compounds have been shown to upregulate expression of antioxidant genes, promote Nrf2-ARE activation, and act as potent inhibitors of the inflammatory process (Ţigu et al. 2021).

Garlic also contains flavonoids that are polyphenolic compounds, which are potential antioxidant agents, such as kaempferol, myricetin, and quercetin (Ramirez et al. 2021). Flavonoids and saponins found in garlic have been investigated for their cytotoxic, antiproliferative, pro-apoptotic, anti-angiogenic, and antimetastatic properties (Ţigu et al. 2021). However, despite their potential, the investigation of these compounds has been ignored or only partially explored. Raw garlic contains higher levels of myricetin than of kaempferol and quercetin (Ramirez et al. 2021). Flavonoids can protect against free radical-induced damage through multiple mechanisms, one of which involves

 DOI: 10.1201/9781003387046-18

their direct scavenging of free radicals. By reacting with the reactive compounds of the radicals, flavonoids stabilize reactive oxygen species. The hydroxyl group in flavonoids is highly reactive and can react with radicals to render them inactive (Khan et al. 2022).

18.2 PHARMACOLOGY OF *ALLIUM SATIVUM*

This chapter provides a summary of garlic's main derivatives and their pharmacological activities, with the aim of discussing the relevant mechanisms of action, including antidiabetic, antioxidant, anti-inflammatory, and anticarcinogenic activities, among others. Several publications have highlighted features of garlic as a rich source of bioactive sulphur-containing and phenolic compounds that demonstrated promising potential aimed at preventing and managing certain diseases, offering a wide range of potential health benefits and therapeutic applications in diabetes, inflammation, oxidative stress, and cancer.

18.2.1 Antidiabetic Activity

Diabetes affects a significant portion of the world's population and is one of the fastest-growing metabolic diseases worldwide, being a chronic disorder of carbohydrate, fat and protein metabolism, characterized by defects in insulin secretion, insulin action, or both (Karimi and Nasab 2014). Garlic has been shown to exhibit hypoglycaemic actions in diabetic patients through a significant reduction of glucose levels in plasma. Furthermore, garlic and its extracts have been demonstrated to be effective in reducing insulin resistance (Padiya and Banerjee 2013). Another study has demonstrated that garlic can help with insulin resistance management (Tesfaye 2021).

Phytochemicals found in garlic extract have been shown to inhibit therapeutic targets for diabetes mellitus, such as α-glucosidase and dipeptidyl peptidase-4 (DPP-4) (Kalhotra et al. 2020). According to a study by Faroughi et al. (2018), compounds such as allyl propyl disulphide, allicin, cysteine sulphoxide, and *S*-allyl cysteine sulph oxide can lower blood glucose levels by several mechanisms, such as preventing the liver from activating insulin, enhancing the secretion of insulin from pancreatic beta cells, isolating insulin from bonded forms, and increasing cell sensitivity to insulin in borderline gestational diabetes mellitus (Faroughi et al. 2018). As a result, there is growing evidence supporting the antidiabetic properties of garlic, adding to which inflammation and oxidative stress are known to play a role in the development of diabetes and its complications (Khan et al. 2022), so that garlic's anti-inflammatory and antioxidant properties can be said to contribute to its antidiabetic potential.

18.2.2 Anti-inflammatory Activity

Inflammation is a natural defence mechanism in the human body that arises in response to harmful stimuli. It is a vital component of the body's innate defence system and is activated when endogenous signalling molecules or foreign pathogens interact directly or indirectly with various types of membrane receptors. Numerous studies have reported that garlic and its related phytochemicals possess anti-inflammatory activity ((El-Saber Batiha et al. 2020; Sasi et al. 2021)).

Garlic has been found to inhibit inflammation by targeting inflammatory mediators such as nitric oxide (NO), tumour necrosis factor-alpha (TNF-α), and interleukin-1 (IL-1.One study demonstrated that ethyl linoleate in garlic could reduce the production of NO and prostaglandin E-2 by downregulating the expression of inducible NO synthase (iNOS) and cyclooxygenase-2 (COX2) in lipopolysaccharide-stimulated RAW 264.7 macrophages (Shang et al. 2019). In another study, it was reported that the 14-kDa garlic protein can inhibit the inflammatory mediators NO, TNF-α, and interleukin IL-1β by interfering with the signalling pathway of the transcription factor nuclear factor-kappa B (NF-κB) in lipopolysaccharide-stimulated J774A.1 macrophages (Shang et al. 2019). NF-κB is an important target molecule of organosulphur compounds from garlic and also functions as a transcription factor that regulates expression of genes responsible for the inflammatory response (Sasi et al. 2021). In a study conducted by Abdel-Daim et al. (2018) (as cited in El-Saber Batiha et al. 2020), DAS demonstrated to exhibit anti-inflammatory effects by reducing the expression of inflammatory cytokines, such as NF-κB, IL-1β, and TNF-α, and by suppressing the generation of reactive oxygen species through the inhibition of the hepatic enzyme CYP-2E1. Overall, the anti-inflammatory effects of garlic organosulphur compounds result from their ability to counteract the activation of pro-inflammatory pathways (El-Saber Batiha et al. 2020).

18.2.3 Antioxidant Activity

Oxidative damage refers to a phenomenon that occurs when there is an imbalance between free radicals and antioxidants in the body. Free radicals are molecules that contain oxygen and have an uneven number of electrons, which makes them highly reactive and unstable (Li et al. 2021). ROS, or reactive oxygen species, is a group of compounds that includes hydrogen peroxide (H_2O_2), superoxide anions (O_2*–), peroxyl (ROO–), hydroxyl (–OH), and alkoxyl (RO–) radicals. Cells generate a variety of free radicals during metabolic events that are necessary for intracellular processes, such as signal transduction, apoptosis, and proliferation, although excessive ROSs disrupt cellular homeostasis resulting in severe oxidative damage to the cellular machinery which can lead to various diseases, including diabetes, chronic inflammation, and cancer (Sasi et al. 2021).

Increasing numbers of studies have found that garlic has strong antioxidant properties; the mechanism of action of garlic might be involved with the increase in antioxidant enzyme activities and the regulation of the Nrf2-ARE pathway (Shang et al. 2019).The DPPH test (1,1-diphenyl-2-picrylhydrazyl) assay can be used to determine the antioxidant activity of garlic.Studies reported that raw garlic exhibited stronger antioxidant activity than cooked garlic when evaluated through the DPPH radical scavenging assay (Shang et al. 2019; E. Subroto et al. 2021). Another study found that garlic and its derivatives has been shown to activate enzymes involved in the reduction of intracellular ROS levels. These findings suggest that garlic may have potential therapeutic applications against the damage caused by ROS to proteins, lipids, and DNA under pathological conditions (Choudhary et al. 2022).

Studies reported that garlic bioactives have free radical scavenging properties and can act as antioxidants, preventing cell and organ damage caused by the oxidation process (Capasso 2013; Subroto et al. 2021); these compounds play an important role in protecting the body against the harmful effects of ROS (Subroto et al. 2021). Garlic and its derivatives can act through different mechanisms one of the ways in which they exert their antioxidant effects is by stimulating the activity of glutathione peroxidase (GPx) and inhibiting the decrease in the ratio of reduced-to-oxidized glutathione. This effect is particularly notable with organosulphur compounds, such as diallyl disulphide (DADS) and diallyl sulphide (DAS); these derivates can increase the activity of glutathione reductase and superoxide dismutase, whereas *S*-allylcysteine (SAC) and *S*-allylmercaptosysteine (SAMC) can enhance the synthesis of reduced glutathione (GSH). Moreover, allicin has a redox potential, meaning that it can be oxidized by ROS and consume phenolic compounds that have potential antioxidant effects, leading to a decrease in the overall antioxidant activity of the garlic extract, known as an antagonistic effect (Subroto et al. 2021). In a study by Patiño-Morales et al. (2021), it was reported that *S*-allylcysteine (SAC) and allicin showed the highest antioxidant activity by inhibiting free radical formation, enhancing the endogenous radical scavenging mechanism, and increasing the cellular antioxidant enzymes (Patiño-Morales et al. 2021). Furthermore, the two allyl derivatives of garlic (DAS and DADS) can modulate the phase II detoxifying enzymes, thus decreasing the intracellular ROS levels. Cultured endothelial cells treated with allicin showed an increase in the expression of phase II detoxifying enzymes, including haem oxygenase-1 (HO-1), thioredoxin reductase 1 and 2, and the glutamate-cysteine ligase modifier subunit, which plays a crucial role in glutathione biosynthesis; this upregulation of enzymes helps prevent damage caused by reactive oxygen species (Patiño-Morales et al. 2021).

Alliin has been shown to reduce osteoclastogenesis induced by the receptor activator of nuclear factor-κB ligand (RANKL) and decrease ROS and NADPH oxidase 1 degeneration. These findings suggest that garlic and its organosulphur compounds could be effective antioxidants and play a crucial role against oxidative stress (Patiño-Morales et al. 2021).

Phenolic compounds found in garlic contain hydroxyl groups that act as hydrogen donors to neutralize free radicals, providing antioxidant effects that protect the body from the damage caused by reactive oxygen species. The antioxidant activity of flavonoids depends on their molecular structure, especially the prenyl group $(CH_3)2C=CH-CH_2$-. Prenyl flavonoid groups have been developed and proven to play a role in the prevention or therapy of diseases associated with free radicals (Subroto et al. 2021). Flavonoid conjugation pathways involve conjugation with mutant glucuronides in the small intestine, followed by binding to albumin and transport to the liver, where they conjugate with sulphate and methyl groups to increase circulation clearance and reduce toxicity. Conjugated flavonoids have been found to inhibit xanthine oxidase enzyme activity and increase antioxidant activity. However, at high concentrations, the antioxidant activity of the phenolic group can be lost, and the antioxidants themselves may even become pro-oxidants (Subroto et al. 2021).

18.2.4 Anticarcinogenic Activity

Inflammation and cancer are intimately interlinked, and reactive oxygen species (ROS) play a fundamental role in both.

Anti-inflammatory and antioxidant use in cancer is a new therapeutic approach to disease treatment, progression, and recurrence. Two highly promising systemic targets common to the development, progression, and recurrence of many common cancers are dysregulated inflammatory and oxidation/reduction pathways; therefore the antioxidant and anti-inflammatory properties can contribute to its anticarcinogenic activity (Tesfaye 2021). Garlic derivatives have been found to influence a growing number of molecular mechanisms involved in carcinogenesis, including DNA adduct formation, scavenging of free radicals, mutagenesis, cell proliferation, cell differentiation, and angiogenesis. Garlic has been shown to reduce the growth rate of cancer cells, specifically by blocking the cell cycle, particularly in the G2/M phase (Capasso 2013), although garlic's antitumour effects are not fully understood at the molecular level and may vary depending on the specific type of cancer being treated as well as the type of garlic compound involved (Patiño-Morales et al. 2021).

According to a study, the mechanisms by which garlic exhibits anticancerous effects include regulating carcinogen metabolism, inhibiting cell growth and proliferation, inducing apoptosis, suppressing angiogenesis, and inhibiting invasion and migration of cancer cells (Shang et al. 2019). Garlic has also been shown to reduce the negative effects of anticancer therapies (Shang et al. 2019).

Another study has suggested that the organosulphur compounds found in garlic can extend their protective anticancer effects in three ways: through direct inhibition of tumour cell metabolism, through inhibition of the initiation and/or promotion phases of carcinogenesis, and through modulation of the host response by augmenting the function of macrophages and T-lymphocytes (Singh and Singh 2008).

Studies have revealed that garlic and its sulphur compounds can reduce the risk of cancer by diminishing the activation of carcinogens; garlic and its organic allyl sulphides can inhibit the generation of nitrosamines, a type of carcinogen produced during cooking and storage. Additionally, garlic allyl sulphides can block DNA alkylation, which is an early step in nitrosamine carcinogenesis (Shang et al. 2019).The observations by Wargovich and Goldberg in 1985 (as cited in Singh and Singh 2008) suggest that diallyl sulphide might inhibit the conversion of procarcinogens to ultimate carcinogens, which is an important factor in the initiation phase of carcinogenesis (Singh and Singh 2008). The review by Capasso (2013) delineated that diallyl sulphide (DAS), diallyl disulphide (DADS), and, more specifically, diallyl trisulphide (DATS) provide significant protection against carcinogenesis (Figure 18.1) (Capasso 2013).

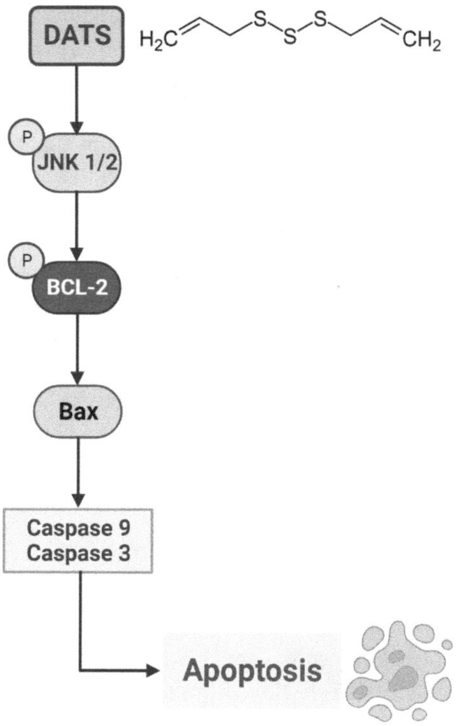

FIGURE 18.1 Proposed molecular mechanism of diallyl trisulphide (DATS) in inducing apoptosis through the mitochondrial pathway.

In glioblastoma cells, treatment with these garlic derivatives triggered the production of reactive oxygen species, inducing apoptosis through the phosphorylation of p38 MAPKs and activation of the redox-sensitive JNK1 pathway (Capasso 2013).

Garlic compounds, such as DATS, can suppress the proliferation of gastric cancer cells and block the cell cycle at the G2/M-phase by activating ataxia-telangiectasia mutated and checkpoint kinase 2, and then inhibiting the phosphorylation of Cdc25C (Ser216) and Cdc2 (Thr14/Tyr15), downregulating the expression of Cyclin B1 and up-regulating p21WAF1 to induce G2/M-phase cell cycle arrest. Moreover, SAC can induce G1/S-phase cell cycle arrest in ovarian cancer cells, and *S*-propargyl-cysteine (SPRC), an analogue of *S*-allylcysteine (SAC), can reduce the proliferation of pancreatic ductal adenocarcinoma cells and induce cell cycle arrest at the G2/M-phase (Shang et al. 2019). The garlic-derived *S*-allymercaptosysteine (SAMC) can suppress the proliferation of hepatocellular carcinoma cells and negatively affect the cell cycle by decreasing the percentage of the S phase and increasing the percentage of the G0/G1 phase. Allicin can inhibit the proliferation of gastric adenocarcinoma cells by inducing cell cycle arrest at the S-phase without affecting normal intestinal cells, and ajoene can restrain the growth of glioblastoma multiforme cancer stem cells and human breast cancer cells (Figure 18.2) (Shang et al. 2019).

You et al. (1988) reported that a significant reduction in stomach cancer risk was found to be associated with increasing consumption of garlic (Singh and Singh 2008).

Garlic has been found to upregulate apoptosis-related genes, including aryl hydrocarbon receptor, hypoxia-inducible factor 1α, and the proto-oncogene c-Jun, which affect the expression of genes related to immunity and cancer in the blood of humans. Bioactive compounds in garlic, such as DATS, SAC, SAMC, alliin, and *S*-propargyl-l-cysteine, induce apoptosis in various cancer cell lines via different pathways. DATS accumulates Bax, p53, and cytochrome c and reduces Bcl-2 expression in gastric cancer cells, while SAC decreases pro-caspase-3, PARP-1, Bcl-2 and increases active caspase-3 and Bax expression in ovarian cancer cells. SAMC induces apoptosis through the

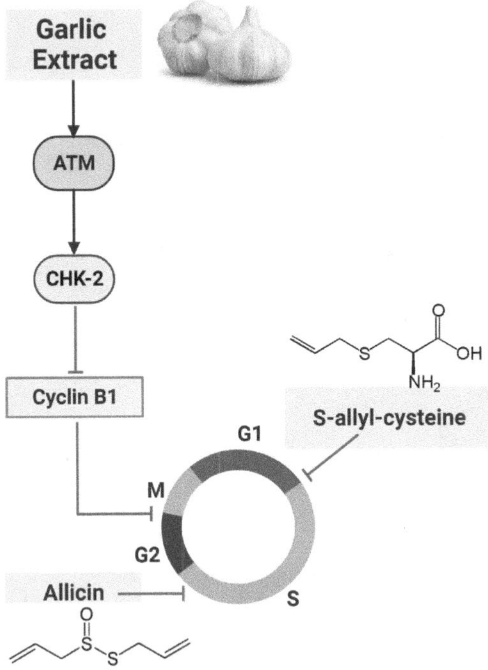

FIGURE 18.2 Proposed apoptotic mechanism of garlic and its active compounds in inhibiting cell cycle progression in cancer cells.

JNK and p38 MAPK pathways in colorectal carcinoma cells, alliin produces ROS and decreases the mitochondrial membrane potential in gastric adenocarcinoma cells, and *S*-propargyl-l-cysteine activates the JNK pathway to induce apoptosis in pancreatic ductal adenocarcinoma cells and limit tumour growth *in vivo* (Shang et al. 2019).

Garlic's bioactive sulphur compounds, such as allicin, diallyl sulphides (DAS), diallyl disulphides (DADS), and diallyl trisulphides (DATS), have been shown to possess antitumour effects, stimulating cytotoxicity, promoting apoptosis, and acting as upregulators of the expression of antioxidant genes and inhibitors of inflammation (Ţigu et al. 2021). Additionally, flavonoids and saponins found in *Allium* species have been investigated for their potential with respect to cytotoxicity, antiproliferation, pro-apoptosis, anti-angiogenesis, and antimetastasis.

Generally, and in various human cancer cell lines, the antitumour mechanisms involve activating apoptosis-related genes, such as caspases (caspase-3, -8, and -9) and Bcl-2. Polyphenols can also induce cell death in a dose-dependent manner through apoptosis (Ţigu et al. 2021). However, allicin has a unique effect on cell signalling pathways, inhibiting cell growth, and inducing apoptosis through the ERK-dependent pathway or by activating both intrinsic and extrinsic apoptotic pathways. Furthermore, allicin can suppress tumour cell progression by inhibiting the HIF pathway (Ţigu et al. 2021).

Garlic-derived compounds have demonstrated antioxidant and antitumour effects on cells derived from neuroblastoma. Allicin was found to induce apoptosis by activating the p38 MAPK pathway, leading to the release of cytochrome c. A report has shown the antitumour effect of *Z*-ajoene through the induction of apoptosis, which was reflected in the increase in annexin V, propidium iodide, and active caspase-3, suggesting a p53-dependent mechanism of action (Figure 18.3). Filomeni et al. (2003) demonstrated that DADS induced apoptosis through the downregulation of the mitochondrial pathway Bcl-2, release of cytochrome c into the cytosol, and activation of caspase-9 in association with the activation of the JNK/c-Jun pathway (Patiño-Morales et al. 2021). DADS induced the disruption of the cytoskeleton explained by tau protein dephosphorylation mediated by the protein

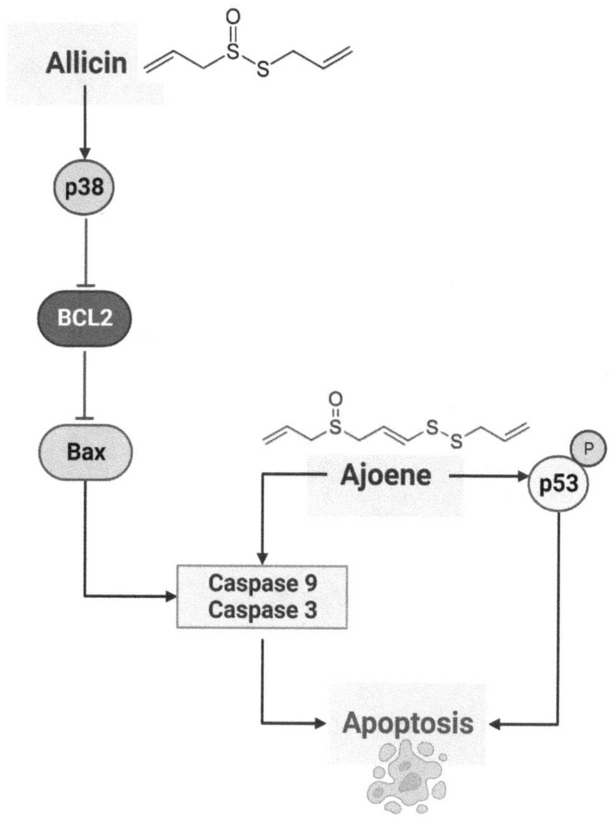

FIGURE 18.3 Proposed apoptotic molecular mechanism of allicin and ajoene.

phosphatase-1 (PP1) in a ROS-dependent pathway after treatment with DADS (Patiño-Morales et al. 2021).

Ajoene increases p53 expression, which is phosphorylated and activates the p53 upregulated modulator of apoptosis (PUMA). In the cytoplasm, PUMA activates pro-apoptotic proteins, such as B-cell lymphoma 2 (Bcl-2)-like protein 4 (Bax), creating a pore in the outer mitochondrial membrane and triggering the release of cytochrome c and the second mitochondrion-derived activator of caspase/direct inhibitor of apoptosis protein (IAP)-binding protein. DAS, DADS, and DATS also increase Bax and decrease Bcl-2, which enhances the effect. Allicin activates p38, which inhibits Bcl-2 and decreases the mitochondrial membrane potential, similar to SAC. Cytochrome c in the cytosol induces the oligomerization of apoptotic peptidase activating factor 1 (Apaf-1). Apaf-1, cytochrome c, and adenosine triphosphate (ATP) form the apoptosome that activates caspase-9. Activated caspase-9 leads to the processing of effector caspases, such as caspase-3. Caspase-3 can also be activated by calpain, which is first activated by the increase in free intracellular Ca^{2+} mediated by DAS and DAD (Patiño-Morales et al. 2021).

Allicin has been shown to be effective against stomach cancer, which is the fifth most common cancer in the world, but its mode of action was unknown for a long time. Several studies have shown that allicin halts the cell cycle during the G2/M phase and induces apoptosis in gastric tumour cells, while normal stomach cells remain unaffected. Moreover, allicin-mediated inhibition of stomach tumourigenesis was also reported in scientific findings (Tudu et al. 2022).

Allicin inhibits TGF-α and its receptor epidermal growth factor receptor (EGFR), resulting in the downregulation of cyclin E and cyclin D1, which can cause cells in the G2 phase to enter the M phase. In addition, allicin can cause DNA damage by generating ROS, leading to the phosphorylation of p53 and p21 proteins, then p21inhibits the cyclin-dependent kinase 4/6 (CDK4/6)-cyclin

D complex, causing the cell to enter G2/M arrest. Allicin has also been shown to have a significant effect on reducing the potential of the outer mitochondrial membrane, this being due to an increase in the proportion of the pro-apoptotic protein B-cell lymphoma 2-like protein 4 (Bax) to the anti-apoptotic protein B-cell lymphoma 2 (Bcl-2) caused by allicin, which can promote cell death through the mitochondrial pathway (Tudu et al. 2022).

According to another study, allicin has been shown to induce the formation of apoptotic bodies, nuclear condensation, and a typical DNA ladder in cancer cells. Additionally, allicin activates caspases-3, -8, and -9 and cleaves poly (ADP-ribose) polymerase (Singh and Singh 2008).

It was also reported that *S*-allylmercaptocysteine (SAMC) also has anti-proliferative effects by directly binding to tubulin and disrupting microtubule assembly, resulting in the arrest of cells in mitosis. This, in turn, activates the JNK1 and caspase-3 signalling pathways, leading to apoptosis (Singh and Singh 2008).

18.3 CONCLUSION

The biological properties of garlic described in this review emphasize the importance of this plant as an ally in human health. Garlic has a significant phytochemical content that has promising and efficient therapeutic potential. This study supports the numerous medicinal properties of garlic and the interdependence of various pathological activities, paying particular attention to the antidiabetic, anti-inflammatory, antioxidant, and anticarcinogenic effects of garlic and its derivatives.

LIST OF ABBREVIATIONS

ADP: Adenosine diphosphate
Apaf-1: Apoptotic protease activating factor 1
ATM: Ataxia telangiectasia mutated
ATP: Adenosine triphosphate
B1: Bradykinin 1
Bax: Bcl-2-associated X protein
Bcl-2: B-cell lymphoma 2
Ca2+: Calcium ion
Cdc2: Cell division control 2
Cdc25C: Cell division cycle 25C
CDK4/6: Cyclin-dependent kinase 4/6
CHK2: Checkpoint kinase 2
COX2: Cyclooxygenase 2
CYP-2E1: Cytochrome P450 2E1
c-Jun: Cellular Jun protein
DADS: Diallyl disulphide
DAS: Diallyl sulphide
DATS: Diallyl trisulphide
DNA: Deoxyribonucleic acid
DPP-4: Dipeptidyl peptidase-4
DPPH: 2,2-diphenyl-1-picrylhydrazyl
ERK: Extracellular signal-regulated kinase
G0/G1: Gap 0/Gap 1
G1/S: Gap 1/Synthesis
G2/M: Gap 2/Mitosis
GPx: Glutathione peroxidase
GSH: Glutathione
H2O2: Hydrogen peroxide

HIF: Hypoxia-inducible factor
HO-1: Haemoxygenase-1
IAP: Inhibitor of apoptosis protein
IL-1: Interleukin-1
IL-1β: Interleukin-1 beta
iNOS: Inducible nitric oxide synthase
JNK: c-Jun N-terminal kinase
JNK1: c-Jun N-terminal kinase 1
JNK1/2: c-Jun N-terminal kinase 1 and 2
MAPK: Mitogen-activated protein kinase
NADPH: Nicotinamide adenine dinucleotide phosphate
NF-κB: Nuclear factor kappa B
NO: Nitric oxide
Nrf2-ARE: Nuclear factor erythroid 2-related factor 2-antioxidant response element
O_2*–: Superoxide anion
p21: Cyclin-dependent kinase inhibitor 1
p38: p38 mitogen-activated protein kinase
p53: Tumour protein p53
p21WAF1: Cyclin-dependent kinase inhibitor 1A
PARP-1: Poly(ADP-ribose) polymerase 1
PP1: Protein phosphatase 1
PUMA: p53 upregulated modulator of apoptosis
RANKL: Receptor activator of nuclear factor kappa-B ligand
ROS: Reactive oxygen species
SAC: *S*-allylcysteine
SAMC: *S*-allymercaptosysteine
Ser216: Serine 216
SPRC: *S*-propargyl-cysteine
Tau: Microtubule-associated protein tau
Thr14: Threonine 14
TNF-α: Tumor necrosis factor alpha
Tyr15: Tyrosine 15
-OH: Hydroxyl
-O2*-: Superoxide
-ROO: Peroxyl
-RO: Alkoxyl
14-kDa: 14 kilodalton

REFERENCES

Bar, Monika, Urszula E. Binduga, and Konrad A. Szychowski. 2022. Methods of isolation of active substances from garlic (*Allium sativum* L.) and its impact on the composition and biological properties of garlic extracts. *Antioxidants* 11(7):1345. doi:10.3390/antiox11071345.

Capasso, Anna. 2013. Antioxidant action and therapeutic efficacy of *Allium sativum* L. *Molecules* 18(1):690–700. doi:10.3390/molecules18010690.

Choudhary, Sunvej, Mohd Ubed Noor, and Md. Sadique Hussain. 2022. Pharmacological properties and phytoconstituents of garlic (*Allium sativum* L.): A review. *Biological Sciences* 02(04). doi:10.55006/biolsciences.2022.2402.

El-Saber Batiha, Gaber, Amany Magdy Beshbishy, Lamiaa G. Wasef, Yaser H. A. Elewa, Ahmed A. Al-Sagan, Mohamed E. Abd El-Hack, Ayman E. Taha, Yasmina M. Abd-Elhakim, and Hari Prasad Devkota. 2020. Chemical constituents and pharmacological activities of garlic (*Allium sativum* L.): A Review. *Nutrients* 12(3):872. doi:10.3390/nu12030872.

Faroughi, Farnaz, Sakineh Mohammad-Alizadeh Charandabi, Yousef Javadzadeh, and Mojgan Mirghafourvand. 2018. Effects of garlic pill on blood glucose level in borderline gestational diabetes mellitus: A randomized controlled trial. *Iranian Red Crescent Medical Journal* 20(7). doi:10.5812/ircmj.60675.

Kalhotra, Poonam, Veera C. S. R. Chittepu, Guillermo Osorio-Revilla, and Tzayhri Gallardo-Velazquez. 2020. Phytochemicals in garlic extract inhibit therapeutic enzyme DPP-4 and induce skeletal muscle cell proliferation: A possible mechanism of action to benefit the treatment of diabetes mellitus. *Biomolecules* 10(2):305. doi:10.3390/biom10020305.

Karimi, Akbar, and Nahid Khayam Nasab. Effect of garlic extract and *Citrus aurantifolia* (lime) juice and on blood glucose level and activities of aminotransferase enzymes in streptozotocin-Induced diabetic rats. 2014. *World Journal of Pharmaceutical Sciences*, 2(8), 821–827. Retrieved from https://wjpsonline.com/index.php/wjps/article/view/effect-garlic-extract-citrus-aurantifolia-diabetic-rats

Khan, Mohd W. A., Ahmed A. Otaibi, Abdulmohsen K. D. Alsukaibi, Eida M. Alshammari, Salma A. Al-Zahrani, Subuhi Sherwani, Wahid A. Khan, Ritika Saha, Smita R. Verma, and Nessar Ahmed. 2022. Biophysical, biochemical, and molecular docking Investigations of anti-glycating, antioxidant, and protein structural stability potential of garlic. *Molecules* 27(6):1868. doi:10.3390/molecules27061868.

Li, Dan, Zongxian Ding, Kaili Du, Ye Xiangshi, and Shixue Cheng. 2021. Reactive oxygen species as a link between antioxidant pathways and autophagy. Edited by Luciana Hannibal. *Oxidative Medicine and Cellular Longevity* (July):1–11. doi:10.1155/2021/5583215.

Padiya, Raju, and Sanjay K. Banerjee. 2013. Garlic as an anti-diabetic agent: recent progress and patent reviews. *Recent Patents on Food, Nutrition and Agriculture* 5(2):105–127. doi:10.2174/18761429113059990002.

Patiño-Morales, Carlos César, Ricardo Jaime-Cruz, Concepción Sánchez-Gómez, Juan Carlos Corona, Estefani Yaquelin Hernández-Cruz, Ivia Kalinova-Jelezova, José Pedraza-Chaverri, Perla D. Maldonado, Carlos Alfredo Silva-Islas, and Marcela Salazar-García. 2021. Antitumor effects of natural compounds derived from *Allium sativum* on neuroblastoma: An overview. *Antioxidants* 11(1):48. doi:10.3390/antiox11010048.

Ramirez, Daniela Andrea, Jorgelina Cecilia Altamirano, and Alejandra Beatriz Camargo. 2021. Multi-phytochemical determination of polar and non-polar garlic bioactive compounds in different food and nutraceutical preparations. *Food Chemistry* 337(February):127648. doi:10.1016/j.foodchem.2020.127648.

Sasi, Minnu, Sandeep Kumar, Manoj Kumar, Sandhya Thapa, Uma Prajapati, Yamini Tak, Sushil Changan, Vivek Saurabh, Shweta Kumari, et al. 2021. Garlic (*Allium sativum* L.) bioactives and its role in alleviating oral pathologies. *Antioxidants* 10(11):1847. doi:10.3390/antiox10111847.

Shang, Ao, Shi-Yu Cao, Xiao-Yu Xu, Ren-You Gan, Guo-Yi Tang, Harold Corke, Vuyo Mavumengwana, and Hua-Bin Li. 2019. Bioactive compounds and biological functions of garlic (*Allium sativum* L.). *Foods* 8(7):246. doi:10.3390/foods8070246.

Singh, Vinay K., and Dinesh K. Singh. 2008. Pharmacological effects of garlic (*Allium sativum* L.). *Annual Review of Biomedical Sciences:*Vol 10. doi: http://dx.doi.org/10.5016/1806-8774.2008.v10p6

Subroto, E., Y. Cahyana, M. Tensiska, F. Filianty, E. Lembong, E. Wulandari, D. Kurniati, R. A. Saputra, and F. Faturachman. 2021. Bioactive compounds in garlic (*Allium sativum* L.) as a source of antioxidants and its potential to improve the immune system: A review. *Food Research* 5(6):1–11. doi:10.26656/fr.2017.5(6).042.

Tesfaye, Azene. 2021. Revealing the therapeutic uses of garlic (*Allium sativum*) and its potential for drug discovery. Edited by Juei Tang Cheng. *The Scientific World Journal* (December):1–7. doi:10.1155/2021/8817288.

Ţigu, Adrian Bogdan, Cristian Silviu Moldovan, Vlad-Alexandru Toma, Anca Daniela Farcaş, Augustin Cătălin Moţ, Ancuţa Jurj, Eva Fischer-Fodor, Cristina Mircea, and Marcel Pârvu. 2021. Phytochemical analysis and *in vitro* effects of *Allium fistulosum* L. and Allium sativum L. extracts on human normal and tumor cell lines: A comparative study. *Molecules* 26(3):574. doi:10.3390/molecules26030574.

Tudu, Champa Keeya, Tusheema Dutta, Mimosa Ghorai, Protha Biswas, Dipu Samanta, Patrik Oleksak, Niraj Kumar Jha, Radha Manoj Kumar, Jarosław Proćków, et al. 2022. Traditional uses, phytochemistry, pharmacology and toxicology of garlic (*Allium sativum*), a storehouse of diverse phytochemicals: A review of research from the last decade focusing on health and nutritional implications. *Frontiers in Nutrition* 9(October):949554. doi:10.3389/fnut.2022.929554.

19 Ginger Rhizome (*Zingiber officinale*)

Sarra Grabsi, Hayette Bouabida, and Djemaa Dris

19.1 INTRODUCTION

Herbs and spices are used to season food and give it flavour as well as taste (Ghafoor et al., 2020; Tanweer et al., 2020). Herbs and spices were widely used in almost all traditional foods and medicines that were used to enhance community health (Ghafoor et al., 2020; Tanweer et al., 2020). The world now uses ginger, which has been cultivated for more than 2,000 years in China, as a well-known condiment and medicine (Jiang et al., 2017). Ginger (*Zingiber officinale* Roscoe) rhizome has been used as a hot spice, home remedy, and as a traditional medicine in many countries, such as Malaysia, Indonesia, and China, for over 2,000 years (Jiang et al., 2017; Srinivasan, 2017; Shahrajabian et al., 2019). This traditional use is in accordance with its chemical content and pharmacological activity to treat symptoms such as inflammation and gastrointestinal disorders; however, it has also been shown to have anti-platelet, antihypertensive, antihyperlipidaemia, anti-hyperuricaemia, antidiabetic, and cancer-preventive properties (Srinivasan, 2017; Suciyati & Adnyana, 2017; Shahrajabian et al., 2019; Kan et al., 2023). Ginger rhizomes are often used in dried, fresh, marinated, or candied form (Zagorska et al., 2023). The phytochemicals in ginger's leaves and blossoms have also been shown to confer health advantages, and their characterization in many foods can make them qualify as nutraceuticals or designer foods. (Tanweer et al., 2020). Ginger rhizome has shown potential use in a variety of areas over the past decades, including herbal medicine, flavouring agents for foods and beverages, and for food and beverage preparation (Jabborova et al., 2021). Overall, this chapter provides valuable information about the ginger rhizome as a plant-based material, in addition to its role in herbal medicine and imparting flavor to our food.

19.2 DESCRIPTION

One of the most recent examples of a medicinal plant, which has many uses, is ginger. It was described for the first time by an English naturalist named William Roscoe in 1807, and Fostor gave it the binomial *Zingiber officinale* (Çatak, 2020), a herbaceous equatorial perennial of the Zingiberaceae family. It is typically cultivated as an annual in agricultural settings (Vasala, 2012). Nigeria accounts for 56.23% of the world's total area covered in ginger, followed by India (23.6%), China (4.47%), Indonesia (3.37%), and Bangladesh (2.32%). (Shahrajabian et al., 2019). This medicinal plant is well described in Figures 19.1 and 19.2.

The ginger plant is upright and has numerous fibrous roots, aerial shoots with foliage, and an underground stem (rhizome). Ginger stems come in two varieties: fleshy and fibrous (Vasala, 2012; Nair, 2019). Ginger is a monocotyledon, $2n = 2x = 22$, belonging to the family Zingiberaceae and the order Zingiberales. This family covers up to 24 genera and around 300 species. The genus *Zingiber* contains about 20 species as well (Moghaddasi & Kashani, 2012; Mao et al., 2019). In the family Zingiberaceae, ginger belongs to the subfamily Zingiberoideae (Vasala, 2012; Mao et al., 2019; Nair, 2019). The rhizome is entirely pale yellow within or with a red external layer. Leafy stems grow to about 50 cm high and 5 mm in diameter, glabrous except for short hairs near the

 DOI: 10.1201/9781003387046-19

FIGURE 19.1 *Zingiber officinale* (ginger).

FIGURE 19.2 Ginger plant.

base of each leaf blade; leaf blades are commonly about 17 cm × 1.8 cm; rather dark green, narrow evenly to the slender tip; ligule broad, thin, glabrous, 5 mm high, slightly bilobed. Scape slender, 12 cm tall, the upper sheaths with or without short leafy tips; inflorescence approximately 4.5 cm long and 15 mm in diameter; bracts approximately 2.5 cm × 1.8 cm, green with pale submarginal band and narrow translucent margin; margins incurved, lower bracts with slender white tip (Nair, 2019) (Figure 19.3).

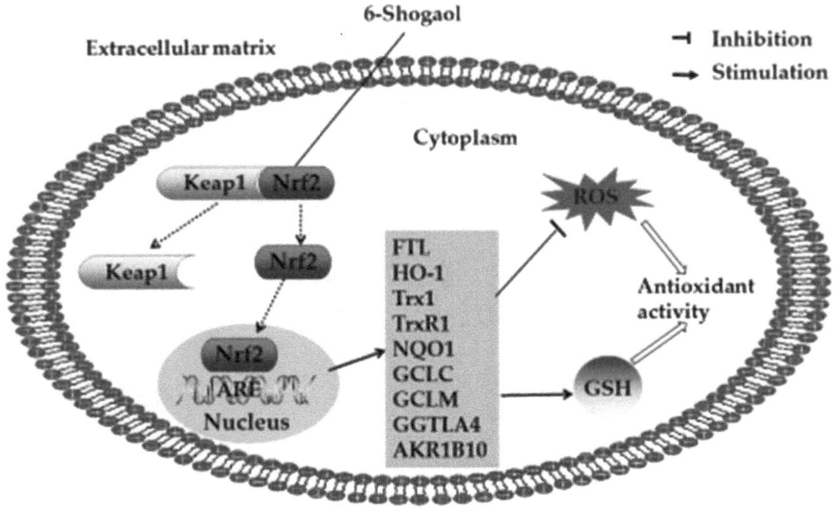

FIGURE 19.3 The potential mechanism for the antioxidant action of 6-shogoal Abbreviations: Nrf2, nuclear factor erythroid 2-related factor 2; Keapl, Kelch-like ECH-associated protein 1; NQO1, nicotinamide adenine dinucleotide phosphate (NADPH) quinone dehydrogenase 1; HO-1, haem oxygenase-1; GCLC, glutamate-cysteine ligase catalytic subunit; GCLM, glutamate-cysteine ligase modifier subunit; Trx1, thioredoxin 1; TrxR1, thioredoxin reductase 1; AKR1B10, Aldo-keto reductase family 1 member B10; FTL, ferritin light chain; GGTLA4, γ-glutamyltransferase-like activity 4; ROS, reactive oxygen species; GSH, reduced glutathione; ARE, antioxidant response element.

19.3 TRADITIONAL KNOWLEDGE

Numerous scientific investigations have recently taken place to support the conventional uses of *Zingiber officinale*. Ginger is a stimulant, carminative, and spicy herb that is frequently used to treat fevers, malaria, and stomachache (Moghaddasi & Kashani, 2012). It is primarily used to treat illnesses such as abdominal pain, anorexia, arthritis, atonic dyspepsia, bleeding, cancer, chest congestion, chickenpox, cholera, chronic bronchitis, cold extremities, colic, colitis, common cold, coughs, cystic fibrosis, diarrhoea, difficulty in breathing, dropsy, fever, and flatulence (Moghaddasi & Kashani, 2012; Shahrajabian et al., 2019; Ghafoor et al., 2020). Ginger and its major components have been utilized in various herbal medicines for a very long time. Sliced or powdered fresh ginger root is used to make ginger tea. It can also be used as a seasoning or tincture.

19.4 BIOACTIVE COMPONENTS AND BIOACTIVITIES OF *ZINGIBER OFFICINALE*

Bioactive components are plentiful in ginger (Mao et al., 2019; Abdullahi et al., 2020). Chemical analysis of ginger shows that it contains over 400 different compounds (Prasad & Tyagi, 2015). Mostly gingerols, shogaols, and paradols constitute ginger's phenolic chemicals (Mao et al., 2019). The primary polyphenols in raw ginger are gingerols, including 6-, 8-, and 10-gingerol. Gingerols can be changed into the corresponding shogaols through heat treatment or extended storage, while shogaols can become paradols after being hydrogenated (Stoner, 2013; Mao et al., 2019). In addition, various other flavonoid compounds, including quercetin, zingerone, gingerenone-A, and 6-dehydrogingerdione, are also present in ginger (Schadich et al., 2016; Ji et al., 2017). Table 19.1 provides information on the nutritional composition of ginger, and the phytochemical profile of ginger is shown in Table 19.2. In addition to having high levels of protein, carbohydrates, and fat, ginger also has a lot of inorganic elements that are vital for human health, like calcium, phosphate, and iron (Çatak, 2020; Shahrajabian et al., 2019).

TABLE 19.1
Nutritional Composition of Ginger (per 100 g)

Constituent	Value
Moisture (%)	15.02±0.04
Protein (g)	5.087±0.09
Fat (g)	3.72±0.03
Insoluble fibre (%)	23.5±0.04
Soluble fibre (%)	25.5±0.04
Carbohydrate (g)	38.35±0.1
Vitamin C (mg)	9.33±0.08
Total carotenoids (mg)	79±0.2
Ash (g)	3.85±0.61
Calcium (mg)	88.4±0.97
Phosphorus (mg)	174±1.2
Iron (mg)	8.0±0.2
Zinc (mg)	0.92±0
Copper (mg)	0.545±0.002
Manganese (mg)	9.13±0.01
Chromium (μg)	70±0

TABLE 19.2
Phytochemical Profile of Ginger

Chemical	Class of Nutrient	Amount
Protein	----	1.8 g
Water	----	78.9 g
Phytosterols	----	15 mg
Carbohydrates	Total carbohydrate	18 g
	Dietary Fibre	2 g
	Sugar	1.7 g
Fats and fatty acids	Total fat	750 mg
	Saturated fat	203 mg
	Monounsaturated fat	154 mg
	Polyunsaturated fat	154 mg
	Omega-3 fatty acids	34 mg
	Omega-6 fatty acids	120 mg
Vitamins	Vitamin C	5 mg
	Vitamin E	260 μg
	Vitamin K	0.1 μg
	Thiamin	25 μg
	Riboflavin	34 μg
	Niacin	750 μg
	Vitamin B6	160 μg
	Folic acid	11 μg
	Pantothenic acid	203 μg
	Choline	28.8 mg

(*Continued*)

TABLE 19.2 (CONTINUED)
Phytochemical Profile of Ginger

Chemical	Class of Nutrient	Amount
Minerals	Calcium	16 mg
	Iron	600 μg
	Magnesium	43 mg
	Phosphorus	34 mg
	Potassium	415 mg
	Sodium	13 mg
	Zinc	340 μg
	Copper	226 μg
	Manganese	229 μg
	Selenium	0.7 μg

19.5 PHARMACOLOGICAL PROPERTIES

19.5.1 Antioxidant Effects

Ginger extracts exhibit strong antioxidant activity on a level with artificial antioxidant preservatives (Moghaddasi & Kashani, 2012). The joint antioxidant effects of linolenic acid and -carotene were responsible for the antioxidant activity of ginger rhizome, leaves, and flower extracts (Tanweer et al., 2020). The antioxidant action of ginger leaves has been evaluated *in vitro* via ferric-reducing antioxidant power (FRAP) and the 2,2-diphenyl-1-picrylhydrazyl (DPPH) assay (Mao et al., 2019; Ghafoor et al., 2020).

The findings showed that dried ginger had the highest antioxidant action, as there were more phenolic compounds in it, being 5.2-, 1.1-, and 2.4-fold higher than that of fresh, stir-fried, and carbonized ginger. The antioxidant capacities of different ginger preparations tended to be as follows: dried ginger > stir-fried ginger > carbonized ginger > fresh ginger (Mao et al., 2019). In accordance with the findings of Zagórska et al. (2023), the bioactive compounds and antioxidant activity of freeze-dried ginger were significantly higher than those of microwave- and air-dried samples, and only slightly lower than those of oven-dried samples. Ingredients responsible for the antioxidant activity of ginger are listed in Table 19.3.

Studies conducted both *in vitro* and *in vivo* have demonstrated that ginger and its therapeutic components, such as 6-shogaol, 6-gingerol, and oleoresin, have strong antioxidant qualities. (Jittiwat & Wattanathorn, 2012; Mao et al., 2019; Shahrajabian et al., 2019). Many research investigations have shown that ginger can effectively ward off oxidative stress (Ahmed et al., 2000; Jittiwat & Wattanathorn, 2012). The potential mechanism for the antioxidant action of the main active compound of ginger (6-shogoal) is shown in Figure 19.3.

TABLE 19.3
Phenolics and Antioxidant Activity of *Zingiber officinale*

Phenolic Class	Amount
Total phenols	840 mg
Tannins	1,51 g
Flavonoids	2,98 g
Total antioxidant activity	73529 μmol/g

By altering Keapl and stopping Nrf2 from being degraded by proteasomes, 6-shogoal causes Nrf2 to translocate into the nucleus and upregulates the expression of Nrf2 target genes. As a result, the concentration of GSH rises whereas that of reactive oxygen species (ROS) falls (Mao et al., 2019).

19.5.2 Anti-inflammatory Effect

The main reason for the beginning and development of many complex illnesses is inflammation including inflammatory skin disorders, cardiovascular inflammation, and other inflammatory diseases (Maione et al., 2016; Zhou et al., 2022). In contrast, the complex mechanism of inflammation, which includes numerous pro-inflammatory mediators and cytokines, has made the creation of new drugs extremely difficult (Zhou et al., 2022). Currently, the first-line anti-inflammatory treatments used are steroids and non-steroidal anti-inflammatory medicines (NSAIDs). Even so, due to their single-target behaviour on only prostaglandins, rather than cytokines, NSAIDs' effectiveness in treating systemic inflammation is constrained and their long-term use is linked to a variety of side effects (Laveti et al., 2013). For this purpose, the hunt for innovative and effective therapeutic targets for inflammation has received a lot of research attention (Zhou et al., 2022). For a long time, medicinal plants have been the primary treatment for different forms of inflammation, and, today, many drugs have been developed for conventional medicine (Maione et al., 2016; Oguntibeju, 2018). Ginger has been shown to have anti-inflammatory properties (Akram et al., 2019). Since ancient times, it has been used as a traditional medicine to treat a variety of illnesses, including inflammatory disorders. Gingerols, shagaols, and paradols, which are the main ingredients in ginger, have anti-inflammatory properties (Akram et al., 2019; Zhou et al., 2022; Guo et al., 2023) . Ginger's phenolic compounds, such as 6-gingerol and 6-shogaol, are thought to be responsible for the reported anti-inflammatory activity, as these compounds are crucial in lowering important pro-inflammatory mediators like the interleukin IL-6 and tumour necrosis factors (TNFs) (Ballester et al., 2022; Lashgari et al., 2022).

The primary polyphenols in raw ginger are gingerols like 6-gingerol, 8-gingerol, and 10-gingerol. Other phenolic compounds found in ginger include quercetin, zingerone, gingerenone-A, and polyphenols; all of these have demonstrated strong antioxidant and anti-inflammatory action (Ballester et al., 2022). Principal bioactive compounds of ginger responsible for its anti-inflammatory action are listed in Table 19.4.

19.5.3 Antimicrobial Effect

Due to antimicrobial resistance, the spread of infectious bacterial, fungal, and viral diseases has posed a serious danger to the general human population (Meade et al., 2021; Rodríguez-Baño et al., 2021). Many herbs and seasonings have been transformed into organic, powerful antimicrobial agents that fight a variety of pathogenic microorganisms (Egamberdieva et al., 2021). Recent reports suggest that ginger has antibacterial, antifungal, and antiviral properties (Camero et al., 2019; Ozkur et al., 2022)

Biofilm formation is an important part of bacterial infection and antimicrobial resistance (Senobar Tahaei et al., 2021). According to one finding, ginger reduced membrane integrity and prevented the development of biofilms, which prevented the spread of a strain of *Pseudomonas aeruginosa* that was resistant to multiple drugs (Mao et al., 2019). Additionally, ginger extract treatment prevented biofilm development by lowering the amount of *bis*-(30-50)-cyclic dimeric guanosine monophosphate in *P. aeruginosa* (Kim et al., 2019; Miari et al, 2020).

Also, the results of the studies by Mao et al. (2019) showed that the ginger crude extract and methanolic fraction reduced the expression of bacterial virulence genes, preventing the development of biofilms, the production of glucan, and the adhesion of *Streptococcus mutans*. Furthermore, an *in vitro* study showed that gingerenone-A and 6-shogaol had a deterrent effect on *Staphylococcus*

TABLE 19.4
Principal Bioactives of the Phytochemicals of Ginger Responsible for its Anti-inflammatory Action

Phenolic Compounds, as Gingerol Analogues			Other Phenolic Compounds	Terpenes
Gingerols	Shogaols	Paradols		
6-Gingerol8-Gingerol10-Gingerol12-Gingerol	6-Shogaol8-Shogaol10-Shogaoldehydro-14-gingerdione6-Gingerdione 10-Gingerdione	6-Paradol8-Paradol10-Paradol	ZingeroneQuercetinGingerenone-A6-Dehydrogingerdione	β-Bisabolene α-Curcumene α-Farnesene β- Sesquiphellandrene Zingiberene

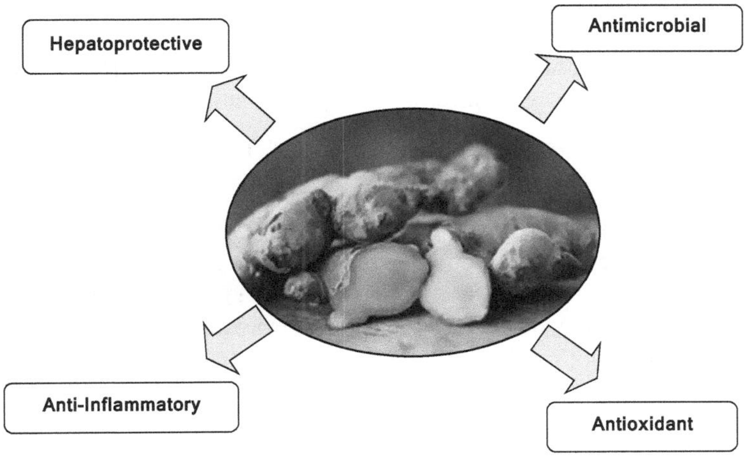

FIGURE 19.4 Pharmacological effects of *Zingiber officinale.*

aureus by preventing the pathogen's 6-hydroxymethyl-7, 8-dihydropterin pyrophosphokinase from being activated. The aqueous ginger extract exhibited a moderate response against *S. aureus* with zone of inhibition of 16.73±1.83 mm at 500 mg/dL (Olaleye et al., 2022; Khatri et al., 2023). However, the virulence of *S. mutans* was reduced by ginger crude extract and methanolic fraction by preventing the glucan production, biofilm formation, and adhesion (Khatri et al., 2023).

19.5.4 Hepatoprotective Effect

Studies conducted *in vitro* showed that ginger's active ingredients can prevent liver cancer (Hamza et al., 2021; Mahomoodally et al., 2021). According to a research, 6-shogaol causes Mahlavu hepatoma cells to die *via* an oxidative stress-mediated, caspase-dependent mechanism. Depletion of the antioxidant reduced glutathione (GSH) has been demonstrated to be a significant mediator of 6-shogaol-induced death in Mahlavu cells (Wu et al., 2015). In addition to these, a study showed that the levels of the antioxidant enzyme superoxide dismutase (SOD), and GSH and glutathione reductase in mouse blood, as well as activities of antioxidant enzymes glutathione-*S*-transferase, glutathione peroxidase, and SOD enzymes in mouse liver, are increased when ginger oil is taken orally for a month (Prasad & Tyagi, 2015). In HepG2 hepatoma cells, ginger extract at a dosage of 250 g/mL significantly alters the morphology of the cells, causing cell shrinkage and chromosome condensation (Vijaya Padma et al., 2007).

Before ROS were produced and cytochrome c was released from the mitochondria, 6-gingerol-induced cathepsin D release occurred. Additionally, liver tissue lysates and mitochondrial lipid peroxidation are said to be protected. The antioxidant capacity of ginger extract can be linked to these protective processes (Ajith at al., 2010; Hosseinzadeh et al., 2017). Figure 19.4 summarizes the pharmacological activities of ginger.

19.6 PRODUCTS OF GINGER RHIZOMES

19.6.1 Fresh Ginger

Compared with other preparations made from the spice, fresh ginger is exceptional for its flavouring because it retains the entire notes of the spice (Vasala, 2012; Bartak et al., 2020) For green ginger, fresh rhizomes with low fibre content but high levels of flavour, pungency, fat, and protein are favoured (Akhlaghi & Najafpour-Darzi, 2023).

19.6.2 Preserved Ginger

Green ginger that is still immature is kept in a sugar solution or vinegar. Through additional preparation steps, ginger in sugar syrup can also be transformed into crystallized ginger (Vasala, 2012; Shahrajabian et al. 2019).

19.6.3 Dry Ginger

The spice industry uses dried raw ginger for the production of ground ginger and the extraction of oil (Vasala, 2012; Zubair, 2020), The process of making dried ginger begins with mature rhizomes that have fully developed aroma, flavour, and pungency. Harvesting is typically done between eight and nine months after sowing (Vasala, 2012; Fueki et al., 2022).

19.6.4 Ginger Oil

The production of ginger oil involves steam distillation of newly ground dry ginger (Mahboubi, 2019). In terms of extraction time and extraction yield, the chemical make-up of hydrolats (aromatic hydrosols) and essential oils, and environmental effect, two applied methods were compared. In comparison with hydrodistillation (1.50%), microwave-assisted hydrodistillation produced a higher extraction yield, 1.85% (Drinić et al., 2021).

19.7 OTHER POTENTIAL USES OF *ZINGIBER OFFICINALE*

19.7.1 Uses in Food Processing

Ginger is an essential food ingredient around the globe due to its reviving aroma, taste, and carminative qualities. For this, a variety of ginger products are used, including fresh ginger, dried ginger flour, oleoresin, and oil (Ibáñez & Blázquez, 2019; Sahamastuti et al., 2019). Its use in food, however, has not been thoroughly investigated and offers a potential field for additional study.

19.8 CONCLUSION

All over the world, the economic value of plants used in medicine is significant. One of them is *Z. officinale*, because it has both nutritional and health benefits. In conclusion, ginger includes a variety of bioactive substances, including paradols, shogaols, and gingerols, and has numerous bioactivities, including anti-inflammatory, anti-oxidant, and antibacterial actions. Furthermore, ginger could be used for the treatment and prevention of a number of illnesses, including liver cancer and microbial infectious diseases. Ginger also has the potential to be an ingredient in functional foods or as nutritional supplements. Future research should be conducted on additional bioactive compounds found in ginger, including their biological activities and associated modes of action. Importantly, carefully planned clinical studies on ginger and its different bioactive compounds are necessary to demonstrate their effectiveness in treating such diseases in humans.

REFERENCES

Abdullahi, A., Khairulmazmi, A., Yasmeen, S., Ismail, I. S., Norhayu, A., Sulaiman, M. R., Ahmed, O. H., & Ismail, M. R. (2020). Phytochemical profiling and antimicrobial activity of ginger (*Zingiber officinale*) essential oils against important phytopathogens. *Arabian Journal of Chemistry*, 13(11): 8012–8025.

Ahmed, R., Seth, V., Pasha, S. T., & Banerjee, B. D. (2000). Influence of dietary ginger (*Zingiber officinale* Rosc.) on oxidative stress induced by malathion in rats. *Food and Chemical Toxicology*, 38(5): 443–450.

Ajith, T. A. (2010). Ameliorating reactive oxygen species-induced *in vitro* lipid peroxidation in brain, liver, mitochondria and DNA damage by *Zingiber officinale* Roscoe. *Indian Journal of Clinical Biochemistry*, 25(1): 67–73.

Akhlaghi, N., & Najafpour-Darzi, G. (2023). Potential applications of ginger rhizomes as a green biomaterial: A review. *International Journal of Engineering*, 36(2): 372–383.

Akram, A., Rasul, A., Waqas, M. K., Irfan, M., Khalid, S. H., Aamir, M. N., Murtaza, G., Ur Rehman, K., Iqbal, M., & Khan, B. A. (2019). Development, characterization and evaluation of *in-vitro* anti-inflammatory activity of ginger extract based micro emulsion. *Pakistan Journal of Pharmaceutical Sciences*, 32(3): 1327–1332.

Ballester, P., Cerdá, B., Arcusa, R., Marhuenda, J., Yamedjeu, K., & Zafrilla, P. (2022). Effect of ginger on inflammatory diseases. *Molecules*, 27(21): 7223.

Bartak, M., Lange, A., Słońska-Zielonka, A., & Cymerys-Bulenda, J. (2020). Homemade onion syrup. Why is it a great common cold remedy? Badania i Rozwój Młodych Naukowców w Polsce, 143.

Camero, M., Lanave, G., Catella, C., Capozza, P., Gentile, A., Fracchiolla, G., Britti, D., Martella, V., Buonavoglia, C., & Tempesta, M. (2019). Virucidal activity of ginger essential oil against caprine alpha herpes virus-1. *Veterinary Microbiology*, 230: 150–155.

Çatak, E. (2020). Botanical description, chemical composition and medical use of *Zingiber officinale* (Roscoe) (ginger). *Theory and Research in Science and Mathematics*, II: 187.

Drinić, Z., Jovanović, M., Pljevljakušić, D., Ćujić-Nikolić, N., Bigović, D., & Šavikin, K. (2021). Microwave-assisted extraction of essential oil from ginger (*Zingiber officinale* Rosc.). *Lekovite Sirovine*, 41: 22–27.

Egamberdieva, D., Jabborova, D., Babich, S., Xalmirzaeva, S., Salakhiddinov, K., & Madazimov, M. (2021). Antimicrobial activities of herbal plants from Uzbekistan against human pathogenic microbes. *Environmental Sustainability*, 4(1): 87–94.

Fueki, T., Nose, I., Liu, Y., Tanaka, K., Namiki, T., & Makino, T. (2022). Oxalic acid in ginger specifically denatures the acrid raphides in the unprocessed dried tuber of *Pinellia ternata*. *Acupuncture and Herbal Medicine*, 2(1): 33–40.

Ghafoor, K., Al Juhaimi, F., Özcan, M. M., Uslu, N., Babiker, E. E., & Ahmed, I. A. M. (2020). Total phenolics, total carotenoids, individual phenolics and antioxidant activity of ginger (*Zingiber officinale*) rhizome as affected by drying methods. *LWT*, 126: 109354.

Guo, M., Shen, Q., Wu, Y., Li, L., Zhang, L., Wang, Y., Zhang, Y., Zhuang, P., & Guo, H. (2023). Multivariate analysis of original identification and chemical markers exploration of Chinese ginger. *Food Science and Biotechnology*: 32: 911–923.

Hamza, A. A., Heeba, G. H., Hamza, S., Abdalla, A., & Amin, A. (2021). Standardized extract of ginger ameliorates liver cancer by reducing proliferation and inducing apoptosis through inhibition oxidative stress/inflammation pathway. *Biomedicine and Pharmacotherapy*, 134: 111102.

Hosseinzadeh, A., Juybari, K. B., Fatemi, M. J., Kamarul, T., Bagheri, A., Tekiyehmaroof, N., & Sharifi, A. M. (2017). Protective effect of ginger (*Zingiber officinale* Roscoe) extract against oxidative stress and mitochondrial apoptosis induced by interleukin-1β in cultured chondrocytes. *Cells, Tissues, Organs*, 204(5–6): 241–250.

Ibáñez, M. D., & Blázquez, M. A. (2019). Ginger and turmeric essential oils for weed control and food crop protection. *Plants*, 8(3): 59.

Jabborova, D., Sayyed, R. Z., Azimov, A., Jabbarov, Z., Matchanov, A., Enakiev, Y., Baazeem, A., El Sabagh, A., Danish, S., & Datta, R. (2021). Impact of mineral fertilizers on mineral nutrients in the ginger rhizome and on soil enzymes activities and soil properties. *Saudi Journal of Biological Sciences*, 28(9): 5268–5274.

Ji, K., Fang, L., Zhao, H., Li, Q., Shi, Y., Xu, C., Wang, Y., Du, L., Wang, J., & Liu, Q. (2017). Ginger oleoresin alleviated γ-ray irradiation-induced reactive oxygen species via the Nrf2 protective response in human mesenchymal stem cells. *Oxidative Medicine and Cellular Longevity*: 1–13.

Jiang, Y., Liao, Q., Zou, Y., Liu, Y., & Lan, J. (2017). Transcriptome analysis reveals the genetic basis underlying the biosynthesis of volatile oil, gingerols, and diarylheptanoids in ginger (*Zingiber officinale* Rosc.). *Botanical Studies*, 58(1): 1–12.

Jittiwat, J., & Wattanathorn, J. (2012). Ginger pharmacopuncture improves cognitive impairment and oxidative stress following cerebral ischemia. *Journal of Acupuncture and Meridian Studies*, 5(6): 295–300.

Kan, C. Y., H'ng, J. X., Goh, A., Smales, F., Tan, E. L., Zhang, S., Pichika, M. R., & Parolia, A. (2023). Effect of sustained systemic administration of ginger (*Zingiber officinale*) rhizome extracts on salivary flow in mice. *International Dental Journal*, 73(1): 63–70.

Khatri, P., Rani, A., Hameed, S., Chandra, S., Chang, C. M., & Pandey, R. P. (2023). Current understanding of the molecular basis of spices for the development of potential antimicrobial medicine. *Antibiotics*, 12(2): 270.

Kim, H. S., Ham, S. Y., Jang, Y., Sun, P. F., Park, J. H., Lee, J. H., & Park, H. D. (2019). Linoleic acid, a plant fatty acid, controls membrane biofouling via inhibition of biofilm formation. *Fuel*, 253: 754–761.

Lashgari, N. A., Momeni Roudsari, N., Khayatan, D., Shayan, M., Momtaz, S., Roufogalis, B. D., Abdolghaffari, A. H., & Sahebkar, A. (2022). Ginger and its constituents: Role in treatment of inflammatory bowel disease. *BioFactors*, 48(1): 7–21.

Laveti, D., Kumar, M., Hemalatha, R., Sistla, R., Naidu, V. G., Talla, V., Verma, V., Kaur, N., & Nagpal, R. (2013). Anti-inflammatory treatments for chronic diseases: A review. *Inflammation & Allergy-Drug Targets (Formerly Current Drug Targets-Inflammation & Allergy)(Discontinued)*, 12(5): 349–361.

Mahboubi, M. (2019). *Zingiber officinale* Rosc. essential oil, a review on its composition and bioactivity. *Clinical Phytoscience*, 5(1): 1–12.

Mahomoodally, M. F., Aumeeruddy, M. Z., Rengasamy, K. R., Roshan, S., Hammad, S., Pandohee, J., Hu, X., & Zengin, G. (2021). Ginger and its active compounds in cancer therapy: From folk uses to nano-therapeutic applications. *In Seminars in Cancer Biology*, 69: 140–149.

Maione, F., Russo, R., Khan, H., & Mascolo, N. (2016). Medicinal plants with anti-inflammatory activities. *Natural Product Research*, 30(12): 1343–1352.

Mao, Q. Q., Xu, X. Y., Cao, S. Y., Gan, R. Y., Corke, H., Beta, T., & Li, H. B. (2019). Bioactive compounds and bioactivities of ginger (*Zingiber officinale* Roscoe). *Foods*, 8(6): 185.

Meade, E., Slattery, M. A., & Garvey, M. (2021). Biocidal resistance in clinically relevant microbial species: A major public health risk. *Pathogens*, 10(5): 598.

Miari, M., Rasheed, S. S., Ahmad, N. H., Itani, D., Abou Fayad, A., & Matar, G. M. (2020). Natural products and polysorbates: Potential Inhibitors of biofilm formation in *Pseudomonas aeruginosa. The Journal of Infection in Developing Countries*, 14(06): 580–588.

Moghaddasi, M. S., & Kashani, H. H. (2012). Ginger (*Zingiber officinale*): A review. *Journal of Medicinal Plants Research*, 6(26): 4255–4258.

Nair, K. P., & Nair, K. P. (2019). The agronomy and economy of ginger. Turmeric (*Curcuma longa* L.) and ginger (*Zingiber officinale* Rosc.) - World's invaluable medicinal spices. *The Agronomy and Economy of Turmeric and Ginger*: 245–315.

Oguntibeju, O. O. (2018). Medicinal plants with anti-inflammatory activities from selected countries and regions of Africa. *Journal of Inflammation Research*, 11: 307–317.

Olaleye, O. N., & Momoh, J. O. (2022). Evaluation of secondary metabolites profiling of ginger (*Zingiber officinale* Roscoe) rhizome using GC-MS and its antibacterial potential on *Staphylococcus aureus* and *Escherichia coli. Microbiology Research Journal International*, 32(7): 7–31.

Ozkur, M., Benlier, N., Takan, I., Vasileiou, C., Georgakilas, A. G., Pavlopoulou, A., Cetin, Z., & Saygili, E. I. (2022). Ginger for healthy ageing: A systematic review on current evidence of its antioxidant, anti-inflammatory, and anticancer properties. *Oxidative Medicine and Cellular Longevity*: 1–16.

Prasad, S., & Tyagi, A. K. (2015). Ginger and its constituents: Role in prevention and treatment of gastrointestinal cancer. *Gastroenterology Research and Practice*: 1–12.

Rodríguez-Baño, J., Rossolini, G. M., Schultsz, C., Tacconelli, E., Murthy, S., Ohmagari, N., & Plant, L. (2021). Antimicrobial resistance research in a post-pandemic world: Insights on antimicrobial resistance research in the COVID-19 pandemic. *Journal of Global Antimicrobial Resistance*, 25: 5–7.

Sahamastuti, A. A. T., Andre, A., Foustine, S., Sumarpo, A., & Hartiadi, L. Y. (2019). Synergistic antibacterial activities of ginger and lemongrass essential oils as an alternative prevention to food-borne disease. *Indonesian Journal of Life Sciences*, 1(2): 54–61.

Sandeep, S. H. A. H. (2017). Commentary on therapeutic role of ginger (*Zingiber officinale*) as medicine for the whole world. *International Journal of Pharmacognosy and Chinese Medicine*, 1(1): 1–3.

Schadich, E., Hlaváč, J., Volná, T., Varanasi, L., Hajdúch, M., & Džubák, P. (2016). Effects of ginger phenylpropanoids and quercetin on Nrf2-ARE pathway in human BJ fibroblasts and HaCaT keratinocytes. *BioMed Research International*: 1–7.

Senobar Tahaei, S. A., Stájer, A., Barrak, I., Ostorházi, E., Szabó, D., & Gajdács, M. (2021). Correlation between biofilm-formation and the antibiotic resistant phenotype in *Staphylococcus aureus* isolates: A laboratory-based study in Hungary and a review of the literature. *Infection and Drug Resistance*: 1155–1168.

Shahrajabian, M. H., Sun, W., & Cheng, Q. (2019). Clinical aspects and health benefits of ginger (*Zingiber officinale*) in both traditional Chinese medicine and modern industry. *Acta Agriculturae Scandinavica, Section B—Soil & Plant Science*, 69(6): 546–556.

Shahrajabian, M. H., Wenli, S. U. N., & Cheng, Q. (2019). Pharmacological uses and health benefits of ginger (*Zingiber officinale*) in traditional Asian and ancient Chinese medicine, and modern practice. *Notulae Scientia Biologicae*, 11(3): 309–319.

Srinivasan, K. (2017). Ginger rhizomes (*Zingiber officinale*): A spice with multiple health beneficial potentials. *PharmaNutrition*, 5(1): 18–28.

Stoner, G. D. (2013). Ginger: Is it ready for prime time? *Cancer Prevention Research*, 6(4): 257–262.

Suciyati, S. W., & Adnyana, I. K. (2017). Red ginger (*Zingiber officinale* Roscoe var *rubrum*): A review. *Review of Education*, 2: 60–65.

Tanweer, S., Mehmood, T., Zainab, S., Ahmad, Z., & Shehzad, A. (2020). Comparison and HPLC quantification of antioxidant profiling of ginger rhizome, leaves and flower extracts. *Clinical Phytoscience*, 6(1): 1–12.

Vasala, P. A. (2012). Ginger. In *Handbook of Herbs and Spices*. Woodhead Publishing, 319–335.

Vijaya Padma, V., Arul Diana Christie, S., & Ramkuma, K. M. (2007). Induction of apoptosis by ginger in HEp-2 cell line is mediated by reactive oxygen species. *Basic and Clinical Pharmacology and Toxicology*, 100(5): 302–307.

Wu, J. J., Omar, H. A., Lee, Y. R., Teng, Y. N., Chen, P. S., Chen, Y. C., Huang, H. S., Lee, K. H., & Hung, J. H. (2015). 6-Shogaol induces cell cycle arrest and apoptosis in human hepatoma cells through pleiotropic mechanisms. *European Journal of Pharmacology*, 762: 449–458.

Zagórska, J., Czernicka-Boś, L., Kukula-Koch, W., Iłowiecka, K., & Koch, W. (2023). Impact of thermal processing on the selected biological activities of ginger rhizome—A review. *Molecules*, 28(1): 412.

Zhou, X., Afzal, S., Wohlmuth, H., Münch, G., Leach, D., Low, M., & Li, C. G. (2022). Synergistic anti-inflammatory activity of ginger and turmeric extracts in inhibiting lipopolysaccharide and interferon-γ-induced proinflammatory mediators. *Molecules*, 27(12): 3877.

Zubair, S., Sawate, A. R., Kshirsagar, R. B., Agarkar, B. S., & Patil, B. M. (2020). Studies on impact of different processing methods on phyto-chemical and antioxidant activity of dried ginger (*Zingiber officinale* Roscoe) rhizome. *Journal of Pharmacognosy and Phytochemistry*, 9(4): 3153–3158.

20 Mustard Seeds (*Brassica* species)

Radhika Sharma, Manik Devgan, Arshdeep Kaur, Tejinder Singh, Usha Rana, Anuj Choudhary, Amit Kumar Kesharwani, Rajeev Rathor, and Sahil Mehta

20.1 INTRODUCTION

The concept of modernization has diverted everyday life towards semi-finished products and ready-made meals to reduce the time expenditure on cooking. Such products are mainly generated based on taste and affordability, with least interest in the product composition (Hu et al., 2021; Li et al., 2022). More recently, the number of people suffering from allergies, obesity, cancer, and cardiovascular diseases has increased at a significantly increased rate. Before buying, a person must read the brand label on the product and care must be given to altering the product trends to move towards healthy nutrition (Rahman et al., 2018; Chand et al., 2021). To enhance or supplement the existing food diets with more nutritive kinds, producers and researchers have been considering less-focused plants for a long time in the kitchen and even they have a large profile of bioactive compounds.

Mustard is a group of related plants that have a high nutritive potential due to their abundance of bioactive ingredients. Several species have been used for mustard cultivation, such as *Sinapsis alba* (white mustard), *Brassica nigra* (black mustard), *Brassica juncea* (brown mustard), *Brassica carinata* (Ethiopian mustard), *Brassica eruca* (rocket mustard), and *Sinapsis arvensis* (wild mustard) (Batool et al., 2022; Bortoloti et al., 2022). Therefore, it is interesting to discuss mustard for its phytochemical potential to enhance nutritional status, its probable health benefits for improving day-to-day life, supplementary advantages such as biodiesel production, livestock feed, and others, and also the role of advanced technology, such as nanotechnology, in developing improved preparations from a pharmaceutical viewpoint.

20.2 ORIGIN AND CLASSIFICATION

Since mustard is suitable for cultivation in both subtropical and temperate climates, it has been grown for centuries in many regions of Eurasia. Central Asia (northwestern India) is thought to be its primary centre of origin, followed by eastern India, central and western China, Myanmar, Iran, and the Near East (Tian & Deng, 2020). A variety of mustard known as *Brassica carinata* A. Braun, commonly referred to as Ethiopian or Abyssinian mustard, has a high content of erucic acid, and is believed to have originated in northern Africa, while kale (*Brassica oleracea* L. [Acephala group]) and turnip (*Brassica napus* L.) interbred to produce oilseed rape or rapeseed (*Brassica napus* L.), an amphidiploid plant. *Alliaria petiolata* (M. Bieb.) Cavara & Grande, an invasive biennial herb with the common name garlic mustard, is native to several nations in temperate and tropical Asia (Shankar et al., 2019). The best-known mustard in Europe is *S. alba* L., which was originally used in approximately 400 BC and is also known as white mustard or yellow mustard. *Brassica integrifolia* is a perennial herb that originates originally from eastern and southern Asia. Black mustard, brown mustard, moutarde noire, and senf are all common names for the annual herb *Brassica nigra* L., which is native to most of Europe. *Brassica rapa* L., also known as oilseed rape (in Canada), kale rape, rapa, rappi, keblock, colza, bird rape, rape mustard, turnip, or bok choy is native to central Asia and Europe (Francisco et al., 2017; Rahman et al., 2018).

DOI: 10.1201/9781003387046-20

20.3 PHYTOCHEMICAL OVERVIEW IN MUSTARD

The phytochemical analysis of the Brassicaceae family to which all the mustards belong reveals that it is composed of active ingredients such as xanthoproteins, alkaloids, terpenoids, cardiac glycosides, tannins, coumarins, flavonoids, steroids, phenols, phlorotannins, sugar, resins, quinines, and saponins (Paul et al., 2019). Because they include tocopherols, carotenoids, polyphenols, and glucosinolates, *Brassica* vegetables are also a valuable source of natural antioxidants (Hedges & Lister, 2006). In general, the Brassicaceae family is high in polyphenolic components and polyphenols. The flavonoids, which mostly consist of flavonols and anthocyanins, and the hydroxycinnamic acids are the most prevalent and varied groups of polyphenols in *Brassica* spp. (Cartea et al., 2010; Bhandari & Kwak, 2015). Due to glucosinolates and phenolic compounds, a distinctive flavour and odour is produced from mustards. Nonetheless, the presence of phenolic chemicals is primarily responsible for the majority of the antioxidative impact associated with eating plant foods, including mustards. Many biological activities, including anticancer, antioxidant, and antibacterial capabilities, are attributed to these bioactives (Singh et al., 2016). Glucosinolates, which are mostly found in the seeds of Brassicaceae plants, are abundant in nature. Glucosinolates are β-thioglucoside *N*-hydroxysulphates, which are a class of secondary metabolites that are high in sulphur content. Glucosinolates activate enzymes that carry out detoxification, apoptosis, and inhibition of the growth of tumour cells, reducing inflammation and adipogenesis and also decreasing oxidative stress (Dinkova-Kostova & Kostov, 2012; Francisco et al., 2017).

Additionally, *Brassica* spp. are a good source of natural antioxidants due to the presence of vitamins A, C, K, B_6, β -carotenes, tocopherols, zeaxanthin, and lutein. The mineral constituents, such as Ca, K, Zn, Na, Fe, Mg, Se, and Mn, are involved in different metabolic processes, and serve as cofactors for enzymes, boost bone and dental health, and stimulate red blood cell production (Jahangir et al., 2009; Francisco et al., 2017). Erucic acid (26.5–36%) dominates the high fat content of mustard seeds, which ranges from 23% to 47%. The quantity of different fatty acids varies among species; for instance, white mustard oil contains a large amount of oleic acid (22%) while linoleic acid is more prevalent in brown and black mustard oils (19.5–22%). All species of mustard have considerable amounts of linolenic acid in their seeds (9–15%) (Sawicka et al., 2007; Grygier, 2022). Sinalbin, sinigrin, neoglucobrassicin, glucoraphanin, gluconapine, glucoiberin, glucobrassicin, and progoitrin are the most prevalent glucosinolates in the Brassicaceae plant family (Sawicka et al., 2007; Grygier, 2022). The pH of the surrounding environment and the presence of ferrous ions affect the breakdown products of the glucosinolate sinigrin, with 1-cyano-2,3-epithiopropane, allyl-isothiocyanate, allyl cyanide and allylthiocyanate being among the breakdown byproducts. Sinigrin is converted into allyl-isothiocyanate in neutral pH mustard seeds, which gives crushed mustard seeds their distinctive aroma (Dai et al., 2015; Grygier, 2022). Mustard seeds contain phenolic chemicals, such as sinapic acid, ferulic acid, and 3,4-di-hydroxybenzoic acid, which are antioxidants. In addition to these substances, rutin is present in black mustard seeds (Rasera et al., 2019).

20.4 POTENTIAL BIOACTIVE USES IN MUSTARDS

20.4.1 Antioxidant Activities

Reactive oxygen species (ROS) are a class of chemicals that are constantly produced by living things under physiological homeostasis, and play essential roles in some biological processes in healthy cells, but accumulate under stressful conditions. The oxidants that are subsequently created can interact with various biological elements, including proteins, lipids, and nucleic acids. Indeed, the majority of oxidants' targets are proteins. All the antioxidant polyphenols have been related to anticancer, antiageing, neuroprotective, antidiabetic, and cardioprotective properties and are recognized to have the optimal structural chemistry for free radical-scavenging actions. Dehydroascorbic acid, a byproduct of ascorbic acid oxidation, has several biological functions in the human body and has been linked to a lower risk of diabetes, cardiovascular diseases, and cancer (Jahangir et al., 2009; Buval et al., 2021).

20.4.2 Food Preservation and Antimicrobial Activities

Due to growing public concern over synthetic preservatives' safety issues and resulting harmful impacts on human health, the preservation of foods using natural preservatives has gained popularity recently. Due to antioxidant and antibacterial properties, substances derived from natural sources have the potential to be employed as natural preservatives. *Brassica* species have attracted interest from scientists for a variety of reasons, including their value in agriculture and nutrition and their potential to naturally preserve food because they are a major source of glucosinolates and have high concentrations of other antioxidant and antimicrobial compounds, essential oils, and polyphenolic compounds. It has been established that glucosinolates and volatile sulphur compounds derived from *S*-alkyl-L-cysteine sulfoxide are responsible for the majority of the antibacterial action of *Brassica* plants (Paul et al., 2012). The potential of *Brassica* plants as a source of chemicals with antibacterial action has already been highlighted in numerous studies. One study confirmed the antibacterial properties of *B. oleracea* active phyto-components concerning *Aspergillus fumigatus*, *Citrobacter divergens*, and *Klebsiella* at various doses (Aires et al., 2012). Table 20.1 summarizes research on the potential antibacterial and antioxidant actions of *Brassica* plants in terms of their preservation qualities.

20.4.3 Anticancer Activities

According to a study (Thangam et al., 2013),, 2-pyrrolidinone, an active substance found in *B. oleracea* leaves, may have anticancer properties. This chemical was anticipated to be used as a lead chemical in the development of fresh anticancer drugs to halt the growth of cancer cells in culture. In HeLa and PC-3 cancer cells, this chemical can block the proliferator cells, leading to a considerable decrease in cell proliferation, cell viability, and significant induction of apoptosis. Also, *in vitro* studies were conducted on the impact of mustard (*B. juncea*) leaf ethanolic extracts on HCT116 colorectal cancer and H1299 non-small-cell lung carcinoma cells (Kwak et al., 2016). In both cell lines, the treatment with mustard leaf extract significantly inhibited the release of basic fibroblast growth factor and vascular endothelial cell growth factor. This study found that the mustard leaf extract exhibits *in vitro* colon cancer-fighting properties (Buval et al., 2021).

20.4.4 Antidepressant Activities

The therapeutic effect was measured using learned helplessness, behavioral despair, the tail-hanging test, and motor activity in mice. It also examined how the extract therapy affected the levels of dopamine, norepinephrine, and serotonin in the serum. The findings demonstrated that *B. juncea*'s antidepressive effects were discernible not only in behavioral tests of diabetic rats and mice but also in biochemical analyses. When compared with the diabetic-depressed baseline, daily mustard extract dose-dependently boosted levels of serum serotonin, norepinephrine, and dopamine in the brain. These findings also suggest that *B. juncea* could be utilized as a nutritional supplement to treat depression brought on by diabetes (Tian & Deng, 2020).

20.4.5 Traditionally Accepted Benefits of Mustard

Seven *Brassica* species, including *B. juncea, B. rapa, B. napus, B. oleracea*, and *B. nigra* are the most-often-utilized plants in traditional medicine. According to Petropoulos et al. (2017), the *Brassica* genus is distinguished by glucosinolates and phenolic chemicals, which are responsible for numerous biological features. So, in various regions of the world, these qualities have been

TABLE 20.1
Pharmacological Activity of Mustard

Plant	Plant Part	Activity	Extract	Pharmacological	Key References
Brassica juncea	Seeds	*Bacillus cereus*, *Staphylococcus aureus*, *Listeria monocytogenes*, *Escherichia coli*	Methanolic	Antibacterial activity	Elansary et al., 2018
	Seeds	Nitric oxide and DPPH scavenging activity and ABTS scavenging activity	Acetone extract (10-1000 g/mL)	Antioxidant activity	Khandayataray & Murthy, 2019
	Seeds	Inhibition zone against *Microsporum ferrogenium* and *Mucor mucaralis*	Methanolic	Antifungal activity	Ogidi et al., 2019
	Leaves	Protease inhibitor activity	Acetone	Antiinflammatory activity	Khandayataray & Murthy, 2019
	Leaves	Decrease of blood glucose in streptozotocin-induced diabetic rats	Methanolic	Antihyperglycaemic activity	Kumar et al., 2019
	Flowers	Cause of death and paralysis of *Phertima posthuma*	n-HSF at the concentration of 10 mg/mL showed paralysis and death	Anthelmintic activity	Uddin et al., 2020
	Flowers	Anxiolytic action in mice was improved	MCE and its fractions at doses of 200 and 400 mg/kg	Anxiolytic activity	Uddin et al., 2020
	Seeds	Anticonvulsant activity against PTZ-induced seizure in mice	Methanolic	Anticonvulsant activity	Buval et al., 2021
	Plant	Nontoxicity against the Madin-Darby canine kidney cells	Aqueous and ethanolic extracts	Cytotoxicity	Buval et al., 2021
	Leaves	Cognitive disorders	Ethanol	Antidiabetic activity	Buval et al., 2021
	Canola meal	Identification of the free radical scavenging activity	Methanol, ethanol, acetone, butanol, chloroform, and hexane	Antioxidant properties	Hussain et al., 2022
	Canola	Highest recovery of antioxidant phenols	Aqueous acetone	Antidiabetic assay	Hussain et al., 2022
Brassica nigra	Flowers	Increase in clot lysis	Addition of 100 μL streptokinase	Thrombolytic activity	Uddin et al., 2020

used in traditional medicinal applications. The seeds and leaves of *B. juncea* are used to treat pain and arthritis/rheumatism problems in Australia (Rahman et al., 2018). On the other hand, its seeds are used in Pakistan (Swat River region of the Khyber-Pakhtunkhwa Province) as a tonic and to treat ascariasis, and the leaves are utilized in India because they have a hypoglycaemic effect by stimulating the production of glycogen. Also, fresh *B. rapa* pollen from various regions of Asia that is rich in phytosterols and are used in Qianlie Kang Pulean tablets, which has positive effects on chronic prostatitis, incontinence, and swollen joints. The leaves and roots of *B. napus* are used to treat abscesses, furuncles, acne, eczema, and chilblains in Romania (Gilca et al., 2018), whereas the leaves and seeds of *B. rapa* are used as a liniment in Australia to treat wounds and soothe sore muscles. Muscle aches, rheumatism, and arthritis pain have all traditionally been treated with mustard seeds and their oil. In India, it is believed that using mustard oil on the scalp has health benefits. Because it contains a lot of essential minerals such as macro and micronutrients, mustard boosts the body's resistance to disease (Divakaran & Babu, 2016). Tocopherols present in mustard oil are antioxidants and help to prevent rancidity, increasing the shelf life of final food products. The key to treating constipation is mucilage, which is a thick, slimy substance found in *Brassica* seeds. In addition, it has a lot of fibre and helps to promote the development of hair. Its pulverized seeds stimulate intestinal secretion, function as a laxative, and stimulate the gastric mucosa. Mustard seeds, which are high in sulphur, are a fantastic way to prevent skin diseases. The seeds' antifungal and antibacterial qualities based on sulphur aid in the treatment of common skin conditions (Paul et al., 2012; Tian & Deng, 2020; Grygier et al., 2022).

20.5 'MUSTARD THE COMPLETE PACKAGE' (ADD-ON AND PURPOSES BEYOND)

In addition to the aforementioned functional (pharmacological) qualities, the mustard crop has the following specific uses (Figure 20.1).

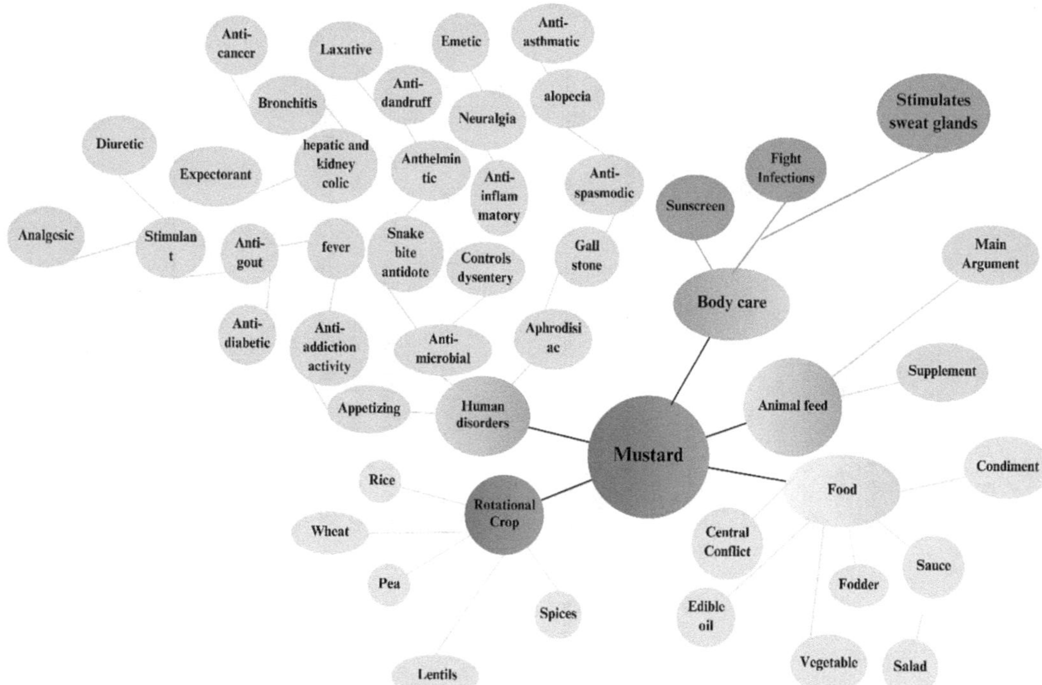

FIGURE 20.1 Multipurpose role of mustard on human disease, animal feed, and body care.

20.5.1 Biodiesel Production

The ongoing need for food, particularly in developing nations, casts a shadow over efforts to produce biodiesel from vegetables or other food crops. We may be able to circumvent this problem by using seed cakes and non-edible mustard seed oils that are particularly high in erucic acid (and therefore are unsuitable for human consumption). Mustard biodiesel is a combined product of first- and second-generation biofuels because it can also be produced from non-edible feedstock (Hoekman et al., 2011). The *Brassica* species most frequently used to produce biodiesel is *Brassica napus*, popularly known as rapeseed or oilseed rape (Sanchez et al., 2013). Mustard feedstock fulfills the two basic demands for an ideal biofuel, i.e., industrial-scale production at minimum cost (Zulqarnain et al., 2021). However, the intricate procedures involved in extracting biofuels from mustard cast a cloud over its enormous potential (Hamza et al., 2020). In order to retrieve a higher level of oil from the mustard feedstocks, several industrial biodiesel synthesis processes, such as physical, chemical, or enzymatic treatments, are used. Some commonly employed processes in biodiesel extraction from mustard are presented in Table 20.2.

TABLE 20.2
Some Commonly Employed Processes in Biodiesel Extraction from Mustard

Process	Efficiency	Modification	Shortcoming	Key References
Steam Distillation	>80%	Pressure variance based on plant material	1. This process requires evenly distributed raw materials for recovery. 2. High capital investment is required	Karmakar et al., 2010
Catalytic cracking	>90%	Providing environments similar to petroleum products	1. Pyrolyzed material contains high amounts of impurities such as sulfur, moisture, etc.	Taufiqurrahmi et al., 2011
Transesterification	~97%	Use of catalysts based on calibrated yield	1. Acid catalysts lead to corrosion 2. High cost Slow reaction	Issariyakul et al., 2014; Kumar et al., 2019
Chemical leaching	85%	Modification in temperature, solvent, mixing in the system	1. Separation of heterogeneous catalysts from the mixture is a major issue	Du et al., 2018; Thangaraj et al., 2019
Enzymatic oil extraction	-	Enzyme immobilization plus chromatography	1. High cost of enzymes 2. Higher incubation time	Ahmad et al., 2019
Mechanical Extraction • (manual pressing or engine-driven screw pressing)	60–65% (manual) 68–80% (engine)	Seed pre-treatment with micro-organisms	1. Only suitable for specific seeds 2. Requires filtration, degumming post extraction	Tang et al., 2020

20.5.2 Soil Amendments, Supplementation, and Bioremediation

Brown mustard can be applied to extract, sequester, or detoxify the heavy metals from polluted sediments. Growing Indian mustard with suitable agronomic interventions has proven beneficial for controlling as well as remediating metal-contaminated soils. The phytoremediation capacity of Indian mustard can be increased by adding organic matter, organic chelates, soil amendments, appropriate cropping methods, intercrops, and fertilizer choices (Rathore et al., 2019). *Alyssum murale, Thlaspi rotundifolium,* and *Dicomanic colifera* and nearly 80 other species of the Brassicaceae have demonstrated hyperaccumulation of several heavy metals, including Ni, Zn, Cr, As, and Pb (Ansari et al., 2015). Goswami and Das (2015) observed a maximum tolerance index for cadmium phytoremediation *via* accumulation in *B. juncea* shoots. Similarly, Indian mustard grown under greenhouse conditions was found to extract nearly 32 mg of lead (Salido et al., 2003). The abundance of water-soluble phenolic fractions and other phytochemicals that can be directly absorbed by plants is increased when decomposed *Brassica* seeds and oilcake are added to the soil. This helps to speed up root growth and overall plant health, which in turn helps to increase host tolerance to nematode invasion.

20.5.3 Rotational Crop

Several mustard species, including *Brassica alba, B. nigra, B. juncea, B. napus,* etc., are popular intercropping species that in conservation agriculture, in combination with efficient nutrient management strategies, can improve soil fertility and hence crop productivity (Pooniya et al., 2021). *Brassica* spp. are disease-suppressive rotation crops that have demonstrated promise for managing soil-borne diseases and enhancing yield in a variety of crop production systems (Larkin & Halloran, 2014). Certain allelochemicals produced by mustard crops have an impact on the germination, growth, reproduction, and survival of neighbouring plants. The phenolics 8-hydroxyquinoline and (±)-catechin, which are found in mustard plants, are effective in chelating nutrients, including nitrogen, phosphorus, etc. These substances are released from the roots and change nitrogen fixation in the soil, assisting in nitrogen accumulation. They also help the soil microbes to break down organic matter and release ammonium nitrogen, which enhances nutrient uptake (Rahman et al., 2018).

20.5.4 Livestock Feed

After collecting the seeds, most of the leftover mustard leaves can be used as animal fodder as the protein content in rapeseed has a well-balanced amino acid profile and is superior to that of cereals in terms of essential amino acids (e.g., lysine, methionine, and tryptophan) (Thanaseelaan, 2013). Moreover, the concentration of certain essential elements. such as calcium, phosphorus, selenium, iron, and manganese, in rapeseed meal is very high (Georgeta, 2009). A significant amount of albumin, glutelin, and globulin are present in the true protein in the meal, which makes up 80–83% of the crude protein concentration, which ranges from 33% to 40%. However, mustard meal digestibility is greatly affected by the nutrition imbalance created by glucosinolates' this could be compensated for by ensuring that the total glucosinolate level is in the range 1–1.5 mmol/kg of diet (Jeroch et al., 2013). Rapeseed meal can be treated to remove its antinutritional components in a variety of ways, including physically, chemically, biologically, and through crop breeding. Myrosinase inactivation and shelling are physical techniques, the former involving thickening, bulking, roasting, and microwave therapy. Myrosinase can be activated during milling with water to decrease glucosinolate content. Drying increases the volatilization of isothiocyanates, whereas extraction reduces the survival rate of glucosinolates in rapeseed powder by 40–80% (Nega & Woldes, 2018).

20.5.6 Food Processing

Mustard seeds exhibit certain pharmacological characteristics, including antifungal, antioxidant, and flavouring properties (Rahman et al., 2018), which make the crop ideal for food processing purposes. The food processing industry uses mustards as a food additive in a variety of applications, including flavouring agents, emulsifiers, adhesives, stabilizers, marinades, biodegradable surfactants, suspending agents, etc. (Wiersema et al., 2016). For instance, mustard seed cakes are very successful in decolourizing raw sugar in the sugar processing industry by converting mustard into powdered active mustard cake form for effective absorption (Singh et al., 2012). Glucosinolate-rich mustard is an excellent antimicrobial agent and is active against problematic pathogens such as *Pseudomonas aeruginosa* and *Staphylococcus aureus* for preventing food spoilage of food products in advanced atmospheric packaging technology (Melrose, 2019). Lecithin (lipophilic fatty acid chains) is an amphiphilic lipid found in mustard oil seeds; due to this characteristic, mustard paste mucilage can be utilized as an emulsifier to help keep the components of salad sauces and mayonnaise from separating and to help suspend oils in water (Patterson et al., 2016).

20.6 USING MUSTARD IN NANOPARTILE SYNTHESIS

Production of green nanoparticles (NPs) such as Au, Ag, CuO, FeO, and Se has increased dramatically in recent years. Numerous biological processes have been used in this context, including those involving bacteria, fungi, and algae (Husen & Iqbal, 2019; Husen & Jawaid, 2020; Husen, 2019, 2023a,b; Husen & Siddiqi, 2023; Khan et al., 2023; Ansari et al., 2023). Metal NPs have been synthesized through multiple chemical and physical techniques including the photochemical process, microwave irradiation, planetary ball mill, electron irradiation, and UV irradiation (Khatami et al., 2016). The majority of reported techniques, however, call for considerable amounts of energy, have low conversion efficiencies, and may possess hazardous chemical leaks, etc. (Figure 20.2).

These issues could be adjudged by green synthesis of NPs i.e. using biological resources (bioresources). Green synthesis of metal and alloy NPs is a low-cost, environmentally friendly method for producing well-characterized NPs on an extensive scale (industrially) (Khatami et al., 2015; Husen et al., 2019; Husen, 2020; Godeto et al., 2023; Mengstu et al., 2023; Bachheti et al., 2023; 1Abate et al., 2023a,b; Zebeaman et al., 2023; Belay et al., 2023). Utilizing bio-resource entities like plant leaves or seed extracts can serve as an environmentally friendly substitute (Soltani et al., 2015). Green NP synthesis from plants has intriguing qualities that make it suitable for use as catalysts, biosensors, conductive dyes, and electronic devices in addition to the antimicrobial properties of such NPs (Singh et al., 2014). In one of the initial efforts, seedlings from *in-vitro* cultured plants were used to create silver NPs from *B. juncea* (Shekhawat & Arya, 2009). Frequent studies established that certain parameters such as pH, $AgNO_3$ concentration, and temperature can all be changed to modify the size of the NPs. The hyperaccumulating characteristics of *B. juncea* provide an edge in the synthesis of copper/zinc NPs, revealing a novel approach to green NP synthesis (Qu et al., 2012).

Ansar et al. (2020), observed the anticancerous, antibacterial, and antioxidant activities of silver NPs synthesized from *B. oleracea* and proved its effectiveness in cancer therapy. Similarly, selenium NPs synthesized with a *B. oleracea* extract can be integrated into affordable, biocompatible dental caries prevention products like toothpaste, gum, and mouthwash (Dhanrajandkumar, 2021). Furthermore, the antibacterial activity of silver NPs synthesized from *B. nigra* was evaluated against *Propionibacterium acnes*, *Pseudomonas aeruginosa* and *Klebsiella pneumoniae*showing synergistic activity in combination with vancomycin (Pandit, 2015). In a nutshell, plant-mediated green synthesis of NPs from *Brassica* spp. has shown promise in extending the use of plant-mediated nanocarriers and provides numerous untapped advantages (Table 20.3).

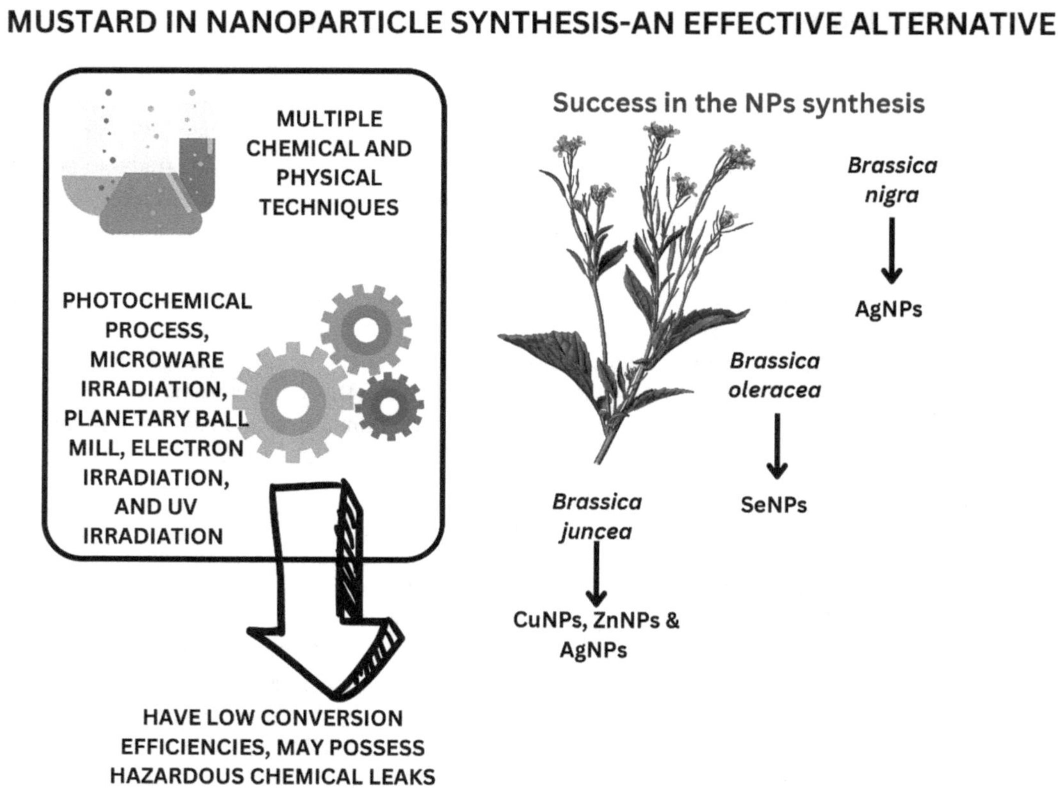

FIGURE 20.2 Mustard as an effective alternative in green nanoparticle synthesis.

20.7 FUTURE PERSPECTIVES AND CONCLUSION

The rapidity of climatic change has resulted in various environmental extremes and posed a major challenge to agricultural productivity. Abiotic stresses cause osmotic (and oxidative) stress, altering the physiological processes and leading to poor growth and development of mustard. Plants activate ROS-detoxifying systems and induce osmotic compatible solutes in response to stress. Numerous genes respond to stress under various stress-tolerance mechanisms. Moreover, breeding programmes are gaining significance in the field of agriculture. The hybridization, transgenic and biotechnological approaches offer tremendous improvements in the desirable traits. The genetic variability in mustard can be considered critical for strengthening abiotic stress tolerance. The genetic × environmental interaction for *B. napus* is the poorest. Therefore, research progress is essential for greater adaptation to wider climatic conditions in the field. Moreover, management practices have an important role to play in tolerating abiotic stress. The impacts of cytokinin, gibberellic acid, and auxin on abiotic stress response need to be studied on the mustard plants Seed treatment with plant growth-promoting substances can improve abiotic stress tolerance, so further research should be focused on the reduction of stress impacts on *Brassica* species to increase the yield and the oil production in the face of global warming.

Brassica has a complex genome and inter-specific hybridization has collected high levels of structural genomic variants and high genetic diversity. Furthermore, the germplasm of mustard crops was built on the traditional and imprecise route of breeding for the long-term, and further improvements need to exploit heterosis with the help of genome-wide selection, omic analysis, gene editing, and speed breeding. It is important to examine the pangenome diversity and complex genomic structures among the *Brassica* members for genome improvement. The decrease in sequencing costs and the increased population size needed to detect the pangenomic variation is

TABLE 20.3
Beneficial Role of Nanoparticles on Mustard Growth and Development

Nanomaterial	Concentration	Crystal Size	Role	Key References
Multi-walled carbon nanotubes	10 mL	30 nm	Promote germination rate, shoot and root growth	Mondal et al., 2011
Ag NPs	500 mg/L	-	Increase root/shoot length, chlorophyll and protein content	Pandey et al., 2014
CuONPs	500 mg/L	-	Antioxidant enzyme activity increased	Nair & Chung, 2015
Fe_3O_4 NPs	500 mg/L	100nm	Increased root–shoot length, chlorophyll, carotenoid and protein content	Praveen et al., 2018
Cu NPs	5.0 ppm	35 nm	Increase length and weight of the seedling	Churilov et al., 2019
SiO_2 NPs	75 µg/ml	80.75 nm	Siliqua length, number of seeds per siliqua increased	Dhingra et al., 2022
CuO NPs	8 mg/L	10 µm	Increased chlorophyll content, net photosynthetic rate, leaf proline content, and antioxidant enzyme activity	Faraz et al., 2022
Zn NPs	20 ppm	30 nm	Increased siliqua number per plant	Kumar et al., 2022
Se NPs	25 mg/L	60.31 nm	Improved germination percentage, shoot length, root length, and total chlorophyll content	Sarkar and Kalita., 2022a
Se NPs	30 mg/L	55.81 nm	Improved percent germination, seedling length, chlorophyll, carbohydrate, and protein content	Sarkar and Kalita., 2022b
ZnO NPs	200 mg/L	200 nm	Enhanced the shoot biomass, chlorophyll content, relative water content, and membrane stability	Geremew et al., 2023

considered to be a pre-requisite. Gene editing and functional genomics for crucial pathways must be further exploited for mustard. The use of high-throughput genomics and phenomics, functional genomics and population genetics along with collecting and evaluating the genetic resources will potentially improve the breeding of this under-exploited crop. This systemic strategy will require additional efforts, such as sharing, exchanging, and integrating data. Such findings will revolutionize mustard genomic improvement to achieve sustainable development of the mustard industry.

REFERENCES

Abate, L., Megra, M.B., Nigussie, Y., Zebeaman, M., Tadesse, M.G., Bachheti, A., Bachheti, R.K. & Husen, A. 2023a. Medicinally important seed extract and seed oil-mediated nanoparticles synthesis and their role in drug delivery and other applications. In: *Secondary Metabolites from Medicinal Plants - Nanoparticles Synthesis and Their Applications* (Eds. Bachheti, R.K. & Bachheti, A). CRC Press, Boca Raton, FL, pp. 199–212. https://doi.org/10.1201/9781003213727

Abate, L., Tadesse, M.G., Bachheti, A., Husen, A. & Bachheti, R.K. 2023b. Medicinal plant-based lignin and its role in nanoparticles synthesis and applications. In: *Secondary Metabolites from Medicinal Plants - Nanoparticles Synthesis and Their Applications* (Eds. Bachheti, R.K. & Bachheti, A.). CRC Press, Boca Raton, FL, pp. 87–110. https://doi.org/10.1201/9781003213727

Ahmad, F.B., Zhang, Z., Doherty, W.O. & O'Hara, I.M. 2019. The prospect of microbial oil production and applications from oil palm biomass. *Biochemical Engineering Journal*, 143: 9–23.

Aires, A., Carvalho, R. & Rosa, E. 2012. Glucosinolate composition of Brassica is affected by postharvest, food processing, and myrosinase activity. *Journal of Food Processing and Preservation*, 3(3): 214–224.

Ansar, S., Tabassum, H. & Aladwan, N.S.M. 2020. Eco-friendly silver nanoparticles synthesis by *Brassica oleracea* and its antibacterial, anticancer and antioxidant properties. *Scientific Reports*, 10(1): 18564.

Ansari, N., Rahman, Q.I., Khan, T., Khan, A., Khan, R., Wagay, J.A. & Husen, A. 2023. Mycogenic-assisted synthesis of nanoparticles and their efficient applications. In: *Advances in Smart Nanomaterials and Their Applications* (Eds. Husen, A. & Siddiqi, K.S.). Elsevier Inc., Cambridge, MA, pp. 569–595.

Ansari, M.K., Ahmad, A., Umar, S., Zia, M.H., Iqbal, M. & Owens, G. 2015. Genotypic variation in phytoremediation potential of Indian mustard exposed to heavy metal stress: A hydroponic study. *International Journal of Phytoremediation*, 17(1–6): 135–144.

Bachheti, R.K., Abate, L., Hunde, Y., Tadesse, M.G., Bachheti, A., Pandey, D.P., Sharma, A., Zebeaman, M. & Husen, A. 2023. Essential oils from medicinal plants and their role in nanoparticles synthesis, characterization, and applications. In: *Secondary Metabolites from Medicinal Plants - Nanoparticles Synthesis and Their Applications* (Eds. Bachheti, R.K. & Bachheti, A.). CRC Press, Boca Raton, FL, pp. 177–198. https://doi.org/10.1201/9781003213727

Batool, M., El-Badri, A.M., Hassan, M.U., Haiyun, Y., Chunyun, W., Zhenkun, Y., Jie, K., Wang, B. & Zhou, G. 2022. Drought stress in *Brassica napus*: Effects, tolerance mechanisms, and management strategies. *Journal of Plant Growth Regulation*, 42: 1–25.

Belay, T., Worku, L.A., Bachheti, R.K., Bachheti, A. & Husen, A. 2023. Nanomaterials: Introduction, synthesis, characterization and applications. *Advances in Smart Nanomaterials and Their Applications* (Eds. Husen, A. & Siddiqi, K.S.). Elsevier Inc., Cambridge, MA, pp. 1–21.

Bhandari, S.R. & Kwak, J.H. 2015. Chemical composition and antioxidant activity in different tissues of *Brassica* vegetables. *Molecules*, 20(1): 1228–1243.

Bortoloti, G.A. & Baron, D. 2022. Phytoremediation of toxic heavy metals by *Brassica* plants: A biochemical and physiological approach. *Environmental Advances*, 8: 100204.

Buval, J., Pandya, D., Pandya, H. & Mankad, A. 2021. Pharmacological activities of *Brassica juncea* l.-A review. *World Journal of Pharmacy and Pharmaceutical Sciences*, 10: 768–782.

Cartea, M.E., Francisco, M., Soengas, P. & Velasco, P. 2010. Phenolic compounds in *Brassica* vegetables. *Molecules*, 16(1): 251–280.

Chand, S., Patidar, O.P., Chaudhary, R., Saroj, R., Chandra, K., Meena, V.K., Limbalkar, O.M., Patel, M.K., Pardeshi, P.P. &Vasisth, P. 2021. *Rapeseed-Mustard Breeding in India: Scenario, Achievements and Research Needs. Brassica Breeding and Biotechnology*, 174.

Churilov, D., Churilova, V., Stepanova, I., Polischuk, S., Gusev, A., Zakharova, O. & Churilov, G. 2019. Size-dependent biological effects of copper nanopowders on mustard seedlings. In *IOP Conference Series: Earth and Environmental Science* (Vol. 392, No. 1, p. 012008). IOP Publishing.

Dai, R. & Lim, L.T. 2015. Release of allylisothiocyanate from mustard seed meal powder entrapped in electrospun PLA-PEO nonwovens. *Food Research International*, 77: 467–475.

Dhanraj, G. & Rajeshkumar, S. 2021. Anticarcinogenic effect of selenium nanoparticles synthesized using *Brassica oleracea. Journal of Nanomaterial*, 2022: 8115585.

Dhingra, P., Sharma, S., Singh, K.H., Kushwaha, H.S., Barupal, J.K., Haq, S. & Kachhwaha, S. 2022. Seed priming with carbon nanotubes and silicon dioxide nanoparticles influence agronomic traits of Indian mustard (*Brassica juncea*) in field experiments. *Journal of King Saud University – Science*, 34(4): 102067.

Dinkova-Kostova, A.T. &Kostov, R.V. 2012. Glucosinolates and isothiocyanates in health and disease. *Trends in Molecular Medicine*, 18(6): 337–347.

Divakaran, M. & Babu, K.N. 2016. *Mustard: Encyclopedia of Food and Health.* Academic Press, Waltham, MA, pp. 9–19.

Du, L., Ding, S., Li, Z., Lv, E., Lu, J. & Ding, J. 2018. Transesterification of castor oil to biodiesel using NaY zeolite-supported La_2O_3 catalysts. *Energy Conservation and Management*, 173: 728–734.

Elansary, H.O., Szopa, A., Kubica, P., Ekiert, H., Ali, H.M., Elshikh, M.S. &ElAnsary, D.O. 2018. Bioactivities of traditional medicinal plants in Alexandria. *Evidence-Based Complementary and Alternative Medicine*, 2018: 1–13.

Faraz, A., Faizan, M., Hayat, S. &Alam, P. 2022. Foliar application of copper oxide nanoparticles increases the photosynthetic efficiency and antioxidant activity in *Brassica juncea*. *Journal of Food Quality, 2022*: 1–10.
Francisco, M., Tortosa, M., Martínez-Ballesta, M.D.C., Velasco, P., García-Viguera, C. & Moreno, D.A. 2017. Nutritional and phytochemical value of Brassica crops from the agri-food perspective. *Annals of Applied Biology*, 170(2): 273–285.
Georgeta, C. 2009. Efficiency of soybean meal replacement by rapeseed meal and/or canola seeds in commercial layer diet. *ArchivaZootechnica*, 12: 27–33.
Geremew, A., Carson, L., Woldesenbet, S., Wang, H., Reeves, S., Brooks, Jr N. & Peace, E. 2023. Effect of zinc oxide nanoparticles synthesized from *Carya illinoinensis* leaf extract on growth and antioxidant properties of mustard (*Brassica juncea*). *Frontiers in Plant Science*, 14: 1108186.
Gilca, M., Tiplica, G.S. & Salavastru, C.M. 2018. Traditional and ethnobotanical dermatology practices in Romania and other Eastern European countries. *Clinics in Dermatology*, 36(3): 338–352.
Godeto, Y.G., Ayele, A., Ahmed, I.N., Husen, A. & Bachheti, R.K. 2023. Medicinal plant-based metabolites in nanoparticles synthesis and their cutting-edge applications: An overview. In: *Secondary Metabolites from Medicinal Plants - Nanoparticles Synthesis and Their Applications* (Eds. Bachheti, R.K. & Bachheti, A.). CRC Press, Boca Raton, FL, pp. 1–34. https://doi.org/10.1201/9781003213727
Goswami, S. & Das, S. 2015. A study on cadmium phytoremediation potential of Indian mustard, *Brassica juncea*. *International Journal of Phytoremediation*, 17(1–6): 583–588.
Grygier, A. 2022. Mustard seeds as a bioactive component of food. *Food Reviews International*, 39: 1–14.
Hamza, M., Ayoub, M., Bin Shamsuddin, R., Mukhtar, A., Saqib, S., Zahid, I., Ameen, M., Ullah, S., Al-Sehemi, A.G. & Ibrahim, M. 2020. A review on the waste biomass derived catalysts for biodiesel production. *Environmental Technology and Innovation*, 21: 101200.
Hedges, L.J. & Lister, C. 2006. Nutritional attributes of brassica vegetables. Report Number: 1618.
Hoekman, S., Broch, A. & Robbins, C. 2011. Review of biodiesel composition, properties and specifications. *Renewable and Sustainable Energy Reviews*, 16: 143–169.
Hu, D., Jing, J., Snowdon, R.J., Mason, A.S., Shen, J., Meng, J. & Zou, J. 2021. Exploring the gene pool of *Brassica napus* by genomics-based approaches. *Plant Biotechnology Journal*, 19(9): 1693–1712.
Husen, A. 2019. Natural product-based fabrication of zinc oxide nanoparticles and their application. In: *Nanomaterials and Plant Potential* (Eds. Husen, A. & Iqbal, M.) Springer International Publishing AG, Cham, pp. 193–291. https://doi.org/10.1007/978-3-030-05569-1_7
Husen, A. 2020. Introduction and techniques in nanomaterials formulation. In: *Nanomaterials for Agriculture and Forestry Applications* (Eds. Husen, A. & Jawaid, M.). Elsevier Inc., Cambridge, MA, pp. 1–14.
Husen, A. 2023a. *Secondary Metabolites Based Green Synthesis of Nanomaterials and Their Applications*. Springer Nature Singapore Pte Ltd. https://doi.org/10.1007/978-981-99-0927-8
Husen, A. 2023b. Nanomaterials from agricultural and horticultural products. Springer Nature Singapore Pte Ltd.. https://link.springer.com/book/9789819934348
Husen, A. & Iqbal, M. 2019. *Nanomaterials and Plant Potential*. Springer International Publishing AG, Cham. https://doi.org/10.1007/978-3-030-05569-1
Husen, A. & Jawaid, M. 2020. *Nanomaterials for Agriculture and Forestry Applications*. Elsevier Inc., Cambridge, MA. https://doi.org/10.1016/C2018-0-02349-X
Husen, A., Rahman, Q.I., Iqbal, M. & Yassin, M.O. 2019. Plant-mediated fabrication of gold nanoparticles and their applications. *In*: *Nanomaterials and Plant Potential* (Eds. Husen, A. & Iqbal, M.) Springer International Publishing AG, Cham, pp. 71–110. https://doi.org/10.1007/978-3-030-05569-1_3
Husen, A. & Siddiqi, K.S. 2023. *Advances in Smart Nanomaterials and Their Applications*. Elsevier Inc., Cambridge, MA. https://doi.org/10.1016/C2021-0-02202-1
Hussain, S., Rehman, A.U., Obied, H.K., Luckett, D.J. & Blanchard, C.L. 2022. Extraction, chemical characterization, in vitro antioxidant, and antidiabetic activity of canola (*Brassica napus* L.) meal. *Separations*, 9(2): 38.
Issariyakul, T. & Dalai, A.K. 2014. Biodiesel from vegetable oils. *Renewable and Sustainable Energy Reviews*, 31: 446–471.
Jahangir, M., Kim, H.K., Choi, Y.H. &Verpoorte, R. 2009. Health-affecting compounds in Brassicaceae. *Comprehensive Reviews in Food Science and Food Safety*, 8(2): 31–43.
Jeroch, H., Kozlowski, K. & Olsztyn. 2013. Improving the nutritional value of poultry feed stuffs: The rapeseed products example. *Lohmann Information*, 48: 29–35.
Karmakar, A., Karmakar, S. & Mukherjee, S. 2010. Properties of various plants and animals' feedstocks for biodiesel production. *Bioresource Technology*, 101(19): 7201–7210.

Khan, T., Tripathi, A.K., Zehra, S., Alvi, A., Ahmad, N., Rahman, Q.I., Ansari, N., Ahmad, N. & Husen, A. 2023. Nano-dimensional materials: An approach towards the biogenic synthesis and their applications. In: *Advances in Smart Nanomaterials and Their Applications* (Eds. Husen, A. & Siddiqi, K.S.). Elsevier Inc., Cambridge, MA, pp. 523–568.

Khandayataray, P. & Murthy, M.K. 2019. Qualitative and quantitative phytochemical screening, antioxidant and anti-inflammatory activities of acetone extract of *Brassica juncea* L. Leaf. *Asian Journal of Research in Biochemistry*, 5: 1–15.

Khatami, M., Azizi, Z., Pourseyedi, S.H. & Najarion, O. 2015. Antibacterial effect of silver Nano particles synthesized by green method against the standard strains *Escherichia coli* K12 and *Escherichia coli* 25922. *Journal of Gorgan University of Medical Sciences*, 17: 1–6.

Khatami, M., Soltani, N. & MeysamPourseyedi, S. 2016. Biogenic synthesis of silver nanoparticles using mustard and its characterization. *International Journal of Nanoscience and Nanotechnology*, 11: 281–288.

Kumar, A., Rana, A.K., Singh, A. & Singh, A. 2019. Bioactivity of methanolic extract of *Brassica juncea* in animal model of diabetes mellitus. *Chinese Herbal Medicines*, 11(4): 434–437.

Kumar, A. S.A., Sakthinathan, G., Vignesh, R., Rajesh Banu, J. & Al-Muhtaseb, A.H. 2019. Optimized transesterification reaction for efficient biodiesel production using Indian oil sardine fish as feedstock. *Fuel*, 253: 921–929.

Kwak, Y., Lee, J. & Ju, J. 2016. Anti-cancer activities of *Brassica juncea* leaves *in vitro*. *EXCLI Journal*, 15: 699–710.

Larkin, R.P. & Halloran, J.M. 2014. Management effects of disease-suppressive rotation crops on potato yield and soilborne disease and their economic implications in potato production. *American Journal of Potato Research*, 91(5): 429–439.

Li, J., Yu, X., Zhang, C., Li, N. & Zhao, J. 2022. The application of CRISPR/Cas technologies to Brassica crops: Current progress and future perspectives. *Abiotech*, 3(2): 146–161.

Melrose. 2019. The glucosinolates: A sulphur glucoside family of mustard anti-tumour and antimicrobial phytochemicals of potential therapeutic application. *Biomedicines*, 7(3): 62.

Mengstu, A., Esubalew, S., Abate, L., Bachheti, R.K., Husen, H. & Bachheti, A. 2023. Flavonoid-mediated nanoparticles synthesis, characterization, and applications. In: *Secondary Metabolites from Medicinal Plants - Nanoparticles Synthesis and Their Applications* (Eds. Bachheti, R.K. & Bachheti, A.). CRC Press, Boca Raton, FL, pp. 35–52. https://doi.org/10.1201/9781003213727

Mondal, A., Basu, R. & Das, S. 2011. Beneficial role of carbon nanotubes on mustard plant growth: An agricultural prospect. *Journal of Nanoparticle Research*, 13(10): 4519–4528.

Nair, P.M.G. & Chung, I.M. 2015. Study on the correlation between copper oxide nanoparticles induced growth suppression and enhanced lignification in Indian mustard (*Brassica juncea* L.). *Ecotoxicology and Environmental Safety*, 113: 302–313.

Nega, T. & Woldes, Y. 2018. Review on nutritional limitations and opportunities of using rapeseed meal and other rape seed by-products in animal feeding. *Journal of Nutritional Health and Food Engineering*, 8(1): 00254.

Ogidi, O.I., George, D.G., Enenebeaku, U.E., Esie, N.G. & Akpan, U.M. 2019. Efficacy evaluation of extracts of *Brassica juncea* (Brown mustard) seeds as potential antimicrobial agent against pathogenic microbes. *Journal of Medicinal Plants*, 7: 263–265.

Pandey, C., Khan, E., Mishra, A., Sardar, M. & Gupta, M. 2014. Silver nanoparticles and its effect on seed germination and physiology in *Brassica juncea* L. (Indian mustard) plant. *Advanced Science Letters*, 20(7): 1673–1676.

Pandit, R. 2015. Green synthesis of silver nanoparticles from seed extract of *Brassica nigra* and its antibacterial activity. *Nusantara Bioscience*, 7(1): 15–19.

Patterson, C. 2016. *Mustard: Protein, Mucilage and Bioactives*. The Saskatchewan Mustard Development Commission, Saskatoon, SK.

Paul, S., Bin Emran, T., Saha, D. & Zahid Hosen, S. 2012. Phytochemical and antimicrobial activity of the plant extracts of *Brassica oleracea* against selected microbes. *Asian Journal of Pharmacy and Medical Science*, 2: 30–34.

Paul, S., Geng, C.A., Yang, T.H., Yang, Y.P. & Chen, J.J. 2019. Phytochemical and health-beneficial progress of turnip (*Brassica rapa*). *Journal of Food Science*, 84(1): 19–30.

Petropoulos, S., Di Gioia, F. & Ntatsi, G. 2017. Vegetable organosulfur compounds and their health promoting effects. *Current Pharmaceutical Design*, 23(19): 2850–2875.

Pooniya, V., Biswakarma, N., Parihar, C.M., Swarnalakshmi, K., Lama, A., Zhiipao, R.R. & Singh, K. 2021. Six years of conservation agriculture and nutrient management in maize–mustard rotation: Impact on soil properties, system productivity and profitability. *Field Crops Research*, 260: 108002.

Praveen, A., Khan, E., Ngiimei, D.S., Perwez, M., Sardar, M. & Gupta, M. 2018. Iron oxide nanoparticles as nano-adsorbents: A possible way to reduce arsenic phytotoxicity in Indian mustard plant (*Brassica juncea* L.). *Journal of Plant Growth Regulation*, 37(2): 612–624.

Qu, J., Luo, C., Cong, Q. & Yuan, X. 2012. A new insight into the recycling of hyperaccumulator: Synthesis of the mixed Cu and Zn oxide nanoparticles sing *Brassica juncea* L. *International Journal of Phytoremediation*, 14(9): 854–860.

Rahman, M., Khatun, A., Liu, L. & Barkla, B.J. 2018. Brassicaceae mustards: Traditional and agronomic uses in Australia and New Zealand. *Molecules*, 23(1): 231.

Rasera, B.G., Hilkner, M.H., de Alencar, S.M. & de Castro, R.J.S. 2019. Biologically active compounds from white and black mustard grains: An optimization study for recovery and identification of phenolic antioxidants. *Industrial Crops and Products*, 135: 294–300.

Rathore, S.S., Shekhawat, K. & Das, A. 2019. Phytoremediation mechanism in Indian mustard (*Brassica juncea*) and its enhancement through agronomic interventions. *Proceedings of National Academy of Sciences*, 89: 419–427.

Salido, A.L., Hasty, K.L., Lim, J.M. & Butcher, D.J. 2003. Phytoremediation of arsenic and lead in contaminated soil using Chinese brake ferns (*Pteris vittata*) and Indian mustard (*Brassica juncea*). *International Journal of Phytoremediation*, 5(2): 89–103.

Sánchez, M., Aracil, J. & Martínez, M. 2013. Current status and prospects of biodiesel production from Brassica species. *International Review of Chemical Engineering (IRECHE)*, 5: 342–350.

Sarkar, R.D. &Kalita, M.C. 2022a. Se nanoparticles stabilized with *Allamanda cathartica* L. flower extract inhibited phytopathogens and promoted mustard growth under salt stress. *Heliyon*, 8(3): e09076.

Sarkar, R.D. &Kalita, M.C. 2022b. Green synthesized Se nanoparticle-mediated alleviation of salt stress in field mustard, TS-36 variety. *Journal of Biotechnology*, 359: 95–107.

Sawicka, B., Kotiuk, E., Bienia, B., Krochmal-marczak, B. & Wójcik, S. 2007. The importance of mustard (Sinapis alba) Indian mustard (*Brassica juncea* var. *sareptana*) and Black Mustard (*Brassica nigra*) in nutrition and phytotherapy. *Acta Scientiarum Polonorum, Technologia Alimentaria*, 2: 17–27.

Shankar, S., Segaran, G., Sundar, R.D.V., Settu, S. & Sathiavelu, M. 2019. Brassicaceae - A classical review on its pharmacological activities. *International Journal of Pharmaceutical Sciences*, 55: 107–113.

Shekhawat, G.S. & Arya, V. 2009. Biological synthesis of ag nanoparticles through in vitro cultures of *Brassica juncea* C. Zern. *Advanced in Materials Research*, 67: 295–299.

Singh Chauhan, E., Tiwari, A., Singh, A. & Anamika Tiwari, C. 2016. Phytochemical screening of red cabbage (*Brassica oleracea*) powder and juice - A comparative study. *Journal of Medicinal Plants Studies*, 4: 196–199.

Singh, K., Bharose, R., Verma, S.K. & Singh, V.K. 2012. Potential of powdered activated mustard cake for decolorising raw sugar. *Journal of the Science of Food and Agriculture*, 93(1): 157–165.

Singh, K., Panghal, M., Kadyan, S., Chaudhary, U. & Parkash, Y.J. 2014. Antibacterial activity of synthesized silver nanoparticles from *Tinospora cordifolia* against multi-drug resistant strains of *Pseudomonas aeruginosa* isolated from burn patients. *Journal of Nanomedicine and Nanotechnology*, 5: 2.

Soltani, N.M., Shahidi-Bonjar, G.H. & Khaleghi, N. 2015. Biosynthesis of gold nanoparticles using *Streptomyces fulvissimus* isolate. *Nanomediciene Journal*, 2: 153–159.

Tang, D.Y.Y., Yew, G.Y., Koyande, A.K., Chew, K.W., Vo, D.V.N. & Show, P.L. 2020. Green technology for the industrial production of biofuels and bioproducts from microalgae: A review. *Environmental Chemistry Letters*, 18(6): 1967–1985.

Taufiqurrahmi, N. & Bhatia, S. 2011. Catalytic cracking of edible and non-edible oils for the production of biofuels. *Energy and Environmental Science*, 4(4): 1087.

Thanaseelaan, V. 2013. Proximate analysis, mineral and amino acid profiles of deoiled rapeseed meal. *International Journal of Food, Agriculture and Veterinary Sciences*, 3: 66–69.

Thangam, R., Suresh, V., Rajkumar, M. Vincent, J.D., Gunasekaran, P. Anbazhagan, C., Kaveri, K. & Kannan, S. 2013. Antioxidant and in vitro anticancer effect of 2-pyrrolidinone rich fraction of *Brassica oleracea* var. capitata through induction of apoptosis in human cancer cells. *Phytotherapy Research* 27(11): 1664–1670.

Thangaraj, B., Solomon, P.R., Muniyandi, B., Ranganathan, R. & Lin, L. 2019. Catalysis in biodiesel production-a review. *Clean Energy*, 3(1): 2–23.

Tian, Y. & Deng, F. 2020. Phytochemistry and biological activity of mustard (*Brassica juncea*): A review. *CyTA-Journal of Food*, 18(1): 704–718.

Uddin, M.S., Millat, M., Islam, M.S., Hussain, M., Uddin, M., Siddiqui, S.A. & Ferdous, M. 2020. Exploration of in vitro thrombolytic, anthelminthic, cytotoxic and in vivo anxiolytic potentials with phytochemical screening of flowers of *Brassica nigra*. *Future Journal of Pharmaceutical Sciences*, 6: 1–9.

Wiersema, J.H. & Leon, B. 2016. *World Economic Plants: A Standard Reference*. CRC Press, Melbourne.

Zebeaman, M., Bachheti, R.K., Bachheti, A., Pandey, D.P. & Husen, A. 2023. Green and cost-effective nanoparticles synthesis from some frequently used medicinal plants and their various applications. In: *Secondary Metabolites from Medicinal Plants - Nanoparticles Synthesis and Their Applications* (Eds. Bachheti, R.K. & Bachheti, A.). CRC Press, Boca Raton, FL, pp. 287–304. https://doi.org/10.1201/9781003213727

Zulqarnain, A.M., Yusoff, M.H.M., Nazir, M.H., Zahid, I., Ameen, M., Sher, F., Floresyona, D., Budi Nursanto, E.A. 2021. A Comprehensive review on oil extraction and biodiesel production technologies. *Sustainability*, 13(2): 788.

21 Nutmeg Seeds (*Myristica fragrans*)

Mansi Shah, Kruti Dave, and Indu Anna George

21.1 INTRODUCTION

Nutmeg is a popular spice across the globe, belonging to the Myristicaceae family which consists of 18 genera and 300 species (Kumar Verma et al., 2021; Malik et al., 2022). The binomial of the nutmeg plant is *Myristica fragrans* (Houtt.) The nutmeg tree is the only plant that produces two spices: one is nutmeg, i.e., the seed, and the other is mace, which is the aril i.e., the protective coating of the seeds.

The documented discovery of nutmeg was in 1512 by the Portuguese in the Banda Islands in the Moluccas to the east of Indonesia and since then it has been traded across the world (Beckerman & Persaud, 2019; Obranović et al., 2020; Ashokkumar et al., 2022). It has been cultivated across the globe around the tropical regions, such as Bhutan, Grenada, Indonesia, India, Laos, Mauritius, Nepal, North Australia, Sri Lanka, South Africa, and the USA, due to its use in global cuisine (Obranović et al., 2020; Barman et al., 2021; Kumar Verma et al., 2021; Ashokkumar et al., 2022). The presence of nutmeg trees has also been reported in the forests of China, Malaya, Penang, Sumatra, and Singapore (Tripathi & Dwivedi, 2015). Bhutan, Guatemala, Indonesia, India, Laos, Nepal, and Sri Lanka produce more than 97% of global nutmeg (Obranović et al., 2020). In India, nutmeg is cultivated in hotter and more humid parts at altitudes above sea level of 750 m and annual rainfall of 150–300 cm, such as Kokan, Madras, and Karnataka (Tripathi & Dwivedi, 2015).

The common trading name of nutmeg and mace spice are mace, magic, muscdier, nux moschata, myristica oil, and muskatbaum (Beckerman & Persaud, 2019) Since nutmeg has been cultivated across the globe since the sixteenth century, it has acquired native names from all over the world. Native names of nutmeg are as follows: Arabic: Josat at-Tib, Bisbasah, Jainsiban; Chinese: rou dou kou; English: nutmeg, mace; French: muscade; Greek: moschokarido; Hindi: jaiphal; Indonesian: Pala; kannada: Jadipattiri; Latin: Merisiniae; Malayalam: Jadi-patri; Persian: bizbaz, charkhun; Sanksrit: Jatiphala; Shami: Warkiyah; Sinhala: Vasavasi; Tamil: Jadi-pattiri; Telugu: Japatri; Unani: Jaiphal, Javitri; and Urdu: Jaiphal, Jouzbuwwa, Maakhan valadin.

Myristica fragrans is a medium-sized (20–25 ft) evergreen tree. It has soft bark with a greyish colour and slender spreading branches. The plant thrives in warm and humid climates with 150–250 cm annual rainfall up to an altitude of 1,000 m above sea level. The leaves of the plant are aromatic, and arranged alternately; they are oblong or globous and acuminate in shape. The colour of the leaves is bright green with a shiny dorsal surface. The flowers are compound or sometimes forked, mostly dioecious, occasionally monoecious with variable sex expression and are situated on small axillary and sub-umbellate racemes. The plant's fruit is spherical in shape and splits into two longitudinal valves. The skin of the fruit is yellowish and has a fleshy pericarp surrounding one seed. The aril of the mace fruit is a crimson, fleshy covering that is intricate and folded and encases the nut when moist. Upon drying, it becomes much harder, takes on a yellowish-brown hue, and becomes quite brittle. The 'nut' (seed) is generally oval or broad-ovate and has a shiny, coarse, dark-brown outer shell. The interior of the shell is pale and smooth. The kernel of the seed is initially soft, oval, and pale brown, but quickly shrivels and develops irregular, vertical furrows on its surface. It is also

DOI: 10.1201/9781003387046-21

FIGURE 21.1 The nutmeg tree, fruits, and seeds. The nutmeg tree (a) The nutmeg fruits (b) The opened fruits showing the fresh mace and nutmeg (c) dried mace (d) the nutmeg in its shell (e) the opened shell (f) the broken nutmeg (g). The authors wish to thank Mr. Yam Bahadur Nepali and Ms. Simone Vaz (Department of Life Sciences, University of Mumbai) for their contribution to Figure 21.1 in this chapter.

rich in oil. The nutmeg tree produces fruit throughout the year, but the period between April and November is the most favourable time for harvesting (Tripathi & Dwivedi, 2015; Obranović et al., 2020; Kumar Verma et al., 2021; Ashokkumar et al., 2022) (Figure 21.1).

21.2 TRADITIONAL USES

The nutmeg seed has two parts: the outer, red aril called mace and an inner, brown kernel called nutmeg, which are both used as spices and are traded in whole or powdered form. However, whole dried form is in greater demand than the powdered form due to its longer shelf life. Drying of nutmeg is done commercially in many countries, although, in Indonesia, farmers still employ a traditional way of drying nutmeg, which involves drying it in the sun for four or five days, depending on the level of moisture, on plastic bags, jute sacks, tarps, or cloth material. The purpose of using a sack or fabric is to keep the temperature stable. Mace is sun dried for a few hours before being aerated. This process is repeated until the mace is dry (Budi et al., 2018).

Nutmeg has a sweet, spicy and nutty taste. Nutmeg has become widespread worldwide since the sixteenth century, and has found its place in many of the native dishes all over the world. Most spices are added to savoury dishes, whereas nutmeg can be an ingredient for sweet pastries and desserts, such as ginger bread (Poliński et al., 2022). Usually, ground nutmeg seeds are used as a flavouring agent in dishes; to retain its flavour, it is usually added at the end of cooking (Budi et al., 2018). Similar to nutmeg, mace also has a sweet, spicy, and nutty taste but is milder than nutmeg and it is also used widely throughout the world in sweet and savoury dishes. In Indonesia, nutmeg is used in stews, sauces, and curries, whereas, in China, it is used in hotpots. Many traditional Indian sweet dishes have nutmeg as an ingredient. Nutmeg is used to give a spicy flavour to the food and is considered to help warm a body. Candied nutmeg seeds are very popular in Malaku as '*pala gula*' and Java as '*pala manis*' (Budi et al., 2018). Fat derived from nutmeg seed, known as nutmeg butter, is used in perfumery, tobacco, and toothpaste (Obranović et al., 2020). Not only seed and mace but the seed kernel is also used in culinary dishes; for example, liquid smoke produced from the nutmeg shell is used in the sea cucumber smoking process to yield the aroma of sea cucumber (Netty Salindeho, 2017)

Apart from their culinary uses, nutmeg and mace have also found application in natural and traditional medicines in Asia, Africa, China, Europe, and America. The essential oil from the mace, and the nutmeg kernel and seeds are also used worldwide as potential medicines (Table 21.1).

The individual dosage of nutmeg recommended by the Ayurveda and Siddha medicinal systems is 0.5–1 g while the Unani system suggests 4–5 g for nutmeg and 1.5 g to 10.5 g for mace (Nifras et al., 2021). Many medicinal formulations are made from the nutmeg in Ayurveda, Unani, and Siddha medicinal systems as listed below (Nifras et al., 2021; Table 21.2).

TABLE 21.1
Various Traditional Medicine Systems have Described Therapeutic Effect of Nutmeg

Ayurveda:	*Deepana* (appetizer, digestant) *Grahi* (antidiarrhoeal), *Vrsya* (aphrodisiac), *Mukhalednasaka, Mukhadaurgandhyasnasaka* (antihallitosis), *Kaphavatapana* (Alleviates Vata and Kapha dosas)	Nifras et al., 2021
Siddha	*Akadduvayvakatri* (carminative), *Kaamamperukki* (aphrodisiac), *Manamootti* (aromatic), *Moorchchaiyundaakki* (sedative/hypnotic), *Uramaakki* (nutrient), and *Veppamundaakki* (stimulant)	Nifras et al., 2021
Unani	*Ishal* (diarrhoea), *Falij* (paralysis), *Wajaul Mafasil* (arthritis), *Amraz-e-Qalb* (cardiac diseases), *Su-e-Hazm* (indigestion), *Zof-e-Bah* (sexual debility)	Nifras et al., 2021
Traditional Chinese medicine	Analgesic, aphrodisiac, antidiarrhoeal, anti-inflammatory, to treat dysmenorrhoea, painful menstruation, abdominal pain, and liver problems	Sharifi-Rad et al., 2017
Tibetan medicine	To manage gastro-intestinal conditions such as colic, nausea, diarrhoea, and flatulence	Naeem et al., 2016

TABLE 21.2
Formulations Using Nutmeg in Ayurveda, Siddha and Unani Medicine

Ayurveda	Jatiphaladi Churna; Jatiphaladi Vati
Siddha	Ashtapayiravak Kulikai; Saampiranippoo Pathangam; Surapunga Vilvathi Ilakam; Ilaku Santhanaathi Thailam; Kapaada Maathirai
Unani	Laboob-e-Kabir; Laboob-e-Asrar; Jawarish-Ood-Shirin; Habb-e-Mumsik.

21.3 PHARMACOLOGICAL USES

Essential oils (EOs) are a chemical concoction of strongly scented aromatic compounds that are derived from aromatic plants. The volatile chemicals found in EOs serve an important environmental purpose by attracting specific insects for plant pollination and protecting plants from harmful bacteria, fungi, insects, and viruses. Thus, they are the most important part of the plant, which is why they are referred to as 'essential'. Geographic location, environment, and stage of maturation are all factors that might affect plants and the concentration and composition of essential oils in them. Plant essential oil chemical compositions vary depending on these factors. EOs are complex mixtures of bioactive substances with a range of structural types, including mono-, sesqui-, and diterpenes, phenol components, sulphur-containing substances, and phenylpropanoid compounds (Mohamed & Alotaibi, 2022). *Myristica fragrans* seeds and mace produce a range of secondary metabolites, the main components of which are phenolic compounds (primarily caffeic, ferulic, and protocatechuic acids, lignans/neolignans, and diarylalkanes) and essential oils (terpenes and phenylpropanoids). Additionally, catechins, epicatechins, flavonoids, and cyanidins, which are polyphenolic pigments, are also found in them (Abourashed & El-Alfy, 2016). Different crude extracts of *Myristica fragrans* seeds and mace containing essential oil have shown many potential pharmacological activities, such as antioxidant, antimicrobial, anti-inflammatory, anticancer, antidiabetic, anti-obesity, and antidepressant properties to name but a few.

21.3.1 Antimicrobial (Antibacterial and Antifungal) Properties

The increasing bacterial resistance to present antibiotics and the exorbitant costs of recent generations of antibiotics may be problems that can be solved thanks to the effective antibacterial capabilities of EOs. The vast array of bioactive elements found in EOs interfere with the defence mechanisms of different bacteria by interacting with their cellular enzymes or cell structures, promoting the rate of microbial cell death. Hence, EOs have emerged as a rich potential source of novel and potent bioactive compounds with a range of therapeutic advantages (Mohamed & Alotaibi, 2022). The essential oil extracted from nutmeg seeds has compounds that possess antimicrobial properties. Major constituents of nutmeg EO are terpene hydrocarbons like pinene, sabinene, camphene, *p*-cymene, terpinene, phellandrene, myrcene, and limonene that make up 60–80% of the essential oil, aromatic ethers like myristicin, elemicin, safrole, and eugenol derivatives that form 15–20% of the oil and oxygenated terpenes like geraniol, linalool, and terpineol that make up the remaining 5–15% of the oil (Rashidian et al., 2022).

21.3.2 Antibacterial Properties

The two main antibacterial components identified in nutmeg seeds are myristic acid and trimyristin, along with others such as α-pinene and β-pinene, which are monoterpene hydrocarbons, carvacrol and γ-cymene that are involved in microbial cell disruption by increasing the cell membrane permeability to protons and potassium ions, destroying the proton motive force followed by ATP pool depletion. (Naeem et al., 2016) Various solvent extractions have been carried out to assess the antibacterial property of the essential oil from nutmeg seed and mace, including both aqueous (Suthisamphat et al., 2020) and organic solvents like methanol (Urabee et al., 2022; Okiki et al., 2023), ethanol (Shafiei et al., 2012; Abourashed & El-Alfy, 2016; Suthisamphat et al., 2020; Okiki et al., 2023), chloroform (Naeem et al., 2016), ethyl acetate (Shafiei et al., 2012; Abourashed & El-Alfy, 2016), and acetone (Urabee et al., 2022), against both Gram-positive bacteria such as *Staphylococcus aureus* (Ibrahim et al., 2020; Rashidian et al., 2022; Ashokkumar et al., 2022; Urabee et al., 2022), *Listeria monocytogenes, Bacillus cereus, Streptococcus iniae*, (Rashidian et al., 2022), *Enterococcus faecalis, Streptococcus mutans* (Rashidian et al., 2022; Ashokkumar et al., 2022; Urabee et al., 2022), *Staphylococcus epidermis* (Ashokkumar et al., 2022; Urabee et al., 2022),

Corynebacterium accolens, Propionibacterium acnes, Staphylococcus aureus ATCC 25923 (Okiki et al., 2023), *Mycobacterium intracellulare* (Ibrahim et al., 2020), *Streptococcus mutans* ATCC 25175, *Streptococcus mitis* ATCC 6249, *Streptococcus salivarius* ATCC 13419 (Shafiei et al., 2012), *Streptococcus salivarius, Streptococcus sanguis* and *Lactococcus casei* (Rashidian et al., 2022), and Gram-negative bacteria such as *Klebsiella pneumoniae* (Rashidian et al., 2022; Ashokkumar et al., 2022; Urabee et al., 2022), *Escherichia coli* (Ibrahim et al., 2020; Rashidian et al., 2022; Urabee et al., 2022;), *Proteus vulgaris* (Rashidian et al., 2022; Ashokkumar et al., 2022), *Pseudomonas aeruginosa* (Ibrahim et al., 2020; Rashidian et al., 2022), *Pasteurella multocida* (Ashokkumar et al., 2022), *Salmonella typhimurium* (Ashokkumar et al., 2022), *Shigella dysenteriae* (Ashokkumar et al., 2022), *Eschericia coli* ATCC 35218 (Okiki et al., 2023), *Aggregatibacter actinomycetemcomitans* ATCC 29522, *Porphyromonas gingivalis* ATCC 33277, *Fusobacterium nucleatum* ATCC 25586 (Shafiei et al., 2012; Rashidian et al., 2022), *Helicobacter pylori* ATCC 43504 (Suthisamphat et al., 2020), *Yersinia ruckeri*, and *Aeromonas hydrophila* (Rashidian et al., 2022).

The results of the *in-vitro* disc diffusion method indicated a wide range of oil extract concentrations that potentially inhibited the growth of test organisms. Based on one study, among the three concentrations (0.5, 1, and 2%) of nutmeg essential oil extracts used, the highest zone of inhibition was obtained for 2% extract against *Streptococcus iniae, Yersinia ruckeri*, and *Aeromonas hydrophila* (Rashidian et al., 2022). Similar observations were obtained from the use of 0.2–0.5% of nutmeg essential oil against *Enterococcus faecalis, Streptococcus mutans* and *Pasteurella multocida* which was able to completely suppress their growth, and this effect could be enhanced by addition of 1% magnesium alumino-metasilicate to the oil in order to target a broader range of bacterial species. The results of another study indicated that, among the concentrations tested (20–100%) of Indonesian nutmeg seed essential oil extracts, 60% concentration gave the highest zone of inhibition against bacterial strains of *Staphylococcus aureus, Staphylococcus epidermis, Shigella dysenteriae*, and *Salmonella typhi* (Ashokkumar et al., 2022). A study using the agar well diffusion method indicated significant antibacterial activity against *Staphylococcus aureus, Proteus vulgaris*, and *Klebsiella pneumoniae* with the use of 25% (v/v) concentration of nutmeg essential oil (Ashokkumar et al., 2022). A different study, testing methanol and acetone extracts of nutmeg seeds, using the Soxhlet extraction method, against the bacterial strains *Staphylococcus aureus, Staphylococcus epidermidis, Escherichia coli*, and *Klebsiella* sp. indicated that, whereas the methanolic extracts remained ineffective, the acetone extract displayed antibacterial activity with a yield of 2.11 g/100 g and a zone of inhibition measuring between 11 and 15 mm in diameter (Urabee et al., 2022).

A few other studies conducted described the antibacterial activity in terms of the Minimum Inhibitory Concentration (MIC value) against the test pathogens. Nutmeg essential oil demonstrated potent antibacterial activity against *Salmonella typhimurium* with a MIC value of 1.6 μg/mL (Ashokkumar et al., 2022). The ethanolic and aqueous extracts of mace when tested against *Helicobacter pylori* ATCC 43504 and six clinical strains, the ethanolic extract was found to be more effective, with MIC values of 125–250 μg/mL as compared with the aqueous extract with a MIC value of 500 μg/mL against the positive control of clarithromycin with an MIC value of 0.5 μg/mL (Suthisamphat et al., 2020). In another comparative study between methanolic and ethanolic extracts of nutmeg seed essential oils, the methanolic extract produced significantly higher susceptibilities than the ethanolic extracts against *Corynebacterium accolens, Propionibacterium acnes, Staphylococcus aureus* ATCC 25923, and *Escherichia coli* ATCC 35218 (Okiki et al., 2023).

21.3.3 Antifungal Properties

Components present in nutmeg seed responsible for its antifungal properties were identified as myristicin (Ibrahim et al., 2020; Seneme et al., 2021), derivatives of neolignans and eugenol (Nagja et al., 2016), β-caryophyllene (Naeem et al., 2016), and lignans like meso-dihydroguaiaretic acid, nectandrin-B and erythron austrobailignan-6 (Naeem et al., 2016; Abourashed & El-Alfy, 2016).

This antifungal potency of the nutmeg seed essential oil was tested against both unicellular fungi, like *Candida glabrata*, *Candida krusei*, (Ibrahim et al., 2020), and *Candida albicans* (Ibrahim et al., 2020; Urabee et al., 2022), and filamentous fungi like *Colletotrichum gloeosporioides*, *Colletotrichum musa*, *Fusarium oxysporum*, *Fusarium semitectum*, *Aspergillus niger*, *Aspergillus glaucus* (Ashokkumar et al., 2022), *Aspergillus flavus*, *Aspergillus ochraceus*, (Seneme et al., 2021), *Aspergillus fumigatus*, and *Cryptococcus neoformans* (Ibrahim et al., 2020). Methods used for extraction of essential oil were hydrodistillation (Ibrahim et al., 2020) and Soxhlet, using organic solvents like methanol and acetone (Urabee et al., 2022).

According to the Soxhlet method's results, acetone solvent had a higher extraction yield than methanol solvent (2.11 g per 100 g), and it also inhibited the growth of *Candida albicans* with a zone of inhibition that measured 15 mm in diameter (Urabee et al., 2022). In the case of the hydrodistillation method, the four selected essential oil distillation time fractions were evaluated, with the fraction obtained at time 120–240 mins showing greater potency against *Cryptococcus neoformans* at an IC_{50} value of 195.206 g/mL than the other three fractions at time intervals of 1–2.5 mins, 10 –30 mins, and 0 –240 mins. Furthermore, the GC-MS profile of the abovementioned 120–240 mins fraction revealed the presence of myristicin as being responsible for the antifungal activity. (Ibrahim et al., 2020). Similarly, another study reported the presence of myristicin in the range of 10–21% could strongly inhibit the growth of *Aspergillus ochraceus* and *Aspergillus flavus* and stated that nutmeg essential oil could act as a food preservative as it protected the food from production of aflatoxins by *Aspergillus flavus* (Seneme et al., 2021). The results of studies conducted on other fungi showed strong inhibition (97–98%) of the growth of *Colletotrichum gloeosporioides* and *Colletotrichum musa*, whereas a moderate inhibition range (60–78%) was observed for *Fusarium oxysporum*, *Fusarium semitectum*, *Aspergillus niger*, and *Aspergillus glaucus* (Ashokkumar et al., 2022).

21.3.4 Antioxidant Properties

The accumulation of free reactive oxygen species (ROS) during metabolic processes causes oxidative stress in a cell, leading to its damage. Antioxidants protect the cell from these free radicals by scavenging or neutralizing them (Warsito, 2021). Different assay methods, such as the deoxyribose assay, Trolox equivalent antioxidant capacity (TEAC) assay, DPPH radical scavenging assay, and ferric reducing antioxidant power (FRAP) assay (Nagja et al., 2016), radioprotective potential assay, hydroxyl radical (OH·) assay, and BHT equivalent (β-carotene/linoleic acid) assay (Warsito, 2021) were employed to evaluate this scavenging property of various natural compounds present in nutmeg. In a comparative study of methanolic extracts of nutmeg seed, mace, and flesh, the DPPH and FRAP test methodologies revealed that nutmeg seed oil had the strongest antioxidant free radical scavenging property whereas mace had none, due to its lower polyphenolic concentration (Abourashed & El-Alfy, 2016). The drying methods influenced the quality of essential oils obtained from seeds. The shade drying method had the greatest antioxidant percentage at 72.82, as opposed to fresh (without drying) at 24.18 and sun-drying at 45.56, respectively. These results were supported by the value of 2.94 ± 0.09 μM/g as Trolox equivalents for DPPH assay and reducing power with an EC_{50} value of 181.4 μg/mL (Warsito, 2021).

The compounds present in nutmeg seeds that possessed this ROS-scavenging property were found to be lignans and their glycosides, β-caryophyllene (Naeem et al., 2016; Warsito, 2021), several terpenoids, like sabinene, myristicin, eugenol (Nagja et al., 2016; Warsito, 2021), tannins and flavonoids (Abourashed & El-Alfy, 2016), elemicin, 4-terpineol, myristicin, *trans*-sabinene hydrate, safrole, and isoeugenol of which the most potent antioxidant component identified was elemicin (Warsito, 2021). Further analysis showed that these potent anti-oxidative capabilities are due to the presence of hydrogen atoms in the allylic or benzylic positions in compounds like β-caryophyllene and eugenol, and the ability of eugenol to boost the activity of enzymes like superoxide dismutase, catalase, glutathione peroxidase, glucose-6-phosphate dehydrogenase, and glutamine transferase.

Also, the breakdown of lignans and their glycosides resulted in structural bioactive components that resemble catechol which are easy electron or phenolic hydrogen donors to acceptors like ROS and hence act as a good antioxidant agent. (Naeem et al., 2016).

21.3.5 Anti-inflammatory Properties

The anti-inflammatory effects of *M. fragrans* were examined with respect to oedema caused by carrageenan in rats and for vascular permeability caused by acetic acid in mice. The results proposed that the anti-inflammatory activity in mace is due to the presence of myristicin that brings about inhibition of chemokines, cytokines, nitrous oxide, and growth factors in double- stranded RNA (dsRNA)-stimulated macrophages *via* the calcium pathway (Lee & Park, 2011). In order to test the effectiveness of aqueous and ethanolic extracts as anti-inflammatory agents, a number of assay methods were employed, such as the inhibition of lipopolysaccharide (LPS)-induced nitric oxide (NO) in a RAW264.7 cell line, the sulphorhodamine B (SRB) assay, the DPPH radical scavenging assay, and the ABTS radical cation decolourization assay. According to the results, the ethanolic extract displayed cytotoxic activity and an inhibitory impact on LPS-induced NO release with IC_{50} values of 82.19 g/mL and 26.06 g/mL, respectively. The EC_{50} values for the DPPH and ABTS antioxidant assays were 13.41 g/mL and 12.44 g/mL, respectively (Suthisamphat et al., 2020).

The methanolic extract, ethanolic extract, and supercritical CO_2 extract from seeds of *M. fragrans*, which are used to treat inflammatory illnesses, had inhibitory effects on the formation of nitric oxide and inflammatory cytokines (Cao et al., 2015; Nagja et al., 2016). The Folin-Ciocalteu and DPPH methods were used to determine the total phenol content and radical scavenging capacities of the acetone extracts of nutmeg seeds, and they were qualitatively compared using the microfluidic paper-based analytical device (PADs) as a new platform. In the DPPH assay, acetone extract displayed a considerable quantity of antioxidant components. The effect of nutmeg acetone extract on the activity of the enzyme cyclooxygenase (COX-2) suggested that nutmeg extract had a greater ability to inhibit COX-2 activity than the anti-inflammatory medicine aspirin (Urabee et al., 2022). Various compounds, such as γ-terpinene, linalool, myristicin, safrole, eugenol, and elemicin present in nutmeg essential oil extracts are known to possess analgesic and anti-inflammatory activity. This activity is associated with its capability to inhibit the expression of blood substance P level and COX-2. The γ-terpinene could inhibit pro-inflammatory mediators and affect the neurotransmission pathways at the SNc level (such as substance P or CGRP). In capsaicin, formalin, and glutamate tests, it may have an antinociceptive effect at low doses affecting the opioid system *via* K^+ ATP channels and the cholinergic system. Eugenol was able to suppress the production of thromboxane B2, prostaglandin, and leukotriene whereas myristicin showed suppression of several inflammatory regulators, such as calcium, NO, the interleukins IL-6 and IL-10, macrophage inflammatory protein (MIP)-1α, MIP-1β, granulocyte-macrophage colony-stimulating factor (GM-CSF), interferon-inducible protein-10, the monocyte chemotactic proteins (MCP)-1, MCP-3, and leukaemia inhibitory factor (LIF, a member of IL-6 family) (Warsito, 2021).

Nutmeg essential oil (NEO) also acted as a potential chronic pain reliever as it was able to alleviate mechanical allodynia, joint swelling, and heat hyperanalgesia of rats stimulated by the complete Freund's adjuvant (CFA) injection. Oral intake of 20 mg/kg NEO preparation decreased swelling, and pain score significantly in comparison with the 30 mg/kg diclofenac-treated group (Warsito, 2021).

21.3.6 Anticancer Properties

Different concentrations of *M. fragrans* essential oil were added to the A-357 epidermal skin cancer cell line and the Michigan Cancer Foundation-7 (MCF-7) breast cancer cell line and exhibited cytotoxic effects. An *in vitro* experiment showed that nutmeg essential oil inhibited the growth of a colon cancer cell line (undifferentiated Caco-2 cells). The Jurkat leukaemia T cell line's cells died as a result of exposure to the methanol extract of *M. fragrans* through a mechanism involving the downregulation of SIRTI mRNA. There was a noticeable reduction in the carcinoma cells of methylcholanthrene-induced carcinogenesis in the uterine cervix of Swiss albino mice after 90 days of receiving a dose of 10 mg/mouse/day. Similar results were seen when *M. fragrans* ethanolic extract was utilized on human cancer cell lines, where more than 70% growth inhibition was seen at a dosage of 100 μg/mL. The chemopreventive property is also seen in the aril of nutmeg (Jannu et al. 1991) Additionally, the oil exhibited potent cytotoxic activity against breast and colorectal malignant carcinoma cell lines. Potential anticancer properties were also observed for machilin A which included inhibition of proliferation of blood lymphocytes, human leukaemia HL-60 cells, and of topoisomerase I and II (Abourashed & El-Alfy, 2016). In order to slow down the growth of tumour cells, antiangiogenesis process is triggered which inhibits the creation of new blood vessels in tumour cells. Such anti-angiogenic activity was observed in *M. fragrans* essential oil with IC_{50} of 77.6 μg/mL compared with *Morinda citrifolia* oil which exhibits IC_{50} of 109.3 μg/mL (Ashokkumar et al., 2022). Melanoma B16-F10 cells were used in a comparative investigation of *M. fragrans* extracts as an anticancer agent that induced apoptosis. *M. fragrans* seed was extracted with ethanol, followed by fractionation in n-hexane, ethyl acetate, and n-butanol. The resazurin reduction assay was used to examine the effects of different doses of treatment on B16-F10 melanoma cells in culture. All fractions evaluated had IC_{50} values of 21.83 μg/mL for the ethanol extract, 21.66 μg/mL for the ethyl acetate fraction, 47.53 μg/mL for the n-hexane fraction, and >1,000 μg/mL for the n-butanol fraction. The ethyl acetate fraction gave the highest IC_{50} value of the various fractions and was able to induce apoptosis *via* caspase-3 proteins which showed results similar to cisplatin, which was used as a positive control (Susianti et al., 2021). The anticancer benefits of several substances found in nutmeg essential oil included myristicin, limonene, eugenol, and 4-terpineol. Myristicin caused the activation of particular intracellular death-related pathways, which in turn triggered caspase-3 activation and the start of apoptotic cytotoxicity in human neuroblastoma SK-N-SH cells. D-limonene, another component, was identified as a possible chemoprevention agent when it was discovered to decrease the overexpression of the c-myc oncoprotein in AKR mice models for hepatocarcinoma caused by *N*-nitrosodiethylamine. Caspase-dependent apoptosis of human melanoma M14 WT cells and M14 adriamycin-resistant cells was caused by 4-terpineol. Eugenol was reported to be able to stimulate apoptosis in osteosarcoma, skin tumours, melanoma, leukaemia, gastric, and mast cells (Warsito, 2021).

21.3.7 Hepatoprotective Properties

Nutmeg possesses hepatoprotective activity, a finding which was demonstrated in a study by giving nutmeg in the diet of mice with damaged livers. The influence of nutmeg essential oil was examined on the activities of hepatic carcinogen-metabolizing enzymes including aryl hydrocarbon hydroxylase, cytochrome P450, acid-soluble sulphydryl, and glutathione-*S* transferase levels in albino mice. The results showed that the essential oil hinders the activities of the host enzymes related to detoxication and activation of xenobiotic components, as well as mutagens and chemical carcinogens. (Naeem et al., 2016) The hepatoprotective activity of myristicin could be due to the inhibition of tumour necrosis factor (TNF)-alpha release from macrophages, as it was able to inhibit the lipopolysaccharide plus D-galactosamine-induced enhancement of serum TNF-alpha concentrations (Nagja et al., 2016) and hepatic DNA fragmentation in mice (Ashokkumar et al., 2022).

Hepatic detoxification system-inducing activities were identified in mace. A study found that the hepatoprotective properties of mace lignan extracted from *M. fragrans* prevented liver damage caused by the drug cisplatin in mice. In another study, it was found that nutmeg aqueous extract had significant hepatoprotectant activity against isoproterenol-induced hepatotoxicity (Nagja et al., 2016).

21.3.8 Hypoglycaemic and Antidiabetic Properties

Since nutmeg (*M. fragrans*) extract is a naturally occurring dual peroxisome proliferator-activated receptor (PPAR) agonist, it may one day be used to treat type 2 diabetes. The PPARs can act as ligand activated transcription factors. Activation of PPAR-γ enhances glucose uptake and sensitivity to insulin (Nicholls and Uno, 2012). In a rat investigation, it was found that pre-treatment with a petroleum ether (60–80°C) extract of *M. fragrans* at a concentration of 200 mg/kg resulted in a considerable drop in blood glucose levels. When a *M. fragrans* petroleum ether extract was administered daily for two weeks to alloxan-induced diabetic rats, the blood sugar level considerably decreased from 326.25 ± 7.05 to 268.0 ± 9.6 mg%. The PPAR α/γ dual agonist activity of the nutmeg extract was notable, although its potency was inferior to that of the PPAR α and PPAR γ full agonist (Nagja et al., 2016). The nutmeg extract was found to activate an AMP-activated protein kinase enzyme which could be the potential therapeutic target for curing the metabolic syndrome, including type-2 diabetes and obesity. Analysis of nutmeg extracts showed the presence of seven active compounds like tetrahydrofuroguaiacin B, 2,5-bis-aryl-3,4-dimethyltetrahydrofuranlignans, fragransin C1, saucernetindiol, nectandrin B, verrucosin, galbacin, and nectandrin A (Naeem et al., 2016). Among the seven isolated compounds, tetrahydrofuroguaiacin, nectandrin A, and nectandrin B exerted strong activation of AMPK in differentiated C2C12 muscle cells at 5 μM concentration (Nguyen et al., 2010). The study also examined the effect of administration of a nectandrin B-rich active fraction in a high fat-induced animal model of obesity (Abourashed & El-Alfy, 2016).

21.3.9 Hypolipidaemic and Platelet Anti-aggregatory Activity/Anti-obesity Activity

A significant reduction in total cholesterol content in heart and liver was observed in albino rats after intake of an ethanolic extract of nutmeg, at an oral dose of 500 mg/kg for 60 days. These results proposed that the seed extracts of *M. fragrans* possess hypolipidaemic and platelet anti-aggregatory activities. With this, low-density lipoproteins (LDL) and very low-density lipoproteins (VLDL) levels were also significantly reduced and there was a slight increase in high-density lipoprotein (HDL) cholesterol level. It also showed prevention of the accumulation of phospholipids, triglycerides and cholesterol in the heart, aorta, and liver, aiding in the dissolution of atheromatous plaque in the aorta. There was a significant increase in the excretion of cholesterol and phospholipids through fecal matter in rabbits fed with nutmeg seed extract (Nagja et al., 2016). A study was conducted on high fat diet induced mice to find the effect of lignans present in nutmeg (Nguyen et al 2010). The tetrahydrofuran (THF) type lignans isolated from *M. fragrans* showed an anti-obesity effect due to the (AMP)-activated protein kinase activation mechanism. It was seen that THF prevented an increase in adipose tissue mass, body weight, LDL, and glucose levels in THF-treated mice (Nagja et al., 2016). The research study was undertaken to examine the effects of mace extract on food intake (FI) capacity and bodyweight (BW) of normal and diet-induced obese rats. The results of this study revealed that the extract could reduce FI capacity and prevent the increase in BW in a dose-dependent manner. A high dose (400 mg/kg) showed the greatest effect. The presence of saponins and tetrahydrofuran in mace extracts were considered to be the reason behind this effect. Saponins can inhibit the pancreatic lipase enzyme; thereby reducing the cholesterol levels. Tetrahydrofuran, which is one of the active chemical components present in mace, also contributed to preventing weight gain by stimulating the AMPK enzyme in differentiated C2C12 cells. Its steroidal compounds reduced the FI capacity for its inhibitory effect on the hunger sensing mechanism in the hypothalamus (Vangoori et al., 2018).

21.3.10 Neurological CNS Effects/Analgesic/Anticonvulsant/Anxiogenic Activities

Components in nutmeg are also known for their effects on the central nervous system (CNS) and have been analyzed for analgesic, anticonvulsant, and anxiogenic activities. The analgesic activity was recorded in the acid-induced writhing animal model by giving a 1 g/kg dosage of an ethanolic extract of nutmeg, whereas anxiogenic effects were observed at a lesser dosage range of 10–100 mg/kg of the n-hexane extracts of nutmeg exerted by the presence of trimyristin (Abourashed & El-Alfy, 2016). Numerous animal models with seizures induced by maximum electroshock, pentylenetetrazole, or strychnine were used in investigations to determine the dose-dependent anticonvulsant efficacy of nutmeg essential oil. The EO was administered *via* the inhalation route and the potential effects were observed on locomotor activity (Abourashed & El-Alfy, 2016). Among all the induction methods, a dosage of 200 μL/kg of nutmeg essential oil showed complete protection against the hind limb tonic extension phase in case of the maximal electroshock seizure test, whereas, in the case of other induction methods, the difference seen was dose dependent (Warsito, 2021). The effects were due to the presence of components like myristicin, safrole, and 4-terpineol (Abourashed & El-Alfy, 2016), pinene analogue, linalool, α-terpineol, eugenol, and citronellol (Warsito, 2021). The mechanism of action is known for these components, from which the pinene analogue acts as an inhibitor of acetylcholinesterase and a modulator of gamma-aminobutyric acid (GABA-A) receptor where it increases the postsynaptic GABA-dependent flow of chloride ions. It also reduces levels of nitrate at the hippocampal region and the striatal content of norepinephrine and dopamine. Neuronal hyperexcitability is modulated by the action of eugenol by blocking the sodium current. Eugenol has both repressive and excitation effects on neurons and its dose-dependent effects are seen where, at a low concentration of 0.5 mM, it shows antiepileptic effect and, at a higher concentration of 2 mM, it induces epileptiform activity (Warsito, 2021).

Myristicin, a key component of nutmeg, has been shown to have anxiolytic effects. A study was conducted to examine the potential interactions between myristicin and the GABA-A receptor in male Sprague-Dawley rats. According to reports, the myristicin group's outcomes did not significantly differ from those of the control group. The information suggests that myristicin may work at GABA-A receptor locations other than the one that benzodiazepines are thought to bind to. These results might be the outcome of another neurotransmitter location in the central nervous system being modulated. The results implied that myristicin regulates the central nervous system by enhancing gross and fine motor activity when combined with flumazenil but did not show any anxiolytic effects (Leiter et al., 2011).

21.3.11 Antidepressant Activity

Nutmeg extracts have been shown to have antidepressant effects in both behavioural despair models as well as reserpine reversal test paradigms (Abourashed & El-Alfy, 2016). The effects of three distinct oral dosages of 5, 10, or 20 mg/kg bodyweight on mice were evaluated using the forced swim test (FST) and the tail suspension test (TST). The most effective dose, which demonstrated the greatest reduction in the immobility of mice as compared with the control, was discovered to be 10 mg/kg. This dosage also demonstrated equal potency to fluoxetine (20 mg/kg) and imipramine (15 mg/kg). Hence, when tested by TST and FST, the nutmeg extract was capable of having a considerable antidepressant effect in mice. The interplay between the dopaminergic, adrenergic, and serotonergic systems and the nutmeg seed extract may be the cause of the antidepressant effect (Nagja et al., 2016; Abourashed & El-Alfy, 2016). When evaluated in animal models of behavioural despair, trimyristin had a depressive impact and potentiated the hypothermia brought on by reserpine. Pre-administration of a serotonin 5-HT2A receptor antagonist prevented the effects that were seen. In the four-point tetrad experiment, the neurobehavioural effects of nutmeg were compared with those of commonly abused substances such as morphine, amphetamine, and 9-tetrahydrocannabinol (THC). The study's findings demonstrated that, depending on the type of the extract and the mode of administration, nutmeg extracts have a variety of assay-related activities. The study showed that the dichloromethane nutmeg extract had some cannabimimetic action in the tetrad experiment when administered at doses of 100 or 300 mg/kg intravenously (Abourashed & El-Alfy, 2016).

21.3.12 Memory-Enhancing Activity

The n-hexane extract of nutmeg seeds, when administrated orally at a low dose of 5 mg/kg of body weight, improved the learning and memory of the young and adult mice. The extract successfully reversed the scopolamine- and diazepam-induced impairment in learning and memory as well (Parle et al., 2004). Oral administration of nutmeg powder and its extract to Wistar albino rats significantly boosted their memory and memory regaining (Jissa et al., 2014; Kumar et al., 2013). This could be attributed to the acetylcholinesterase inhibitors present in nutmeg. Acetylcholinesterase inhibitors are among the approved drugs to treat Alzheimer's disease. During an *in vitro* study, the hydroalcoholic extract of nutmeg exhibited 50% inhibition of acetylcholinesterase at the concentrations of 100–150 μg/mL (Mukherjee et al., 2007).

21.3.13 Cardiovascular Effects

Nutmeg extracts have been shown to have a cardioprotective effect against myocardial infarction (MI). The effect of daily therapy with 100 mg/kg bodyweight/day of aqueous nutmeg extract for 30 days on isoproterenol-induced MI in adult male rats was studied. The findings point to nutmeg consumption having a possible cardiovascular preventive role, since pre-treatment with nutmeg extract provided protection against isoproterenol effects on blood glucose, plasma lipids, and histological myocardial changes (Abourashed & El-Alfy, 2016). According to a follow-up study, nectandrin B activated AMP protein kinase (AMPK) in vascular smooth muscle cells (VSMC) and reduced VSMC proliferation and neointima formation, two processes that are critical in the establishment of vascular occlusive diseases. An in-depth examination of the anti-proliferative effect of VSMCs revealed that AMPK activation resulted in the synthesis of p53 and p21, which in turn reduced Rb phosphorylation and E2 transcription factor 1 (E2F1) expression, suppressing the development of Pin1 (Kang et al., 2013). The sequence of events prevents intimal hyperplasia. These findings imply that nectandrin B may be useful in the treatment or prevention of a variety of occlusive vascular diseases (Abourashed & El-Alfy, 2016).

21.3.14 Osteoblast Proliferation Stimulation Activity

Lignans are categorized as phytoestrogens because they bind to the oestrogen receptor and then display oestrogenic activity. In both preclinical and clinical investigations, a number of phytoestrogens, particularly isoflavones, have shown favorable therapeutic effects against bone loss (Abourashed & El-Alfy, 2016). Compounds from *M. fragrans* have been shown to induce osteoblastic differentiation (Nagja et al., 2016). One of nutmeg's lignan components, machilin A, can promote osteoblast development by triggering p38 mitogen-activated protein (MAP) kinase pathway (Lee et al., 2009). Machilin A dramatically increases alkaline phosphatase (ALP) activity, a frequently used marker for promoting differentiation, in the early stages of osteoblast differentiation. Machilin A similarly induced considerable bone mineralization and late-stage differentiation. The anabolic activity in the bones that was seen was dose dependent (Abourashed & El-Alfy, 2016). Additional lignans from *M. fragrans* that have anabolic activity in bone metabolism include safrole, myristargenol, mace lignan, nectandrin B, mesodihydroguaiaretic acid, machilin F, licarin B, and licarin A. These lignans likewise exhibit similar effects on bone metabolism (Nagja et al., 2016).

21.3.15 Immuno-Modulatory and Radio-Protective Activities

The essential oil of nutmeg protected plasmid (pBR322) DNA against the dose rate of 7 Gy min^{-1} of gamma rays (Adiani et al., 2015). The alcoholic extract of nutmeg seeds, when administered to Swiss albino mice at the concentration of 10 mg/kg bodyweight in saline for three consecutive days, showed a higher survival rate than the control group (Sharma & Kumar, 2007). The nutmeg-derived macelignan reduced splenocyte proliferation in response to the polyclonal T cell mitogen concanavalin A and protected splenocytes against radiation-induced intracellular ROS generation in a dose-dependent manner (Checker et al., 2008).

21.3.16 Anti-malarial Activity

The nutmeg oil was subjected to distillation to obtain fractions and every fraction was tested against *Plasmodium falciparum*, the mosquito-borne parasite causing malaria. The fraction obtained at 1–2.5 min, containing the highest concentrations of α-pinene, β-pinene, and sabinene, was the most effective against *P. falciparum* (Ibrahim et al., 2020). Ethanolic extract of nutmeg also exhibited inhibition of *P. falciparum* (IC_{50} < 10 mg/mL) during *in vitro* testing (Thiengsusuk et al., 2013).

21.3.17 Anti-diarrhoeal Activity

The hexane-soluble fraction of the methanolic extract of dried nutmeg fruits displayed anti-secretory activities in the ileum of rabbits and guinea pigs against *Escherichia coli* enterotoxins (Gupta et al., 2008). The crude nutmeg suspension and petroleum ether extract reduced the quantity of loose stool and increased the latency period, exhibiting sedative properties and anti-diarrhoeal efficacy (Grover et al., 2002).

21.3.18 Anti-cariogenic Activity

Mace lignan (20 μg/mL in 1 minute), purified from methanolic extract of nutmeg, inactivated *Streptococcus mutans,* an oral pathogen that causes dental caries. Compared with other naturally occurring anti-cariogenic drugs, mace lignan was discovered to have a relatively low MIC (Chung et al., 2006).

21.3.19 Pesticide Activity (Insecticidal)

Nutmeg essential oil is proven to be a promising, environmentally friendly, and stable insecticide or insect repellent. The active components of nutmeg essential oil, namely α-pinene, β-caryophyllene, eugenol, elemicin, limonene, methyleugenol, methylisoeugenol, myristicin, sabinene, and safrole, are reported to be responsible for its insecticidal property (Abourashed & El-Alfy, 2016; Nagja et al., 2016; Warsito, 2021).

Nutmeg essential oil has exhibited larvicidal and adult toxicity against various species of mosquitoes, namely adult female *Anopheles gambiae*, which is a vector for the malarial parasite (LD_{50} 10.5 to 19.0 μg/g), adult female *Aedes aegypti*, a vector for dengue, chikungunya, Zika, Mayaro, and yellow fever viruses (LD_{50} 19.1 and 33.3 μg/g) (Norris et al., 2015; Lim et al., 2019). Cellulose nanoparticle-based Pickering emulsions of nutmeg essential oil exhibited 100% larvicidal activity against Asian tiger mosquito, *Aedes albopictus*, which is a vector for dengue and chikungunya viruses (Seo et al., 2021).

Nutmeg essential oil also exhibited larvicidal activity against pulse beetle *Callosobruchus chinensis* and the mushroom-infesting fly *Lycoriella ingénue* (Chaubey, 2008; Mahdi, 2016). Furthermore, the oil has also shown insecticidal activity against adult insects which are vectors for many diseases or are harmful for stored food such as *Plodia interpunctella* (LC_{50} 6.65 μL/L air), *Lasioderma serricorne* (LD_{50} 19.3 μg/adult), *Musca domestica* and *Chrysomya albiceps (Du et al., 2014; Soni et al., 2016; Cossetin et al., 2021).* Exposure to nutmeg essential oil for 24 hours eliminated 100% of the population of *Callosobruchus maculatus* (pulse beetle) (Adedire, 2002; Alibabaie & Safaralizadeh, 2015), while another study showed that methyl eugenol extracted from nutmeg essential oil attracted *Bactrocera carambolae*, *B. dorsalis,* and *B. umbrosa*, all species of fruit fly, thus exhibiting potential as a fruit fly attractant in horticulture (Susila et al., n.d.).

The aqueous decoction made from nutmeg seeds proved to be lethal for cockroaches. The hexane-soluble fraction of a methanolic extract of nutmeg seed exhibited insecticidal activity against adult female *Blatella germanica* (Jung et al., 2007). Nutmeg powder effectively eliminated alfalfa leaf cutting bee, a vector for the parasitoid pest *Pteromalus venustus*. It also reduced the female fertility and sexual function of the alfalfa leaf cutting bee (Ong et al., 2020), reducing the pest infestation.

21.4 INDUSTRIAL APPLICATIONS

The bioactive characteristics of nutmeg have given rise to several new applications for this spice. The essential oils and powder have been used to augment the properties of finished products. A few of these current applications are discussed below.

21.4.1 Food Industry

21.4.1.1 Food Preservation

The deterioration of food products could be a result of storage conditions, changes in their chemical composition, microbial load, and sensory properties (Šojić et al., 2015).

One contributing factor to chemical degradation would be the oxidation of lipids. The current synthetic antioxidants have been considered potentially hazardous to human health, prompting the search for natural antioxidants. A study conducted on cooked sausages that incorporated 20 ppm of nutmeg essential oil in the sausage batter concluded that the oil resulted in lower lipid oxidation, reduced the growth of mesophilic bacteria, and improved the aroma of the product (Šojić et al., 2015).

Another study compared the effect of spices, including nutmeg and mace, on the *Lactobacillus curvatus* LTH 1174 starter culture that is used in the production of fermented sausages. Nutmeg significantly reduced the bacterial biomass of *L. curvatus*, along with the pathogenic strain of *E. coli* O157 and *Listeria monocytogenes*. It also decreased the production of curvacin A (a bacteriocin). Mace, on the other hand, had only a minor effect on bacteriocin production (Verluyten et al., 2004).

Nitrites are found in large amounts in cured meats. A study was conducted on the efficacy of eight spice microparticles in the scavenging of these nitrites. The scavenging activity of the nutmeg microparticles was shown to be moderate (Wu et al., 2020).

21.4.1.2 Packaging

Packaging of food products is an essential part of delaying and preventing deterioration of the product. Most packaging relies on plastics. The concerns of the effect of plastics on the ecosystem have encouraged a search for alternative, natural, and effective packaging (Bharti et al., 2020).

Freshly picked fruits and vegetables are subjected to active packaging which mimic the natural cuticular barrier and extends the shelf life of the product. Alginate-based edible coating with nutmeg essential oil and cinnamon essential oil improved the shelf life of the coated samples by 15 days (Santos et al., 2018)

Essential oils incorporated into cellulose ester films modified its properties. The nutmeg essential oil, like other essential oils tested in the study, showed a strong plasticizing effect on acetate and decreased water vapour permeability (Biswas et al., 2020).

Tapioca starch containing nutmeg essential oil could be cast into films that were colourless, glassy, flexible, and transparent. These films were used to wrap chicken nuggets made in an experimental lab. Although the tensile strength of the film decreased with the addition of essential oils, it was able to delay degradation of the meat product, increase oxidative stability of the product, delay detectable counts of Gram-negative psychrophilic bacteria, and reduce yeast and mould counts (Bharti et al., 2020).

Gelatin, a protein, can be used to produce flexible films that are good barriers to oxygen and carbon dioxide but are ineffective against water vapour. Addition of oily compounds to this film formed films that had greater resistance to water with additional advantages, such as antibacterial and antifungal activity and enhancement of sensory properties (26Figueroa-Lopez et al., 2018a). These authors also tested nutmeg oleoresins, among other oleoresins, and found that these oleoresins reduced permeability to water vapour and lowered film solubility while introducing antimicrobial activity to the film (Figueroa-Lopez et al., 2018b). A further study examined the effect of clove, nutmeg, and black pepper oleoresins incorporated into biocomposite films based on gelatin and microcrystalline cellulose that were used to package bread. All of these treatments suppressed *S. aureus* and *E. coli* growth, reduced mould and yeast development, and reduced oil and vapor permeability by 20 to 30% (26Figueroa-Lopez et al., 2018a).

21.4.1.3 Anti-biofilm Formation and Quorum Sensing

Staphylococcus aureus can form biofilms on handmade soy products leading to potential health issues. D-amino acids have been used to impair biofilm formation and nutmeg oil essential oil was selected as an antimicrobial agent. A combination of D-amino acids and nano-encapsulated nutmeg essential oil was included in electro-spun nanofibrous polyethylene oxide membranes. The result was an excellent wrapper with antibiofilm and antibacterial activity (Cui et al., 2017).

Chromobacterium violaceum (strains 12472 and CV026) and *Pseudomonas aeruginosa* are biosensor bacteria capable of producing biofilm on food packaging *via* AHL-mediated quorum sensing. Essential oil from nutmeg and mace have shown promising results in inhibiting the quorum sensing and biofilm formation under *in vivo* conditions (Bai & Vittal, 2014).

Licarin A (a neolignan) and its derivative found in nutmeg exhibited potential inhibitory effect against rapidly growing mycobacterial strains. Further inhibition was improved by structural changes in licarin A and its derivatives at the phenolic sub-unit and alkenyl side chain (Alvarenga et al., 2020)

21.4.2 Health Care Products and Cosmetics

Malabaricone C, derived from an ethanolic extract of nutmeg powder, inhibited the activity of 5-lipoxygenase which catalyzes the first two steps of the biosynthesis of leukotrienes from arachidonic acid as well as ameliorating the symptoms of psoriasis-like skin inflammation in mice (Tsukayama et al., 2022).

Ethanolic extract of nutmeg prevents atopic dermatitis induced by house dust mites (*Dermatophagoides farinae*) in an experimental model of mice by decreasing trans-epidermal water loss, erythema, and the generation of serum immunoglobulin E, interleukin (IL)-4, and interferon-γ by auxillary lymph node cells. It also decreased epidermal thickness and inflammatory cell infiltration in mouse skin (Chung et al., 2012).

Essential oil of nutmeg exhibited potential anti-photoageing effect by reducing wrinkle score in mice due to UVB exposure and its skin irritation score was slight, making it a potential candidate for anti-photoageing (Rahmi et al., 2021).

Alkaline lignin myristicin and three neolignans derived from nutmeg essential oil were effective against *Cutibacterium acnes* and *Staphylococcus aureus*, bacteria responsible for acne. Furthermore, these four compounds don't induce any skin irritation and showed anti-inflammatory and anti-acne effect in Wistar rats (Lee et al., 2020).

There is a growing demand for plant extracts to replace the synthetic principles that are used for oral hygiene in toothpastes and mouthwash, prompting a study on the effectiveness of nutmeg as a potential replacement. Ethyl acetate and ethanol extracts of the seed flesh, seed and mace of the nutmeg fruit was tested against three facultative anaerobic Gram-positive cariogenic bacteria (*Streptococcus mutans* ATCC 25175, *Streptococcus mitis* ATCC 6249, and *Streptococcus salivarius* ATCC 13419) and three obligate periodontopathic anaerobes (*Aggregatibacter actinomycetemcomitans* ATCC

29522, *Porphyromonas gingivalis* ATCC 33277 and *Fusobacterium nucleatum* ATCC 25586). All extracts showed antibacterial activity with the flesh being the most effective among the parts of the nutmeg fruit tested. The authors also mentioned that there were reports of macelignan or malabaricone C isolated from the nutmeg seed having greater antibacterial activity than the crude extracts tested in their study (Shafiei et al., 2012).

21.4.3 Nanotechnology

Phytochemical-mediated synthesis of nanoparticles is a rapid and well accepted method to produce bioactive nanoparticles. Copper nanoparticles and copper silver nanoparticles synthesis *via* nutmeg oleoresin exhibited antimicrobial and anti-inflammatory activities (Madhumitha et al., 2021; Uthayasankar et al., 2021; Vijayakumar et al., 2021).

Super magnetic nanoparticles with superior antimicrobial activities were generated by green synthesis *via* nutmeg oil (Mousavi et al., 2019), while the mace-mediated silver nanoparticles also showed antimicrobial and cytotoxicity activities (Rizwana et al., 2021). Copper nanoparticles synthesized *via* nutmeg oleoresin exhibited anti-microbial activities against *Staphylococcus aureus*, *Enterococcus faecalis*, *Streptococcus mutans*, and *Candida albicans*, responsible for oral pathogenicity, thus proving the potential for these nanoparticles in oral hygiene (Madhumitha et al., 2021).

21.4.4 Veterinary Applications

Bacterial infections in the aquaculture industry, including those of the common carp, result in severe economic losses. Natural feed additives reduce the requirement for medical treatments during culture. Commercial fish feed supplemented with 2% nutmeg extracts fed to an aquaculture population of the common carp (*Cyprinus carpio*) showed improved growth performance, increased total protein, albumin, and globulin levels, higher skin mucus antibacterial effects, and improved survival rates when challenged with *Aeromonas hydrophila* (Rashidian et al., 2022).

Korean chicken was fed diets that contained 250 ppm or 500 ppm nutmeg oil. Although there was not a significant difference in most of the parameters tested, the serum aspartate aminotransferase, malondialdehyde concentration (a measure of oxidative stress) and rectal temperature were decreased in birds exposed to elevated heat conditions for three hours. Further studies are needed to determine the formulation of nutmeg oil necessary to confer the ability to tolerate heat in chicken (Hartanto et al., 2019).

Zebrafish fed with a meal containing 3% nutmeg powder resulted in a significantly lower survival rate than the control. However, these fish also showed higher protein levels in both mucus and serum, higher total immunoglobulin levels in the serum, increased superoxide dismutase activity and decreased mortality following a low-temperature shock (Vakili et al., 2023).

21.4.5 Insect-Repellent and Insecticidal Applications

Nutmeg essential oil was extracted in ethanol and used under greenhouse conditions against *Bemisia tabaci* (whitefly) as an alternative to synthetic pesticides. It proved to be a successful fumigation agent against adult whiteflies and displayed contact toxicity to the nymphs at a concentration of 10 mg/mL. Its action as a repellent lasted for 48 h. It also prevented oviposition for 48 h (Wagan et al., 2017).

Nutmeg oil was tested for its mosquito-repellent activity against *Aedes aegypti* and it was found that the repellent activity was similar to the positive control of the synthetic insect repellent diethyltoluamide (DEET) for the first two hours but the protection was lost after six hours (Lim et al., 2019). The mosquito-repellent activity of essential oils could be compromised by the uncontrolled release of the essential oil. Nano emulsions of nutmeg oil (20%) using glycerol, Montanov 82 (surfactant), and deionized water were tested for their efficacy as a mosquito repellent against *Aedes aegypti* and compared with the positive control (DEET). Some of the nano emulsions produced were stable, non-toxic to mouse fibroblast cells, and had a prolonged insect-repellent effect. One such formulation offered 100% protection at 0 h and 81.09% protection at 8 h (Mohd Narawi et al., 2020).

The nutmeg essential oil was found to be effective against *Musa domestica* larvae using immersion bioassays. The oil showed larval mortality at a concentration of 5% and reduced adult fly emergence (20Cossetin et al., 2021b).

Farmers in pursuit of economical and natural methods for post-harvest protection of stored products against pests have explored the use of underutilized plant materials. A petroleum ether extract of nutmeg seeds was effective as a biocontrol agent against a bruchid beetle pest (*Callosobruchus maculatus*) of cowpea seeds. Its mode of action is attributed to asphyxiation (Adedire, 2002).

21.4.6 Applications beyond Pharmacy and Spice

The nutmeg seeds and mace not only proved to have pharmacological properties, but they have also exhibited various other properties, such as mechanical and thermal insulation, anti-corrosion, and dye removal.

Polyurethane (PU) foams were reinforced with nutmeg filler to improve mechanical, thermal, anti-microbial, and anti-ageing properties. The nutmeg filler composite PU foams developed had better compression strength, flexural strength, improved fire resistance, stabilization, and anti-bacterial properties (Członka et al., 2020).

Green corrosion inhibitors are those plant formulations that inhibit corrosion. Several plants have shown this kind of activity. The fruit extract of nutmeg had an 87.81% inhibition efficiency (Haldhar & Saxena, 2018).

Shell type wastes have a lignin content that limits its use as animal feed (Thomas & Thalla, 2022). The nutmeg seed shell contains 9.5% hemicelluloses, 28.63% cellulose, and 50.14% lignin that prompted its conversion into biobriquettes that have a higher calorific value than wood and are relatively economical (Idris et al., 2018).

Activated carbon is the adsorbent of choice for dye removal from the wastewater from textile, paper, and leather industries. The nutmeg seed shell biochar proved to be a cost-effective value-added product that was more efficient that the conventional adsorbents used to remove the dye Remazol brilliant blue R (Thomas & Thalla, 2022).

21.5 CONCLUSION AND FUTURE REMARKS

Nutmeg seeds have proven their potential as spices since as early as the sixteenth century. Gradually, its uses in cuisine and medicine have spread worldwide, given its various pharmaceutical properties, finding its place in many of the traditional medicines and dishes. Recent advances in scientific research have recognized numerous pharmaceutical applications of nutmeg seeds and its essential oils. Many industrial products of health care and cosmetics employ nutmeg as a raw product. Recent research in nanotechnology and green products will further enhance the use of nutmeg in industries such as food, packaging, pesticides and other novel applications.

REFERENCES

Abourashed, E. A., & El-Alfy, A. T. (2016). Chemical diversity and pharmacological significance of the secondary metabolites of nutmeg (*Myristica fragrans* Houtt.). *Phytochemistry Reviews*, *15*(6), 1035–1056. Springer Netherlands. https://doi.org/10.1007/s11101-016-9469-x

Adedire, C. O. (2002a). Use of nutmeg *Myristica fragrans* (Houtt.) powder and oil for the control of cowpea storage bruchid, *Callosobruchus maculatus* Fabricius / Verwendung von Pulver und Ölaus Samen der Muskatnuss (*Myristica fragans* Houtt.) zur Bekämpfung von *Callosobruchus maculatus* Fabricius, einem Lagerschädling an Kundebohnen, *109*(2), 193–199. https://www.jstor.org/stable/43216195

Adiani, V., Gupta, S., Chatterjee, S., Variyar, P. S., & Sharma, A. (2015). Activity guided characterization of antioxidant components from essential oil of Nutmeg (*Myristica fragrans*). *Journal of Food Science and Technology*, *52*(1), 221–230. https://doi.org/10.1007/S13197-013-1034-7/FIGURES/4

Alibabaie, M., & Safaralizadeh, M. H. (2015). Fumigant toxicity of nutmeg seed essential oil (*Myristica fragrans* Houtt.) (MF, Myristicaceae) on Cowpea Weevil, *Callosobruchus maculatus* F. (Coleoptera: Bruchidae). *New Horizons in Insect Science: Towards Sustainable Pest Management*, 127–133. https://doi.org/10.1007/978-81-322-2089-3_13

Alvarenga, D. J., Matias, L. M. F., Oliveira, L. M., Leão, L. P. M. de O., Hawkes, J. A., Raimundo, B. V. B., Castro, L. de F. D., Campos, M. M. A. de, Siqueira, F. dos S., Santos, T. dos, & Carvalho, D. T. (2020). Exploring how structural changes to new Licarin A derivatives effects their bioactive properties against rapid growing mycobacteria and biofilm formation. *Microbial Pathogenesis*, *144*. https://doi.org/10.1016/j.micpath.2020.104203

Ashokkumar, K., Simal-Gandara, J., Murugan, M., Dhanya, M. K., & Pandian, A. (2022a). Nutmeg (*Myristica fragrans* Houtt.) essential oil: A review on its composition, biological, and pharmacological activities. *Phytotherapy Research*, *36*(7), 2839–2851. https://doi.org/10.1002/PTR.7491

Bai, A. J., & Vittal, R. R. (2014). Quorum sensing inhibitory and anti-biofilm activity of essential oils and their *in vivo* efficacy in food systems. *Food Biotechnology*, *28*(3), 269–292. https://doi.org/10.1080/08905436.2014.932287

Barman, R., Bora, P. K., Saikia, J., Kemprai, P., Saikia, S. P., Haldar, S., & Banik, D. (2021). Nutmegs and wild nutmegs: An update on ethnomedicines, phytochemicals, pharmacology, and toxicity of the Myristicaceae species. *Phytotherapy Research*, *35*(9), 4632–4659. https://doi.org/10.1002/PTR.7098

Beckerman, B., & Persaud, H. (2019). Nutmeg overdose: Spice not so nice. *Complementary Therapies in Medicine*, *46*, 44–46. https://doi.org/10.1016/j.ctim.2019.07.011

Bharti, S. K., Pathak, V., Alam, T., Arya, A., Singh, V. K., Verma, A. K., & Rajkumar, V. (2020a). Materialization of novel composite bio-based active edible film functionalized with essential oils on antimicrobial and antioxidative aspect of chicken nuggets during extended storage. *Journal of Food Science*, *85*(9), 2857–2865. https://doi.org/10.1111/1750-3841.15365

Bharti, S. K., Pathak, V., Alam, T., Arya, A., Singh, V. K., Verma, A. K., & Rajkumar, V. (2020b). Materialization of novel composite bio-based active edible film functionalized with essential oils on antimicrobial and antioxidative aspect of chicken nuggets during extended storage. *Journal of Food Science*, *85*(9), 2857–2865. https://doi.org/10.1111/1750-3841.15365

Biswas, A., Do Socorro Rocha Bastos, M., Furtado, R. F., Kuzniar, G., Boddu, V., & Cheng, H. N. (2020). Evaluation of the properties of cellulose ester films that incorporate essential oils. *International Journal of Polymer Science*. https://doi.org/10.1155/2020/4620868

Budi, V., Sihotang, L., Yang, G., Chi, X., & Huang, L. (2018). The ethnobotanical survey of clove, pepper, and nutmeg and their utilization by Chinese and Indonesian people. *Journal of Tropical Biology & Conservation (JTBC)*, *15*, 15â-27. https://doi.org/10.51200/JTBC.V15I0.1472

Cao, G. Y., Xu, W., Yang, X. W., Gonzalez, F. J., & Li, F. (2015). New neolignans from the seeds of *Myristica fragrans* that inhibit nitric oxide production. *Food Chemistry*, *173*, 231. https://doi.org/10.1016/J.FOODCHEM.2014.09.170

Chaubey, M. K. (2008). Fumigant toxicity of essential oils from some common spices against pulse beetle, *Callosobruchus chinensis* (Coleoptera: Bruchidae). *Journal of Oleo Science*, *57*(3), 171–179. https://doi.org/10.5650/JOS.57.171

Checker, R., Chatterjee, S., Sharma, D., Gupta, S., Variyar, P., Sharma, A., & Poduval, T. B. (2008). Immunomodulatory and radioprotective effects of lignans derived from fresh nutmeg mace (*Myristica fragrans*) in mammalian splenocytes. *International Immunopharmacology*, *8*(5), 661–669. https://doi.org/10.1016/J.INTIMP.2008.01.012

Chung, H. C., Kim, M. S., Mun, S. Y., Sa, B. K., Chung, J. Y., Kim, D. U., & Hwang, J. K. (2012). Effect of oral administration of nutmeg extract on American house dust mite (*Dermatophagoides farinae*) extract-induced atopic dermatitis-like skin lesions in NC/Nga mice. *Food Science and Biotechnology*, *21*(2), 559–564. https://doi.org/10.1007/s10068-012-0071-8

Chung, J. Y., Choo, J. H., Lee, M. H., & Hwang, J. K. (2006). Anticariogenic activity of macelignan isolated from *Myristica fragrans* (nutmeg) against *Streptococcus mutans*. *Phytomedicine*, *13*(4), 261–266. https://doi.org/10.1016/j.phymed.2004.04.007

Cossetin, L. F., Santi, E. M. T., Garlet, Q. I., Matos, A. F. I. M., De Souza, T. P., Loebens, L., Heinzmann, B. M., & Monteiro, S. G. (2021a). Comparing the efficacy of nutmeg essential oil and a chemical pesticide against *Musca domestica* and *Chrysomya albiceps* for selecting a new insecticide agent against synantropic vectors. *Experimental Parasitology*, *225*, 108104. https://doi.org/10.1016/J.EXPPARA.2021.108104

Cossetin, L. F., Santi, E. M. T., Garlet, Q. I., Matos, A. F. I. M., De Souza, T. P., Loebens, L., Heinzmann, B. M., & Monteiro, S. G. (2021b). Comparing the efficacy of nutmeg essential oil and a chemical pesticide against *Musca dom*estica and *Chrysomya albiceps* for selecting a new insecticide agent against synantropic vectors. *Experimental Parasitology*, *225*. https://doi.org/10.1016/j.exppara.2021.108104

Cui, H., Yuan, L., Li, C., & Lin, L. (2017). Control of *Staphylococcus aureus* on soya bean products by D-amino acids/nutmeg essential oil-co-loaded nanofilms. *International Journal of Food Science and Technology*, *52*(11), 2393–2403. https://doi.org/10.1111/ijfs.13523

Członka, S., Strąkowska, A., Kairytė, A., & Kremensas, A. (2020). Nutmeg filler as a natural compound for the production of polyurethane composite foams with antibacterial and anti-agingproperties. *Polymer Testing*, *86*, 106479. https://doi.org/10.1016/j.polymertesting.2020.106479

Du, S. S., Yang, K., Wang, C. F., You, C. X., Geng, Z. F., Guo, S. S., Deng, Z. W., & Liu, Z. L. (2014). Chemical constituents and activities of the essential oil from *Myristica fragrans* against cigarette beetle *Lasioderma serricorne*. *Chemistry and Biodiversity*, *11*(9), 1449–1456. https://doi.org/10.1002/CBDV.201400137

Figueroa-Lopez, K. J., Andrade-Mahecha, M. M., & Torres-Vargas, O. L. (2018a). Development of antimicrobial biocomposite films to preserve the quality of bread. *Molecules*, *23*(1). https://doi.org/10.3390/molecules23010212

Figueroa-Lopez, K. J., Andrade-Mahecha, M. M., & Torres-Vargas, O. L. (2018b). Spice oleoresins containing antimicrobial agents improve the potential use of bio-composite films based on gelatin. *Food Packaging and Shelf Life*, *17*, 50–56. https://doi.org/10.1016/j.fpsl.2018.05.005

Grover, J. K., Khandkar, S., Vats, V., Dhunnoo, Y., & Das, D. (2002). Pharmacological studies on *Myristica fragrans* – Antidiarrheal, hypnotic, analgesic and hemodynamic (blood pressure) parameters. *Methods and Findings in Experimental and Clinical Pharmacology*, *24*(10), 675–680. https://doi.org/10.1358/MF.2002.24.10.802317

Gupta, S., Yadava, J. N. S., Mehrotra, R., & Tandon, J. S. (2008). Anti-diarrhoeal profile of an extract and some fractions from *Myristica fragrans* (nutmeg) on Escherichia coli enterotoxin-induced secretory response. *International Journal of Pharmacognosy*, *30*(3), 179–183. https://doi.org/10.3109/13880209209053990

Haldhar, R., Prasad, D., & Saxena, A. (2018). Myristica fragrans extract as an eco-friendly corrosion inhibitor for mild steel in 0.5 M H2SO4 solution. *Journal of Environmental Chemical Engineering*, *6*(2), 2290–2301. https://doi.org/10.1016/j.jece.2018.03.023

Hartanto, S., Ko, H. S., Jee, S. H., Kang, J. U., Seo, J. S., Kang, Y. H., Kim, H. N., & Ohh, S. J. (2019). Effect of dietary nutmeg oil on heat-stress tolerance-related parameters in Korean native chicken reared under hot temperature. *Journal of Animal Physiology and Animal Nutrition*, *103*(4), 1160–1167. https://doi.org/10.1111/jpn.13113

Ibrahim, M. A., Cantrell, C. L., Jeliazkova, E. A., Astatkie, T., & Zheljazkov, V. D. (2020). Utilization of nutmeg (*Myristica fragrans* Houtt.) seed hydrodistillation time to produce essential oil fractions with varied compositions and pharmacological effects. *Molecules*, *25*(3). https://doi.org/10.3390/molecules25030565

Idris, R., Tawiequrrohman Yuliansyah, A., & Purwono, S. (2018). Development of biobriquette from nutmeg seed shells. *E3S Web of Conferences*, *42*. https://doi.org/10.1051/e3sconf/20184201016

Jannu, L.N., Hussain, S.P. and Rao, A.R. (1991). Chemopreventive action of mace (Myristica fragrans, Houtt.) on DMBA-induced papillomagenesis in the skin of mice. *Cancer Letters*, *56*(1), 59–63. https://doi.org/10.1016/0304-3835(91)90194-m

Jissa, G., Sai-Sailesh, K., & Mukkadan, J. (2014). Oral administration of nutmeg on memory boosting and regaining in Wistar albino rats. *Bali Medical Journal*, *3*(1), 3–10. https://balimedicaljournal.id/index.php/BaliMedJ/article/view/111

Jung, W.-C., Jang, Y.-S., Hieu, T. T., Lee, C.-K., & Ahn, Y.-J. (2007). Toxicity of *Myristica fragrans* seed compounds against *Blattella germanica* (Dictyoptera: Blattellidae). *Journal of Medical Entomology*, *44*(3), 524–529. https://doi.org/10.1093/JMEDENT/44.3.524

Kang, K.W., Lee, J.W., Lim, S.C., Hien, T.T., Im, J.H., Oh, W.K., Lee, M.Y., Ji, Y.H., and Kim, Y.G. (2013). Protective effect of nectandrin B, a potent AMPK activator on neointima formation: Inhibition of Pin1 expression through AMPK activation, *British Journal of Pharmacology*, *168*, 932–945. https://doi.org/10.1111/j.1476-5381.2012.02228.x

Kumar, S., Jissa, G., Setty, P., & Mukkadan, J. (2013). A comparative study on effect of oral administration of turmeric and nutmeg on memory boosting and regaining in Wistar albino rats. *Journal of Universal College of Medical Sciences*, *1*(3), 46–52. https://doi.org/10.3126/JUCMS.V1I3.8765

Kumar Verma, N., Singh, A. K., & Maurya, A. (2021). *EAS Journal of Pharmacy and Pharmacology Abbreviated Key Title: EAS J Pharm Pharmacol Myristica fragrans (Nutmeg): A Brief Review*. https://doi.org/10.36349/easjpp.2021.v03i05.004

Lee, C. J., Huang, C. W., Chen, L. G., & Wang, C. C. (2020). (+)-Erythro-Δ80-7S,8R-Dihydroxy-3,30,50-Trimethoxy-8-O-40-Neolignan, an anti-acne component in degreasing *Myristica fragrans* Houtt. *Molecules*, *25*(19). https://doi.org/10.3390/molecules25194563

Lee, J. Y., & Park, W. (2011). Anti-inflammatory effect of myristicin on RAW 264.7 macrophages stimulated with polyinosinic-polycytidylic acid. *Molecules (Basel, Switzerland)*, *16*(8), 7132–7142. https://doi.org/10.3390/MOLECULES16087132

Lee, S. U., Ki, S. S., Shi, Y. R., Yong, K. M., & Seong, H. K. (2009). Machilin A isolated from Myristica fragrans stimulates osteoblast differentiation. *Planta Medica*, *75*(2), 152–157. https://doi.org/10.1055/S-0028-1112197/ID/13

Leiter, Hitchcock, Godwin, Johnson, Sedgwick, Jones, McCall, & Ceremuga. (n.d.). Evaluation of the anxiolytic properties of myristicin, a component of nutmeg, in the male Sprague-Dawley rat. *AANA Journal*, April 2011. www.aana.com/aanajournalonline.aspx

Lim, V., Mohd Narawi, M., Chiu, H. I., Tung, W. H., Tan, J. J., & Lee, C. K. (2019a). Selected essential oils as repellents against *Aedes aegypti*: Validation of the bioconstituents using gas chromatography. *Journal of Essential Oil Bearing Plants*, *22*(4), 1058–1073. https://doi.org/10.1080/0972060X.2019.1661796

Madhumitha, B., Roy, A., Kumar, S., & Thangavelu, L. (2021a). Green synthesis of copper nanoparticles using nutmeg oleoresin and its antimicrobial activity against oral pathogens. *Journal of Complementary Medicine Research*, *12*(3), 21. https://doi.org/10.5455/jcmr.2021.12.03.02

Mahdi, S. H. (2016). Ovicidal and repellent effects of some spice powders against the *Callosobruchus chinensis* L. and *C. maculatus* (F.). *Bangladesh Journal of Zoology*, *44*(1), 51–59. https://doi.org/10.3329/BJZ.V44I1.30176

Malik, T., Sharma, R., Panesar, P. S., Gehlot, R., Tokusoglu, O., Dhull, S. B., Vural, H., & Singh, A. (2022). Nutmeg nutraceutical constituents: *In vitro* and *in vivo* pharmacological potential. *Journal of Food Processing and Preservation*, *46*(6). https://doi.org/10.1111/JFPP.15848

Mohamed, A. A., & Alotaibi, B. M. (2022). Essential oils of some medicinal plants and their biological activities: A mini review. *Journal of Umm Al-Qura University for Applied Sciences*. https://doi.org/10.1007/s43994-022-00018-1

Mohd Narawi, M., Chiu, H. I., Yong, Y. K., Mohamad Zain, N. N., Ramachandran, M. R., Tham, C. L., Samsurrijal, S. F., & Lim, V. (2020). Biocompatible nutmeg oil-loaded nanoemulsion as phyto-repellent. *Frontiers in Pharmacology*, *11*. https://doi.org/10.3389/fphar.2020.00214

Mousavi, S. M., Hashemi, S. A., Ramakrishna, S., Esmaeili, H., Bahrani, S., Koosha, M., & Babapoor, A. (2019). Green synthesis of supermagnetic Fe3O4–MgO nanoparticles via Nutmeg essential oil toward superior anti-bacterial and anti-fungal performance. *Journal of Drug Delivery Science and Technology*, *54*. https://doi.org/10.1016/j.jddst.2019.101352

Mukherjee, P. K., Kumar, V., & Houghton, P. J. (2007). Screening of Indian medicinal plants for acetylcholinesterase inhibitory activity. *Phytotherapy Research: PTR*, *21*(12), 1142–1145. https://doi.org/10.1002/PTR.2224

Naeem, N., Rehman, R., Mushtaq, A., & Ghania, B. (2016). Nutmeg: A review on uses and biological properties. *IJCBS*, *9*, 107–110. www.iscientific.org/Journal.html

Nagja, T., Vimal, K., & Sanjeev, A. (2016). MYRISTICA FRAGRANS: A COMPREHENSIVE REVIEW. *International Journal of Pharmacy and Pharmaceutical Sciences*, *8*(2), 27–30.

Nguyen, P.H., Le, T.V.T., Kang, H.W., Chae, J., Kim, S.K., Kwon, K., Seo, D.B., Lee, S.J., Oh, W.K., (2010). AMP-activated protein kinase (AMPK) activators from Myristica fragrans (nutmeg) and their anti-obesity effect, *Bioorganic & Medicinal Chemistry Letters*, *20* (14) 4128–4131. DOI: 10.1016/j.bmcl.2010.05.067

Nicholls, S. J., & Uno, K. (2012). Peroxisome proliferator-activated receptor (PPAR a/c) agonists as a potential target to reduce cardiovascular risk in diabetes. *Diabetes & Vascular Disease Research*, *9*(2), 89–94. https://doi.org/10.1177/1479164112441477.

Nifras, M., Fasmila, M., Fatheena, J., & Muthalib, A. M. (2021). A comprehensive review on therapeutic uses of *Myristica fragrans* in traditional systems of medicine. *International Journal of Unani and Integrative Medicine*, *5*(2), 111–117. http://www.unanijournal.com

Norris, E. J., Gross, A. D., Dunphy, B. M., Bessette, S., Bartholomay, L., & Coats, J. R. (2015). Comparison of the insecticidal characteristics of commercially available plant essential oils against Aedes aegypti and *Anopheles gambiae* (Diptera: Culicidae). *Journal of Medical Entomology*, *52*(5), 993–1002. https://doi.org/10.1093/JME/TJV090

Obranović, M., Bryś, J., Repajić, M., Balbino, S., Škevin, D., Bryś, A., Tonković, P., Medved, A. M., Uzelac, V. D., & Kraljić, K. (2020). Fatty acid and sterol profile of nutmeg (*Myristica fragrans*) and star anise (*Illicium verum*) extracted using three different methods. *The 1st International Electronic Conference on Food Science and Functional Foods, 33*. https://doi.org/10.3390/foods_2020-07687

Okiki, P. A., Nwobi, C. P., Akpor, O. B., Adewole, E., & Agbana, R. D. (2023). Assessment of nutritional and medicinal properties of nutmeg. *Scientific African, 19*, e01548. https://doi.org/10.1016/J.SCIAF.2023.E01548

Ong, M., Chomistek, N., Dayment, H., Goerzen, W., & Baines, D. (2020). Insecticidal Activity of Plant Powders against the Parasitoid, *Pteromalus venustus*, and Its Host, the Alfalfa Leafcutting Bee. *Insects, 11*(6), 359. https://doi.org/10.3390/INSECTS11060359

Parle, M., Dhingra, D., & Kulkarni, S. K. (2004). Improvement of mouse memory by *Myristica fragrans* seeds. *Journal of Medicinal Food, 7*(2), 157–161. https://doi.org/10.1089/1096620041224193

Poliński, S., Topka, P., Tańska, M., Kowalska, S., Czaplicki, S., & Szydłowska-Czerniak, A. (2022). Effect of grinding process parameters and storage time on extraction of antioxidants from ginger and nutmeg. *Molecules, 27*(21). https://doi.org/10.3390/molecules27217395

Rahmi, D., Yunilawati, R., Jati, B. N., Setiawati, I., Riyanto, A., Batubara, I., & Astuti, R. I. (2021). Antiaging and skin irritation potential of four main Indonesian essential oils. *Cosmetics, 8*(4). https://doi.org/10.3390/cosmetics8040094

Rashidian, G., Shahin, K., Elshopakey, G. E., Mahboub, H. H., Fahim, A., Elabd, H., Prokić, M. D., & Faggio, C. (2022a). The dietary effects of nutmeg (*Myristica fragrans*) extract on growth, hematological parameters, immunity, antioxidant status, and disease resistance of common carp (*Cyprinus carpio*) against *Aeromonas hydrophila*. *Journal of Marine Science and Engineering, 10*(3). https://doi.org/10.3390/jmse10030325

Rizwana, H., Bokahri, N. A., Alkhattaf, F. S., Albasher, G., & Aldehaish, H. A. (2021). Antifungal, antibacterial, and cytotoxic activities of silver nanoparticles synthesized from aqueous extracts of mace-arils of *Myristica fragrans*. *Molecules, 26*(24). https://doi.org/10.3390/molecules26247709

Salindeho, N. (2017). Application of nutmeg shell pyrolysis-based liquid smoke for sea cucumber (Holothuria scabra) processing. *International Journal of ChemTech Research, 10*(12), 278–283.

Santos, S. M. Dos, Malpass, G. R. P., Okura, M. H., & Granato, A. C. (2018). Edible active coatings incorporated with *Cinnamomum cassia* and *Myristica fragrans* essential oils to improve shelf-life of minimally processed apples. *Ciencia Rural, 48*(12). https://doi.org/10.1590/0103-8478cr20180447

Seneme, E. F., Dos Santos, D. C., Silva, E. M. R., Franco, Y. E. M., & Longato, G. B. (2021). Pharmacological and therapeutic potential of myristicin: A literature review. *Molecules, 26*, 5914. https://doi.org/10.3390/MOLECULES26195914

Seo, S. M., Lee, J. W., Shin, J., Tak, J. H., Hyun, J., & Park, I. K. (2021). Development of cellulose nanocrystal-stabilized Pickering emulsions of Massoia and nutmeg essential oils for the control of *Aedes albopictus*. *Scientific Reports*, 1–12. https://doi.org/10.1038/s41598-021-91442-6

Shafiei, Z., Shuhairi, N. N., Fazly, Md., Shah Yap, N., Harry Sibungkil, C. A., & Latip, J. (2012). Antibacterial activity of *Myristica fragrans* against oral pathogens. *Evidence-Based Complementary and Alternative Medicine*. https://doi.org/10.1155/2012/825362

Sharifi-Rad, J., Sureda, A., Tenore, G. C., Daglia, M., Sharifi-Rad, M., Valussi, M., Tundis, R., Sharifi-Rad, M., Loizzo, M. R., Oluwaseun Ademiluyi, A., Sharifi-Rad, R., Ayatollahi, S. A., & Iriti, M. (2017). Biological activities of essential oils: From plant chemoecology to traditional healing systems. *Molecules, 22*, 70. https://doi.org/10.3390/MOLECULES22010070

Sharma, M., & Kumar, M. (2007). Radioprotection of Swiss albino Mice by *Myristica fragrans* Houtt. *Journal of Radiation Research, 48*(2), 135–141. https://doi.org/10.1269/JRR.0637

Šojić, B., Tomović, V., Kocić-Tanackov, S., Škaljac, S., Ikonić, P., Džinić, N., Živković, N., Jokanović, M., Tasić, T., & Kravić, S. (2015). Effect of nutmeg (*Myristica frag*rans) essential oil on the oxidative and microbial stability of cooked sausage during refrigerated storage. *Food Control, 54*, 282–286. https://doi.org/10.1016/j.foodcont.2015.02.007

Soni, R., Sharma, G., & Jasuja, N. D. (2016). Essential oil yield pattern and antibacterial and insecticidal activities of *Trachyspermum ammi* and *Myristica fragrans*. *Scientifica*. https://doi.org/10.1155/2016/1428194

Susianti, S., Lesmana, R., Salam, S., Julaeha, E., Pratiwi, Y. S., Sylviana, N., Goenawan, H., Kurniawan, A., & Supratman, U. (2021). The effect of nutmeg seed (*M. fragrans*) extracts induces apoptosis in melanoma maligna cells (B16-F10). *Indonesian Biomedical Journal, 13*(1), 68–74. https://doi.org/10.18585/inabj.v13i1.1424

Susila, W., Supartha, W., Kadek, I., Yudha, W., Dhyana, U., & Bali, P. (n.d.). *Putu angga Wiradana*. Retrieved April 20, 2023, from https://www.researchgate.net/publication/353752310

Suthisamphat, N., Dechayont, B., Phuaklee, P., Prajuabjinda, O., Vilaichone, R. K., Itharat, A., Mokmued, K., & Prommee, N. (2020). Anti-*Helicobacter pylori*, anti-inflammatory, cytotoxic, and Antioxidant Activities of Mace Extracts from Myristica fragrans. *Evidence-Based Complementary and Alternative Medicine*. https://doi.org/10.1155/2020/7576818

Thiengsusuk, A., Chaijaroenkul, W., & Na-Bangchang, K. (2013). Antimalarial activities of medicinal plants and herbal formulations used in Thai traditional medicine. *Parasitology Research*, *112*(4), 1475–1481. https://doi.org/10.1007/S00436-013-3294-6/METRICS

Thomas, T., & Thalla, A. K. (2022). Nutmeg seed shell biochar as an effective adsorbent for removal of Remazol brilliant blue reactive dye: Kinetic, isotherm, and thermodynamic study. *Energy Sources, Part A: Recovery, Utilization and Environmental Effects*, *44*(1), 893–911. https://doi.org/10.1080/15567036.2022.2050850

Tripathi, I. P., & Dwivedi, N. (2015). Pharmacognostical standardization of nutmeg seeds (*Myristica fragrans* Houtt.)-a traditional medicine. *International Journal of Pharmaceutical Sciences and Research*, *6*(7), 3096. https://doi.org/10.13040/IJPSR.0975-8232.6(7).3096-02

Tsukayama, I., Kawakami, Y., Tamenobu, A., Toda, K., Maruoka, S., Nagasaki, Y., Mori, Y., Sawazumi, R., Okamoto, K., Kanzaki, K., Ito, H., Takahashi, Y., Miki, Y., Yamamoto, K., Murakami, M., & Suzuki-Yamamoto, T. (2022). Malabaricone C derived from nutmeg inhibits arachidonate 5-lipoxygenase activity and ameliorates psoriasis-like skin inflammation in mice. *Free Radical Biology and Medicine*, *193*(1), 1–8. https://doi.org/10.1016/j.freeradbiomed.2022.09.028

Urabee, M. C., Abdulsattar, J. O., & Nasif, Z. N. (2022). Phytochemical profile, antimicrobial, antioxidant activity and cyclooxygenase 2 inhibitory properties of nutmeg (*Myristica fragrans*) seeds extract. *Egyptian Journal of Chemistry*, *65*(1), 317–326. https://doi.org/10.21608/EJCHEM.2021.78192.3831

Uthayasankar, D., Roy, A., Kumar, S., & Thangavelu, L. (2021). Nutmeg oleoresin assisted copper sulphide nanoparticles and its anti-inflammatory activity. *Journal of Complementary Medicine Research*, *12*(4), 87. https://doi.org/10.5455/jcmr.2021.12.04.13

Vakili, F., Roosta, Z., Safari, R., Raeisi, M., Hossain, Md. S., Guerreiro, I., Akbazadeh, A., & Hoseinifar, S. H. (2023). Effects of dietary nutmeg (*Myristica fragrans*) seed meals on growth, non-specific immune indices, antioxidant status, gene expression analysis, and cold stress tolerance in zebrafish (Danio rerio). *Frontiers in Nutrition*, *9*. https://doi.org/10.3389/fnut.2022.1038748

Vangoori, Y., Dakshinamoorthi, A., Prabhakar Rao, R., David, D. C., & Anantha Babu, K. (2018). Effect of *Myristica fragrans* extract on food intake and body weight in experimental models. *Journal of Clinical and Diagnostic Research*, *12*(2), FF01–FF05. https://doi.org/10.7860/JCDR/2018/28944.11169

Verluyten, J., Leroy, F., & De Vuyst, L. (2004). Effects of different spices used in production of fermented sausages on growth of and curvacin a production by *Lactobacillus curvatus* LTH 1174. *Applied and Environmental Microbiology*, *70*(8), 4807–4813. https://doi.org/10.1128/AEM.70.8.4807-4813.2004

Vijayakumar, G., Kesavan, H., Kannan, A., Arulanandam, D., Kim, J. H., Kim, K. J., Song, H. J., Kim, H. J., & Rangarajulu, S. K. (2021). Phytosynthesis of copper nanoparticles using extracts of spices and their antibacterial properties. *Processes*, *9*(8). https://doi.org/10.3390/pr9081341

Wagan, T. A., Wang, W., Hua, H., & Cai, W. (2017). Chemical constituents and toxic, repellent, and oviposition-deterrent effects of ethanol-extracted *Myristica fragrans* (Myristicaceae) oil on *Bemisia tabaci* (Hemiptera: Aleyrodidae). *Florida Entomologist*, *100*(3), 594–601. https://doi.org/10.1653/024.100.0317

Warsito, M. F. (2021). A review on chemical composition, bioactivity, and toxicity of *Myristica fragrans* Houtt. essential oil. *Indonesian Journal of Pharmacy*, *32*(3), 304–313. https://doi.org/10.22146/IJP.1271

Wu, T., Zhao, J., Ding, M., Zhang, T., Tao, N., Wang, X., & Zhong, J. (2020). Preparation of selected spice microparticles and their potential application as nitrite scavenging agents in cured tilapia muscle. *International Journal of Food Science and Technology*, *55*(9), 3153–3161. https://doi.org/10.1111/ijfs.14579

22 Saffron Flower (*Crocus sativus*)

Noureddine Chaachouay, Ouafae Benkhnigue, and Lahcen Zidane

22.1 INTRODUCTION

The use of medicinal plants dates back to ancient times and has been a crucial part of traditional medicine worldwide (Chaachouay et al. 2020a,b; Hassan 2015; Husen 2021, 2022a,b; 2023; Rahman & Husen 2021, 2022, 2023a,b). These plants contain compounds that have different medicinal effects, such as antibacterial, anti-inflammatory, and pain-relieving properties. Many of these plants are still used for their medicinal properties, either in their raw form or as processed extracts or supplements. Their use is gaining renewed interest as an alternative or complementary therapy (Chaachouay et al. 2021a,b,c,d; Chang 2000).

Saffron (*Crocus sativus* L.) is a spice derived from the dried stigmas ('threads') of the crocus flower, native to Southwest Asia (Maggi et al. 2011). It is considered one of the world's most valuable and expensive spices due to the labour-intensive process of harvesting it, with just a small number of saffron threads needed to flavour a dish. For centuries, saffron has been a popular ingredient in traditional medicine and cuisine across diverse cultures, including Indian, Arabic, African, Persian, and Mediterranean cuisines. It imparts a distinct flavor, aroma, and yellow color to dishes and is commonly used in savory dishes like paella, bouillabaisse, risotto, and various traditional rice dishes (Mousavi & Bathaie 2011). It's also used in sweets and liqueurs, such as Moroccan delight and saffron cakes.

Saffron has a long history of usage in traditional medicine, and has been used to cure various ailments, such as depression, respiratory issues, and digestive problems. Modern research supports some of these uses and has found potential health advantages such as antioxidant and anti-inflammatory properties (Abdullaev 1993). It is also being studied for its potential value in treating depression and memory impairment. In addition to its culinary applications, saffron has also been used as a dye and fragrance in textiles, soaps, and traditional perfumes (Yildirim et al. 2020). It has been a popular ingredient in cosmetics, including face masks and creams, due to its anti-inflammatory and antioxidant effects. Saffron has also been used in religious ceremonies, such as Hindu and Buddhist rituals, and is believed to bring good luck and ward off evil in some cultures (Hutchings 2004).

In recent years, saffron has become increasingly popular as a gourmet ingredient in modern cuisine, with top chefs using it to enhance flavours in dishes and cocktails. The high cost of saffron has not deterred its growing popularity, and it is widely available in dried and powdered forms. Saffron is also sold in capsules and extracts, making it more accessible to those interested in its potential health benefits (Licón et al. 2010; Chaachouay et al. 2020a,b,c, Benkhigue et al. 2022). Whether used in cooking, traditional medicine, or modern health practices, saffron remains a highly valued and versatile ingredient. This chapter offers a comprehensive description and the most current data on the saffron flower's traditional and pharmacological properties and other potential benefits.

Saffron is a flavourful ingredient obtained from the autumn crocus or saffron crocus plant, a species of the Iridaceae family. Local common names of *Crocus sativus* L. are: Arabic: Zafaran; English: Saffron flower, Saffron crocus; French: Safran; German: Safran; Hindi: Kesar; Italian: Zafferano; Malayalam: Kashmiram; Portuguese: Açafrão; Spanish: Azafrán; Turkish: Safran.

DOI: 10.1201/9781003387046-22

22.2 SYNONYMS OF *CROCUS SATIVUS* L.

Crocus sativus L. includes eleven synonyms, which are as follows (The Plant List 2013):

(1) *Crocus autumnalis* Sm.; **(2)** *Crocus officinalis* (L.) Honck.; **(3)** *Crocus orsinii* Parl.; **(4)** *Crocus pendulus* Stokes; **(5)** *Crocus sativus* var. *cashmerianus* Royle; **(6)** *Crocus sativus* var. *officinalis* L.; **(7)** *Crocus sativus* var. *orsinii* (Parl.) Maw; **(8)** *Crocus sativus* subsp. *orsinii* (Parl.) K.Richt.; **(9)** *Crocus setifolius* Stokes; **(10)** *Geanthus autumnalis* Raf.; and **(11)** *Safran officinarum* Medik.

22.3 BOTANICAL DESCRIPTION OF *CROCUS SATIVUS*

Crocus sativus is a small stemless perennial plant up to 15 cm tall, with an underground corm 3–5 cm in diameter, encircled by a delicately reticulate-fibrous tunic, and rooted at the base (Saxena 2010). It is a triploid species (2n = 3x = 24) that cannot form seeds; hence, it spreads only by corms. A corm is a bulbous tuberous stem that has a reproductive purpose since it re-grows each year after shedding its aerial portion throughout the winter. Each year, each corm produces two or more new corm (Rashed-Mohassel 2006). The crocus corm contains adventitious roots, and the plant's leaves and flowers will develop from the corm (Figure 22.1). There can be between four and ten leaves in a saffron plant, very narrow and long, about 2 mm in width and 10–20 cm in length. The leaves are stemless, sheathed, with thin ciliate margins, and dark green in colour. A longitudinal white line characterizes the leaves in the middle (Mathew 1999).

Crocus flowers are solitary and terminal, purple, with six petals, three stamens, and a style ending in three red-orange stigmas (Figure 22.2).

The stigmas, also known as strands, saffron pistils, or saffron threads, are long, reaching between 20 and 50 mm (Figure 22.3). Saffron flowers appear in September or October and are open for only two or three days (Kafi et al. 2006).

FIGURE 22.1 The corm of *Crocus sativus* L. shows the regeneration of new cormlets.

FIGURE 22.2 The flowers of *Crocus sativus* L. from the region of Taliouine, Morocco.

FIGURE 22.3 The stigmas of a saffron flower.

22.4 ORIGIN AND GEOGRAPHIC DISTRIBUTION OF *CROCUS SATIVUS*

The origins of *C. sativus* are unknown; some believe the eastern Mediterranean is its centre of origin (Winterhalter & Straubinger 2000), while others believe it originated in Iran. According to archaeological and historical evidence, the cultivation of saffron dates back to antiquity (2,000 to 1,500 years BC) (Grilli Caiola 2003). *C. sativus* is indigenous to the eastern Mediterranean region, Central Asia, and Iran. It has been widely grown since ancient times in Morocco, Palestine, Iraq, Egypt, the United Arab Emirates, Turkey, Afghanistan, Pakistan, Azerbaijan, India, China, Italy, Japan, Spain, Iran, Greece, Switzerland, and France, with Australia recently starting its cultivation (Dar et al. 2017; Maggi et al. 2011). With an area of 115,000 hectares and an annual yield of 405 tons, Iran is now regarded as the globe's top producer of saffron, accounting for more than 85% of international production. Next come India and Greece. Morocco is the world's fourth biggest producer, producing 6.8 tons in 2018 over an area of around 1,800 hectares (Golmohammadi 2019;

Mzabri et al. 2019). In Morocco, the saffron-producing areas are situated in the towns of Taznakht and Taliouine (Aboudrare et al. 2014; Mzabri et al. 2021; Chaachouay & Zidane 2021a,b). Because of its importance, Morocco has placed a particular focus on the growth and development of the saffron sector as part of the Green Morocco Plan. This involves adopting modern production methods and identifying ideal regions for saffron cultivation.

22.5 ECOLOGY OF *CROCUS SATIVUS*

Saffron thrives in climates that are temperate and moderately arid. In the Mediterranean regions where saffron is produced, yearly precipitation seldom surpasses 400 mm (Maleki et al. 2017). It is a cold-tolerant species that can withstand frost and snow but requires a warm and dry summer for optimal growth and flowering. Saffron is primarily propagated by dividing its corms and does not typically produce seeds. The flowers of saffron crocus are pollinated by insects, such as bees and butterflies, but self-pollination is also possible. Saffron is considered a low-maintenance plant, but it may require regular irrigation during dry periods and mulching to conserve soil moisture (Kumar et al. 2022). The leaves of the plant die back after flowering, and the bulbs remain dormant until the following growing season (Saxena 2010; Sönmez et al. 2018). Saffron is not known to be susceptible to many pests or diseases, although bulb mites and fungal pathogens can attack it in some areas. Saffron is a hardy and adaptable species well-suited to cultivation in various climates and soils.

22.6 PHYTOCHEMISTRY OF SAFFRON FLOWER

Saffron is the dried stigmas and styles of the *C. sativus* flower. Its phytochemistry is complex, and many compounds have been isolated, identified, and studied (Anjum et al. 2015; Mounira et al. 2015). The saffron flower contains various chemical compounds, including carbohydrates, phenylpropanoids, triterpenoid saponins, mucilages, gums, ashes, minerals (potassium, calcium, iron, manganese, zinc, etc.), vitamins, fats, fibres and pigments, the latter including carotene, lycopene, crocetin, crocin, and zigzantin, according to phytochemical analyses (Table 22.1). The primary

TABLE 22.1
Compounds of Saffron Flower by Chemical Class

Metabolites	Chemical Class	References
Primary metabolites	Amino acids, carbohydrates, riboflavin, fatty acids, vitamin B_6, vitamin A, vitamin E, vitamin B_2, vitamin C, thiamin, folate, niacin, sugars.	Abdullaev 2002; Bagri et al. 2017; Hashemi & Erim 2016; Popović-Djordjević et al. 2021
Flavonoids	Astragalin, tamarixetin, kaempferol, populin, vitexin, isorhamnetin, myricetin, quercetin, naringenin, taxifolin, orientin, rhamnetin, and dihydro-kaempferol	Ahrazem et al. 2015; Baba et al. 2015; Montoro et al. 2008; Sun et al. 2020; Termentzi & Kokkalou 2008
Triterpenoid saponins	Azafrines 1 and 2	Ahrazem et al. 2015; Rubio-Moraga et al. 2011; S. Sharma & Kumar 2022
Carotenoids	Lycopene, zeaxanthin, α- and β-carotene, crocetin	José Bagur et al. 2017
Phenolic compounds	Catechol, caffeic acid, syringic acid, gentisic acid, methylparaben, pyrogallol, coumaric acid, chlorogenic acid, salicylic acid, *p*- cinnamic acid, gallic acid, vanillin, *t*-ferulic acid, *p*-hydroxy benzoic acid.	Belyagoubi et al. 2021; Esmaeili et al. 2011; Hadizadeh et al. 2003
Minerals	Magnesium, calcium, sodium, potassium, copper, iron, phosphorus, zinc, manganese, selenium.	Fahim et al. 2012; Moghaddasi 2010

biomolecules in saffron threads are crocin, picrocrocin, crocetin, and safranal. Furthermore, it contains small amounts of oil-soluble pigments and anthocyanin pigments, as well as β-carotene, α-carotene, and zegxantin (Anjum et al. 2015; Bhat & Broker 1953; Esmaeili et al. 2011; Mounira et al. 2015; Nørbæk & Kondo 1998; Pandita 2021; Sami et al. 2022). In addition, phytochemical investigations have revealed the presence of flavonoids, such as anthocyanins, bioflavonoids, quercetin, hesperidin, luteolin, tannins, and rutin, in the petals of the flowers (Gohari et al. 2013). Crocin, a degraded carotenoid molecule, is responsible for the color of the threads. The flavour is derived from carotenoid oxidation products, mostly safranal, and the peculiar taste is derived from picrocrocin and its glucoside. Particular essential oils intensify the scent of saffron. Safranal, the primary component of saffron, is a carboxaldehyde volatile molecule comprising about 70% of the total essential oil (Pandita 2021). It is produced through the de-glucosylation of picrocrocin. There are also some terpene aldehydes and ketones in the essential oil. The most prevalent component is safranal and 2,6,6-trimethyl 1,3-cyclohexadiene-1-carboxaldehyde. Additionally, cineol and pinene have been found as terpene derivatives (Ahrazem et al. 2015).

22.7 TRADITIONAL USES OF SAFFRON THREADS

Saffron has a long and rich history of traditional uses in various cultures. Below are some of the most common traditional benefits of saffron.

22.7.1 Phytotherapy

In herbal medicine, saffron has been employed to treat conditions such as digestive issues, respiratory problems, mood improvement, memory enhancement, menstrual cramps, and more. It has also been used as an antidepressant and anxiolytic due to its potential effects on mood (Hosseinzadeh 2014; Javadi et al. 2013; Mousavi & Bathaie 2011; Mzabri et al. 2019; Schmidt et al. 2007; Yousefi & Shafaghi 2020; Chaachouay et al. 2023a,b). Despite its historical use, it's essential to be cautious when using saffron for medicinal purposes and to speak with a specialist before starting any new treatment. This is because herbal remedies, including saffron, may interact with other medications and may not be safe for everyone (Chaachouay et al. 2020; Shi & Klotz 2012).

22.7.2 Cosmetics

Saffron has been used in cosmetic products due to its potential skin-lightening effects and as an ingredient in perfumes due to its distinctive fragrance (Mzabri et al. 2019). It has been used in traditional beauty remedies to lighten the skin, improve its overall appearance, and reduce hyperpigmentation (Figure 22.4). Saffron is also used in traditional hair care remedies, such as hair oils and masks, to improve the health and appearance of the hair (Bahrami et al. 2020; Dawalbhakta & Telang 2017). It is important to note that, although some cosmetic uses of saffron have some scientific evidence to support them, others require further investigation and evidence (Chaachouay et al. 2021). Before using saffron in cosmetics, it is recommended to consult with a dermatologist or skincare professional to determine if it is appropriate for your individual needs and skin type Figure 22.4).

22.7.3 Food and Culinary Uses

Saffron is a highly prized spice used in cooking for its characteristic flavour and vibrant colour. It is commonly used in traditional dishes such as bouillabaisse, biryani, tajine, paella, and risotto Milanese, adding a touch of sophistication to these classic recipes (Mzabri et al. 2019; Chaachouay et al. 2022a,b). Saffron can also be used to flavour soups, stews, and sauces, as well as marinate meats, such as fish and chicken, before cooking. In addition to its flavour, saffron is used as a natural

FIGURE 22.4 Some of the local products made from saffron flower extract.

food dye to colour baked goods, rice dishes, and sweets. It is a popular ingredient in Mediterranean cultures and Middle Eastern cooking, which is often added to tea to make a soothing and flavourful beverage. Saffron can also enhance the flavour of dairy products, such as cheese, yogurt, and milk. In Moroccan and Indian cuisine, it is used in rice dishes, desserts, and traditional drinks like tea (Figure 22.5) and kesar chai (Akaberi et al. 2019; Almodóvar et al. 2018; Fujii et al. 2022; Guclu et al. 2020; Mousavi & Bathaie 2011; Benkhigue et al. 2023; Orch et al. 2021). Despite its many culinary uses, it is essential to remember that saffron is a potent spice and a little goes a long way. Its intense flavour and colour can easily overpower a dish, so it should be used sparingly. Overall, saffron is a versatile ingredient that adds depth and complexity to many dishes.

22.7.4 Perfumery

In perfumery, saffron is used to enhance a fragrance's overall scent and create a more complex and layered aroma (Basker & Negbi 1983; Shokrpour 2019). The aroma of saffron is warm, slightly sweet, and somewhat bitter, with hints of hay, honey, and a metallic note. Saffron is also used in perfumery to add a touch of luxury and sophistication to fragrances. Its high cost and limited availability make it a sought-after ingredient in the perfume industry and often used in premium fragrances. Using saffron in perfumery adds a unique, warm, and spicy note to aromas. Its high cost and rarity make it a valuable ingredient in high-end, premium perfumes (Charles & Charles 2013; Mousavi & Bathaie 2011; Mzabri et al. 2019).

22.7.5 Beauty

Saffron is used in traditional beauty remedies, such as face masks and hair treatments, due to its potential benefits for the skin and hair (Dai et al. 2021; Fekrat 2003; Koocheki & Khajeh-Hosseini 2019; Mzabri et al. 2019). It is important to note that, although scientific studies have validated some

FIGURE 22.5 Moroccan tea flavoured with the stigmas of saffron.

traditional uses of saffron, others still require further investigation and evidence. However, saffron's rich cultural and historical significance continues to make it a valued and much-sought-after ingredient worldwide.

22.7.6 Religious and Cultural

Saffron has a long history of benefit in religious and cultural practices, with references extending back thousands of years. In ancient Persia, it was used in religious ceremonies as an offering to the gods (Koocheki & Khajeh-Hosseini 2019; Motamedi et al. 2020; Mousavi & Bathaie 2011; Yildirim et al. 2020). In Hinduism, saffron is considered a symbol of purity and is often used in wedding ceremonies to signify the couple's commitment to each other. In Buddhism, saffron is operated to dye the robes of Buddhist monks and is sometimes offered as a gift to the Buddha (Narayan 2009). In Christianity, saffron was used in medieval times to dye priests' robes and as an ingredient in religious incense. Saffron continues to hold significant cultural and religious significance in many societies today. Its vibrant colour and unique flavour make it a sought-after ingredient in many cuisines worldwide.

22.7.7 Textile Dye

Saffron has been used as a dye for textiles for centuries due to its rich golden color and ability to provide long-lasting hues. Using saffron to dye fabric involves extracting the colour from the saffron strands and mixing it with a mordant, a substance that helps the dye bond to the material (Barani 2016; Mortazavi et al. 2012). The fabric is then immersed in the dye mixture, penetrating the fibres and producing a vibrant, long-lasting colour. Saffron was especially popular for dyeing silk and wool in medieval Europe and the Middle East and was often used to create luxurious garments for royalty and the wealthy (Mzabri et al. 2019). Today, saffron dye is still used in traditional textiles, particularly in parts of Morocco, Iran, Turkey, and India, where it is prized for its beautiful color and cultural significance (Lachguer et al. 2021). Additionally, modern dyeing techniques have made it possible to produce synthetic saffron dyes that are more cost-effective and easier to use. Still, artisans who value its natural beauty and cultural heritage continue to practise the traditional method of using true saffron.

22.8 PHARMACOLOGICAL PROPERTIES OF SAFFRON FLOWER

The stigmas of the *C. sativus* flower are employed to produce the spice saffron. With the advent of *in vivo* and *in vitro* tests, many pharmacological characteristics of saffron have been documented. Saffron has been utilized for therapeutic reasons for millennia.

22.8.1 Antibacterial Properties

The antibacterial effects of saffron have been investigated in different laboratory and animal studies, which suggest that it may have potential as a natural antibacterial agent. Some of the compounds found in saffron, such as crocetin and crocin, have been demonstrated to have antibacterial activities against a range of bacteria, including *Salmonella typhi*, *Staphylococcus aureus*, and *Escherichia coli* (Bellachioma et al. 2022a; Esmaeelian et al. 2021; Jafari-Sales & Pashazadeh 2020; Lachguer et al. 2022; Zara et al. 2021).

22.8.2 Antifungal Properties

Saffron has been shown to possess antifungal properties against various fungal species, including *Aspergillus niger*, *Candida albicans*, and *Trichophyton mentagrophytes*. The active compounds responsible for these antifungal effects include crocin, crocetin, and safranal (Abu-Izneid et al. 2022; Lachguer et al. 2022; Naim et al. 2022; Rasool et al. 2021). It should be highlighted, however, that further investigation is necessary to demonstrate the full efficacy and potential of saffron as an antifungal agent. Additionally, saffron's antifungal effects may differ depending on the fungal strain and the extraction method used.

22.8.3 Antioxidant Activity

Saffron contains ingredients such as safranal, crocin, and crocetin with potent antioxidant properties (Baba et al. 2015; Bellachioma et al. 2022b; Lahmass et al. 2018; Salem et al. 2022). Antioxidants protect cells from damage produced by free radicals and destructive chemicals. Free radicals may induce oxidative stress and contribute to the development of different illnesses, such as cancer and cardiovascular disease (Makhlouf et al. 2011; Muzaffar et al. 2016). The antioxidant activities of saffron may help protect against oxidative stress and provide multiple health benefits (Cerdá-Bernad et al. 2022; Rikabad et al. 2019). Therefore, further research is required to completely comprehend the scope of these advantages and the underlying processes.

22.8.4 Antidepressant Effect

The antidepressant effect of saffron has been studied in several clinical trials and has been found to positively impact symptoms of depression. Some studies suggest that saffron may be as valuable as some antidepressant drugs in treating mild to moderate depression (Khan et al. 2020; Lopresti & Drummond 2014). The exact mechanism by which saffron has an antidepressant effect is not yet well understood. Still, it is thought to involve the regulation of neurotransmitters in the brain, such as serotonin, norepinephrine, and dopamine (Kamalipour et al. 2010; V. Kumar et al. 2022). It's important to note that, whereas saffron may have potential as an antidepressant, more research is required to fully understand its efficacy and safety and determine the optimal dose and method of administration. Additionally, saffron should not be used as a substitute for conventional antidepressant treatment. One should consult with a health care professional before using saffron for medical purposes.

22.8.5 Anti-inflammatory Effect

The anti-inflammatory effect of saffron has been demonstrated in research studies, suggesting that it may have the potential as a natural treatment for inflammation-related conditions (Amin & Hosseinzadeh 2015; Khan et al. 2020; Salem et al. 2022). Inflammation is a natural body response to injury or infection, but chronic inflammation can develop different illness, such as cardiovascular disease, arthritis, and certain cancers. The anti-inflammatory effect of saffron may be due to its ability to inhibit the production of pro-inflammatory cytokines and other inflammatory mediators (Hazman & Ovalı 2015; Poma et al. 2012; Zhang et al. 2014). Some studies have also shown that saffron can reduce pain and swelling (Abdullaev 2006; Hosseinzadeh & Younesi 2002). It's important to note that, although saffron may have potential as an anti-inflammatory agent, it should not be used as a substitute for conventional medical treatment.

22.8.6 Antinociceptive Activity

The antinociceptive activity of saffron refers to its ability to reduce pain perception. This effect has been observed in several animal studies, which suggest that saffron may have the potential as a natural pain reliever (Amin & Hosseinzadeh 2015; Qadir et al. 2020; Shahi et al. 2016a). The exact mechanisms behind the antinociceptive effect of saffron still need to be completely understood. Still, it affects the regulation of neurotransmitters and other signalling molecules involved in pain perception. Some analyses have also indicated that saffron has anti-inflammatory effects, which may contribute to its antinociceptive effect (Nassiri-Asl & Hosseinzadeh 2015; Zhang et al. 2014).

22.8.7 Anticancer Properties

The potential anticancer properties of saffron have been the subject of several laboratory and animal investigations, indicating that it may have a role in preventing and treating cancer (Ege et al. 2020; Fernández 2006; Lambrianidou et al. 2020). Some of the compounds found in saffron, such as crocin and crocetin, have been revealed to have antiproliferative and apoptotic properties on cancer cells, meaning they can inhibit the growth and induce the death of cancer cells (Abdullaev 2002; Chermahini et al. 2010; Gezici 2019). In addition, saffron has been found to have antioxidant and anti-inflammatory properties, which may also contribute to its potential as an anticancer agent. However, it's necessary to report that the evidence for the anticancer properties of saffron is still limited and mainly comes from laboratory and animal studies. More investigation is needed to define the efficacy and safety of using saffron for cancer prevention and therapy in humans.

22.8.8 Anxiolytic Effect

The anxiolytic effect of saffron refers to its ability to reduce anxiety. This effect has been observed in several human studies, which suggest that saffron may have the potential as a natural anxiolytic agent (Nanda & Madan 2021; Sharma & Kumar 2022). The specific processes behind saffron's anxiolytic action are not entirely known; however, it is believed to include the control of neurotransmitters and other signalling molecules in mood and anxiety regulation. Some studies have also shown that saffron has antidepressant and anticonvulsant properties, which may contribute to its anxiolytic effect (Abu-Izneid et al. 2022; Nassiri-Asl & Hosseinzadeh 2015; Pitsikas 2016).

22.8.9 Other Pharmacological Properties

The bioactive components in saffron have been shown to possess many pharmacological effects that may help protect against oxidative stress and inflammation in the brain. These properties may make saffron a potential treatment for disorders such as Parkinson's disease, Alzheimer's disease, and age-related cognitive decline (Ahmad et al. 2005; D'Onofrio et al. 2021; Namgyal & Sarwat 2020; Sharma et al. 2020; Skladnev & Johnstone 2017). In addition, saffron has been found to have blood pressure-lowering effects and to improve blood flow, making it a promising alternative for treating cardiovascular conditions (Kamalipour & Akhondzadeh 2011; Shahi et al. 2016b; Singletary 2020). Therefore, further investigation is required to fully understand saffron's cardiovascular health benefits and to determine safe and effective doses.

22.9 CONCLUSION AND FUTURE PERSPECTIVES

The saffron flower is a highly valued plant due to the unique characteristics and widespread use of the spice. It is a species of *Crocus* that grows in Mediterranean regions and in areas with similar climates and is known for its deep red stigmas and styles that are used in various applications. The plant has a high concentration of phytochemicals, including kaempferol, crocin, crocetin, safranal, and picrocrocin, contributing to its specific taste and colour. Using saffron as a condiment and colourant stretches back thousands of years, and it has been used in herbal medicine for its many healing benefits. Recent scientific studies have confirmed its pharmacological effects, including antioxidant, anxiolytic, anti-inflammatory, antinociceptive, and neuroprotective effects, making it a promising candidate for developing new drugs. The saffron plant is also known for its potential health actions, including curing depression, anxiety, Alzheimer's disease, and cardiovascular problems. Furthermore, it has been shown to have anti-tumour effects and to improve heart health. In conclusion, the saffron flower is a valuable and versatile plant with a rich history and a promising future in botany, phytochemistry, and pharmacology. Its unique phytochemicals and pharmacological properties make it a beneficial ingredient in developing new drugs. Its traditional use and potential health benefits make it a valuable plant for future investigation. To better understand the impact and mechanism of saffron on human health, more comprehensive epidemiological studies, laboratory tests, and clinical trials are needed to evaluate its medicinal properties.

REFERENCES

Abdullaev, F. (2006). Biological properties and medicinal use of saffron (*Crocus sativus* L.). In *II International Symposium on Saffron Biology and Technology 739* (pp. 339–345).

Abdullaev, F. I. (1993). Biological effects of saffron. *Biofactors (Oxford, England)*, *4*(2), 83–86.

Abdullaev, F. I. (2002). Cancer chemopreventive and tumoricidal properties of saffron (*Crocus sativus* L.). *Experimental Biology and Medicine*, *227*(1), 20–25.

Aboudrare, A., Aden, A.-H., & Lybbert, T. J. (2014). Importance Socio-économique du Safran pour les Ménages des Zones de Montagne de la Région de Taliouine-Taznakht au Maroc. *Revue Marocaine des Sciences Agronomiques et Vétérinaires*, *2*(1), 5–14.

Abu-Izneid, T., Rauf, A., Khalil, A. A., Olatunde, A., Khalid, A., Alhumaydhi, F. A., Aljohani, A. S., Sahab Uddin, M., Heydari, M., & Khayrullin, M. (2022). Nutritional and health beneficial properties of saffron (*Crocus sativus* L): A comprehensive review. *Critical Reviews in Food Science and Nutrition*, *62*(10), 2683–2706.

Ahmad, A. S., Ansari, M. A., Ahmad, M., Saleem, S., Yousuf, S., Hoda, M. N., & Islam, F. (2005). Neuroprotection by crocetin in a hemi-parkinsonian rat model. *Pharmacology, Biochemistry, and Behavior*, *81*(4), 805–813.

Ahrazem, O., Rubio-Moraga, A., Nebauer, S. G., Molina, R. V., & Gomez-Gomez, L. (2015). Saffron: Its phytochemistry, developmental processes, and biotechnological prospects. *Journal of Agricultural and Food Chemistry*, *63*(40), 8751–8764.

Akaberi, M., Boghrati, Z., Amiri, M. S., & Emami, S. A. (2019). Saffron: The golden spice. *Science of Spices and Culinary Herbs-Latest Laboratory, Pre-Clinical, and Clinical Studies*, *1*, 1–29.

Almodóvar, P., Prodanov, M., Arruñada, O., & Inarejos-García, A. M. (2018). Affron® eye, a natural extract of saffron (Crocus sativus L.) with colorant properties as novel replacer of saffron stigmas in culinary and food applications. *International Journal of Gastronomy and Food Science*, *12*, 1–5.

Amin, B., & Hosseinzadeh, H. (2015). Analgesic and anti-inflammatory effects of *Crocus sativus* L. (saffron). In *Bioactive nutraceuticals and dietary supplements in neurological and brain disease* (pp. 319–324). Elsevier.

Anjum, N., Pal, A., & Tripathi, Y. C. (2015). Phytochemistry and pharmacology of saffron, the most precious natural source of colour, flavour and medicine. *SMU Medical Journal*, *2*, 335–347.

Baba, S. A., Malik, A. H., Wani, Z. A., Mohiuddin, T., Shah, Z., Abbas, N., & Ashraf, N. (2015). Phytochemical analysis and antioxidant activity of different tissue types of *Crocus sativus* and oxidative stress alleviating potential of saffron extract in plants, bacteria, and yeast. *South African Journal of Botany*, *99*, 80–87.

Bagri, J., Yadav, A., Anwar, K., Dkhar, J., Singla-Pareek, S. L., & Pareek, A. (2017). Metabolic shift in sugars and amino acids regulates sprouting in saffron corms. *Scientific Reports*, *7*(1), 11904.

Bahrami, A., Mirmohammadsadegh, S.-S., Zare-Hoseini, S.-F., Mousavi-Tabari, R., & Saket, S. (2020). Saffron in the ancient history of Iran. In *Saffron* (pp. 23–34). Elsevier.

Barani, H. (2016). Optimization of wool fibers dyeing with saffron petal to produce blue shade. *Journal of Color Science and Technology*, *10*(4), 259–266.

Basker, D., & Negbi, M. (1983). Uses of saffron. *Economic Botany*, *37*(2), 228–236.

Bellachioma, L., Marini, E., Magi, G., Pugnaloni, A., Facinelli, B., Rocchetti, G., Martinelli, E., Lucini, L., Morresi, C., & Bacchetti, T. (2022a). Phytochemical profiling, antibacterial and antioxidant properties of *Crocus sativus* flower: A comparison between tepals and stigmas. *Open Chemistry*, *20*(1), 431–443.

Bellachioma, L., Marini, E., Magi, G., Pugnaloni, A., Facinelli, B., Rocchetti, G., Martinelli, E., Lucini, L., Morresi, C., & Bacchetti, T. (2022b). Phytochemical profiling, antibacterial and antioxidant properties of *Crocus sativus* flower: A comparison between tepals and stigmas. *Open Chemistry*, *20*(1), 431–443.

Belyagoubi, L., Loukidi, B., Belyagoubi-Benhammou, N., Gismondi, A., Di Marco, G., D'Agostino, A., Canini, A., Benmahieddine, A., Rouigueb, K., & Ben Menni, D. (2021). Valorization of Algerian saffron: Stigmas and flowers as source of bioactive compounds. *Waste and Biomass Valorization*, *12*(12), 6671–6683.

Benkhnigue, O., Chaachouay, N., Khamar, H., El Azzouzi, F., Douira, A., & Zidane, L. (2022). Ethnobotanical and ethnopharmacological study of medicinal plants used in the treatment of anemia in the region of Haouz-Rehamna (Morocco). *Journal of Pharmacy & Pharmacognosy Research*, *10*(2), 279–302.

Benkhnigue, O., Khamar, H., Bussmann, R. W., Chaachouay, N., & Zidane, L. (2023). Ethnobotanical and ethnopharmacological study of medicinal plants used in treating some liver diseases in the Al-Haouz Rehamna region (Morocco). *Ethnobotany Research and Applications*, *25*, 1–32.

Bhat, J. V., & Broker, R. (1953). Riboflavine and thiamine contents of saffron, *Crocus sativus* Linn. *Nature*, *172*(4377), 544–544.

Cerdá-Bernad, D., Clemente-Villalba, J., Valero-Cases, E., Pastor, J.-J., & Frutos, M.-J. (2022). Novel insight into the volatile profile and antioxidant properties of *Crocus sativus* L. flowers. *Antioxidants*, *11*(9), 1650.

Chaachouay, N., Benkhnigue, O., Fadli, M., El Ayadi, R., & Zidane, L. (2019a). Ethnobotanical study of medicinal plants used to treat osteoarticular diseases in the Moroccan Rif, Morocco. *Journal of Pharmacy & Pharmacognosy Research*, *7*(6), 454–470.

Chaachouay, N., Benkhnigue, O., Fadli, M., El Ibaoui, H., El Ayadi, R., & Zidane, L. (2019b). Ethnobotanical and ethnopharmacological study of medicinal and aromatic plants used in the treatment of respiratory system disorders in the Moroccan Rif. *Ethnobotany Research and Applications*, *18*, 1–16. https://doi.org/10.32859/era.18.22.1-17

Chaachouay, N., Benkhnigue, O., Fadli, M., El Ibaoui, H., & Zidane, L. (2019c). Ethnobotanical and ethnopharmacological studies of medicinal and aromatic plants used in the treatment of metabolic diseases in the Moroccan Rif. *Heliyon*, *5*(10), e02191. https://doi.org/10.1016/j.heliyon.2019.e02191

Chaachouay, N., Benkhnigue, O., & Zidane, L. (2020a). Ethnobotanical study aimed at investigating the use of medicinal plants to treat nervous system diseases in the Rif of Morocco. *Journal of Chiropractic Medicine*, *19*(1), 70–81. https://doi.org/10.1016/j.jcm.2020.02.004

Chaachouay, N., Douira, A., Hassikou, R., Brhadda, N., Dahmani, J., Belahbib, N., Ziri, R., & Zidane, L. (2020b). *Etude floristique et ethnomédicinale des plantes aromatiques et médicinales dans le Rif (Nord du Maroc)* [Phdthesis, Département de Biologie - Université Ibn Tofail - Kénitra]. https://tel.archives-ouvertes.fr/tel-03376377.

Chaachouay, N., Douira, A., & Zidane, L. (2021a). Herbal medicine used in the treatment of human diseases in the Rif, Northern Morocco. *Arabian Journal for Science and Engineering*, 47, 131–153. https://doi.org/10.1007/s13369-021-05501-1

Chaachouay, N., Orch, H., & Zidane, L. (2021b). Cystitis treatment with phytotherapy within the Rif, Northern Morocco. *Future Journal of Pharmaceutical Sciences*, 7(81), 1–9. https://doi.org/10.1186/s43094-021-00226-2

Chaachouay, N., Benkhnigue, O., Douira, A., & Zidane, L. (2021c). Poisonous medicinal plants used in the popular pharmacopoeia of the Rif, northern Morocco. *Toxicon, 189*, 24–32. https://doi.org/10.1016/j.toxicon.2020.10.028

Chaachouay, N., Douira, A., & Zidane, L. (2021d). COVID-19, prevention and treatment with herbal medicine in the herbal markets of Salé Prefecture, North-Western Morocco. *European Journal of Integrative Medicine, 42*, 101285. https://doi.org/10.1016/j.eujim.2021.101285

Chaachouay, N., & Zidane, L. (2021a) The Symbolic efficacy of plants in rituals and socio-religious ceremonies in Morocco, Northwest of Africa. *Journal of Religious & Theological Information, 21*: 34–53. https://doi.org/10.1080/10477845.2021.1942433

Chaachouay, N., & Zidane, L. (2021b). Neurological Phytotherapy by Indigenous People of Rif, Morocco. In *Therapy Approaches in Neurological Disorders*. IntechOpen. 122–137. https://doi.org/10.5772/intechopen.97175

Chaachouay, N., Azeroual, A., Bencharki, B., Douira, A., & Zidane, L. (2022a). Ethnoveterinary medicines plants for animal therapy in the Rif, North of Morocco. *South African Journal of Botany, 147*, 176–191. https://doi.org/10.1016/j.sajb.2021.12.037

Chaachouay, N., Azeroual, A., Douira, A., & Zidane, L. (2022b). Ethnoveterinary practices of medicinal plants among the Zemmour and Zayane tribes, Middle Atlas, Morocco. *South African Journal of Botany, 151*, 826–840. https://doi.org/10.1016/j.sajb.2022.11.009

Chaachouay, N., Azeroual, A., Ansari, M. K. A., & Zidane, L. (2023a). Use of Plants as Medicines and Aromatics by Indigenous Communities of Morocco: Pharmacognosy, Ecology and Conservation. In *Genomics, Transcriptomics, Proteomics and Metabolomics of Crop Plants* (Ed. Husen, A., & Ahmad, A.) (pp. 353–363). Academic Press. https://doi.org/10.1016/B978-0-323-95989-6.00017-6

Chaachouay, N., Belhaj, S., El Khomsi, M., Benkhnigue, O., & Zidane, L. (2023b). Herbal remedies used to treat digestive system ailments by indigenous communities in the Rif region of northern Morocco. *Vegetos*. https://doi.org/10.1007/s42535-023-00606-4

Chang, J. (2000). Medicinal herbs: Drugs or dietary supplements? *Biochemical Pharmacology, 59*(3), 211–219.

Charles, D. J., & Charles, D. J. (2013). *Antioxidant properties of spices, herbs and other sources* (pp. 509–520).

Chermahini, S. H., Majid, F. A. A., Sarmidi, M. R., Taghizadeh, E., & Salehnezhad, S. (2010). Impact of saffron as an anti-cancer and anti-tumor herb. *African Journal of Pharmacy and Pharmacology, 4*(11), 834–840.

Dai, R.-C., Nabil, W. N. N., & Xu, H.-X. (2021). The history of saffron in China: From its origin to applications. *Chinese Medicine and Culture, 4*(4), 228–234.

Dar, R. A., Shahnawaz, M., Malik, S. B., Sangale, M. K., Ade, A. B., & Qazi, P. H. (2017). Cultivation, distribution, taxonomy, chemical composition and medical importance of *Crocus sativus*. *The Journal of Phytopharmacology, 6*(6), 356–358.

Dawalbhakta, M., & Telang, M. (2017). Patents on therapeutic and cosmetic applications of bioactives of *Crocus sativus* L. and their production through synthetic biology methods: A review. *Recent Patents on Biotechnology, 11*(1), 3–19.

D'Onofrio, G., Nabavi, S. M., Sancarlo, D., Greco, A., & Pieretti, S. (2021). *Crocus sativus* L. (Saffron) in Alzheimer's disease treatment: Bioactive effects on cognitive impairment. *Current Neuropharmacology, 19*(9), 1606.

Ege, B., Yumrutas, O., Ege, M., Pehlivan, M., & Bozgeyik, I. (2020). Pharmacological properties and therapeutic potential of saffron (*Crocus sativus* L.) in osteosarcoma. *Journal of Pharmacy and Pharmacology, 72*(1), 56–67.

Esmaeelian, M., Jahani, M., Feizy, J., & Einafshar, S. (2021). Effects of ultrasound-assisted and direct solvent extraction methods on the antioxidant and antibacterial properties of saffron (*Crocus sativus* L.) corm extract. *Food Analytical Methods, 14*(1), 74–87.

Esmaeili, N., Ebrahimzadeh, H., Abdi, K., & Safarian, S. (2011). Determination of some phenolic compounds in *Crocus sativus* L. corms and its antioxidant activities study. *Pharmacognosy Magazine*, *7*(25), 74.

Fahim, N. K., Janati, S. F., & Feizy, J. (2012). Chemical composition of agriproduct saffron (*Crocus sativus* L.) petals and its considerations as animal feed. *Gida*, *37*(4), 197–201.

Fekrat, H. (2003). The application of crocin and saffron ethanol-extractable components in formulation of health care and beauty care products. In *I International Symposium on Saffron Biology and Biotechnology 650* (pp. 365–368).

Fernández, J.-A. (2006). Anticancer properties of saffron, *Crocus sativus* Linn. *Advances in Phytomedicine*, *2*, 313–330.

Fujii, S., Morita, Y., Ohta, T., Uto, T., & Shoyama, Y. (2022). Saffron (*Crocus sativus* L.) as a valuable spice and food product: A narrative review. *Longhua Chinese Medicine*, *5*, 18.

Gezici, S. (2019). Comparative anticancer activity analysis of saffron extracts and a principle component, crocetin for prevention and treatment of human malignancies. *Journal of Food Science and Technology*, *56*(12), 5435–5443.

Gohari, A. R., Saeidnia, S., & Mahmoodabadi, M. K. (2013). An overview on saffron, phytochemicals, and medicinal properties. *Pharmacognosy Reviews*, *7*(13), 61.

Golmohammadi, F. (2019). Saffron as a main cash, medical and resistive plant for sustainable economy and livelihood of rural people in dried regions of Iran. *Black Sea Journal of Agriculture*, *2*(3), 156–163.

Grilli Caiola, M. (2003). Saffron reproductive biology. In *I International Symposium on Saffron Biology and Biotechnology 650* (pp. 25–37).

Guclu, G., Kelebek, H., & Selli, S. (2020). Saffron (*Crocus sativus* L.): Its aroma and key odorants. In *Saffron* (pp. 69–82). Elsevier.

Hadizadeh, F., Khalili, N., Hosseinzadeh, H., & Khair-Aldine, R. (2003). Kaempferol from saffron petals. *Science Direct Working Paper, S1574-0331, 04*.

Hashemi, P., & Erim, F. B. (2016). Analysis of vitamin B2 in saffron stigmas (Crocus sativus L) by capillary electrophoresis coupled with laser-induced fluorescence detector. *Food Analytical Methods*, *9*(8), 2395–2399.

Hassan, H. M. A. (2015). A short history of the use of plants as medicines from ancient times. *Chimia*, *69*(10), 622–622.

Hazman, Ö., & Ovalı, S. (2015). Investigation of the anti-inflammatory effects of safranal on high-fat diet and multiple low-dose streptozotocin induced type 2 diabetes rat model. *Inflammation*, *38*(3), 1012–1019.

Hosseinzadeh, H. (2014). Saffron: A herbal medicine of third millennium. In *Jundishapur Journal of Natural Pharmaceutical Products* (Vol. 9, Numéro 1, pp. 1–2). Briefland.

Hosseinzadeh, H., & Younesi, H. M. (2002). Antinociceptive and anti-inflammatory effects of Crocus sativus L. stigma and petal extracts in mice. *BMC Pharmacology*, *2*, 1–8.

Hutchings, J. (2004). Colour in folklore and tradition—The principles. *Color Research and Application: Endorsed by Inter-Society Color Council, The Colour Group (Great Britain), Canadian Society for Color, Color Science Association of Japan, Dutch Society for the Study of Color, The Swedish Colour Centre Foundation, Colour Society of Australia, Centre Français de la Couleur*, *29*(1), 57–66.

Husen, A. (2021). *Traditional herbal therapy for the human immune system*. CRC Press. https://doi.org/10.1201/9781003137955.

Husen, A. (2022a). *Herbs, shrubs and trees of potential medicinal benefits*. CRC Press. https://doi.org/10.1201/9781003205067.

Husen, A. (2022b). *Exploring poisonous plants: Medicinal values, toxicity response and other potential uses*. CRC Press. https://doi.org/10.1201/b23017.

Husen, A. (2023). *Antidiabetic medicinal plants and herbal treatments*. CRC Press. https://doi.org/10.1201/b23347.

ITIS (2022). *Integrated Taxonomic Information System (ITIS) On-Line Database*. www.itis.gov, CC0 https://doi.org/10.5066/F7KH0KBK; http://itis-web-prod.snafu.cr.usgs.gov.

Jafari-Sales, A., & Pashazadeh, M. (2020). Antibacterial effect of methanolic extract of saffron petal (*Crocus sativus* L.) on some standard gram positive and gram negative pathogenic bacteria *in vitro*. *Current Perspectives on Medicinal and Aromatic Plants (CUPMAP)*, *3*(1), 1–7.

Javadi, B., Sahebkar, A., & Emami, S. A. (2013). A survey on saffron in major Islamic traditional medicine books. *Iranian Journal of Basic Medical Sciences*, *16*(1), 1.

José Bagur, M., Alonso Salinas, G. L., Jiménez-Monreal, A. M., Chaouqi, S., Llorens, S., Martínez-Tomé, M., & Alonso, G. L. (2017). Saffron: An old medicinal plant and a potential novel functional food. *Molecules*, *23*(1), 30.

Kafi, M., Koocheki, A., & Rashed, M. H. (2006). *Saffron (Crocus sativus): Production and processing.* Science Publishers.

Kamalipour, M., & Akhondzadeh, S. (2011). Cardiovascular effects of saffron: An evidence-based review. *The Journal of Tehran Heart Center, 6*(2), 59.

Kamalipour, M., Jamshidi, A. H., & Akhoundzadeh, S. (2010). *Antidepressant effect of Crocus sativus: An evidence-based review.*

Khan, A., Muhamad, N. A., Ismail, H., Nasir, A., Khalil, A. A. K., Anwar, Y., Khan, Z., Ali, A., Taha, R. M., & Al-Shara, B. (2020). Potential nutraceutical benefits of in vivo grown saffron (*Crocus sativus* L.) as analgesic, anti-inflammatory, anticoagulant, and antidepressant in mice. *Plants, 9*(11), 1414.

Koocheki, A., & Khajeh-Hosseini, M. (2019). *Saffron: Science, technology and health.* Woodhead Publishing Limited.

Kumar, A., Devi, M., Kumar, R., & Kumar, S. (2022). Introduction of high-value *Crocus sativus (saffron) cultivation in non*-traditional regions of India through ecological modelling. *Scientific Reports, 12*(1), 11925.

Kumar, V., Kaur, N., & Wadhwa, P. (2022). Clinical evidence on the effects of saffron (*Crocus sativus* L.) in anxiety and depression. *World Journal of Traditional Chinese Medicine, 8*(2), 181.

Lachguer, K., El Merzougui, S., Boudadi, I., Laktib, A., Ben El Caid, M., Ramdan, B., Boubaker, H., & Serghini, M. A. (2022). Major phytochemical compounds, *in vitro* antioxidant, antibacterial, and antifungal activities of six aqueous and organic extracts of *Crocus sativus* L. Flower Waste. *Waste and Biomass Valorization*, 1–17.

Lachguer, K., El Ouali, M., Essaket, I., El Merzougui, S., Cherkaoui, O., & Serghini, M. A. (2021). Eco-friendly dyeing of wool with natural dye extracted from Moroccan *Crocus sativus* L. flower waste. *Fibers and Polymers, 22*(12), 3368–3377.

Lahmass, I., Ouahhoud, S., Elmansuri, M., Sabouni, A., Elyoubi, M., Benabbas, R., Choukri, M., & Saalaoui, E. (2018). Determination of antioxidant properties of six by-products of *Crocus sativus* L.(saffron) plant products. *Waste and Biomass Valorization, 9*(8), 1349–1357.

Lambrianidou, A., Koutsougianni, F., Papapostolou, I., & Dimas, K. (2020). Recent advances on the anticancer properties of saffron (*Crocus sativus* L.) and its major constituents. *Molecules, 26*(1), 86.

Licón, C., Carmona, M., Llorens, S., Berruga, M. I., & Alonso, G. L. (2010). Potential healthy effects of saffron spice (*Crocus sativus* L. stigmas) consumption. *Functional Plant Science and Biotechnology, 4*(2), 64–73.

Lopresti, A. L., & Drummond, P. D. (2014). Saffron (*Crocus sativus*) for depression: A systematic review of clinical studies and examination of underlying antidepressant mechanisms of action. *Human Psychopharmacology: Clinical and Experimental, 29*(6), 517–527.

Maggi, L., Carmona, M., Kelly, S. D., Marigheto, N., & Alonso, G. L. (2011). Geographical origin differentiation of saffron spice (*Crocus sativus* L. stigmas)–preliminary investigation using chemical and multi-element (H, C, N) stable isotope analysis. *Food Chemistry, 128*(2), 543–548.

Makhlouf, H., Saksouk, M., Habib, J., & Chahine, R. (2011). Determination of antioxidant activity of saffron taken from the flower of *Crocus sativus* grown in Lebanon. *African Journal of Biotechnology, 10*(41), 8093–8100.

Maleki, F., Kazemi, H., Siahmarguee, A., & Kamkar, B. (2017). Development of a land use suitability model for saffron (*Crocus sativus* L.) cultivation by multi-criteria evaluation and spatial analysis. *Ecological Engineering, 106*, 140–153.

Mathew, B. (1999). Botany, taxonomy and cytology of *Crocus sativus* L. and its allies. *Saffron: Crocus sativus L, 1730.*

Moghaddasi, M. S. (2010). Saffron chemicals and medicine usage. *Journal of Medicinal Plants Research, 4*(6), 427–430.

Montoro, P., Tuberoso, C. I., Maldini, M., Cabras, P., & Pizza, C. (2008). Qualitative profile and quantitative determination of flavonoids from *Crocus sativus* L. petals by LC-MS/MS. *Natural Product Communications, 3*(12), 1934578X0800301215.

Mortazavi, S. M., KAMALI, M. M., Safi, S., & Salehi, R. (2012). *Saffron petals, a by-product for dyeing of wool fibers.*

Motamedi, M., Gharaei, F., & Saket, S. (2020). Saffron and religion. In *Saffron* (pp. 15–21). Elsevier.

Mounira, L., Bernardo, M., Luigi, C. P., Guido, F., Fatima, G., Khadija, B., Abdelmjid, Z., & Luisa, P. (2015). Phytochemical composition of Moroccan saffron accessions by headspace solid-phase-microextraction. *American Journal of Essential Oils and Natural Products, 2*(4), 1–7.

Mousavi, S. Z., & Bathaie, S. Z. (2011). Historical uses of saffron: Identifying potential new avenues for modern research. *Avicenna Journal of Phytomedicine*, *1*(2), 57–66.
Muzaffar, S., Rather, S. A., Khan, K. Z., & Akhter, R. (2016). Nutritional composition and in-vitro antioxidant properties of two cultivars of Indian saffron. *Journal of Food Measurement and Characterization*, *10*(1), 185–192.
Mzabri, I., Addi, M., & Berrichi, A. (2019). Traditional and modern uses of saffron (*Crocus sativus). Cosmetics*, *6*(4), 63.
Mzabri, I., Charif, K., Rimani, M., Kouddane, N., Boukroute, A., & Berrichi, A. (2021). History, biology and culture of *Crocus sativus*: Overview and perspectives. *Arabian Journal of Chemical and Environmental Research*, *8*(1), 1–28.
Mzabri, I., Chetouani, M., Charif, K., Addi, M., Nkengurutse, J., Mimouni, J., Kouddane, N., & Berrichi, A. (2019). The agro-morphological behavior of saffron under the environmental conditions of four areas in Eastern Morocco. *Materials Today: Proceedings*, *13*, 1062–1069.
Naim, N., Fauconnier, M.-L., Ennahli, N., Tahiri, A., Baala, M., Madani, I., Ennahli, S., & Lahlali, R. (2022). Chemical composition profiling and antifungal activity of saffron petal extract. *Molecules*, *27*(24), 8742.
Namgyal, D., & Sarwat, M. (2020). Saffron as a neuroprotective agent. In *Saffron* (pp. 93–102). Elsevier.
Nanda, S., & Madan, K. (2021). The role of Safranal and saffron stigma extracts in oxidative stress, diseases and photoaging: A systematic review. *Heliyon*, *7*(2), e06117.
Narayan, B. (2009). *Fascinating Hindutva: Saffron politics and Dalit mobilisation*. SAGE Publications Ltd.
Nassiri-Asl, M., & Hosseinzadeh, H. (2015). Neuropharmacology effects of saffron (*Crocus sativus*) and its active constituents. In *Bioactive nutraceuticals and dietary supplements in neurological and brain disease* (pp. 29–39). Elsevier.
Nørbæk, R., & Kondo, T. (1998). Anthocyanins from flowers of *Crocus* (Iridaceae). *Phytochemistry*, *47*(5), 861–864.
Orch, H., Chaachouay, N., Douiri, E. M., Faiz, N., Zidane, L., & Douira, A. (2021). Use of medicinal plants in dermato-cosmetology: An ethnobotanical study among the population of Izarène. *Jordan Journal of Pharmaceutical Sciences*, 323–340.
Pandita, D. (2021). Saffron (Crocus sativus L.): Phytochemistry, therapeutic significance and omics-based biology. In *Medicinal and aromatic plants* (pp. 325–396). Elsevier.
Pitsikas, N. (2016). Constituents of saffron (*Crocus sativus* L.) as potential candidates for the treatment of anxiety disorders and schizophrenia. *Molecules*, *21*(3), 303.
Poma, A., Fontecchio, G., Carlucci, G., & Chichiricco, G. (2012). Anti-inflammatory properties of drugs from saffron crocus. *Anti-Inflammatory and Anti-Allergy Agents in Medicinal Chemistry (Formerly Current Medicinal Chemistry-Anti-Inflammatory and Anti-Allergy Agents)*, *11*(1), 37–51.
Popović-Djordjević, J. B., Kostić, A. Ž., & Kiralan, M. (2021). Antioxidant activities of bioactive compounds and various extracts obtained from saffron. In *Saffron* (pp. 41–97). Elsevier.
Qadir, S., Bashir, S., & John, R. (2020). Saffron—Immunity system. In *Saffron* (pp. 177–192). Elsevier.
Rashed-Mohassel, M. H. (2006). Saffron botany. In *Saffron (Crocus sativus)* (pp. 21–34). CRC Press.
Rahman, S., & Husen, A. (2022). *Asparagus racemosus* (Shatavari) and *Dioscorea villosa* (Wild yams). In *Herbs, Shrubs, and Trees of Potential Medicinal Benefits* (Ed. Husen, A.). CRC Press.
Rahman, S., & Husen, A. (2021). Potential role of medicinal plants in the cure of liver and kidney diseases. In *Non-timber Forest Products* (Ed. Husen, A., Bachheti, R.K., & Bachheti A.) (pp. 229–25). Springer International Publishing AG. https://doi.org/10.1007/978-3-030-73077-2_10.
Rahman, S., & Husen, A. (2023a). *Trigonella foenum-graecum* L.: A Useful Medicinal Plant for Diabetes and Other Diseases. In *Antidiabetic medicinal plants and herbal treatments* (Ed. Husen, A.). CRC Press.
Rahman, S., & Husen, A. (2023b). Role of green chiretta [*Andaphis paniculata* (Burm. F.)] Nees in diabetes and other ailments. In *Antidiabetic medicinal plants and herbal treatments* (Ed. Husen A.). CRC Press.
Rasool, A., Mir, M. I., Zulfajri, M., Hanafiah, M. M., Unnisa, S. A., & Mahboob, M. (2021). Plant growth-promoting and antifungal asset of indigenous rhizobacteria secluded from saffron (*Crocus sativus* L.) rhizosphere. *Microbial Pathogenesis*, *150*, 104734.
Rikabad, M. M., Pourakbar, L., Moghaddam, S. S., & Popović-Djordjević, J. (2019). Agrobiological, chemical and antioxidant properties of saffron (*Crocus sativus* L.) exposed to TiO2 nanoparticles and ultraviolet-B stress. *Industrial Crops and Products*, *137*, 137–143.
Rubio-Moraga, Á., Gerwig, G. J., Castro-Díaz, N., Jimeno, M. L., Escribano, J., Fernández, J.-A., & Kamerling, J. P. (2011). Triterpenoid saponins from corms of *Crocus sativus*: Localization, extraction and characterization. *Industrial Crops and Products*, *34*(3), 1401–1409.

Salem, M., Shaheen, M., Tabbara, A., & Borjac, J. (2022). Saffron extract and crocin exert anti-inflammatory and anti-oxidative effects in a repetitive mild traumatic brain injury mouse model. *Scientific Reports*, *12*(1), 5004.

Sami, M., Rashid, A., Akram, M., Ibadi, A. K., & Mbaye, E. H. S. (2022). *Phytochemical analysis and therapeutic properties of medicinal plants Crocus sativus: Saffron.*

Saxena, R. B. (2010). Botany, Taxonomy and Cytology of *Crocus sativus* series. *AYU (An International Quarterly Journal of Research in Ayurveda)*, *31*(3), 374.

Schmidt, M., Betti, G., & Hensel, A. (2007). Saffron in phytotherapy: Pharmacology and clinical uses. *Wiener Medizinische Wochenschrift (1946)*, *157*(13–14), 315–319.

Shahi, T., Assadpour, E., & Jafari, S. M. (2016). Main chemical compounds and pharmacological activities of stigmas and tepals of 'red gold'; saffron. *Trends in Food Science and Technology*, *58*, 69–78.

Sharma, B., Kumar, H., Kaushik, P., Mirza, R., Awasthi, R., & Kulkarni, G. T. (2020). Therapeutic benefits of saffron in brain diseases: New lights on possible pharmacological mechanisms. In *Saffron* (pp. 117–130). Elsevier.

Sharma, S., & Kumar, D. (2022). Chemical composition and biological uses of *Crocus sativus* L.(saffron). In *Edible plants in health and diseases: Volume II: phytochemical and pharmacological properties* (pp. 249–277). Springer.

Shi, S., & Klotz, U. (2012). Drug interactions with herbal medicines. *Clinical Pharmacokinetics*, *51*(2), 77–104.

Shokrpour, M. (2019). Saffron (*Crocus sativus* L.) breeding: Opportunities and challenges. In *Advances in Plant Breeding Strategies: Industrial and Food Crops: Volume 6* (pp. 675–706).

Singletary, K. (2020). Saffron: Potential health benefits. *Nutrition Today*, *55*(6), 294–303.

Skladnev, N. V., & Johnstone, D. M. (2017). Neuroprotective properties of dietary saffron: More than just a chemical scavenger? *Neural Regeneration Research*, *12*(2), 210.

Sönmez, Ç., Taghİloofar, A. H., & Bayram, E. (2018). Determination of some agro-morphological characters among different originated saffron (*Crocus sativus* L.) populations under Bornova ecological conditions. *Ege Üniversitesi Ziraat Fakültesi Dergisi*, *55*(3), 335–339.

Sun, C., Nile, S. H., Zhang, Y., Qin, L., El-Seedi, H. R., Daglia, M., & Kai, G. (2020). Novel insight into utilization of flavonoid glycosides and biological properties of saffron (*Crocus sativus* L.) flower byproducts. *Journal of Agricultural and Food Chemistry*, *68*(39), 10685–10696.

Termentzi, A., & Kokkalou, E. (2008). LC-DAD-MS (ESI+) analysis and antioxidant capacity of *Crocus sativus* petal extracts. *Planta Medica*, *74*(05), 573–581.

The Plant List. (2013). *The Plant List.* Version 1.1. Published on the Internet. http://www.theplantlist.org/ (accessed January 1).

Winterhalter, P., & Straubinger, M. (2000). Saffron—Renewed interest in an ancient spice. *Food Reviews International*, *16*(1), 39–59.

Yildirim, M. U., Sarihan, E. O., & Khawar, K. M. (2020). Ethnomedicinal and traditional usage of saffron (*Crocus sativus* L.) in Turkey. In *Saffron* (pp. 21–31). Elsevier.

Yousefi, M., & Shafaghi, K. (2020). Saffron in Persian traditional medicine. In *Saffron* (pp. 393–404). Elsevier.

Zara, S., Petretto, G. L., Mannu, A., Zara, G., Budroni, M., Mannazzu, I., Multineddu, C., Pintore, G., & Fancello, F. (2021). Antimicrobial activity and chemical characterization of a non-polar extract of saffron stamens in food matrix. *Foods*, *10*(4), 703.

Zhang, Y., Wang, X., Ma, L., Dong, L., Zhang, X., Chen, J., & Fu, X. (2014). Anti-inflammatory, antinociceptive activity of an essential oil recipe consisting of the supercritical fluid CO_2 extract of white pepper, long pepper, cinnamon, saffron and myrrh *in vivo*. *Journal of Oleo Science*, *63*(12), 1251–1260.

23 Turmeric Rhizome and Leaves (*Curcuma longa*)

Mayur Mukut Murlidhar Sharma, Divya Kapoor, Atul Loyal, and Pankaj Sharma

23.1 INTRODUCTION

Turmeric (*Curcuma longa* L.; syn.: *Curcuma domestica* Valeton) belongs to the Zingiberaceae (ginger) family and is a plant whose rhizomes are used as a spice in most south Asian households. It can be grown in subtropical and tropical environments with a temperature range of 25–35°C. The plant's morphology is shrub-like at 3–5 ft height with oblong leaves and funnel-shaped flowers. This plant is also known as yellow roots (Gupta et al., 2013). It grows in well-drained sandy or clay loam soils, having a pH of 4.5–7.5 with good organic status, where it flourishes. Turmeric has a long history of being used as a spice and a food flavour enhancer in Asian cuisines. It is especially added to rice, yogurt, and other food items to improve the flavour and colour tone of the food item. In curries, it is mostly used in combination with other spices. Its use in ancient traditional medicine of Asian countries is well documented, and its use in Indian systems dates back 4000 years to the Vedic culture. Alongside its culinary significance, initially turmeric was of great religious importance. It probably reached China by 700 AD, East Africa by 800 AD, West Africa by 1200 AD, and Jamaica in the eighteenth century. In 1280, Marco Polo defined this spice as having qualities like that of saffron. According to the Sanskrit medical treatise and the Ayurvedic and Unani systems as well, turmeric has a profound and long history regarding its medicinal use in India, with Susruta's Ayurvedic Compendium from 250 BC recommending a turmeric-based ointment to alleviate the effects of poisoned food (Prasad et al., 2011). Apart from the Indian and South Asian culture, its use has also been prominent in other Asian cultures. For instance, in Persian, it is known as *Zard choobe*. The use of turmeric as a household remedy has been known for ages. It is sought to have antibacterial, anti-viral, and anti-inflammatory properties. The use of turmeric as a tonic also has a long history. Its use has been recognized in alleviating the effects of a wide range of diseases, such as stomach disorders, arthritis, and liver-associated ailments (Tayyem et al., 2006).

Curcumin (1,7-bis(4-hydroxy-3-methoxyphenyl)-1*E*,6*E*-heptadiene-3,5-dione or diferuloyl methane) is a polyphenol derived from turmeric. Curcumin is a yellow tincture, which can be obtained from the rhizome of the plant. The yellow colour of turmeric is attributed to the curcumin complexes. Generally, turmeric contains three different types of curcumin, namely curcumin I, curcumin II, and curcumin III. The presence of curcumin in turmeric is around 8% of the dry weight. The curcumin complexes are not soluble in water or ether, but can be dissolved in other organic solvents, such as alcohol. Apart from its anti-inflammatory and antimicrobial properties, turmeric is also known to possess antioxidant, neuroprotective, and anticancer properties. The active ingredient curcumin is considered to be a potent remedy for the process of oxidation and inflammation. The other active compounds found in turmeric are volatile oils. The oils found in turmeric comprise some fatty acids that are known to have antifungal, antimutagenic, and antibacterial activities. This chapter will describe the medicinal benefits conferred by turmeric, including its leaves and rhizomes.

DOI: 10.1201/9781003387046-23

23.2 NOMENCLATURE OF TURMERIC

In North India, turmeric is commonly called '*haldi*', a word derived from the Sanskrit word *haridra*, and in the south it is called 'manjal', a word that is frequently used in ancient Tamil literature. The name turmeric derives from the Latin term *terra merita* (meritorious earth), referring to the color of ground turmeric, which resembles a mineral pigment. It is known as *terre merite* in French and simply as "yellow root" in many languages. In many cultures, its name is based on the Latin word *curcuma*. In Sanskrit, turmeric has at least 53 different names, namely anestha (not offered for sacrifice or homa), bhadra (auspicious or lucky), bahula (plenty), dhirgharaja (long in appearance), gandhaplashika (which produces good smell), gauri (to make fair), gharshani (to rub), haldi (that draws attention to its bright color), haridra (dear to hari, Lord Krishna), harita (greenish), hemaragi (exhibits golden color), hemaragini (gives the golden color), hridayavilasini (gives delight to heart, charming), jayanti (one that wins over diseases), jawarantika (which cures fevers), kanchani (exhibits golden color), kaveri (harlot), krimighni or kashpa (killer of worms), kshamata (capability), laxmi (prosperity), mangalprada (who bestows auspiciousness), mangalya (auspicious), mehagni (killer of fat), nisha (night), nishakhya (known as night), nishawa (clears darkness and imparts color), patwaluka (perfumed powder), pavitra (holy), pinga (reddish-brown), pinja (yellow-red powder), pita (yellow), pitika (which gives yellow color), rabhangavasa (which dissolves fat), ranjani (which gives color), ratrimanika (as beautiful as moonlight), shifa (fibrous root), shobhna (brilliant color), shiva (gracious), shyama (dark colored), soubhagaya (lucky), survana (golden color), survanavara (which exhibits golden color), tamasini (beautiful as night), umavara (Parvati, wife of Lord Shiva), vairagi (who remains free from desires), varavarnini (which gives fair complexion), varna datri (enhancer of body complexion), varnini (which gives color), vishagni (killer of poison), yamini (night), yoshitapriya (beloved of wife), and yuvati (young girl).

23.3 ACTIVE INGREDIENTS AND CHEMICAL CONSTITUENTS OF TURMERIC

There are more than one hundred chemical compounds in turmeric and most of these compounds have been identified for their biological activities (Figure 23.1 and Table 23.1). For instance, 20 organic compounds are known to have antibiotic properties and around 14 of them are known to exhibit anticancer activities. Furthermore, 12 constituents of turmeric have been demonstrated to possess antitumour properties and ten of them are known to mediate anti-inflammatory activities. Overall, 326 biological activities of turmeric have been identified (Ahmad et al., 2020). The major constituent of the rhizome is a volatile oil comprising turmerone, and, in addition to that, there are some additional colouring agents, termed curcuminoids. Curcuminoids comprise of curcumin demethoxycurcumin, 5'-methoxycurcumin, and dihydrocurcumin, which are known to be natural antioxidants (Ruby et al., 1995). In a standard form, turmeric contains moisture (>9%), curcumin (5–6.6%), extraneous matter (<0.5% by weight), mould (<3%), and volatile oils (<3.5%). Volatile oils include d-α-phellandrene, d-sabinene, cineol, borneol, zingiberene, and sesquiterpenes (Ohshiro et al., 1990). There are a variety of sesquiterpenes, like germacrone, termerone; ar-(+)-, α-, and β-termerones; β-bisabolene; α-curcumene; zingiberene; β-sesquiphellanderene; bisacurone; curcumenone; dehydrocurdione; procurcumadiol; bis-acumol; curcumenol; isoprocurcumenol; epiprocurcumenol; procurcumenol; zedoaronediol; and curlone. Among all of them, many are specific to the type of turmeric species. The constituents responsible for the aroma of turmeric are turmerone, arturmerone, and zingiberene. The rhizomes are also stated to contain four different polysaccharides or ukonans, in addition to stigmasterol, β-sitosterol, cholesterol, and 2-hydroxymethyl anthraquinone (Prasad et al., 2011).

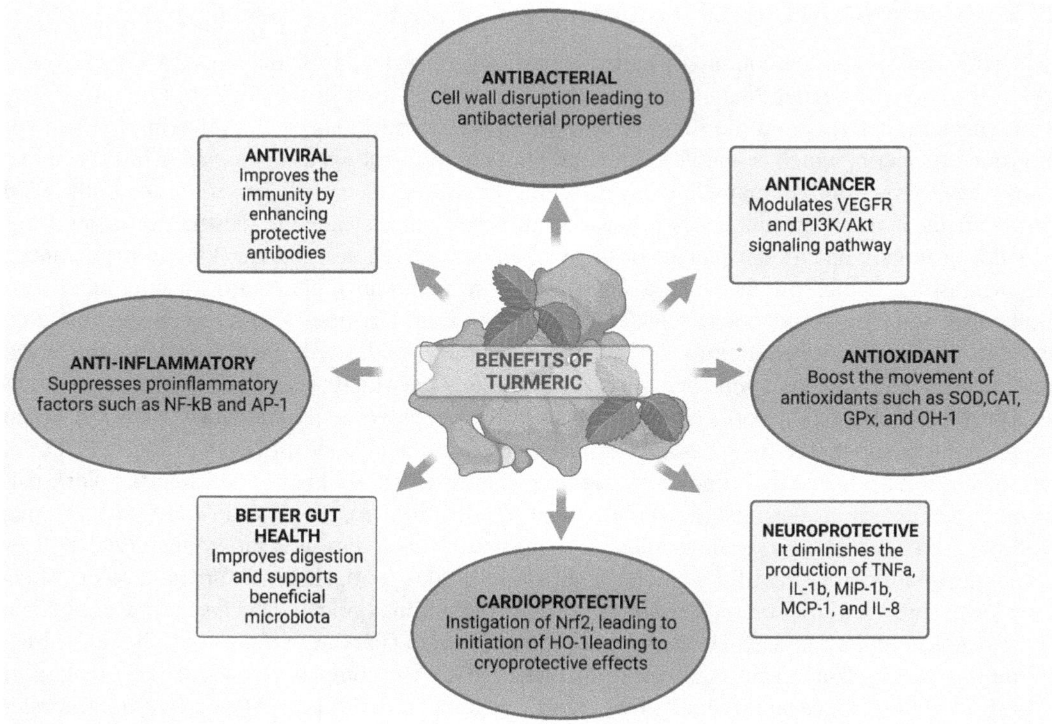

FIGURE 23.1 A portrayal depicting different pharmacological attributes of turmeric.

TABLE 23.1
List of Different Pharmacological Attributes of Turmeric as Reported in the Literature

Serial no.	Activity	Organism or Cell Line	Reference
1	Anticancerous	Human glioblastoma cell lines (U373, U251, D54, and T98G)	Bilal et al., 2021
2	Anticancerous	Human breast cancer cell lines (MCF-7, ZR-75 and MDA-MB 231)	Vemuri et al., 2017
3	Anti-inflammatory and antioxidant	*In vivo* AIA rat model	Ramadan et al., 2011
4	Anti-inflammatory	*In vivo* Wistar rats	Meizarini et al., 2018
5	Anti-inflammatory	RAW 264.7 macrophages	Chuysinuan et al., 2023
6	Neuroprotective	*In vivo* mice	Mythri et al., 2011
7	Neuroprotective	Neural stem cells	Ostuka et al., 2022
8	Hepatoprotective	*In vivo* Wistar rats	Ibrahim et al., 2020
9	Hepatoprotective	*In vivo* Wistar rats	Harmeet et al., 2011
10	Cardioprotective	*In vivo* Wistar rats	Fasooto et al., 2023

23.4 TURMERIC AS A TRADITIONAL MEDICINAL

Turmeric has been part of the folk medicine in different parts of the world since time immemorial, where it has been used in therapeutic preparations over the centuries for treating different ailments in households and also by professional traditional medicinal people. In Ayurvedic practices, turmeric is thought to have numerous medicinal properties comprising maintenance of the vigour of the body, discharging of gas, ousting of worms, improving digestion, managing menstruation, eliminating gallstones, and treating arthritis. In many countries, it is used to treat wounds, cuts, and burns and is also applied to prevent festering of any possible bacterial infection. Turmeric's use is also predominantly regarding the gastro-intestinal discomfort. The anti-inflammatory activities of turmeric makes them a good fit to ease gastro-related inflammatory problems such as irritable bowel syndrome. In Pakistan and Afghanistan, turmeric is used to wash lesions and encourage their recovery by smearing it on a piece of burnt cloth that is positioned above a wound. Indians use turmeric, in addition to its Ayurvedic prescription, to cleanse blood and to treat skin disorders. Turmeric paste is used by women in some regions of India to get rid of unwanted hair. There has been a practice to apply turmeric paste to the bride and groom's body before marriage in some parts of the South-East Asia such as India, Pakistan, and Bangladesh. It is believed that the bride and groom will get a better glow on their skin after application of turmeric paste.

Additionally, it is believed that turmeric also keeps bad bacteria away from them, to ensure their good health before the all-important event of the wedding. Furthermore, turmeric has also been included in sunscreens by many companies, owing to the skin-friendly properties possessed by turmeric. In Ayurvedic medicine, turmeric is a well-documented treatment for various respiratory conditions (e.g., asthma, bronchial hyperactivity, and allergy), as well as for liver disorders, anorexia, rheumatism, diabetic wounds, runny nose, cough, and sinusitis (Araujo et al., 2001). In traditional Chinese medicine, turmeric is used to treat diseases associated with abdominal pain. From ancient times, as prescribed by Ayurveda, turmeric has been used to treat sprains and swelling. In both Ayurvedic and traditional Chinese medicine, turmeric is considered a bitter digestive and a carminative (Aggarwal et al., 2004). Unani practitioners also use turmeric to expel phlegm or kapha, as well as to open blood vessels in order to improve blood circulation. It can be incorporated into foods, including rice and bean dishes, to improve digestion and reduce gas and bloating. It is a cholagogue, stimulating bile production in the liver and encouraging excretion of bile *via* the gallbladder, which improves the body's ability to digest fats. Sometimes, turmeric mixed with milk or water is taken to treat intestinal disorders as well as for colds and sore throats.

23.5 MEDICINAL AND PHARMACOLOGICAL PROPERTIES OF TURMERIC

23.5.1 Anti-inflammatory Properties

Multiple reports have mentioned that turmeric possesses anti-inflammatory properties, and is a potential natural medicine to treat a number of inflammatory diseases. Curcumin, the most active ingredient found in turmeric, governs the anti-inflammatory reactions by suppressing the pro-inflammatory transcription factors such as NF-κB and AP-1 (Sharifi-Rad et al., 2020). Nuclear kappa-B factor is an important regulator of inflammatory responses. Its inhibition is necessary to suppress the inflammatory responses; for instance, when a curcumin analogue was tested *in vivo* against mouse RAW 264.7 macrophage cells, the activity of NF-κB drastically decreased which led to the consequences that the curcumin exhibited both anti-inflammatory and anticancer activities (Olivera et al., 2012). The AP-1 is also an important factor in the JNK inflammatory pathway. The inhibition of AP-1 has also been reported to diminish inflammatory activities (Surh et al., 2000). Turmeric also plays anti-inflammatory activities by decreasing the secretion of pro-inflammatory cytokines. In a recent study conducted by Gorabi et al. (2021), the meta-analysis of different studies available in the literature strongly suggests the diminishing of different anti-inflammatory cytokines, with the most affected cytokines being IL-1 and TNF-α. In another study, the treatment with curcumin of acute pancreatitis significantly affected the cytokines and inhibited the activities of NF-κB and AP-1. The cytokine which were greatly influenced were chemokines, TNF-α and IL-6 (Gulcubuk et al., 2013). Anandakumar et al. 2014 studied the anti-inflammatory activities of curcumin in different acute and chronic inflammatory models. The carrageenan-induced paw oedema and xylene-induced ear oedema models were used to test the acute inflammation response towards turmeric. Both the ear and paw oedema were significantly reduced, which proves the role of turmeric (rich in curcumin and other active compounds) in decreasing acute inflammation. The same study also demonstrated reduction in the weight of granuloma in the cotton pellet-induced granuloma model which further highlights the reduction in chronic inflammation by turmeric. In cases of acute inflammation, taking curcumin orally was discovered to be as effective as using cortisone or phenylbutazone. Consuming *C. longa* orally notably reduced the swelling associated with inflammation. It is also believed that the anti-inflammatory features of *C. longa* could be associated with its capacity to hamper the production of inflammatory prostaglandins resulting from arachidonic acid and to modify the activity of neutrophils in circumstances of inflammation (Labban et al., 2014). Turmeric also compromises inflammation by decreasing the activity of enzymes like 5-lipoxygenase and COX-2 and -5 and by stalling the mitogen-activated protein kinases (MAPK) and routes involved in the formation of nitric oxide synthase (NOS) enzymes. Additionally, curcuminoids inhibit the action of LOX, COX, phospholipases, leukotrienes, prostaglandins, thromboxane, nitric oxide, elastase, hyaluronidase, collagenase, monocyte chemoattractant protein-1, interferon-inducible protein, TNF, and interleukin-12. They also attenuate the synthesis of prostaglandins and inhibit the formation of leukotrienes through the lipoxygenase pathway (Bundy et al., 2004).

23.5.2 Anticancer Properties

Studies on animals reveal curcumin's ability to impact on each of the three phases of cancer growth: the initiation, promotion, and progression. In the initiation and promotion stages, curcumin influences transcription factors that control the detoxification of carcinogens in phases I and II (Garg et al., 2008). Moreover, it lessens the activity of inflammatory signalling molecules, transcription factors triggered by free radicals, and the production of arachidonic acid through cyclooxygenase and lipoxygenase pathways. Additionally, it acts as a scaavenger for free radicals (Hong et al., 2004). The angiogenesis in any type of cancer is well known. Any type of cancer tissue mass can create the blood vessels to feed the growing tumourous mass. The creation of these blood vessels is initiated by the activation of pro-angiogenic factors. Inhibition of these factors will result in suppression of cancer growth. Curcumin displays anticancer activities by inhibiting angiogenic factors stimulators, such as VEGF and basic fibroblast growth factor (Yance et al., 2006). Astinfeshan et al. (2019) findings showed that curcumin has the ability to inhibit the formation of new blood vessels by influencing the modulation of the VEGFR and the PI3K/Akt signalling pathways. Furthermore, the study exposed curcumin's ability to downregulate the expression of MMP-2 and MMP-9, while concurrently increasing levels of tissue inhibitor metalloproteinase-1. This dual action plays a role in maintaining the stability and consistency of the extracellular matrix. Moreover, curcumin has the ability to induce apoptosis in cancerous cells *via* a pathway that relies on p53. Remarkably, p53 stands out as a critical protein in suppressing tumour development, having influence over cell growth, programmed cell death, and DNA repair (Kandoth et al., 2013). Additionally, it was demonstrated that the proapoptotic impact of curcumin is conditional upon the occurrence of miR-192-5p and miR-215, both of which actuate the p53 pathway in cases of non-small-cell lung cancer (Ye et al., 2015). In the study conducted by Chiu and Su (2009), it was shown that the administration of curcumin to the MDA-MB-231 triple-negative breast cancer cell line resulted in interference with CDKs/cyclin complexes. These complexes are vital for the advancement of the cell cycle. Also, the levels of cyclin D1, an important factor for moving through the G1/S phase of the cell cycle, were reduced. Given that increased cyclin D1 expression is associated with an increased incidence of breast cancers, this interruption brought about a halt in the cell cycle at the G1 stage. Curcumin has attracted consideration for its remarkable efficacy against cancer conditions characterized by extreme Ras activity. Notably, Cao et al. (2015) revealed that curcumin beneficially blocks the propagation of AGS gastric cancer cells by lowering Ras protein levels while boosting ERK activity. Curcumin's influence extends to decreasing the PI3K/Akt signalling pathway in cancer models, as established by studies by Rana et al. (2015) and Kasi et al. (2016). Furthermore, targeting the Wnt/ß-catenin signalling pathway represents a favourable target in cancer therapy. This pathway's hyperactivity, which is associated with human cancers, can be blocked by curcumin, which activates cell cycle arrest at the G2/M phase by tempering the Wnt/ß-catenin signalling pathway. Turmeric leaf extracts has also been shown to have anti-proliferative activities making it another potential anticancer treatment. Additionally, the turmeric leaf oil (TLO) is also being studied to determine its effect against cancer progression. For instance, Behera et al. (2023) studied turmeric leaf oil as a surfactant against breast cancer cell lines. According to the results of the *in-vitro* cytotoxicity study, TLO nanoemulgel resulted in 28.56% cell viability after 24 hours and an IC_{50} value of 59.84 g/mL. In another study, turmeric leaf extract inhibited the proliferation of HepG2 (hepatocellular carcinoma), HT-29 (colon carcinoma), MDA-MB-231 (non-hormone-dependent breast cancer), MCF-7 (hormone-dependent breast cancer) and HeLa (cervical cancer) cells; 50% inhibitory values (IC_{50}) were obtained for MDA-MB-231, HepG2, and HT-29 (Hueh et al., 2011).

23.5.3 Hepatoprotective Effect

Several reasons, including alcohol, medication, environmental toxins, parasites, and elements of our diet, can induce both short-term and long-term impairment to the liver, leading to situations like liver fibrosis, non-alcoholic steatohepatitis, non-alcoholic liver disease, and possibly cirrhosis. A noteworthy body of research has been devoted to examining curcumin due to its ability to defend the liver from damage (Tung et al., 2017). In a study by Choudhury et al. (2016), they established that administering curcumin *via* injection (at a concentration of 8.98 μM) to Swiss albino rats with CCl_4-induced liver damage led to a lessening in NADH oxidase levels and an increase in glutathione reductase (GR) and glutathione-S-transferase (GST) levels, as well as an increase in succinate dehydrogenase activity. Likewise, in another experiment encompassing the sameidentical kind of liver injury, the administration of curcumin (at a dosage of 200 mg/kg) to Sprague-Dawley rats caused an expansion of hepatic glutathione levels and a decrease in lipid peroxidase levels, along with reduced activities of both alanine transaminase (ALT) and aspartate aminotransferase (AST), as reported by Lee et al. (2016). When exposed to liver damage caused by drug-related factors like streptozotocin and paracetamol misuse, curcumin has confirmed the capability to alleviate these antagonistic effects in mice. Research conducted by Afrin et al. (2015) discovered that administering curcumin to Sprague Dawley rats with streptozotocin-induced diabetes decreased the level of TNFα, IL-1ß, MAPK, and apoptosis signal-regulating kinase 1 (ASK1) within liver tissues. In cases of non-alcoholic steatohepatitis, encouraged by a mixture of low-dose streptozotocin and a high-fat diet, Afrin et al. (2017) established that curcumin treatment displayed the ability to reduce oxidative stress, inflammation, and lipogenesis. Additionally, curcumin was detected to ease fibrosis and hamper HMGB1-NF-κB translocation and signalling. In mice suffering from alcoholic fatty liver, the introduction of curcumin displayed a positive influence on hepatocyte necroptosis. It also acted to quash the pathway induced by ethanol exposure, to hinder glyoxylate, dicarboxylate, and pyruvate metabolisms, bringing about modifications in antioxidant signalling pathways, and augment the expression of detoxifying genes. These conclusions were attained through the stimulation of the ERK/p38-MAPK pathway (Guo et al., 2017).

23.5.4 Neuroprotective Effects

Neurological conditions like Parkinson's disease (PD), Alzheimer's disease (AD), major depression, and epilepsy have a noteworthy influence on millions of people worldwide, and their manifestation is on the upsurge. Neuroinflammation, described by persistent inflammation, brings about modifications in neuronal metabolism that lead to the weakening of neurons. This progression is exaggerated by the initiation of microglia and astrocytes in states of neuroinflammation. These astrocytes are accountable for the discharge of proinflammatory cytokines like TNFα and IL-1, promote aggravating neuronal decrease. Drawing from prevailing research, curcumin has arisen as a potential therapeutic choice for numerous neurological ailments, comprising dementia, AD, PD, multiple sclerosis, and Huntington's disease (HD). This is accredited to its antioxidative, anti-inflammatory, and anti-protein accumulation properties (Salehi et al., 2020). As an illustration, studies have established curcumin's ability to block the generation of inflammatory cytokines and prostaglandins in triggered microglia and astrocytes (Cianciulli et al., 2016). Furthermore, curcumin diminishes the production of TNFα, IL-1β, macrophage inflammatory protein (MIP-1β), monocyte chemoattractant protein (MCP-1), and IL-8 inside microglial and astrocytic cells (Chen et al., 2014). In Alzheimer's disease (AD), the fundamental cause leads from the buildup of amyloid-ß (Aß) peptide plaques, a consequence of microgliosis, astrocytosis, and the presence of proinflammatory components inside the brain. Particularly, curcumin has occurred as a possible alleviator of AD symptoms through a range of mechanisms. It can adjust the transferring of amyloid precursor protein, thereby inhibiting the production of Aß peptides. Curcumin's interaction with Aß peptides impacts their accumulation propensities, while concurrently reducing the hyperphosphorylation of tau protein and improving its effective clearance. Furthermore, curcumin plays a part in weakening Aß-induced toxicity by impeding JNK-3 phosphorylation. Its influence on cholesterol levels is associated with a reduced risk of AD, and it supports the protective function of the blood-brain barrier through the up-regulation of OH-1 expression. Moreover, curcumin impedes acetylcholinesterase action, contributes to cell signalling through the initiation of Wnt pathways, and adds to the lessening of both inflammation and oxidative impairment within the brain (Goozee et al., 2016; Ganesh et al., 2017). Within the realm of multiple sclerosis, an autoimmune and inflammatory disorder that principally marks young adults and women through the expansion of demyelinating lesions, curcumin has established remarkable neuroprotective effects through different mechanisms. These incorporate its roles as an antioxidant, an anti-inflammatory agent, and an inhibitor of excessive cell propagation. Remarkably, curcumin has shown the ability to affect numerous molecular targets, comprising transcriptional factors (such as NF-κB, Nrf2, AP-1, STAT-1, -3, -4), enzymes (COX-2, iNOS, OH-1, LOX, XO), inflammatory cytokines (like chemokine ligands, interleukins, TNFα), proteins (such as caspase-3, -9, Bcl-2, prostaglandin, CRP, myosin light chain), protein kinases (like AK, JNK, JAK, MAPK), and growth factors and receptors (such as TLRs, chemokine receptors, TGF-α, TGF-ß). These multilayered interactions jointly contribute to curcumin's possible therapeutic influence on various scleroses (Qureshi et al., 2018).

23.5.5 Cardioprotective Effect

Cardiovascular (CV) diseases is one of global health's major challenges and are related to substantial rates of morbidity and mortality. Research findings have determined that curcumin achieves effectiveness in providing barriers against such cardiovascular ailments. The advantageous properties of curcumin with respect to cardiovascular health are primarily based on protection against several conditions, such as atherosclerosis, cardiac hypertrophy, heart failure, aortic aneurysm, stroke, myocardial infarction, and diabetic cardiovascular complications (Salehi et al., 2020b). One important characteristic of curcumin's molecular action implicates the instigation of Nrf2, leading to the initiation of heme oxygenase-1 (HO-1), which is involved in achieving cytoprotective and anti-inflammatory effects against oxidative stress (Pittala et al., 2018). For instance, Sunagawa et al. (2018) revealed that curcumin functions as a natural inhibitor of p300 histone acetyltransferase (HAT), whereas Monfoulet et al. (2017) recognized curcumin's capability to augment endothelial function and diminish TNFα-induced monocyte union in endothelial cells through the inhibition of NF-κB. In addition, Yao et al. (2016) recognized that curcumin weakens the expression of the angiotensin II type 1 receptor (AT1R), thus achieving protection against cardiovascular diseases. Interestingly, curcumin accomplishes this by reducing the binding potential of the *AT1R* gene promoter with the specificity protein 1 (SP1). Likewise, Cao et al. (2018) provided evidence that curcumin could possibly alleviate chronic heart failure by increasing the activity of p38 MAPK, JNK, and ASK1.

23.5.6 Antioxidant Properties

Curcumin's antioxidative effects have attracted widespread attention in the scientific literature. Multiple *in vitro* and *in vivo* studies have examined this aspect, accrediting curcumin's antioxidant potential to its chemical composition, which integrates features like carbon-carbon double bonds, β-diketo groups, and phenyl rings with hydroxyl and *o*-methoxy groups (Menon and Sudheer, 2007). The mechanism of its antioxidant competence lies in its capability to intermingle with various reactive oxygen species (ROS), comprising superoxide radicals, hydrogen peroxide, and nitric oxide (NO) radicals, while also curbing lipid peroxidation (Ak and Gulcin, 2008). This latter function stems from its ability to boost the activity of critical antioxidant enzymes, such as superoxide dismutase (SOD), catalase (CAT), glutathione peroxidase (GPx), and HO-1. Moreover, curcumin raises glutathione levels by upregulating glutathione transferase and their analogous mRNA levels. By impeding ROS-producing enzymes, like LOX, COX, and xanthine oxidase, curcumin further exhibits its antioxidant properties. Its lipophilic nature also succeeds as a chain-breaking antioxidant, making it capable of foraging peroxyl radicals (Priyadarsini et al., 2003). The antioxidant effects of turmeric leaves (TL) come from distinct bioactive compounds, such as curcumin, total phenolic compounds, and flavonoids. Some studies have also focused specifically on the turmeric leaves to evaluate their antioxidant effects. Kim et al. (2021) investigated the antioxidant effects of a water extract of TL (TLE) through *in vitro* Vero cell and *in vivo* zebrafish models when subjected to H_2O_2-induced oxidative stress. The assessment of intracellular ROS generation and sub-G1 phase cell proportions serve as key indicators of TLE's antioxidant potential. The results demonstrated that TLE exhibited remarkable intracellular ROS scavenging activity, effectively counteracting the deleterious effects of H_2O_2-induced oxidative stress. Moreover, the dose-dependent reduction in the proportion of cells in the sub-G1 phase further underscored the significant antioxidant activity of TLE. These findings not only highlight the potential of TLE as a potent antioxidant agent but also suggest its prospective application in mitigating oxidative stress-induced cellular damage. The robust antioxidant properties of TLE observed in both cellular and organismal models warrant further exploration and validation, potentially leading to its utilization in therapeutic interventions aimed at ameliorating oxidative stress-related conditions.

23.6 CONCLUSION

Turmeric (*Curcuma longa* L.) is a plant with considerable historical and cultural implications and is widely documented for its culinary uses and therapeutic properties across numerous traditional medicinal systems. Its names reveal its varied roles, from spice to skincare. The active compound, curcumin, characterizes turmeric with its distinct colour and a countless number of medicinal benefits. Turmeric has been lauded for its extraordinary anti-inflammatory traits, established through its impact on key transcription factors and cytokines, with potential to mediate diverse inflammatory disorders. Its anticancer properties have attracted considerable attention, with effects over different stages of cancer growth and inhibiting angiogenesis. The hepatoprotective prospective of turmeric has been demonstrated to safeguard the liver against various insults, including drug-induced impairment and fatty liver situations. The neuroprotective properties of turmeric, ascribed to its antioxidative and anti-inflammatory characteristics, position it as a encouraging candidate for addressing neurodegenerative diseases such as Alzheimer's and Parkinson's diseases. Furthermore, turmeric displays cardioprotective activities, impacting factors linked to cardiovascular diseases, and potentially improving endothelial utility. The antioxidative properties of curcumin accentuate the ability of curcumin to stabilize oxidative stress by mediating reactive oxygen species, boosting antioxidant enzyme activity, and quenching harmful radicals. This multifaceted antioxidant mechanism contributes to its shielding effects against oxidative damage and lipid peroxidation. Integrating turmeric into daily routines, whether as a spice in food preparation or through dietary supplements, offers a plethora of potential health benefits. However, it is important to note that, while the research conducted thus far is promising, further studies are needed to illuminate the mechanisms underlying turmeric's medicinal properties and to establish ideal dosages for different therapeutic situations.

REFERENCES

Afrin R, Arumugam S, Rahman A, Wahed MI, Karuppagounder V, Harima M, Suzuki H, Miyashita S, Suzuki K, Yoneyama H, Ueno K. Curcumin ameliorates liver damage and progression of NASH in NASH-HCC mouse model possibly by modulating HMGB1-NF-κB translocation. *International Immunopharmacology*. 2017 Mar 1;44:174–82.

Afrin R, Arumugam S, Soetikno V, Thandavarayan RA, Pitchaimani V, Karuppagounder V, Sreedhar R, Harima M, Suzuki H, Miyashita S, Nomoto M. Curcumin ameliorates streptozotocin-induced liver damage through modulation of endoplasmic reticulum stress-mediated apoptosis in diabetic rats. *Free Radical Research*. 2015 Mar 4;49(3):279–89.

Aggarwal BB, Takada Y, Oommen OV. From chemoprevention to chemotherapy: Common targets and common goals. *Expert Opinion on Investigational Drugs*. 2004 Oct 1;13(10):1327–38.

Ahmad RS, Hussain MB, Sultan MT, Arshad MS, Waheed M, Shariati MA, Plygun S, Hashempur MH. Biochemistry, safety, pharmacological activities, and clinical applications of turmeric: A mechanistic review. *Evidence-Based Complementary and Alternative Medicine*. 2020 May10;2020:7656919

Ak T, Gülçin I. Antioxidant and radical scavenging properties of curcumin. *Chemico-Biological Interactions*. 2008 Jul 10;174(1):27–37.

Anandakumar S, Joseph JA, Bethapudi B, Agarwal A, Jung EB. Anti-inflammatory effects of turmeric (Curcuma longa L.) extract on acute and chronic inflammation models. Journal of the Korean Society of Food Science and Nutrition. 2014;43(4):612–7.

Araujo CA, Leon LL. Biological activities of *Curcuma longa* L. *Memórias do Instituto Oswaldo Cruz*. 2001;96(5):723–8.

Astinfeshan M, Rasmi Y, Kheradmand F, Karimipour M, Rahbarghazi R, Aramwit P, Nasirzadeh M, Daeihassani B, Shirpoor A, Gholinejad Z, Saboory E. Curcumin inhibits angiogenesis in endothelial cells using downregulation of the PI3K/Akt signaling pathway. *Food Bioscience*. 2019 Jun 1;29:86–93.

Behera MP, Devi J, Mohitkar SP, Dhir V, Marndi SS, Sharma D, Goda C. An innovative breast cancer treatment using turmeric leaf oil encapsulated Nanoemulgels. *Rivista Italiana di Filosofia Analitica Junior*. 2023 Jul 29;14(1):775–83.

Bilal I, Xie S, Elburki MS, Aziziaram Z, Ahmed SM, Jalal Balaky ST. Cytotoxic effect of diferuloylmethane, a derivative of turmeric on different human glioblastoma cell lines. *Cellular, Molecular and Biomedical Reports*. 2021 Apr 1;1(1):14–22.

Bundy R, Walker AF, Middleton RW, Booth J. Turmeric extract may improve irritable bowel syndrome symptomology in otherwise healthy adults: A pilot study. *Journal of Alternative and Complementary Medicine*. 2004 Dec 1;10(6):1015–8.

Cao AL, Tang QF, Zhou WC, Qiu YY, Hu SJ, Yin PH. Ras/ERK signaling pathway is involved in curcumin-induced cell cycle arrest and apoptosis in human gastric carcinoma AGS cells. *Journal of Asian Natural Products Research*. 2015 Jan 2;17(1):56–63.

Cao Q, Zhang J, Gao L, Zhang Y, Dai M, Bao M. Dickkopf3 upregulation mediates the cardioprotective effects of curcumin on chronic heart failure. *Molecular Medicine Reports*. 2018 May 1;17(5):7249–57.

Chen N, Geng Q, Zheng J, He S, Huo X, Sun X. Suppression of the TGF-β/Smad signaling pathway and inhibition of hepatic stellate cell proliferation play a role in the hepatoprotective effects of curcumin against alcohol-induced hepatic fibrosis. *International Journal of Molecular Medicine*. 2014 Oct 1;34(4):1110–6.

Chiu TL, Su CC. Curcumin inhibits proliferation and migration by increasing the Bax to Bcl-2 ratio and decreasing NF-κBp65 expression in breast cancer MDA-MB-231 cells. *International journal of molecular medicine*. 2009 Apr 1;23(4):469–75.

Choudhury ST, Das N, Ghosh S, Ghosh D, Chakraborty S, Ali N. Vesicular (liposomal and nanoparticulated) delivery of curcumin: A comparative study on carbon tetrachloride–mediated oxidative hepatocellular damage in rat model. *International Journal of Nanomedicine*. 2016 May 18; 11: 2179–93.

Chuysinuan P, Pengsuk C, Lirdprapamongkol K, Thanyacharoen T, Techasakul S, Svasti J, Nooeaid P. Turmeric herb extract-incorporated biopolymer dressings with beneficial antibacterial, antioxidant and anti-inflammatory properties for wound healing. *Polymers*. 2023 Feb 22;15(5):1090.

Cianciulli A, Calvello R, Porro C, Trotta T, Salvatore R, Panaro MA. PI3k/Akt signalling pathway plays a crucial role in the anti-inflammatory effects of curcumin in LPS-activated microglia. *International Immunopharmacology*. 2016 Jul 1;36:282–90.

Fasooto TT, Ogundana OR, Adebayo AF, Olawade DB, Olugbogi EA, Fapohunda O, Akinmoladun AC. *In-silico* and *in-vivo* comparative evaluation of the cardioprotective potential of yellow turmeric and white turmeric in MN-induced cardiac oxidative Stress. SSRN. 2023 Feb 1;357(1):196–205.

Ganesh P, Karthikeyan R, Muthukumaraswamy A, Anand J. A potential role of periodontal inflammation in Alzheimer's disease: A review. *Oral Health and Preventive Dentistry*. 2017 Jan 1;15(1):7–12.

Garg R, Gupta S, Maru GB. Dietary curcumin modulates transcriptional regulators of phase I and phase II enzymes in benzo[a]pyrene-treated mice: Mechanism of its anti-initiating action. *Carcinogenesis*. 2008 May; 29(5): 1022–32.

Goozee KG, Shah TM, Sohrabi HR, Rainey-Smith SR, Brown B, Verdile G, Martins RN. Examining the potential clinical value of curcumin in the prevention and diagnosis of Alzheimer's disease. *British Journal of Nutrition*. 2016 Feb;115(3):449–65.

Gorabi AM, Razi B, Aslani S, Abbasifard M, Imani D, Sathyapalan T, Sahebkar A. Effect of curcumin on proinflammatory cytokines: A meta-analysis of randomized controlled trials. *Cytokine*. 2021 Jul 1;143:155541.

Gulcubuk A, Haktanir D, Cakiris A, Ustek D, Guzel O, Erturk M, Karabagli M, Akyazi I, Cicekci H, Altunatmaz K, Uzun H. Effects of curcumin on proinflammatory cytokines and tissue injury in the early and late phases of experimental acute pancreatitis. *Pancreatology*. 2013 Jul 1;13(4):347–54.

Guo C, Ma J, Zhong Q, Zhao M, Hu T, Chen T, Qiu L, Wen L. Curcumin improves alcoholic fatty liver by inhibiting fatty acid biosynthesis. *Toxicology and Applied Pharmacology*. 2017 Aug 1;328:1–9.

Gupta SC, Sung B, Kim JH, Prasad S, Li S, Aggarwal BB. Multitargeting by turmeric, the golden spice: From kitchen to clinic. *Molecular Nutrition and Food Research*. 2013 Sep;57(9):1510–28.

Harmeet S, Bedi PS, Singh B. Hepatoprotective activity of turmeric and garlic against 7–12, dimethylbenzanthracene induced liver damage in Wistar albino rats. *European Journal of Medicinal Plants*. 2011;1(4):162–70.

Hong J, Bose M, Ju J, Ryu JH, Chen X, Sang S, Lee MJ, Yang CS. Modulation of arachidonic acid metabolism by curcumin and related β-diketone derivatives: Effects on cytosolic phospholipase A 2, cyclooxygenases and 5-lipoxygenase. *Carcinogenesis*. 2004 Sep 1;25(9):1671–9.

Hueh Zan C, Rahmat A, Abdah, A, Alitheen BM. N, Othman F, Ee Cheng Lian G. Anti-proliferative effects of pandan leaves (*Pandanus amarylfolius*), kantan flower (*Etlingera elatior*) and turmeric leaves (*Curcuma longa*). *Nutrition and Food Science*. 2011 Jul 19;41(4):238–41.

Ibrahim J, Kabiru AY, Abdulrasheed-Adeleke T, Lawal B, Adewuyi AH. Antioxidant and hepatoprotective potentials of curcuminoid isolates from turmeric (*Curcuma longa*) rhizome on CCl4-induced hepatic damage in Wistar rats. *Journal of Taibah University for Science*. 2020 Jan 1;14(1):908–15.

Kandoth C, McLellan MD, Vandin F, Ye K, Niu B, Lu C, Xie M, Zhang Q, McMichael JF, Wyczalkowski MA, Leiserson MD. Mutational landscape and significance across 12 major cancer types. *Nature*. 2013 Oct 17;502(7471):333–9.

Kasi PD, Tamilselvam R, Skalicka-Woźniak K, Nabavi SF, Daglia M, Bishayee A, Pazoki-Toroudi H, Nabavi SM. Molecular targets of curcumin for cancer therapy: An updated review. *Tumor Biology*. 2016 Oct;37(10):13017–28.

Kim S, Kim M, Kang MC, Lee HH, Cho CH, Choi I, Park Y, Lee SH. Antioxidant effects of turmeric leaf extract against hydrogen peroxide-induced oxidative stress *in vitro* in vero cells and in vivo in zebrafish. *Antioxidants*. 2021 Jan 14;10(1):112.

Labban L. Medicinal and pharmacological properties of Turmeric (*Curcuma longa*): A review. *International Journal of Biomedical and Pharmaceutical Sciences.* 2014;5(1):17–23.

Lee HY, Kim SW, Lee GH, Choi MK, Jung HW, Kim YJ, Kwon HJ, Chae HJ. Turmeric extract and its active compound, curcumin, protect against chronic CCl 4-induced liver damage by enhancing antioxidation. *BMC Complementary and Alternative Medicine.* 2016 Dec;16:1–9.

Meizarini A, Riawan W, Puteri A. Anti-inflammatory properties of a wound dressing combination of zinc oxide and turmeric extract. *Veterinary World.* 2018 Jan;11(1):25.

Menon VP, Sudheer AR. Antioxidant and anti-inflammatory properties of curcumin. Adv Exp Med Biol. 2007;595:105-25..

Monfoulet LE, Mercier S, Bayle D, Tamaian R, Barber-Chamoux N, Morand C, Milenkovic D. Curcumin modulates endothelial permeability and monocyte transendothelial migration by affecting endothelial cell dynamics. *Free Radical Biology and Medicine.* 2017 Nov 1;112:109–20.

Mythri RB, Veena J, Harish G, Rao BS, Bharath MS. Chronic dietary supplementation with turmeric protects against 1-methyl-4-phenyl-1, 2, 3, 6-tetrahydropyridine-mediated neurotoxicity in vivo: Implications for Parkinson's disease. *British Journal of Nutrition.* 2011 Jul;106(1):63–72.

Ohshiro M, Kuroyanagi M, Ueno A. Structures of sesquiterpenes from *Curcuma longa. Phytochemistry.* 1990 Jan 1;29(7):2201–5.

Olivera A, Moore TW, Hu F, Brown AP, Sun A, Liotta DC, Snyder JP, Yoon Y, Shim H, Marcus AI, Miller AH. Inhibition of the NF-κB signaling pathway by the curcumin analog, 3, 5-Bis (2-pyridinylmethylidene)-4-piperidone (EF31): Anti-inflammatory and anti-cancer properties. *International Immunopharmacology.* 2012 Feb 1;12(2):368–77.

Ostuka S, Kawamura M, Fujino S, Nakamura F, Arai D, Fusetani N, Nakao Y. Coronarin D. A metabolite from the wild turmeric, *Curcuma aromatica*, promotes the differentiation of neural stem cells into astrocytes. *Journal of Agricultural and Food Chemistry.* 2022 Mar 4;70(10):3300–9.

Pittala V, Vanella L, Salerno L, Romeo G, Marrazzo A, Di Giacomo C, Sorrenti V. Effects of polyphenolic derivatives on heme oxygenase-system in metabolic dysfunctions. *Current Medicinal Chemistry.* 2018 Apr 1;25(13):1577–95.

Prasad S, Aggarwal BB. *Turmeric, the Golden Spice: From Traditional Medicine to Modern Medicine. In: In: Benzie IFF, Wachtel-Galor S, editors. Herbal Medicine: Biomolecular and Clinical Aspects.* 2nd ed. Boca Raton (FL): CRC Press/Taylor & Francis; 2011. Chapter 13.

Priyadarsini KI, Maity DK, Naik GH, Kumar MS, Unnikrishnan MK, Satav JG, Mohan H. Role of phenolic OH and methylene hydrogen on the free radical reactions and antioxidant activity of curcumin. *Free Radical Biology and Medicine.* 2003 Sep 1;35(5):475–84.

Qureshi M, Al-Suhaimi EA, Wahid F, Shehzad O, Shehzad A. Therapeutic potential of curcumin for multiple sclerosis. *Neurological Sciences.* 2018 Feb;39(2):207–14.

Ramadan G, Al-Kahtani MA, El-Sayed WM. Anti-inflammatory and anti-oxidant properties of *Curcuma longa* (turmeric) versus *Zingiber officinale (ginger) rhizomes in rat adjuvant-induced arthritis. Inflammation.* 2011 Aug;34(4):291–301.

Rana C, Piplani H, Vaish V, Nehru B, Sanyal SN. Downregulation of PI3-K/Akt/PTEN pathway and activation of mitochondrial intrinsic apoptosis by diclofenac and curcumin in colon cancer. *Molecular and Cellular Biochemistry.* 2015 Apr;402(1–2):225–41.

Ruby AJ, Kuttan G, Babu KD, Rajasekharan KN, Kuttan R. Anti-tumour and antioxidant activity of natural curcuminoids. *Cancer Letters.* 1995 Jul 20;94(1):79–83.

Salehi B, Calina D, Docea AO, Koirala N, Aryal S, Lombardo D, Pasqua L, Taheri Y, Marina Salgado Castillo C, Martorell M, Martins N. Curcumin's nanomedicine formulations for therapeutic application in neurological diseases. *Journal of Clinical Medicine.* 2020 Feb 5;9(2):430.

Salehi B, Del Prado-Audelo ML, Cortés H, Leyva-Gómez G, Stojanović-Radić Z, Singh YD, Patra JK, Das G, Martins N, Martorell M, Sharifi-Rad M. Therapeutic applications of curcumin nanomedicine formulations in cardiovascular diseases. *Journal of Clinical Medicine.* 2020 Mar 10;9(3):746.

Sharifi-Rad J, Rayess YE, Rizk AA, Sadaka C, Zgheib R, Zam W, Sestito S, Rapposelli S, Neffe-Skocińska K, Zielińska D, Salehi B. Turmeric and its major compound curcumin on health: Bioactive effects and safety profiles for food, pharmaceutical, biotechnological and medicinal applications. *Frontiers in Pharmacology.* 2020 Sep 15;11:1021.

Sunagawa Y, Funamoto M, Sono S, Shimizu K, Shimizu S, Genpei M, Miyazaki Y, Katanasaka Y, Morimoto E, Ueno M, Komiyama M. Curcumin and its demethoxy derivatives possess p300 HAT inhibitory activity and suppress hypertrophic responses in cardiomyocytes. *Journal of Pharmacological Sciences.* 2018 Apr 1;136(4):212–7.

Surh YJ, Han SS, Keum YS, Seo HJ, Lee SS. Inhibitory effects of curcumin and capsaicin on phorbol ester-induced activation of eukaryotic transcription factors, NF-\kappaB and AP-1. *BioFactors*. 2000 Jan 1;12(1–4):107–12.

Tayyem RF, Heath DD, Al-Delaimy WK, Rock CL. Curcumin content of turmeric and curry powders. *Nutrition and Cancer*. 2006 Jul 1;55(2):126–31.

Tung BT, Hai NT, Son PK. Hepatoprotective effect of Phytosome curcumin against paracetamol-induced liver toxicity in mice. *Brazilian Journal of Pharmaceutical Sciences*. 2017 Apr 20;53(1):e16136.

Vemuri SK, Banala RR, Subbaiah GP, Srivastava SK, Reddy AG, Malarvili T. Anti-cancer potential of a mix of natural extracts of turmeric, ginger and garlic: A cell-based study. *Egyptian Journal of Basic and Applied Sciences*. 2017 Dec 1;4(4):332–44.

Yance Jr DR, Sagar SM. Targeting angiogenesis with integrative cancer therapies. *Integrative Cancer Therapies*. 2006 Mar;5(1):9–29.

Yao Y, Wang W, Li M, Ren H, Chen C, Wang J, Wang WE, Yang J, Zeng C. Curcumin exerts its anti-hypertensive effect by down-regulating the AT1 receptor in vascular smooth muscle cells. *Scientific Reports*. 2016 May 5;6(1):25579.

Ye M, Zhang J, Zhang J, Miao Q, Yao L, Zhang J. Curcumin promotes apoptosis by activating the p53-miR-192-5p/215-XIAP pathway in non-small cell lung cancer. *Cancer Letters*. 2015 Feb 1;357(1):196–205.

Index

A

B

C

D

E

F

G

O

P

Q

R

S

T

Kolibri 360 GmbH
Ferdinand-Jühlke-Straße 7
99095 Erfurt, Deutschland
produktsicherheit@kolibri360.de